QUANTUM NONLOCALITY AND REALITY

50 Years of Bell's Theorem

Combining twenty-six original essays written by an impressive line-up of distinguished physicists and philosophers of physics, this anthology reflects some of the latest thoughts by leading experts on the influence of Bell's Theorem on quantum physics.

Essays progress from John Bell's character and background, through studies of his main work, on to more speculative ideas, addressing the controversies surrounding the theorem, investigating the theorem's meaning and its deep implications for the nature of physical reality. Combined, they present a powerful comment on the undeniable significance of Bell's Theorem for the development of ideas in quantum physics over the past 50 years.

Questions surrounding the assumptions and significance of Bell's work still inspire discussion in field of quantum physics. Adding to this with a theoretical and philosophical perspective, this balanced anthology is an indispensable volume for students and researchers interested in the philosophy of physics and the foundations of quantum mechanics.

MARY BELL is a physicist and the widow of John Bell, with whom she frequently collaborated. She held several positions working on particle acceleration design, notably with the Atomic Energy Research Establishment in Harwell, Oxfordshire, and several accelerator divisions at CERN.

SHAN GAO is an Associate Professor at the Institute for the History of Natural Sciences, Chinese Academy of Sciences. He is the founder and managing editor of the *International Journal of Quantum Foundations*. He is the author of several books and the editor of the recent anthology *Protective Measurement and Quantum Reality: Towards a New Understanding of Quantum Mechanics*. His research focuses on the foundations of quantum mechanics and history of modern physics.

QUANTUM NONLOCALITY AND REALITY

50 Years of Bell's Theorem

Edited by

MARY BELL

SHAN GAO
Chinese Academy of Sciences

CAMBRIDGE
UNIVERSITY PRESS

CAMBRIDGE
UNIVERSITY PRESS

University Printing House, Cambridge CB2 8BS, United Kingdom

Cambridge University Press is part of the University of Cambridge.

It furthers the University's mission by disseminating knowledge in the pursuit of
education, learning and research at the highest international levels of excellence.

www.cambridge.org
Information on this title: www.cambridge.org/9781107104341

© Cambridge University Press 2016

First published 2016

Printed in the United States of America by Sheridan Books, Inc.

A catalogue record for this publication is available from the British Library

ISBN 978-1-107-10434-1 Hardback

Contents

Contributors

Stephen L. Adler
The Institute for Advanced Study, Princeton

Yakir Aharonov
Tel Aviv University and Chapman University

Jean Bricmont
Université catholique de Louvain

Harvey R. Brown
University of Oxford

Eliahu Cohen
Tel Aviv University

Bernard d'Espagnat
University of Paris–Orsay (deceased)

Gordon N. Fleming
Pennsylvania State University

Olival Freire Jr.
Universidade Federal da Bahia

Shan Gao
Chinese Academy of Sciences

Marco Genovese
Istituto Nazionale di Ricerca Metrologica

GianCarlo Ghirardi
University of Trieste and Abdus Salam International Centre for Theoretical Physics

Sheldon Goldstein
Rutgers University

Richard A. Healey
University of Arizona

Basil J. Hiley
University of London

Gregg Jaeger
Boston University

Tim Maudlin
New York University

Wayne C. Myrvold
University of Western Ontario

Michael Nauenberg
University of California, Santa Cruz

Travis Norsen
Smith College

Philip Pearle
Hamilton College

Roger Penrose
University of Oxford

Osvaldo Pessoa Jr.
University of São Paulo

Daniel Rohrlich
Ben-Gurion University of the Negev

Henry P. Stapp
University of California

Christopher G. Timpson
University of Oxford

Roderich Tumulka
Rutgers University

Lev Vaidman
Tel Aviv University

Andrew Whitaker
Queen's University Belfast

H. Dieter Zeh
Heidelberg University

Preface

I am very pleased to see, in this volume, the papers written in John's memory. He had many interests and it is good to see such a variety of authors. I thank all of them for their work. I am sure that John would have enjoyed the book.

Mary Bell

Preface

In 1964, John Stewart Bell published an important result that was later called Bell's theorem [1]. It states that certain predictions of quantum mechanics cannot be accounted for by any local realistic theory. Bell's theorem has been called "the most profound discovery of science" [2]. By introducing a notion of quantum nonlocality, it not only transforms the study of the foundations of quantum mechanics, but also paves the way to many quantum technologies that have been developed in the last decades. It can be expected that Bell's work will play a more important role in the physics of the future. Admittedly, there are still controversies on the underlying assumptions and deep implications of Bell's theorem. This also poses a challenge to us in understanding quantum theory, as well as the physical world at the most fundamental level.

This book is an anthology celebrating the 50th anniversary of Bell's theorem. It contains 26 original essays written by physicists and philosophers of physics, reflecting the latest thoughts of leading experts on the subject. The content includes recollections of John Bell, an introduction to Bell's nonlocality theorem, a review of its experimental tests, analyses of its meaning and implications, investigations of the nature of quantum nonlocality, discussions of possible ways to avoid nonlocality, and last but not least, analyses of various nonlocal realistic theories. The book is accessible to graduate students in physics. It will be of value to students and researchers with an interest in the philosophy of physics and especially to physicists and philosophers working on the foundations of quantum mechanics.

This book is arranged in four parts. The first part introduces John Bell, the great Northern Irish physicist, and it also contains a few physicists' treasured recollections of him. In Chapter 1, Andrew Whitaker introduces the Irish tradition of physics, Bell's Belfast background, and his studies of physics at Queen's University Belfast. Whitaker argues that there may exist a connection between Bell's work and the Irish tradition, especially concerning his views on the apparent incompatibility between quantum nonlocality and special relativity. In Chapter 2, Michael Nauenberg, who coauthored with Bell the paper "The moral aspect of quantum mechanics," recollects his encounters with Bell at SLAC during 1964–5 when Bell discovered the theorem, as well as his later interactions with Bell. In Chapter 3, GianCarlo Ghirardi recalls his repeated interactions with Bell in the last four years of Bell's life, including their deep discussions about the elaboration and interpretation of collapse theories, Bell's contributions to the development of this approach, and Bell's clearcut views on the locality issue. These discussions of Bell's character and background, as well as the recollections of him from other physicists, will help readers understand better the background and significance of Bell's theorem.

The second part of this book then introduces Bell's theorem, its later developments, and its experimental tests. In Chapter 4, Jean Bricmont first gives a pedagogical introduction to Bell's original theorem, in particular to how it establishes the existence of nonlocal effects. He also discusses several misunderstandings of Bell's result and explains how Bohm's theory allows people to understand what nonlocality is. In Chapter 5, Roderich Tumulka further analyzes the various statements that have been claimed to be assumed in the derivation of Bell's inequality. He gives reasons that some assumptions such as realism and determinism are not made in the derivation, and others such as locality are indeed refuted by experimental violations of Bell's inequality. Moreover, he also briefly analyzes the relationship between nonlocality and relativity. He argues that the GRW flash theory demonstrates that it is possible to retain relativity and give up locality, and to have nonlocal influence without direction.

In Chapter 6, Harvey Brown and Christopher Timpson analyze the change in the notion of nonlocality in Bell's papers between 1964 and 1990, and discuss the relevance of the modern Everettian stance on nonlocality. They argue that violating the 1964 locality condition gives rise to action at a distance, while violating the local causality of 1976 need not, and Bell came more and more to recognise that there need be no straightforward conflict between violation of either of his locality conditions and the demands of relativity. In their view, the significance of Bell's theorem can be fully understood by taking into account that a fully Lorentz-covariant version of quantum theory, free of action at a distance, can be articulated in the Everett interpretation. In Chapter 7, Marco Genovese presents the experimental progress in testing Bell's inequalities and discusses the remaining problems for a conclusive test of the inequalities, such as eliminating the detection loophole and spacelike loophole. In the Appendix added in proof, he also briefly introduces the latest loophole-free tests of Bell inequalities. One of the main theoretical developments that follow Bell's work is the Greenberger–Horne–Zeilinger (GHZ) theorem, also known as Bell's theorem without inequalities. In Chapter 8, Olival Freire Jr. and Osvaldo Pessoa Jr. introduce the history of the creation of this theorem and analyze its scope.

The third part of this book further investigates the nature of quantum nonlocality, as well as possible ways to avoid the nonlocality implication of Bell's theorem. In Chapter 9, Henry Stapp gives a new proof of Bell's inequality theorem, which avoids the hidden-variable assumption and the assumption of "outcome independence." The proof places no conditions on the underlying process, beyond the macroscopic predictions of quantum mechanics. In Chapter 10, Bernard d'Espagnat analyzes the nature of the premises assumed in Bell's proof, such as the nature of causality. Based on this analysis, he argues that a theory compatible with the quantum predictions is not necessarily nonlocal, and assuming realism is by far the safest way to establish nonlocality on truly firm grounds.

In Chapter 11, Richard Healey analyzes Bell's assumptions about probability in his formulations of local causality. He argues that probability does not conform to these assumptions when quantum mechanics is applied to account for the particular correlations that Bell argues are locally inexplicable. By assuming a pragmatist view of quantum mechanics, he also gives an explanation of nonlocalized quantum correlations. The explanation involves no superluminal action and there is even a sense in which it is local, but it is in tension with the requirement that the direct causes and effects of events be nearby. In Chapter 12, Lev Vaidman argues that the lesson we should learn from Bell's inequalities is not that quantum mechanics requires some kind of action at a distance, but that it leads us to believe in parallel worlds. In the many-worlds interpretation of quantum mechanics, Bell's proof of action at a distance fails, since it requires a single world to ensure that measurements have single outcomes. In his view, although there is no action at a distance in the many-worlds interpretation, it still has nonlocality, and the core of the nonlocality is entanglement, which is manifested in the connection between the local Everett worlds of the observers. In Chapter 13, Travis Norsen analyzes the many-worlds theory and the quantum Bayesian interpretation, which purport to avoid the nonlocality implication of Bell's theorem. By investigating each theory's grounds for claiming to explain the EPR–Bell correlations locally, he argues that the two theories share a common for-all-practical-purposes (FAPP) solipsistic character, and this undermines such theories' claims to provide a local explanation of the correlations. In Norsen's view, this analysis reinforces the assertion that nonlocality really is necessary to coherently explain the empirical data.

In Chapter 14, Wayne Myrvold investigates the possibility of the compatibility of nonlocality with relativity. He argues that the nonlocality required to violate the Bell inequalities need not involve action at a distance, and the distinction between forms of nonlocality makes a difference when it comes to compatibility with relativistic causal structure. Concretely speaking, although parameter dependence involves a departure from relativistic causal structure, nonlocal theories that satisfy parameter independence and exhibit only outcome dependence, such as collapse theories, can satisfy the compatibility with relativistic causal structure at a truly fundamental level. In Chapter 15, Gordon Fleming argues that in collapse theories the "elements of reality" can retain their Lorentz invariance or frame independence if the hyperplane dependence of their localization is recognized and the conflation of hyperplane dependence with frame dependence is avoided. He also criticizes a view of the nature of Lorentz transformations presented by Asher Peres and co-workers that conflicts with the view employed by him in the argument. In Chapter 16, Shan Gao presents a new analysis of quantum nonlocality and its apparent incompatibility with relativity. First, he gives a simpler proof of nonlocality in standard quantum mechanics, which also avoids the controversial assumption of counterfactual definiteness. Next, he argues that the new proof may imply the existence of a preferred Lorentz frame. After arguing for the detectability of the preferred frame, he further shows that the frame can be detected in a recently suggested model of energy-conserved wave function collapse. Moreover, he analyzes possible implications of quantum nonlocality for simultaneity of events. Last, he also discusses a possible mechanism of nonlinear quantum evolution and superluminal signaling.

In Chapter 17, Daniel Rohrlich shows that maximally nonlocal "superquantum" (or "PR-box") correlations, unlike quantum correlations, do not have a classical limit consistent with relativistic causality, and by deriving Tsirelson's bound from the three axioms of relativistic causality, nonlocality, and the existence of a classical limit, this result can be generalized to all stronger-than-quantum nonlocal correlations. Moreover, he argues that local retro-causality offers us an alternative to nonlocality. In Chapter 18, Yakir Aharonov and Eliahu Cohen analyze entanglement and nonlocality in the framework of pre-/postselected ensembles with the aid of weak measurements and the two-state-vector formalism that admits local retrocausality. In addition to the EPR–Bohm experiment, they revisit the Hardy and Cheshire Cat experiments, whose entangled pre- or postselected states give rise to curious phenomena. Moreover, they also analyze even more peculiar phenomena suggesting "emerging correlations" between independent pre- and postselected ensembles of particles, which can be viewed as a quantum violation of the classical "pigeonhole principle."

The last part of this book introduces and analyzes various nonlocal realistic theories, including Bohm's theory, collapse theories, and twistor theory. In Chapter 19, Tim Maudlin analyzes Bell's theory of local beables, which not only underpins all analyses of the significance of violations of Bell's inequalities, but also stands on its own as a contribution to the foundations of physics. In particular, he discusses Bohm's theory, GRW theory, and Bell's Everett(?) theory, which are different nonlocal realistic theories of local beables. In his view, the analysis of local beables also highlights a challenge for the orthodox Everettian position. In Chapter 20, H. Dieter Zeh compares and discusses various realistic interpretations of quantum mechanics, which were either favored or neglected by John Bell in the context of his nonlocality theorem.

In Chapter 21, Basil Hiley presents the background of Bohm's theory that led Bell to a study of quantum nonlocality from which Bell's inequalities emerged. He recalls the early experiments done at Birkbeck with the aim of exploring the possibility of "spontaneous collapse," a way suggested by Schrödinger to avoid the conclusion that quantum mechanics was grossly nonlocal. He also reviews some of the work that Bell did that directly impinged on his own investigations into the foundations of quantum mechanics, and reports some new investigations toward a more fundamental theory, such as the Clifford algebra approach to quantum mechanics. In Chapter 22, Sheldon Goldstein further analyzes Bell's supportive views on Bohm's theory in more detail. He points out that these views are not nearly as well appreciated as they should be. Moreover, he also briefly discusses nonlocality and the "big question" about Lorentz invariance.

In addition to Bohm's theory, Bell was also a staunch supporter of collapse theories. In Chapter 23, Philip Pearle reminisces on his handful of interactions with Bell on such theories through letters between them. He also discusses quantum nonlocality and some of its implications within the framework of the CSL (continuous spontaneous localization) model of dynamical collapse. In Chapter 24, Stephen Adler first recalls intersections of his research interests with those of Bell. He then argues that the noise needed in collapse theories most likely comes from a fluctuating complex part in the classical spacetime metric; that is, wave function collapse is driven by complex-number-valued "spacetime foam."

In Chapter 25, Roger Penrose argues that quantum nonlocality must be gravitationally related, as it comes about only with quantum state reduction, this being claimed to be a gravitational effect. He also outlines a new formalism for curved space–times, palatial twistor theory, which seems to be able to accommodate gravitation fully, providing a nonlocal description of the physical world. In Chapter 26, Gregg Jaeger considers Bell's critique of standard quantum measurement theory and some alternative treatments wherein Bell saw greater conceptual precision, such as collapse theories. He also makes further suggestions as to how to improve conceptual precision.

No doubt, the discussions about the significance and implications of Bell's theorem still have a long way to run. This anthology is a tribute to John Bell, and I hope it will arouse more researchers' interest in his profound work and its ramifications. I thank all contributors for taking the time to write the new essays in this anthology. I am particularly grateful to Martinus Veltman, Kurt Gottfried, and Mary Bell for their help and support. I also wish to express my warm thanks to all participants of the John Bell Workshop 2014, which was organized by the *International Journal of Quantum Foundations* and in which the drafts of the essays of this anthology were discussed [3]. I thank Simon Capelin of Cambridge University Press for his kind support as I worked on this project, and the referees who gave helpful suggestions on how the work could best serve its targeted audience. Finally, I am deeply indebted to my parents, QingFeng Gao and LiHua Zhao, my wife Huixia, and my daughter Ruiqi for their unflagging love and support.

References

[1] J.S. Bell, On the Einstein–Podolsky–Rosen paradox. *Physics* **1**, 195–200, 1964.
[2] H.P. Stapp, Bell's theorem and world process, *Nuovo Cimento B* **29**, 270–76, 1975.
[3] John Bell Workshop 2014, www.ijqf.org/groups-2/bells-theorem/forum/.

Shan Gao

Part I

John Stewart Bell: The Physicist

1

John Bell – The Irish Connection

ANDREW WHITAKER

John Bell lived in Ireland for only 21 years, but throughout his life he remembered his Irish upbringing with fond memories, pride and gratitude.

1.1 Irish Physics and the Irish Tradition

Ireland has a very respectable tradition in physics, and particularly mathematical physics [1]. As early as the eighteenth century, Robert Boyle is usually given credit for establishing the experimental tradition in physics [2, 3]. Through the nineteenth century, luminaries such as William Rowan Hamilton [4], James MacCullagh [5], Thomas Andrews [6], George Francis FitzGerald [7, 8] and Joseph Larmor [9] all made very important contributions to the establishment of physics as an intellectual discipline in its own right, while it should not be forgotten that although the academic careers of George Gabriel Stokes [10], William Thomson [Lord Kelvin] [11] and John Tyndall [12] were spent in England or Scotland, all had Irish origins which they never forgot.

An important aspect of Irish mathematical physics in the nineteenth century was the existence of the so-called Irish tradition, which consisted of studying the wave theory of light and the ether. It was a tradition that was to be important for John Bell and it was to transcend what may be described as the most important event in nineteenth century physics – James Clerk Maxwell's theory of electromagnetism, and his conclusion that light was an electromagnetic wave [13]. The work of MacCullagh and Hamilton was performed in the 1830s and 1840s, while Maxwell's mature work was not to emerge until the 1860s.

MacCullagh improved the theory of Christian Huygens and Augustine Fresnel, being able first to derive the laws of reflection and refraction of light at the surfaces of crystals and metals, and then to write down equations for a light-bearing ether that justified his previous work. He was also able to handle the phenomenon of total internal reflection, and his model of the ether involved an effect that he called 'rotational elasticity', which was easy to describe mathematically but difficult to picture physically [14].

MacCullagh is probably not very well known even among physicists. One physicist, though, who was well aware of his achievements was Richard Feynman. In his famous lectures [15], he discusses MacCullagh's work in his very first chapter of his volume on electromagnetism, where he argues that one should not look for a mechanical model for electric and magnetic fields.

Feynman comments that 'It is interesting that the correct equations for the behavior of light were worked out by MacCullagh in 1839. But people said to him: "Yes, but there is no real material whose mechanical properties could possibly satisfy those equations, and since light is an oscillation that must vibrate in *something*, we cannot believe this abstract equation business." If people had been more open-minded, they might have believed in the right equations for the behavior of light a lot earlier than they did.'

Hamilton was broadly a contemporary of MacCullagh, though, of course, much better known, principally for his theory of quaternions and his invention of what is now known as 'Hamiltonian mechanics'. His work on optics used the same general methods as that on mechanics, and he also studied the wave surface in detail, making the important prediction of conical refraction, a prediction soon verified by another Irish physicist, Humphrey Lloyd.

FitzGerald and Larmor, whose main work was performed after Maxwell's discoveries, in the 1880s and 1890s, respectively, were thus, of course, working in a completely different scientific context than MacCullagh and Hamilton. They were to become central members of the group often known as the 'Maxwellians' [16, 17], the other main members being Oliver Lodge, Oliver Heaviside and Heinrich Hertz; the last was the first to generate and detect the electromagnetic waves predicted by Maxwell.

While Maxwell's work was undoubtedly brilliant in the extreme, Yeang describes it as 'promising but esoteric and somehow puzzling' and he argues that it was largely FitzGerald, Lodge and Heaviside who transformed his theory into a 'fully-fledged research programme', while Larmor was able to supplement the field-based paradigm of Maxwell with a microphysics in which charge and current 'regained the status of fundamental physical entities'. This blended in neatly with J.J. Thomson's 'discovery' of the electron [18] and the new age in physics thus entered. As is well known, Hendrik Lorentz came to conclusions similar to Larmor's in the same period.

It is interesting to note the extent to which the work of FitzGerald and Larmor was based on the much earlier ideas of MacCullagh and Hamilton, particularly the former. O'Hara [5] writes that '[MacCullagh's] work was received with scepticism by many contemporaries ... His dynamical theory did, however, find supporters, particularly among the Anglo-Irish, decades after his death.'

An important idea of FitzGerald, his vortex ether model, was based wholly on Mac-Cullagh's model of the ether. Working by analogy with MacCullagh, he was able to interpret the potential and kinetic energy of the ether in terms of the energy of the field, and use this to derive the laws of reflection and refraction [17]. FitzGerald was also able to use MacCullagh's model of the ether to explain the propagation of radiant heat and electromagnetic radiation in general [14].

Larmor drew on the ideas of both Hamilton and MacCullagh [14]. He followed Hamilton in his statement that 'The great *desideratum* for any science is its reduction to the smallest number of dominating principles. This has been achieved for dynamical science mainly by Sir William Rowan Hamilton of Dublin.'

In his famous book *Aether and Matter*, Larmor [19] developed MacCullagh's model of the ether to include electric charges. For Larmor the ether was not a material medium; rather he visualised electric particles or 'electrons' as moving in the ether and following Maxwell's laws, while an electron itself was not a material particle but a nucleus of intrinsic strain in the ether [14]. Larmor was so impressed by MacCullagh's ideas that he described him as one of the great figures of optics.

Their work on the ether led both FitzGerald and Larmor to become involved with the search for signs of the motion of the earth through the ether, and it was in this context that Bell might have considered himself to be an honorary member of the Irish tradition. As is well known, FitzGerald and also Lorentz postulated a contraction of the length of the object along the direction of its motion through the ether that would serve to make its motion through the ether unobservable.

Larmor went rather further by discussing the equilibrium conditions of his electrons or intrinsic strains for different states of motion. The relations he obtained between the various conditions are what are now known as the Lorentz transformations. Larmor was thus the first person to write these down, although his argument showed only that they were correct to second order; at the time he did not realise that they were correct to all orders.

The ideas of FitzGerald and Larmor were of interest to Einstein early in the following century, although, of course, his stance was notably different; as we shall see, they were also of interest to John Bell much later in the century.

Let us now turn to the first half of the twentieth century before Bell entered university, and note again the significant contributions of Irish physicists during this period, such as John Synge [20], John Desmond Bernal [21] and Ernest Walton [22]. It should also be remembered that the Dublin Institute for Advanced Studies gave sanctuary to Erwin Schrödinger [23] and Walter Heitler [24] in World War II, and Cornelius Lanczos [25] a little after the war.

The majority of those mentioned so far were associated with Dublin, and particularly with Trinity College Dublin, which was founded as early as 1592, but after 1849, when Queen's College Belfast [from 1908 Queen's University Belfast] had been founded, a substantial contribution was made to physical science there as well. Kelvin, Andrews and Larmor were linked to Belfast, as well as such worthy figures as Peter Guthrie Tait [26] and James Thomson [27].

It is also interesting that, a few years before Bell entered Queen's University Belfast, just before World War II in fact, there was quite a remarkable flowering of potential talent in physics and mathematics at the university [28]. Three of the rather small group became Fellows of the Royal Society (FRS) and were also knighted. These were William McCrea, then Head of Mathematics, who became a very well-known astronomer, Harrie Massey, Head of Mathematical Physics, who after the war was the UK's leading atomic and space physicist, and David Bates, who was to set up a large and influential group working on atomic physics and geophysics in Belfast itself.

Samuel Francis Boys also became an FRS; James Hamilton became one of the world's leading particle physicists and had a period as Head of NORDITA, the Nordic

[Scandinavian] Institute for Theoretical Physics, an international organisation set up by Niels Bohr, while Richard Buckingham became the world's leading expert on the application of computers in education, and the first Chair of the Technical Committee for Education of the International Federation for Information Processing.

1.2 John Bell – The Belfast Background

As would be expected, this group had largely dispersed to carry out war work before John Bell was to arrive at Queen's, but nevertheless it is clear that traditions and standards were high. Indeed, in 1949, when John was interviewed for his first scientific job at the UK atomic energy station at Harwell, a little concerned that he might not be considered as seriously as a graduate of Oxbridge or one of the leading London colleges, the main question was not whether he would be given a job – that was soon taken for granted – but which of the panel members, Klaus Fuchs or Bill Walkinshaw, should gain his services. Of course this was partly, perhaps largely, because of John's ability, but obviously his education must have been of a high standard as well. (Fuchs as the senior person obtained Bell for his atomic reactor group, but within months he was unmasked as the 'atomic spy' and almost by default Bell ended up with Walkinshaw and accelerators.)

Yet there must have been differences between studying at a major university and at a university which, whatever its merits, was rather remote and had a much greater emphasis on medicine than on physical science. However, before considering this, and in particular how it would have affected John's studies, let us think about his background and especially any effect his Irish roots may have had on his intellectual development and aspirations.

In the 1930s, the Bell family were not well off. This should certainly not be exaggerated – they were certainly no worse off than hundreds of thousands of others in the industrial cities of the United Kingdom. Indeed, shrewd management by his mother Annie meant that the family did not go without any essentials, and could even scrape up a few luxuries, such as second-hand bicycles for the children. And John found education up to the age of 11 enjoyable and interesting. Fortunately it was also free!

After that age everything was much more difficult, and it was rather more difficult in Northern Ireland than in the rest of the United Kingdom. It was only to be a few years before free secondary education was to become law in 1947, actually three years after it had done so in England. In 1939, though, it was not only not free but expensive, and although John passed the actual entrance examination for grammar school with ease, he sat the scholarship examination in every grammar school without success.

In Northern Ireland there seemed to be little leeway even for the obviously extremely talented. Each grammar school had its own junior or 'preparatory' department, where pupils paid to study up to age of 11, and the school was likely to award post-11 scholarships to those who had already paid into the system, rather than to the poorer child without perhaps such a posh accent.

While it must be admitted that different parts of the United Kingdom had their own policies and practices, an interesting comparison might be with Fred Hoyle [29, 30], a distinguished scientist of the future, though, like John Bell, with parents who were not very well off. Hoyle did obtain a scholarship to Bingley Grammar School in Yorkshire, admittedly only after an appeal, and thereafter Yorkshire Education Committee financed him throughout his university studies.

In contrast, John Bell obtained a small amount of support, sufficient only to attend the much less prestigious Technical School, and he was lucky to obtain that – he never knew where it came from. The course allowed him to matriculate at Queen's, but again he needed a small amount of money – obtained as a grant from the Cooperative Society – to be able to attend the first year at the university. He had actually finished school a year too young to commence at the university, and had been fortunate enough to get some experience and earn some money by working as a laboratory attendant in the very department where he hoped go become a student, but, as we have seen, it was touch and go whether this would be possible.

So at the time of his entry to university, John Bell had taken advantage of friendly and helpful teachers, and been very much encouraged by his mother. However, he had been sorely discouraged by the grammar schools of Belfast, and, to a lesser extent, by the problem of university entrance. His father's attitude – that the natural course was to leave school as early as possible and get a job [31] – must have been discouraging. (It must be admitted, though, that this attitude would probably have been far more common among working people at the time than that of his mother, which was that it was wise to get all the education that was possible.)

Also, helpful as all his teachers had been, and he was keen to praise them all [31], his personal intellectual wanderings must have by far transcended them. His library-based analysis of philosophy, physics and the writings of such figures as Cyril Joad, H.G. Wells, George Bernard Shaw and Bertrand Russell was to play a considerable part in determining the type of person he was to become, both in intellectual interests and in personal standards. Just as an example, his choice to become a vegetarian, following the example of Shaw, would have been exceptionally unusual in young men at that period.

1.3 Determination – A Protestant Characteristic?

It is clear that, at this time in his life, and despite his quiet demeanour and good behaviour, he had the ability to overcome obstacles and to decide on his own course of action, not being put off by a good deal of opposition. It is possible to suggest that these traits were encouraged by the political and social background in which he was situated.

It may not be too much of a simplification of Irish history to say that while, from the end of the seventeenth century, the minority Protestant community had exercised full domination over the majority Catholics, from the 1860s with renewed talk of Home Rule, and particularly from 1885, with the conversion of William Gladstone, leader of the Liberal Party, to the cause of Home Rule, the Protestant community had become more and more nervous and defensive about the political situation. Home Rule would have given Ireland a very limited form of devolution, but broadly the Protestants would have none of it [32].

By 1922, though, 26 of Ireland's 32 counties had gained a degree of freedom much greater than that offered by Home Rule, with the 'Irish Free State' being a self-governing dominion within the British empire of the same nature as Canada, though an oath of allegiance to the crown was still required. Protestants in these counties obviously had to come to terms with their minority status. However, the six counties in the northeast of the island, where there was a Protestant: Catholic ratio of roughly 2:1, had gained their own form of Home Rule, a 'Province' of 'Northern Ireland', with a Parliament in Belfast which had control of security and internal matters, but also with representation in and subservient to the British Parliament.

The Protestants in Northern Ireland had a permanent majority in the Parliament and took the opportunity to strengthen the sectarian nature of the Province by gerrymandering electoral boundaries and considerable use of discrimination, though it has to be said that the minority community assisted the process by refusal to take part in the new political institutions.

Without entering into rights and wrongs, it is undoubtedly the case that the Protestant community in Northern Ireland felt themselves to be beleaguered from two directions: from outside, by the government of the remaining 26 counties, which never recognised Northern Ireland, which described its own territory as the full 32 counties of the island, and which maintained the right, in principle, to subsume the six northeastern counties, however little it might do in practice; and from inside, from the recalcitrant minority, who also refused to accept the present constitutional position, and who would, from time to time, turn to passive resistance and guerrilla activities. They also suspected that, when the chips were down, they would receive precious little support from the rest of the United Kingdom.

While it will be readily admitted that the minority population in the north felt *themselves* to be a beleaguered minority *within* Northern Ireland, it is only to be expected that the typical psyche of the Northern Irish Protestant tended to be rather defensive and determined, unafraid of being in a minority on the island, keen to assert rights, keen to maintain the dominance of the majority community in the province, and especially keen to eliminate any influence of the Catholic church!

John Bell was certainly not interested in Protestantism as such – his wife Mary [33] has reported that he was an atheist most of his life. Though feeling a broad affinity to the Protestant community, he was very open to fruitful political discussion between the two communities, and while at Queen's had interesting theological discussions with a Catholic friend.

Nevertheless it is at least plausible to surmise that he took over some of the better qualities of his community – determination, honesty, stickability, and the willingness of take on a minority cause even at the risk of some ridicule. Fortunately he did not (except in very special and very exceptional cases) take on the pugnacity which often goes with these praiseworthy attributes.

Two other ways in which he may have assimilated qualities from his community may be tentatively suggested. First is perhaps a tendency not to promote one's own achievements – not so much a desire for privacy or to limit attention to oneself, but more a preference for the attention to be drawn by others. The second is the tendency to ignore or wish away medical symptoms, just to hope that they will go away if left alone – anything better than attending the doctor.

It may well be argued that these two attitudes are by no means confined to Northern Irish Protestants, which is certainly true, the second perhaps applying to most men everywhere! However, there does seem to be a certain resonance with the stoical undemonstrative community in which John Bell was brought up, and it is very easy to relate these to John's own life.

1.4 Studying Physics at Queen's

Let us now return to John's studies at Queen's University Belfast. Queen's was very much a provincial university and definitely had been taken to the hearts of the local people. The noted poet Philip Larkin worked in the library at Queen's from 1950 to 1955 before his long period as (Chief) Librarian at the University of Hull [34], and his Belfast days were actually the happiest and the most successful creatively of his life; he very much enjoyed both the city and the university. At Leicester, where he had worked before, he felt that the University was regarded 'if at all, as an accidental impertinence'. In Belfast, on the other hand, the university had local character and was fully integrated into the community. 'Your doctor, your dentist, your minister, your solicitor would all be Queen's men, and would probably know each other. Queen's stood for something in the city and in the province … It was accepted for what it was.'

When Bell was at Queen's, the number of students was around 1500, but of these roughly half were studying medicine or dentistry [35]. There were about 300 in engineering, around 250 in arts, and half as many in science, with the remainder in law, agriculture and theology. Physics and mathematical physics each had about five students in any year, but medical and engineering students had to study physics in their first year, so that pushed up the number taking the subject in that year enormously. On the teaching side and across all subjects there were around 30 professors and around 120 others – readers, lecturers, clinical lecturers, assistants and demonstrators.

In physics at this period there were one or two temporary assistant lecturers or demonstrators, usually graduates of Queen's often working for PhDs, who did much of the first-year laboratory work at lower levels, but there were two permanent members of staff in experimental physics, Professor Karl George Eméleus and Dr Robert Harbinson Sloane. Both were known by their second names – George and Harbinson.

Professor Eméleus had worked under Rutherford at Cambridge, where he took some of the first pictures of particle tracks and helped to develop the Geiger counter. He had spent a brief period working on plasma physics with Edward Appleton at King's College London before he was appointed to a lectureship at Queen's in 1927, and promoted to professor in 1933. He was an excellent research worker, publishing over 250 papers in his field of conduction of electricity through plasmas at low temperatures. In 1929, he wrote the definitive monograph on the topic, *The Conduction of Electricity through Gases* [36], and this was reprinted in 1936 and 1951.

He may have been disappointed that he was not elected as an FRS, unlike his brother, Harry Julius Emeléus, who was Professor of Inorganic Chemistry at Cambridge University from 1945, but the reason must surely be that he dedicated his life to undergraduate teaching. He delivered the great majority of the lectures in the department, and was well known for his ability to present the most difficult topics in a clear and indeed captivating way. Though a rather formal person, he was eminently concerned and helpful with the problems and requirements of his students.

Dr Sloane was equally dedicated to his students. A protégé of Emeléus, he carried out research in the same area and he was particularly skilled in constructing the highly sophisticated equipment that was required. In Bell's day his teaching was mostly restricted to laboratory work at the higher levels and also a course for the final year students, of which more shortly. Sloane was an excellent physicist, becoming a reader, equivalent to an associate professor elsewhere, and, like Emeléus, a Member of the Royal Irish Academy.

What did the limited size of the department, in numbers of both students and lecturing staff, mean for John Bell and the other students? One might also mention the geographical separation of Northern Ireland from the rest of the United Kingdom and the political division from southern Ireland. Such separation does not imply actual remoteness – physically it was easy enough to get to, for instance, Glasgow, Liverpool or Dublin. Nevertheless there was a certain parochialism. Extremely few students would come to Queen's from outside Northern Ireland, while it seems quite probable that, for example, John Bell might never have left the province until he left to work at Harwell.

The first point, an entirely positive one, would be that, with such a small number of students, coupled with the helpfulness of Emeléus and Sloane, the students were in the favourable position of being able to solicit help and obtain answers to worrying questions. An excellent example of this occurred with John Bell even before he became a student, in his year as a laboratory assistant. Recognising his great interest and ability, the staff lent him books, and allowed him to attend the first year lectures and take the exams at the end of the year. This allowed him to have a year over at the end of the course, which was extremely beneficial for him, as it allowed him to add a second degree in mathematical physics to the one he already had in experimental physics. He obtained first-class honours in both degrees.

Another important example for him, from Professor Eméleus in particular, was the importance of developing a wide-ranging understanding of physics. This was to be very important in his research career. When he began work on particle accelerators at Harwell, his task was to trace the paths of particles through the various arrays of electric and magnetic fields in any particular design. His deep understanding of electromagnetism allowed him to make substantial contributions to the work immediately, and, when the all-important 'strong focussing' was discovered [37], he was able to write a seminal report [38]. Similarly, when he returned to work on accelerators and the Unruh effect in the 1980s, he was able to make superb use of his knowledge of thermodynamics [39]. Last, as we shall see shortly, he understood in depth not only Einstein's ideas on relativity but also the historical background of the subject.

With the limited size of the department, was there any evidence of lack of breadth or depth in the teaching? In his discussions with Jeremy Bernstein [31], John did raise two issues. The first concerned Dr Sloane's final-year course on atomic spectra. The atomic energy levels and hence the spectrum of atomic hydrogen can, of course, be calculated exactly by quantum theory, but this is not the case for any larger atom. Rather the useful, and indeed quite instructive, 'vector model' has to be used to give results that are meaningful but certainly not exact.

John's complaint was that rather than merely illustrating this model to give the general idea, which is certainly worth understanding, the course laboured through 'all the atoms in the periodic table'. He suggested that this was because atomic spectra were a speciality at Belfast as a result of an interest in the physics of the upper atmosphere. John, of course, would have much preferred to have some discussion of the conceptual or philosophical background to quantum theory!

Leslie Kerr [40], who was a contemporary of John's at Queen's and later a close friend for some years, has explained that these criticisms are a little one-sided. In fact, earlier in the course, Professor Eméleus had given an excellent course on theoretical physics, including not only quantum theory but also statistical mechanics, thermodynamics and so on. Then half of Dr Sloane's course in the final year covered atomic spectra, but the other half dealt in alternate years with radio-frequency spectroscopy and accelerators. So while Leslie was prepared to admit that John Bell was not the only student to find the material on atomic spectra rather tedious, the other two topics dealt with very current research, and it is very creditable for the department that they appeared on an undergraduate syllabus.

Probably no physics department in the world at the time would have provided lectures on what John Bell would have really liked – the ideas behind quantum theory – and it was in attempting to discuss these that he had his other difficulty with Queen's physics. This was in the form of an argument with Dr Sloane. It will be understood that Eméleus would have graduated before the arrival of Heisenberg–Schrödinger quantum theory, while Sloane would have learned about the theory from Eméleus. Both would have been quite capable of dealing with the mathematical formalism, and also of giving the standard account of the conceptual problems of quantum theory and the Copenhagen 'solutions'.

But Dr Sloane in particular was unable to satisfy John Bell's concerns about the theory – he would be the first of many! John knew, of course, that the position might be well defined, in which case the momentum was poorly defined, or vice versa. But who or what determines what case you actually have? From the books and courses that John had studied, it seemed that it was practically a subjective matter; the observer might choose what they wished – something he did not believe! Unfortunately discussion with Dr Sloane gave him no enlightenment – quite the reverse, in fact – he found Sloane's attempts at explanations incoherent and accused him of dishonesty!

In time he grasped the answer to his question – the criterion, at least according to any orthodox interpretation of quantum theory, was in fact provided by the previous

measurement. Having realised what the orthodox interpretation would say, it was not difficult for John to recognise that he did not like it. Physics, he felt, should not be just about
the results of measurements – it should say something about a real universe. The result of
practically thirty years of frustration was his late famous paper 'Against "measurement"'
[41].

To come back to the argument with Dr Sloane, it is certainly possible to argue that it
showed a certain parochialism in the presentation of quantum theory at Queen's. Be that as
it may, it may also be thought that the experience was extremely beneficial for John Bell. It
showed him that an intelligent man and excellent physicist such as Dr Sloane could claim to
understand and indeed agree with the Copenhagen argument, and yet, as John saw it, could
not defend the position in an intellectually responsible way. Perhaps John not being able to
appreciate Bohr's arguments was not a demonstration of his own intellectual limitations.
Indeed, perhaps it showed a genuine weakness in the established position.

This may be contrasted [42] with the experience of Euan Squires as described in [43].
Squires' life intersected with that of John Bell at several points. He was five years younger
than John, and in his student days in the 1950s, like John, he was concerned about the
standard approach to quantum theory. Unlike John, though, as we shall see, he did not
feel encouraged at that time to pursue his interest. In the late 1950s he worked for a short
period with John at Harwell, and they wrote an important paper on the nuclear optical model
together. Much later, after a successful career in elementary particle physics, Squires finally
did get involved in the foundations of quantum theory, and wrote some interesting papers.

Squires was a research student at Manchester University, and at this time the Professor
of Theoretical Physics there was Léon Rosenfeld [44], who was by this time Bohr's closest
collaborator and generally accepted spokesman. At Christmas the students sang, to the tune
of 'The boar's head in hand bear I':

> At Bohr's feet I lay me down
> For I have no theories of my own.
> His principles perplex my mind
> But he is so very kind.
> Correspondence is my cry:
> I don't know why; I don't know why.

And Squires went on to say that they were afraid to ask. Even if they had dared to ask
Rosenfeld, the answer would almost certainly have been too contorted conceptually actually
to understand, but too erudite to allow the students to think that the failure in understanding
was anybody's fault but their own. It could be argued that John Bell was very lucky that he
was studying in Belfast rather than Manchester.

Before the discussion of student days is concluded, it should be mentioned that in his
extra year studying mathematical physics at the end of his course, he was a student of Peter
Paul Ewald, who had been involved at the very beginning of X-ray crystallography. Ewald's
dynamical theory of X-ray interference was highly influential for Max von Laue's original
ideas, and he was responsible for many of the central concepts of X-ray crystallography,

such as the reciprocal lattice and the sphere of reflection. Ewald had to leave Germany in 1937, and, as Bell said, he was 'washed up on the shores of Ireland' as a refugee from the Nazis [31]. In 1979, John was to write to Ewald that ' ... your bad fortune in being cast away in Belfast was my very good luck ...'[45].

Indeed it was certainly extremely beneficial for John that Ewald was in Belfast. He spent half his time in this year doing a project with the older man, and very much impressed him – Leslie Kerr has suggested that there was an idea that this work might be published, though it seems that nothing came of this. John very much appreciated the considerable time spent with Ewald, who, compared with Professor Emeléus and Dr Sloane, was extremely unstuffy and would talk about anything! This was an intensity of interaction probably only possible in a small university and an intimate department. It is certain that Ewald, comparatively isolated in the university, would very much have enjoyed his discussions with such an intelligent student as John Bell. And during these conversations, John must have learned a great deal about carrying out research at the highest level, and it must have reinforced his desire to undertake a career in physics research.

As a general comment, Mary Bell says that John 'was extremely satisfied with his education at Queen's University' [46].

1.5 John Bell and the Protestant 'Virtues'?

Let us move on to some of John Bell's relatively early work and note the quality mentioned above of unwillingness, at least in print, to push or personalise his ideas. A prime example of this is his important paper 'Time reversal in field theory' [47], in which he proves the CPT theorem. This was a crucial paper in John's career. He had been offered a year away from Harwell carrying out research at the University of Birmingham under Rudolf Peierls, and this paper was the fruit of this year, which he clearly hoped would begin to establish his name among quantum field theorists and the wider theoretical physics community.

It is true that he realised that in any race for priority he was behind Wolfgang Pauli and Gerhart Lüders [31], both of whose work is actually cited in John's paper, and of course the theorem is justifiably known as the Pauli–Lüders theorem to this day.

Nevertheless the presentation is remarkable. It reads exactly like a chapter in a textbook – beautifully clear and elegant, but without the slightest attempt to draw attention to the particular contributions that he had made. One could almost describe it as extreme self-abasement! Jackiw and Shimony [48] comment that 'Bell is hardly ever credited for his independent derivation, presumably because he was not in the circle of formal field theorists (Wolfgang Pauli, Eugene Wigner, Julian Schwinger, Res Jost, and others) who appropriated and dominated this topic.' But it must be said that John Bell made it easy for them to ignore him!

Fortunately and deservedly, Peierls himself recognised the great merit in the paper, and was indeed vastly impressed by John in general, and arranged for him to move on from accelerators to a new group specialising in theoretical physics when he moved back to Harwell.

The same type of comment may be made about John's two great quantum mechanics papers of the mid-1960s. The first (in order of writing, not publication), 'On the problem of hidden variables in quantum mechanics', was sent to *Reviews of Modern Physics* rather than a journal of current research, and introduced with the words 'The realization that von Neumann's proof is of limited relevance has been gaining ground since the 1952 work of Bohm. However it is far from universal' [49]. There could scarcely have been a more deliberate attempt to downplay the significance of the argument.

The second, 'On the Einstein–Podolsky–Rosen paradox' [50], was sent to the journal *Physics*, a new journal that lasted only a few issues. His official reason was that the journal did not, like many established journals, require page charges. Since he was visiting the United States he would have had to ask his hosts to pay these. While this argument makes sense, the choice of journal did allow him to put his head only briefly and rather surreptitiously over the parapet. It was almost as if he was saying – here is the argument, but it is up to others to see if it is worthy of study. He may have been lucky that in the fullness of time, Abner Shimony and John Clauser did pick up the argument and ran with it.

1.6 Did Bell Belong to the Irish Tradition?

Let us now move on to the aftermath of his famous theorem and discussion of how the nonlocality might be discussed or perhaps even 'explained away'. The best-known method is that of Abner Shimony [51], which argues that, because the nonlocality involved with Bell's theorem cannot be used to send signals, there may be 'peaceful coexistence' between quantum theory and special relativity.

However, as Jackiw and Shimony [48] report, John Bell was 'scornful' of such a proposal, as he considered 'signalling' to be as ill defined and/or anthropomorphic as 'measurement' [52].

His preferred or at least 'cheapest' solution [53] was to return to the Irish electromagnetic tradition of Fitzgerald and Larmor, with whose names he coupled those of Lorentz and Henri Poincaré.

The first sign of this view was the publication of his paper 'How to teach special relativity' in 1976 [54]. In this article, he argues that, without making any reservation about the power and precision of Einstein's approach to special relativity, he thinks that there is an advantage in emphasising the continuity with the earlier years rather than just the radical break.

Making use of what he calls Fitzgerald contraction and Larmor time-dilation, following very much what he says is the approach of Lorentz, he shows that the laws of physics in one frame may account for all physical phenomena, including the observations made by moving observers. It is thus totally allowable to define a unique rest frame of the universe, but with the proviso that the laws of physics, rather perversely, forbid this rest frame to be identified, as, for example, in the Michelson–Morley experiment.

There is no mention of quantum theory in this paper, so at first it seems surprising that John Bell included it in his 1987 collection, *Speakable and Unspeakable in Quantum*

Mechanics [55]. It was also included in the quantum mechanics section when Mary Bell, Kurt Gottfried and Martinus Veltmann put together the collected works in 1995 [45].

There is some clarification, though, in an interview John gave to Paul Davies in a radio programme in 1985 [53]. He saw the outstanding clash between the nonlocality predicted from his theorem and relativity as follows. Nonlocality says that in one particular frame, something may go faster than light. Then, if, as Einstein postulated, all inertial frames are equivalent, there must be inertial frames in which things go backwards in time, with all the problems for causality that are created in that way.

If, however, one goes back to relativity before Einstein – Larmor, Fitzgerald, Lorentz, Poincaré – there is a unique rest frame of the ether, though the laws of physics (perversely, as has been said) do not allow this frame to be identified. In this ether frame, there is what John Bell calls a 'real causal sequence'. It is true that in frames moving with respect to this ether frame, things may 'seem' to go backwards in time, but this according to John, is just an 'optical illusion'.

Just before his death, John spoke on the same theme, though with a more explicit Irish connection, in Dublin. His talk was abridged by Denis Weaire [56]. In a talk specifically on Fitzgerald, he stressed that Fitzgerald's idea of contraction was by no means, as often believed, ad hoc. It was based on the work of Heaviside, who calculated the electrical field of a moving charge, demonstrating a contraction; Fitzgerald surmised that matter should behave in the same way. With the addition of his time dilation, Larmor was able to establish Lorentz invariance. John also mentions that Fitzgerald was the first, in response to a query of Heaviside, to conjecture that no object may travel faster than light.

Again he is insistent on stressing that, in contrast to some relatively common beliefs, 'pre-Einstein relativity' is correct and so certainly cannot be proved wrong by experiment. His most interesting point is that in relativistic arguments, Einstein tells you what happens by use of a general principle, but Fitzgerald and Larmor tell you *how* it happens using detailed dynamics.

While Bell goes beyond Einstein's philosophy, he does not question his formalism. However, one does gain some freedom of choice by doing so, and Franco Selleri [57] follows this path by denying Einstein's postulate that the speed of light is the same in all inertial frames. Specifically, he discusses the one-way speed of light; in a frame moving with respect to the ether frame, the speed of light depends on whether the light is travelling in the direction of the motion or against it, though the two-way speed will be c. Denial of Einstein's postulate leads to a requirement for a full discussion of clock synchronisation. A brief treatment is given by Whitaker [58].

1.7 A Sad Note

As a brief note on a sad topic, we may note Mary Bell's [33] remark that John Bell had frequent migraine attacks through his life. As suggested before, he displayed a certain reluctance to consult his doctor, though ironically he did have a full checkup a short time before he died.

1.8 A Light Note

On a lighter note, it will of course be recognised that the Irish joke (and its cousins) are the lowest form of wit. However, it appears that John Bell liked them! One of his favourites was 'Green side up. Green side up', which relates to an Irishman laying turf.

More relevant to quantum theory is his explanation of his discovery of Bell's inequalities. It is well known that disagreement with the inequalities occurred only for certain ranges of angles of correlation, which did not include 0 degrees or 90 degrees. Einstein and Bohr had considered correlations only at 0 degrees and 90 degrees, while John, as an Irishman, said that he had considered correlations at 37 degrees [59].

References

[1] McCartney, Mark and Whitaker, Andrew (eds.) 2003, *Physicists of Ireland: Passion- and Precision* (Institute of Physics, Bristol).

[2] More, Louis Trenchard 1944, *The Life and Works of the Honourable Robert Boyle* (Oxford University Press, Oxford).

[3] Conant, James Bryant 1970, Robert Boyle's experiments in pneumatics, in *Harvard Case Studies in Experimental Science*, James Bryan Conant (ed.) (Harvard University Press, Cambridge, MA), vol. 1, pp. 1–63 (originally published 1948).

[4] Hankins, Thomas Leroy 1980, *Sir William Rowan Hamilton* (John Hopkins University Press, Baltimore).

[5] O'Hara, James Gabriel 2003, James MacCullagh 1809–1847, in McCartney and Whitaker 2003, pp. 69–76.

[6] Burns, Duncan Thorburn 2003, Thomas Andrews 1813–1885, in McCartney and Whitaker 2003, pp. 77–84.

[7] Weaire, Denis 2008, Kelvin and Fitzgerald, great Irish physicists, in *Kelvin: Life, Labours and Legacy*, Raymond Flood, Mark McCartney and Andrew Whitaker (eds.) (Oxford University Press, Oxford), pp. 86–93.

[8] Weaire, Denis (ed.) 2009, *George Francis Fitzgerald* (Living Edition Publishers, Vienna).

[9] Warwick, Andrew 2003, *Masters of Theory: Cambridge and the Rise of Mathematical Physics* (University of Chicago Press, Chicago). **231**, 479–95; also in [45, pp. 129–45].

[10] Wilson, David B. 1987, *Kelvin and Stokes: A Comparative Study in Victorian Physics* (Adam Hilger, Bristol).

[11] Smith, Crosbie and Wise, M. Norton 1989, *Energy and Empire: ABiographical Study of Lord Kelvin* (Cambridge University Press, Cambridge).

[12] Brock, William, McMillan, Norman, and Mollan, R. Charles (eds.) 1981, *John Tyndall: Essays on a Natural Philosopher* (Royal Dublin Society, Dublin).

[13] Flood, Raymond, McCartney, Mark and Whitaker, Andrew (eds.) 2014, *James Clerk Maxwell: Perspectives on His Life and Work* (Oxford University Press, Oxford).

[14] Flood, Raymond 2003, Joseph Larmor 1857–1942, in McCartney and Whitaker 2003, pp. 151–60.

[15] Feynman, Richard, Leighton, Robert and Sands, Matthew 1964, *The Feynman Lectures on Physics* (Addison-Wesley, Boston), vol. 2.

[16] Hunt, Bruce 1991, *The Maxwellians* (Cornell University Press, Ithaca).

[17] Yeang, Chin-Pang 2014, The Maxwellians: The reception and further development of Maxwell's electromagnetic theory, in *James Clerk Maxwell: Perspectives on His Life and Work*, Raymond Flood, Mark McCartney and Andrew Whitaker (eds.) (Oxford University Press, Oxford), pp. 204–22.

[18] Davis, Edward A. and Falconer, Isabel 1997, *J.J. Thomson and the Discovery of the Electron* (Taylor and Francis, Abingdon).

[19] Larmor, Joseph 1900. *Aether and Matter* (Cambridge University Press, Cambridge).

[20] Florides, Petros S. 2003, *John Synge 1897–1995*, in McCartney and Whitaker 2003, pp. 208–20.

[21] Brown, Andrew 2005, *J.D. Bernal: The Sage of Science* (Oxford University Press, Oxford).

[22] Cathcart, Brian 2005, *The Fly in the Cathedral* (Penguin, Harmondsworth).

[23] Moore, Walter 1989, *Schrödinger: Life and Thought* (Cambridge University Press, Cambridge).

[24] Glass, David 2003, Walter Heitler 1904–81, in McCartney and Whitaker 2003, pp. 238–47.

[25] Gellai, Barbara 2003, Cornelius Lanczos 1893–1974, in McCartney and Whitaker 2003, pp. 198–207.

[26] Knott, Cargill Gilston 1911, *Life and Scientific Work of Peter Guthrie Tait* (Cambridge University Press, Cambridge). Available at https://archive.org/stream/lifescientificwo00knotuoft#page/n15/mode/2up.

[27] Whitaker, Andrew 2015, James Thomson, engineer and scientist: The path to thermodynamics, pp. 93–133; *and* James and William Thomson: The creation of thermodynamics, pp. 67–91 in *Kelvin, Thermodynamics and the Natural World*, M.W. Collins, R.C. Dougall and C. Koenig (eds.) (WIT Press, Southampton).

[28] Whitaker, Andrew 2016, *John Stewart Bell and Twentieth-Century Physics: Vision and Integrity* (Oxford University Press, Oxford, to be published).

[29] Burbidge, Geoffrey 2003, Sir Fred Hoyle, *Biographical Memoirs of Fellows of the Royal Society* **49**, 213–47.

[30] O'Connor and Robertson, E. F. 2003, Sir Fred Hoyle, *Mactutor History of Mathematics*. Available at www-history.mcs.st-and.ac.uk/Biographies/Hoyle.html.

[31] Bernstein, Jeremy 1991, *Quantum Profiles* (Princeton University Press, Princeton, NJ).

[32] Beckett, James C. 1966, *The Making of Modern Ireland 1603–1923* (Faber, London).

[33] Bell, Mary 2002, Some reminiscences, in [60, pp. 3–6].

[34] Motion, Andrew 1993, *Philip Larkin: A Writer's Life* (Faber, London).

[35] Walker, Brian and McCreary, Alf 1995, *Degrees of Excellence: The Story of Queen's, Belfast, 1845–1995* (Institute of Irish Studies, Queen's University, Belfast).

[36] Eméleus, Karl George 1929, *The Conduction of Electricity through Gases* (Methuen, London).

[37] Mladjenović 1998, *The Defining Years in Nuclear Physics 1932–1960s* (Institute of Physics, Bristol).

[38] Bell, John S. 1953, A new focussing principle applied to the proton linear accelerator, *Nature* **171**, 167–8.

[39] Bell, John S. and Leinaas, Jon M. 1983, Electrons as accelerated thermometers, *Nuclear Physics B* **212**, 131–50.

[40] Kerr, Leslie 1998, Recollections of John Bell (unpublished memoir).

[41] Bell, John S. 1990a, Against 'measurement', *Physics World* **3** (August), 33–40; also in [45, pp. 902–9; 55 (second ed. only), pp. 213–31].

[42] Whitaker, Andrew 2002, John Bell in Belfast: Early years and education, in [60, pp. 7–20].

[43] Squires, Euan 1986, *The Mystery of the Quantum World* (Institute of Physics, Bristol).

[44] Jacobsen, Anja Skaar 2012, *Léon Rosenfeld: Physics, Philosophy, and Politics in the Twentieth Century* (World Scientific, Singapore).

[45] Bell, John S. 1995, *Quantum Mechanics, High Energy Physics and Accelerators* (Mary Bell, Kurt Gottfried and Martinus Veltman (eds.)) (World Scientific, Singapore).

[46] Bell, Mary 2000, Biographical notes on John Bell, in [61], pp. 3–5.

[47] Bell, John S. 1955, Time reversal in field theory, *Proceedings of the Royal Society A* **231**, 479.

[48] Jackiw, Roman and Shimony, Abner 2002, The depth and breadth of John Bell's physics, *Physics in Perspective* **4**, 78–116.

[49] Bell, John S. 1966, On the problem of hidden variables on quantum mechanics, *Reviews of Modern Physics* **38**, 447–52; also in [45, pp. 695–700; 55, pp. 1–13].

[50] Bell, John S. 1964, On the Einstein–Podolsky–Rosen paradox, *Physics* **1**, 195–200; also in [45, pp. 701–6; 55, pp. 14–21].

[51] Shimony, Abner 1978, Metaphysical problems in the foundations of quantum mechanics, *International Philosophical Quarterly* **18**, 3–17.

[52] Bell, John S. 1990b, La nouvelle cuisine, in A. Sarlemijn and P. Kroes (eds.), *Between Science and Technology* (North-Holland, Amsterdam), pp. 97–115; also in [45, pp. 910–28; 55 (second ed. only), pp. 232–48].

[53] Davies, Paul C.W. and Brown, Julian R. 1986, *The Ghost in the Atom* (Cambridge University Press, Cambridge). The interview with John Bell is on pp. 45–57.

[54] Bell, John S. 1976, How to teach special relativity, *Progress in Scientific Culture* **1**, no. 2, pp. 1–13; also in [45, pp. 755–63; 55, pp. 67–80].

[55] Bell, John S. 1987, *Speakable and Unspeakable in Quantum Mechanics* (Cambridge University Press, Cambridge).

[56] Bell, John S. 1992, George Francis Fitzgerald (Denis Weaire (ed.)), *Physics World* **5** (September), 31–5; also in [45, pp. 929–33].

[57] Selleri, Franco 2002, Bell's spaceships and special relativity, in [60], pp. 413–28.

[58] Whitaker, Andrew 2012, *The New Quantum Age* (Oxford University Press, Oxford).

[59] Hey, Tony and Walters, Patrick 2003, *The New Quantum Universe* (Cambridge University Press, Cambridge). Also private communication from Tony Hey.

[60] Bertlmann, Reinhold and Zeilinger, Anton (eds.) 2002, *Quantum [Un]speakables* (Springer, Berlin).

[61] Ellis, John and Amati, Daniele (eds.) 2000, *Quantum Reflections* (Cambridge University Press, Cambridge).

2

Recollections of John Bell

MICHAEL NAUENBERG

It is a pleasure to contribute to this anthology some of my recollections of John Bell.

I first met him at SLAC, where we were visitors during 1964–5 when he was on leave from CERN, and I was on leave from the Columbia University Physics Department. Soon I found that we had a common interest in the foundations of quantum mechanics, and we had lively discussions on this subject. We were concerned with the "reduction of the wave packet" after an experiment in quantum mechanics was completed, and wrote a tongue-in-cheek article on this subject for a *festschrift* in honor of Viki Weisskopf [1], stating that

> We emphasize not only that our view is that of a minority but also that current interest in such questions is small. The typical physicist feels that they have long been answered, and that he will fully understand just how, if ever he can spare twenty minutes to think about it.

Now, 50 years later, this topic, known as the *measurement problem*, has become the source of numerous articles representing many different viewpoints. At the time, however, I confess that I did not realize that the main flaw in John von Neumann's presentation of this problem was his assumption that a measuring device can be represented by a pointer with only two quantum states. Actually, a measuring device must be able to *record* the outcome of an experiment, and for this purpose it has to have an enormous number of quantum states, leading to an irreversible macroscopic transition in the device. A correct discussion of the measurement problem was given, for example, by Nico van Kampen [2], but recently he told me that he was unable to persuade Bell, who continued to be concerned about this problem for the rest of his life. Indeed, shortly before he died, he wrote a diatribe [3] entitled "Against Measurement," where he continued to argue that this problem constituted a fundamental flaw in quantum mechanics. Kurt Gottfried, who referred to this problem as Bell's second major theme, called his article "a fervent jeremiad summarizing a lifetime of reflection on what he saw as the fatal flaws of the orthodox theory" [4].

Concerning his disagreement with van Kampen, Bell wrote,

> Let us look at one more good book, namely Physica A 153 (1988), and more specifically at the contribution: 'Ten theorems about quantum mechanical measurements', by N. G. van Kampen. This paper is distinguished especially by its robust common sense. The author has no patience with '... such mind-boggling fantasies as the many world interpretation ...'. He dismisses out of hand the notion of

von Neumann, Pauli, Wigner that 'measurement' might be complete only in the mind of the observer: '... I find it hard to understand that someone who arrives at such a conclusion does not seek the error in his argument'. For vK '... the mind of the observer is irrelevant ... the quantum mechanical measurement is terminated when the outcome has been macroscopically recorded ...'. Moreover, for vK, no special dynamics comes into play at 'measurement': '... The measuring act is fully described by the Schrödinger equation for object system and apparatus together. The collapse of the wavefunction is a consequence rather than an additional postulate ...'. [5]

Bell concluded that van Kampen's kinematics "is of the de Broglie-Bohm 'hidden variable' dual type," but this claim is not justified. One of the most eminent physicists of the twentieth century, Rudolph Peierls, a pioneer in the development of quantum mechanics, who had supervised Bell when he was his graduate student, also disagreed with him on this issue. In a reply to Bell's article entitled "In Defence of Measurement," he wrote,

He [Bell] regarded it as necessary to have a clearly formulated presentation of the physical significance of the theory without relying on ill-defined concepts. I agree with him that this is desirable, and, like him, I do not know of any textbook which explains these matters to my satisfaction. I agree in particular that the books he quoted do not give satisfactory answers (I assume that they are fairly quoted; I have not re-read them). But I do not agree with John Bell that these problems are very difficult. I think it is easy to give an acceptable account, and in this article I shall try to do so. I shall not aim at a rigorous axiomatic, but only at the level of the logic of the working physicist. [6]

But this "logic" is what Bell called proof by FAPP=for all practical purposes.

In 1984 John received an invitation from Cambridge University Press to publish a collection of his papers on quantum philosophy, and he asked me whether I had any objections to including in it our joint paper on the measurement problem. By then I had lost interest in this subject, but I agreed when he wrote to me that "one of those [papers] I still like is the one that you and I wrote together for the Viki Weisskopf book." A copy of his letter is reproduced in Fig. 2.1 because it reveals his characteristic sense of humor when he ended with "... you might not want to remind people that you got involved with this." He was right.

In the late 1960s, I often met Bell during my frequent visits to CERN. We regularly had lunch with other physicists at the CERN cafeteria, and in retrospect, it is remarkable that during that time, Bell never brought up his seminal theorem that quantum mechanics cannot be reproduced by local hidden variables. I think that a plausible reason was that some of our common friends and lunch companions at the time, who included Martinus Veltman and Jack Steinberger (both later received the Nobel Prize for their contributions to particle physics), did not express any interest in this subject. Recently, I asked Veltman for this recollections, and he responded,

Frankly, I never discussed hidden variables with John. In 1963 both John and I were at SLAC, where he wrote his famous paper. I was not interested in hidden variables at all; John once told me that he was trying to definitely silence Jauch who was apparently trying to convince John about hidden variables. I remember saying that I gladly left it to him, and we collaborated on other things .... [7]

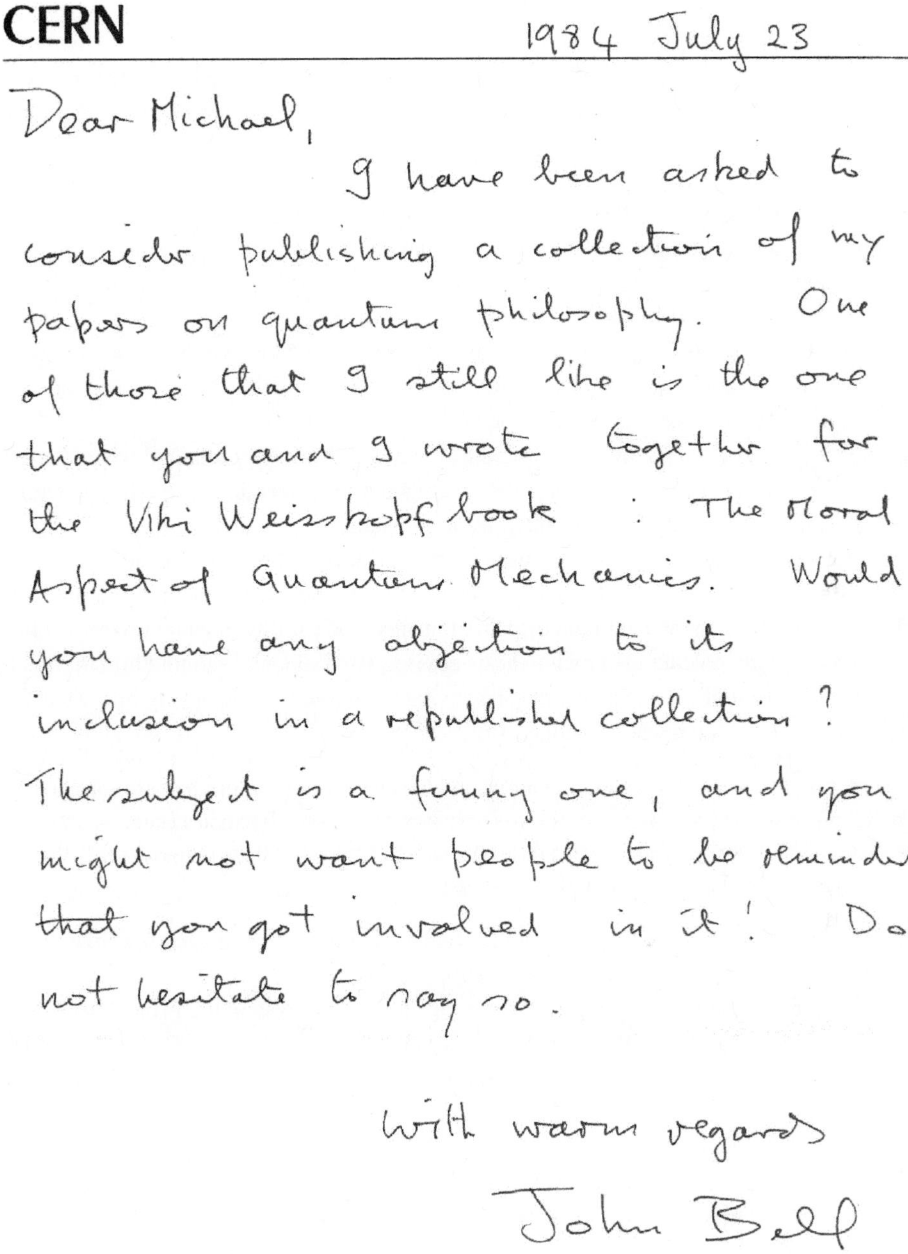

CERN   1984 July 23

Dear Michael,

I have been asked to consider publishing a collection of my papers on quantum philosophy. One of those that I still like is the one that you and I wrote together for the Viki Weisskopf book : The Moral Aspect of Quantum Mechanics. Would you have any objection to its inclusion in a republished collection? The subject is a funny one, and you might not want people to be reminded that you got involved in it! Do not hesitate to say no.

With warm regards

John Bell

Figure 2.1 A letter from John Bell to the author.

But Bell's comments in high-energy physics were always appreciated. For example, Veltman and I discussed with him some issues related to current algebra and gauge variance that later led to an article by Bell on this subject [8]. I also recall that at the time we were both interested in the decay of the neutral K meson, and Bell wrote an excellent review article on this subject with Steinberger [9].

In the 1980s, Bell and I attended a meeting of the American Academy of Arts and Sciences in Boston, MA. One of the presentations was given by Alain Aspect on his experiment on two-photon spin correlations of an entangled state with total spin zero demonstrating that these correlations were in agreement with quantum mechanics, and in violation of Bell's inequality based on local hidden variables [10]. Bell had bought two tickets to a special retrospective exhibit of Renoir's paintings at the Boston Art Museum, and since his wife Mary could not come, he invited me to join him. While walking along the galleries he told me that a few years earlier, Alain had come to visit him at CERN to discuss his plan for an experiment to test his inequality. Bell asked him whether he had tenure, and when Aspect responded that his position was secure, he encouraged him to pursue his experiment, which closed a *loophole* in an earlier experiment by John Clauser.

Clauser was a graduate student in astronomy at Columbia University when he stumbled on Bell's paper in the physics library and became very excited about the possibility of carrying out an experiment that might disprove the universal validity of quantum mechanics. But he was strongly discouraged by his thesis adviser, Pat Taddeus, who told him that such an experiment would be a waste of time. Then Clauser wrote to Bell, asking him what the thought about doing such an experiment. Bell responded,

In view of the general success of quantum mechanics, it is very hard for me to doubt the outcome of such experiments. However, I would prefer these experiments, in which the crucial concepts are very directly tested, to have been done, and the results on record. Meanwhile there is always the chance of an unexpected result, which would shake the world. [11]

Clauser also visited Richard Feynman at Cal Tech to discuss his proposed experiment, but Feynman also told him that he would be wasting his time.

Earlier, a former student of Wigner, Abner Shimony, who was in the Princeton philosophy department, had also become interested in an experimental test of Bell's inequality. When he read an abstract on this subject that Clauser had submitted to an APS meeting, he immediately contacted him. Afterwards, together with his student, Michael Horne, and with Francis Pipkin's student at Harvard, Richard Holt, they collaborated on a paper that discussed in detail how such an experimental test could be performed [12]. After graduating from Columbia University, Clauser obtain an astrophysics post doctoral position with Charles Townes at UC Berkeley and succeed in convincing him that his proposed atomic physics experiment was worth doing. As luck would have it, Gene Commins, with his student Carl Kocher, had carried out an experiment measuring the polarization correlation between two entangled photons in a state of spin zero from the decay of calcium atoms. But they had only considered two special cases where the polarization analyzers where either aligned or orthogonal to each other [13]. Together with Commins' graduate student,

Stuart Freedman, Clauser continued their experiment by building polarization analyzers that could be rotated at an arbitrary relative angle and carried out the first experimental test of Bell's inequality. As is well known, they obtained results in good agreement with the predictions of quantum mechanics [14]. Twelve years later, after Aspect presented his results, Feynman also changed his mind about the relevance of this experimental test, and after a seminar that Aspect gave at Cal Tech on his experiment, Feynman wrote to him that "… your talk was excellent" [15].

The last time I saw Bell was at the end of the summer of 1989, when he invited me to present my work on the motion of wavepackets in a Coulomb potential [16] at the weekly CERN theory seminar. Schrödinger had been unable to solve this problem, and he maintained an incorrect interpretation of the absolute square of such a wavepacket as the charge or mass density of a quantum particle, disagreeing with Born's correct probability interpretation. During this talk, I emphasized that initially the wavepacket gave a distribution very similar to that of a classical ensemble of particles rotating around the center of force. But the particles in such an ensemble spread, and when the head of the packet catches up with its tail, which occurs after a well-defined interval of time that depends on the mean principal quantum number of the wavepacket, this quantum–classical correspondence ceases to be valid due to wave interference in quantum mechanics. After the seminar, I suggested to Bell that he submit a proposal to the Santa Barbara Institute of Theoretical Physics (ITP) for a workshop on problems in the foundations of quantum mechanics. He agreed, provided that I would do the leg work. At the end of our conversation he walked out with me through the long corridors of CERN until we reached the parking lot. That was the last time I saw him; he died unexpectedly a few months later.

Outside the community of experts in his field, recognition of Bell's seminal insights has been slow in coming, but this is changing now. In particular, Belfast, the Irish city where he was born and first educated, has dedicated the entire month of November 2014 to celebrate the 50th anniversary of the publication of his famous theorem. To honor Bell, Belfast also plans to rename various streets and buildings with his name.

References

[1] J.S. Bell and M. Nauenberg, The moral aspect of quantum mechanics, in *Preludes in Theoretical Physics*, edited by A. de Shalit, H. Feshbach, and L. Van Hove (North Holland, Amsterdam 1966); reproduced in J.S. Bell, *Speakable and Unspeakable in Quantum Mechanics* (Cambridge University Press, New York, 1987), pp. 22–28.

[2] N.G. van Kampen, Ten theorems about quantum mechanical measurements, *Physica A* **153**, 97 (1988).

[3] I use the term *diatribe* in the sense of "an ironic or satirical criticism".

[4] M. Bell, K. Gottfried and M. Veltman (eds.), *Quantum Mechanics, High Energy Physics and Accelerators: Selected Papers of John S. Bell* (World Scientific, Singapore, 1995), p. 8.

[5] J.S. Bell, Against measurement, *Physics World*, 33 August (1990).

[6] R. Peierls, In defence of measurement, *Physics World*, 19 January (1991).

[7] M. Veltman, private communication.

[8] J.S. Bell, Current algebra and gauge variance, *Nuovo Cim.* **50**, 129 (1967).

[9] J.S. Bell and J. Steinberger, Weak interaction of kaons, in L. Wolfenstein (ed.), *Oxford International Symposium Conference on Elementary Particles*, Sept. 1965, pp. 42–57.

[10] A. Aspect, Proposed experiment to test the non separability of quantum mechanics, *Phys. Rev. D* **14**, 1944 (1976); A. Aspect, P. Grangier, and G. Roger, Experimental tests of realistic local theories via Bell's theorem, *Phys. Rev. Lett.* **47**, 460 (1981); Experimental realization of Einstein-Podolsky-Rosen-Bohm Gedankenexperiment: A new violation of Bell's inequality, *Phys. Rev. Lett.* **49**, 91 (1982).

[11] L. Gilder, *The Age of Entanglement: When Quantum Physics Was Reborn* (Knopf, New York, 2008), p. 256.

[12] J.S. Clauser, M.A. Horne, A. Shimony and R. Holt, Proposed experiment to test local hidden-variable theories, *Phys, Rev. Lett.* **23**, 880 (1969).

[13] C.A. Kocher and E.D. Commins, Polarization correlation of photons emitted in an atomic cascade, *Phys. Rev. Lett.* **18**, 575 (1967).

[14] S.J. Freedman and J.F. Clauser, Experimental test of local hidden-variable theories, *Phys. Rev. Lett.* **28**, 938 (1972).

[15] A. Aspect, private communication.

[16] M. Nauenberg, Quantum wavepackets on Kepler elliptical orbits, *Phys. Rev. A* **140**, 1133 (1989); Wave packets: Past and present, in J.A. Yeazell and T. Uzer (eds.), *The Physics and Chemistry of Wave Packets*, (Wiley, New York, 2000), pp. 1–30.

3

John Bell: Recollections of a Great Scientist and a Great Man

GIANCARLO GHIRARDI

3.1 Introduction

This contribution to the book in honour of J. S. Bell will probably differ from the remaining ones, in particular since only a part of it will be devoted to specific technical arguments. In fact I have considered it appropriate to share with the community of physicists interested in the foundational problems of our best theory the repeated interactions I had with him in the last four years of his life, the deep discussions in which we have been involved in particular in connection with the elaboration of collapse theories and their interpretation, and the contributions he gave to the development of this approach at a formal level, as well as championing it on repeated occasions.[1] In brief, I intend to play here the role of one of those lucky persons who became acquainted with him personally, who has exchanged important views with him, who has learned a lot from his deep insight and conceptual lucidity, and, last but not least, whose scientific work has been appreciated by him.

Moreover, due to the fact that this book intends to celebrate the 50th anniversary of the derivation of the fundamental inequality which bears his name, I will also devote a small part of the text to recalling his clearcut views about the locality issue, views that I believe have not been grasped correctly by a remarkable part of the scientific community. I will analyze this problem in quite general terms at the end of the paper.

3.2 Some of Bell's Scientific Achievements

Bell received bachelor's degrees in experimental physics and in mathematical physics at Queen's University of Belfast in the years 1948 and 1949, respectively, and a PhD in physics at the University of Birmingham. We all know very well that already at that time he was absolutely unsatisfied with the conceptual structure of quantum mechanics and with the way in which it was taught. This is significantly expressed by the statement he made during an interview to Jeremy Bernstein: *I remember arguing with one of my professors, a Doctor Sloane, about that. I was getting very heated and accusing him, more or less, of dishonesty. He was getting very heated too and said, 'You're going too far.'*

---

[1] I have also added various pictures taken from letters by John, and/or illustrating important moments of our interactions during his last years.

He began his career working at the Atomic Energy Research Establishment at Harwell, Oxfordshire, but he soon joined the accelerator design group at Malvern. There he met Mary Ross, whom he married in 1954. To summarize the enormous relevance of this event it seems sufficient to mention that when writing the preface of his collected works on quantum mechanics he stated, *I here renew very especially my warm thanks to Mary Bell. When I look through these papers again I see her everywhere.* Subsequently they moved to CERN, the Centre for European Nuclear Research in Geneva, and John worked almost exclusively on particle physics and on accelerator design. However, quantum theory was his hobby, perhaps his obsession. And it made him famous. Much more about this in what follows.

Concerning his first scientific activity, let me stress that modelling the paths of charged particles through accelerators in these days before electronic computers became available required a rigorous understanding of electromagnetism, and the insight and judgment to make the necessary mathematical simplifications to render the problem tractable on mechanical calculators, while retaining its essential physical features. Bell's work was masterly. We cannot avoid mentioning that in this period he gave some clear indications concerning the effect of *strong focusing*, which has played such a relevant role in accelerator science.

In 1953 we find him, during a year's leave of absence, at Birmingham University with Rudolf Peierls. In this period he did work of paramount importance [1], producing his version of the CPT theorem, independent of Gerhard Lüders and Wolfang Pauli, who got all the credit for it.[2] Subsequently, both John and Mary moved to CERN. Here they spent almost all their careers.

In 1967 he produced another important piece [2] in elementary particle theory: he pointed out that many successful relations following from current algebra, and in fact current algebra itself, can be seen as a consequence of gauge invariance. However, by far the most important work by John in the field of elementary particle theory was the 1969 one [3] with Jackiw in which they identified what has become known as the Adler–Bell–Jackiw anomaly in quantum field theory. This work solved an outstanding problem in the theory of elementary particles, and over the subsequent thirty years the study of such anomalies became important in many areas of particle physics.

3.3 Bell and the Foundational Problems

3.3.1 A Brief Picture

In spite of the extremely relevant papers mentioned in the previous section, I cannot forget that, as I have stated above, quantum theory was his hobby, perhaps his obsession. Actually when in 1963–4 he left CERN for SLAC, he concentrated his attention almost exclusively on the foundations of this theory. It is not surprising that the most relevant theoretical paper

---

[2] In M. Veltman's words, *John's article was conceived independently. It is of course very different, and perhaps today more relevant, than the rather formal field theory arguments of Lüders. In his introduction John acknowledged the paper of Lüders, and never thought to even suggest that his work had been done "independently".*

which attracted him was the celebrated EPR paper [4], a work which, taking for granted the local nature of physical processes, challenges, in the authors' intentions, the completeness of quantum theory. This conclusion entails that Einstein can be regarded as *the most profound advocate of the hidden variables*, as Bell, quoting A. Shimony, made clear in Ref. [5]. So Bell shifted his attention to hidden variable theories, and more specifically, to what [6] had been for him *a revelation*: Bohmian mechanics. I would like to stress that this theory was extremely interesting to him for two main reasons: *the elimination of indeterminism* but *more important . . . the elimination of any need for a vague division of the world into 'system' on the one hand, and 'apparatus' or 'observer' on the other.*

Two problems were strictly related to this new perspective. First of all, J. von Neumann had proved [7] that no deterministic completion of quantum mechanics was possible, in principle. How could this be reconciled with the existence and consistency of Bohm's theory? The essential contribution of John, after his arrival to SLAC, was to show [8] that von Neumann's argument was based on a logically unnecessary necessary assumption.[3] The second problem arose from the fact that Bohmian mechanics exhibited a very peculiar feature: it was basically nonlocal. John tried hard to work out a similar theory which was free of nonlocality, but he did not succeed. So he entertained the idea that one might prove that nonlocal features would characterize any theory whatsoever which reproduces the quantum predictions. With this in mind he conceived and wrote his fundamental paper [9] in which he derived his celebrated and revolutionary inequality.

I would like to stress that I consider this a result which makes John one of the greatest physicists of the past century, since he has made crystal clear something which nobody had ever conceived and which implies a radical change in our views about the world around us: Nature is nonlocally causal!

3.3.2 The Shifty Split

The other problem of quantum theory which, as remarked above, had worried John since his university times is the one of its resorting to two different dynamical evolution principles: the first one described by the *linear and deterministic* Schrödinger equation and the second one taking place when measurements are performed and described by the projection postulate of J. von Neumann, which is fundamentally *nonlinear and stochastic*. And this is not the only point. The crucial fact is that there is nothing in the theory which marks in any sense the borderline between the ranges of applicability of the two dynamical principles just mentioned. This fact specifically worried John and he referred to it as *the shifty split*. His position is wonderfully summarized by his sentence:

---

[3] The publication of the paper he wrote on this subject, for various reasons, was delayed until 1966. In connection with this paper two remarks are in order. First, in 1935 Grete Herman had already proved that von Neumann's argument was circular. Her contribution has been completely ignored by the scientific community, which made systematic reference to von Neumann's book as "The Gospel." Second, it is important to remember that in his paper, while relating his derivation to Gleason's theorem, Bell stresses that in the case of dispersion-free states, to avoid a contradiction one must give up the assumption that *the measurement of an observable must yield the same value independent of what other measurements may be made simultaneously*. In brief, he has also clarified the unavoidably contextual nature of any deterministic completion of quantum mechanics, a point of remarkable relevance which I have not put into evidence in the main text since I will not make reference to it in what follows.

There is a fundamental ambiguity in quantum mechanics, in that nobody knows exactly what it says about any particular situation, for nobody knows exactly where the boundary between the wavy quantum world and the world of particular events is located . . . every time we put that boundary – we must put it somewhere – we are arbitrarily dividing the world into two pieces, using two quite different descriptions . . .

This fundamental question, which had been the subject of a long-lasting and vivid debate between the founders of the theory, is a theme to which Bell returned continuously from the second half of the sixties up to his last days. But this is not the whole story: he also analyzed many of the proposed *solutions* to this problem and he proved that almost all of them are characterized by imprecise, vague, verbal assumptions aimed at avoiding to face the real contradiction which, in a quantum view, occurs between the *waviness of quanta* and *our definite perceptions.*

Actually, it was just this position which led him to pay specific attention, first of all (i.e., from 1964 on) to Bohmian mechanics [10] and its variants, and, second, in the last five years of his life, to the collapse model [11] that we (Rimini, Weber and myself) presented for the first time in 1984 in a very concise form and then discussed in great detail [12] in 1985. These facts allow me to pass now to the real core of my contribution: describing the many interactions we had and how useful they have been for me and my colleagues.

3.4 John's Interest in Our Work

Obviously we knew John and his fundamental contributions very well, and we considered him by far the greatest scientist in our field (and not only in our field).

Everybody can imagine our surprise and pleasure, after we had sent the preprint of our paper to the CERN library, to receive a letter from John starting with the following sentence (Fig. 3.1):

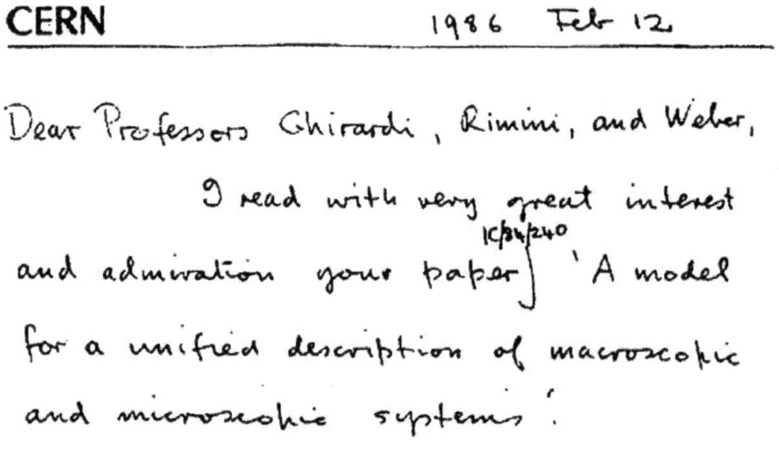

Figure 3.1 The first letter we got from John.

After this significant appreciation, the scientist Bell appears with an absolutely appropriate remark (Fig. 3.2):

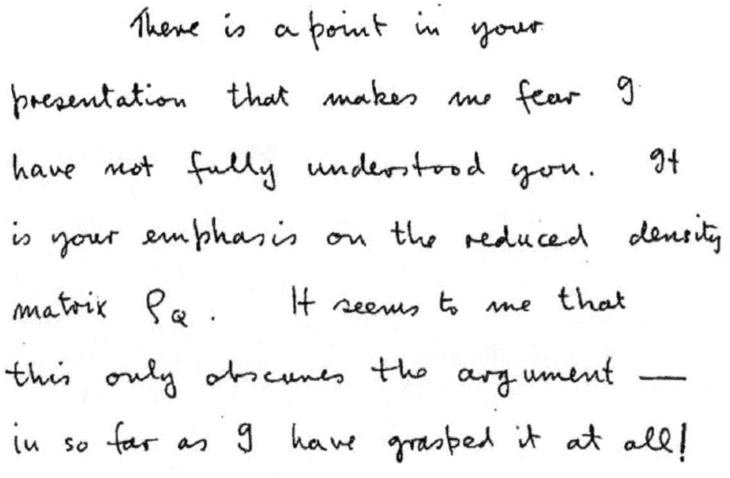

There is a point in your presentation that makes me fear I have not fully understood you. It is your emphasis on the reduced density matrix ρ_Q. It seems to me that this only obscures the argument — in so far as I have grasped it at all!

Figure 3.2 A remark by John.

The problem is that of ensemble versus individual reductions. There is no doubt that our work dealt with individual reductions (and we mentioned this briefly in various parts of the paper), but, to perform the complete calculation of the dynamics of a free particle, we resorted to statistical operator language. John's letter continues with other general remarks and concludes with a sentence (Fig. 3.3) which clearly shows how great and open-minded John was as a man:

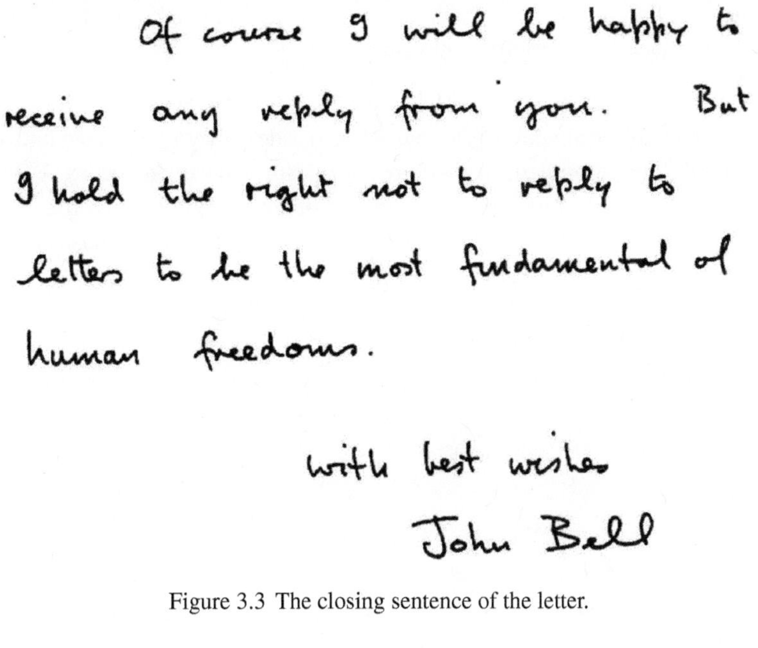

Of course I will be happy to receive any reply from you. But I hold the right not to reply to letters to be the most fundamental of human freedoms.

with best wishes
John Bell

Figure 3.3 The closing sentence of the letter.

3.5 Collapse or GRW Models

Before proceeding, a short summary of what we had done in 1984–5, i.e., to present a proposal to solve the measurement problem, is in order. Our approach is based on the idea that [13]

Schrödinger's equation is not always true.

Actually, we suggested modifying the standard evolution equation by adding nonlinear and stochastic terms which strive to induce WPR at the appropriate level, leading to states which correspond to definite macroscopic outcomes. The theory, usually referred as the GRW theory, is a rival theory of quantum mechanics and is experimentally testable against it. Its main merit is that it qualifies itself as a precise example of a unified theory governing all natural processes, in full agreement with quantum predictions for microscopic processes, and inducing the desired objectification of the properties of macroscopic systems. Let us be precise about it.

- The first problem to be faced is the choice of the so-called preferred basis: if one wants to objectify some properties, which ones have to be privileged? The natural choice is that of choosing the position basis, as suggested by Einstein [14]:

  A macrobody must always have a quasi-sharply defined position in the objective description of reality.

- The second problem, and the more difficult, is to embody in the scheme a triggering mechanism implying that the modifications to the standard theory are absolutely negligible for microsystems while they have remarkable (and appropriate) effects at the macroscopic level.

The theory is based on the following assumptions:

- Let us consider a system of N particles and let us denote as $\psi(\mathbf{r}_1, \ldots, \mathbf{r}_N)$ the configuration space wave function. The particles, besides obeying the standard Hamiltonian evolution, are subjected, at random times with a mean frequency λ, to random and spontaneous localization processes around appropriate positions. If a localization affects the ith particle at point $\mathbf{x}$, the wave function is multiplied by a Gaussian function $G_i(\mathbf{x}) = (\frac{\alpha}{\pi})^{3/4} exp[-\frac{\alpha}{2}(\mathbf{r}_i - \mathbf{x})^2]$.
- The probability density of a localization taking place for particle i and at point $\mathbf{x}$ is given by the norm of the function $G_i(\mathbf{x})\psi(\mathbf{r}_1, \ldots, \mathbf{r}_N)$. This implies that localizations occur with higher probability where, in the standard theory, there is a high probability of finding the particle.
- Obviously, after the localization has occurred, the wave function has to be normalized again.

It is immediate that a localization, when it occurs, suppresses the linear superposition of states in which the same particle is well localized at different positions separated by a distance larger than $1/\sqrt{\alpha}$.

However, the most important feature of the model is its trigger mechanism. To understand its basic role, let us consider the superposition of two macroscopical pointer states, $|H\rangle$ and $|T\rangle$, corresponding to two macroscopically different locations of the pointer's c.o.m. Taking into account that the pointer is "almost rigid" and contains on the order of Avogadro's number of microscopic constituents, one immediately realizes that a localization of any one of them suppresses the other term of the superposition: the pointer, after the localization of one of its constituents, is definitely either Here or There.

With these premises we can choose the values of the two constants (which Bell considered as new constants of nature) of the theory: the mean frequency of the localizations λ and their accuracy $1/\sqrt{\alpha}$. These values have been taken (with reference to the processes suffered by nucleons, since it is appropriate to make the frequency λ proportional to the mass of the particles) to be

$$\lambda = 10^{-16}\,\mathrm{s}^{-1}, \quad \frac{1}{\sqrt{\alpha}} = 10^{-5}\,\mathrm{cm}. \tag{3.1}$$

It follows that a microscopic system suffers a localization, on the average, every hundred million years. This is why the theory agrees to an extremely high level of accuracy with quantum mechanics for microsystems. On the other hand, due to the trigger mechanism, one of the constituents of a macroscopic system, and, correspondingly, the whole system, undergoes a localization every 10^{-7} seconds.

A few comments are in order:

- The theory makes it possible to locate the ambiguous split between micro and macro, reversible and irreversible, quantum and classical. The transitions between the two regimes are governed by the number of particles which are well localized at positions further apart than 10^{-5} cm in the two states whose coherence is going to be dynamically suppressed.
- The theory is testable against quantum mechanics, and various proposals in this sense have been put forward [15–20]. The tests are difficult to perform with current technology, but the model clearly identifies appropriate sets of mesoscopic processes which might reveal the limited validity of the superposition principle.
- Most of the physics does not depend separately on the two parameters of the theory, but only on their product $\alpha\lambda$, and a change of few orders of magnitude of its value will already conflict with experimentally established facts. So, in spite of its appearing ad hoc, if one chooses to make the positions objective (we mention that one can prove that making variables involving the momenta objective leads to a nonviable theory), not much arbitrariness remains.

An interesting feature of the theory deserves a comment. Let us make reference to a discretized version of the model. Suppose we are dealing with many particles and, accordingly, we can disregard the Schrödinger evolution of the system because the dominant effect is the collapse. Suppose that we divide the universe in to elementary cells of volume $10^{-15}\,\mathrm{cm}^3$, the volume related to the localization accuracy. Denote as $|n_1, n_2, \ldots\rangle$ a state in which there are n_i particles in the ith cell and let us consider the superposition of two states $|n_1, n_2, \ldots\rangle$

and $|m_1, m_2, \ldots\rangle$ which differ in the occupation numbers of the various cells. It is then quite easy to prove that the rate of suppression of one of the two terms is governed by the quantity

$$\exp\left\{-\lambda t \sum_i (n_i - m_i)^2\right\}, \tag{3.2}$$

the sum running over all cells of the universe.

It is interesting to remark that the above equation, where $\lambda = 10^{-16}\,\mathrm{s}^{-1}$, if one is interested in time intervals on the order of the perceptual times (i.e., about $10^{-2}\,\mathrm{s}$), implies that the universal dynamics characterizing the theory does not allow the persistence for perceptual times of a superposition of two states which differ in the fact that 10^{18} nucleons (Planck's mass) are differently located in the whole universe. This remark establishes some interesting connections between the collapse models and the important suggestion by Penrose [21], who, to solve the measurement problem by following the quantum gravity line of thought, has repeatedly claimed that it is the Planck mass which should define the boundary between the wavy quantum universe and the one in which the superposition principle fails and, in particular, the world of our definite perceptions emerges.

3.6 John Bell and the GRW Model

3.6.1 Our First Personal Contacts

The first time I personally met John was at the Imperial College (London) at the centenary celebration [13] of Schrödinger. Before this event John wrote to us the letter shown in Fig. 3.4, in which he anticipated that he was going to discuss our work. After having delivered his talk he immediately exhibited his great generosity by telling us: You should have delivered it in place of me!

At that time our proposal had still a big problem: it did not preserve the (anti)-symmetrization principle for identical constituents. I remember that during the official dinner of the meeting we (Rimini and I) discussed this problem with him, and he wrote some formulae on a paper napkin. In particular, we discussed seriously reductions in which the dynamics strives to make objective the number of particles in an appropriate volume. The collapse process, which is accounted for by the operator which one should apply to the state vector when a stochastic process occurs, would then have taken the form

$$|\Psi\rangle \to e^{-\beta[N(\mathbf{r})-n]^2}|\Psi\rangle, \quad N(\mathbf{r}) = \int_V a^\dagger(\mathbf{r})a(\mathbf{r})d\mathbf{r}. \tag{3.3}$$

Here $a^\dagger(\mathbf{r})$ and $a(\mathbf{r})$ are the creation and annihilation operators for a particle at point $\mathbf{r}$ and V is the characteristic localization volume. As usual, the state vector has then to be normalized. The reader will easily realize that this process suppresses superpositions of states with different numbers of particles in V, and that, being expressed in terms of the creation and annihilation operators, it automatically respects the symmetry conditions for identical constituents.

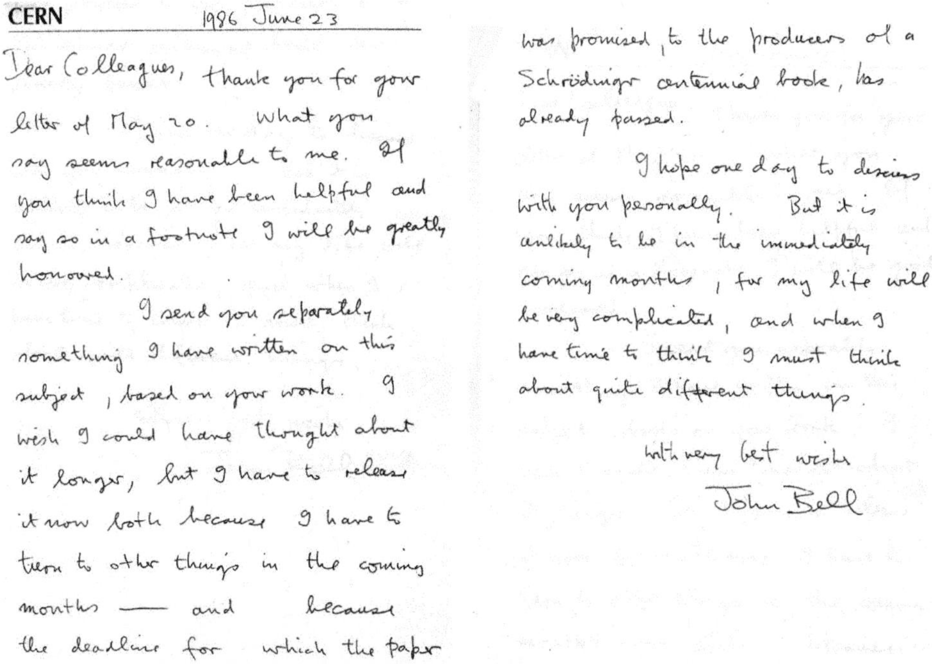

CERN 1986 June 23

Dear Colleagues, thank you for your letter of May 20. What you say seems reasonable to me. If you think I have been helpful and say so in a footnote I will be greatly honoured.

I send you separately something I have written on this subject, based on your work. I wish I could have thought about it longer, but I have to release it now both because I have to turn to other things in the coming months — and because the deadline for which the paper was promised, to the producers of a Schrödinger centennial book, has already passed.

I hope one day to discuss with you personally. But it is unlikely to be in the immediately coming months, for my life will be very complicated, and when I have time to think I must think about quite different things.

With very best wishes

John Bell

Figure 3.4 John's letter before the conference in London.

The proposal represents a very obvious and simple way to overcome the problem we were facing, and from a physical point of view it leads to results quite similar to those of the original collapse model [12]. However, we were not fully satisfied with it because it required the introduction of a new parameter in to the theory. In fact, besides the frequency of the stochastic processes and the localization volume V, it involved the parameter β governing the rate of suppression of states with different number of particles in this volume. We were not very keen to add new phenomenological parameters to the theory. In contrast, John did not feel so uneasy in doing so. In fact, when I met him a few months later in Padu, he told me, *if you do not write the identical constituents paper I will write it!*

3.6.2 Enter Philip Pearle

John Bell played also an important role in the subsequent development of collapse theories. P. Pearle, for a long time, had argued [22] that the measurement problem had to be solved by resorting to a stochastic modification of Schrödinger's equation, but he had not been able to identify an appropriate preferred basis and a dynamics implying the trigger mechanism. In 1986, Pearle wrote to Bell, asking whether he could spend a sabbatical year at CERN interacting with him on foundational problems. Bell replied suggesting that he come to Trieste, and wrote a letter to us (Fig. 3.5) concerning this matter. It is interesting to see the reasons that John puts forward for his proposal, reasons which, on one side, make clear

ORGANISATION EUROPÉENNE POUR LA RECHERCHE NUCLÉAIRE

EUROPEAN ORGANIZATION FOR NUCLEAR RESEARCH

CERN   CH-1211 GENÈVE 23   SUISSE/SWITZERLAND
Téléphone: GENÈVE (022) 83 6111   Telex: 419000 CER CH - Télégramme: CERNLAB-GENÈVE

SIÈGE: GENÈVE, SUISSE

13 November 1986

Professor G.C. Ghirardi
Dipartimento di Fisica Teorica
Università di Trieste
Strada Costiera 11
I-34014 - TRIESTE

Professor A. Rimini
Dipartimento di Fisica Nucleare
e Teorica
Università di Pavia
Via A. Bassi 6
I-27100 - PAVIA

Dear Colleagues,

   I enclose a copy of a letter from Philip Pearle. As you will see he
has a sabbatical year coming, and is considering where to go. He is
interested in the problem of quantum mechanics, so CERN would not be a
good place. There would be only me to talk to, and most of my time goes
on efforts more appropriate for this laboratory. He would feel isolated
and frustrated here. It would be a bit frustrating for me too, for even
without anyone to talk to here I have difficulty keeping away from the
quantum question.

   So, as he himself suggests, I pass his letter on to you, in case you
might consider inviting him. I have a very good impression of Philip
Pearle. Before your own work, his work on spontaneous wave packet
reduction seemed to me the most serious in the literature. You could get
an idea of his work from Phys.Rev. D33 (1986) 2240, and the references
there.

   Thank you for the paper of Benatti et al., which I enjoyed reading.

   With best wishes,

                                    Yours sincerely,

                                    John Bell

                                    J.S. Bell

Figure 3.5 John's letter supporting the visit of P. Pearle.

Figure 3.6 Erice's conference in 1989.

how involved he was in other problems and, at the same time, reveal his continuous desire to deal with foundational issues.[4] Obviously we reacted immediately, and so Philip spent a reasonably long period in Trieste interacting with me and Renata Grassi. Subsequently he spent a few months in Pavia, invited by A. Rimini, who had been given a chair there. This stay of Pearle gave rise to an extremely useful collaboration between us. But, more importantly, this contact allowed him to grasp the precise spirit and the technical details of our work, so that, integrating the new ideas in to the stochastic evolution equations he was investigating, he produced the elegant version [23] of collapse model which became known as CSL (continuous spontaneous localization).

This proposal, which physically has effects quite similar to those of ours, is formulated in a much more elegant way than GRW and satisfies the quantum requirement for systems with identical constituents, yielding a solution to our problem without requiring further parameters.

3.6.3 The Erice 89 Meeting

The subsequent relevant event at which we met was the Conference *Sixty-two years of uncertainty* organized by A. Miller in August 1989 at Erice. Readers involved in foundational problems will have no difficulty in identifying some important scientists and philosophers who were present by looking at the photo of the participants as shown in Fig. 3.6.

---

[4] It seems appropriate to recall that one time Bell said, *I am a Quantum Engineer, but on Sundays I Have Principles.*

During this meeting, the problem of the interpretation (today one would say the ontology) of collapse models has seen an interesting development. To make things clear, I will begin by mentioning that in his presentation of the GRW theory at Schrödinger's centenary conference Bell had proposed a very specific interpretation strictly connected with his firm conviction concerning the necessity of making clear what are the "beables" of any scientific theory. His proposal has been denoted recently [24] as the *flash ontology*. I summarize it by resorting to the precise words he used in *Are There Quantum Jumps?*:

The collapse processes *are the mathematical counterparts in the theory to real events at definite places and times in the real world . . . (as distinct from the observables of other formulations of quantum mechanics, for which we have no use here). A piece of matter is then a galaxy of such events.*

As it is stated clearly, the stochastic processes characterizing the theory are taken as the very basic elements of its ontology. However, at Erice, Bell seemed to have changed his mind by attributing an absolutely privileged role to the wave function of a many-particle system in the full configuration space:

The GRW-type theories have nothing in their kinematics but the wave function. It gives the density (in a multidimensional configuration space!) of stuff. To account for the narrowness of that stuff in macroscopic dimensions, the linear Schrödinger equation has to be modified, in this GRW picture, by a mathematically prescribed spontaneous collapse mechanism.

I must confess that this is the only point on which I disagreed with John. I am firmly convinced, like many people who are interested in collapse theories, that they need an interpretation. Limiting all considerations exclusively to the wave function in the $3N$-dimensional configuration space does not lead to a clear picture. One needs to connect the mathematical entities with the reality of the world we live in and with our perceptions about it. For this reason, we [25] proposed what is currently known as the mass-density ontology (as opposed to the flash ontology), which, at the nonrelativistic level, represents a meaningful way to make sense of the implications of the theory. What the theory is assumed to be about is the mass-density distribution in real 3-dimensional space at any time, defined as

$$m(\mathbf{x}, t) = \sum_i m_i \int d\mathbf{r}_1 d\mathbf{r}_2 \dots d\mathbf{r}_N |\psi(\mathbf{r}_1, \mathbf{r}_2, \dots, \mathbf{r}_N)|^2 \delta(\mathbf{r}_i - \mathbf{x}). \qquad (3.4)$$

The mass density interpretation is still at the center of a lively debate involving, among others, philosophers of science.[5] However, John stuck, from then on, to the idea that his wave-function ontology is the appropriate one, as I will briefly describe in the next subsection.

---

[5] After having completed our paper [25] we sent a copy of it to Bas van Fraassen, who answered with the following e-mail: *Dear GianCarlo, your message was almost the first I found when I returned here after the holidays, and it makes me very glad. I have talked with my students about your paper and also brought it up in David Albert's seminar (which he was giving here for the fall term). We all agreed that your paper addresses the most important issue about how to relate QM to the macroscopic phenomena in a truly fundamental and new way. . . . I will explain below why I see this as part of a consensus with discussions about other interpretations of QM. But there is this difference: that you have given, in your discussion of appropriate and inappropriate topologies, an important and even (to my mind) very convincing rationale for this solution.*

Figure 3.7 John and me.

To conclude this part referring to the Erice meeting, allow me to present a personal photo taken at Erice (Fig. 3.7). This is my preferred image among those referring to my professional career, since it calls to my mind the exciting moments of that event, and has for me a deep emotional impact, as any reader will easily understand.

3.6.4 More on Bell's Ontology

It goes without saying that I had various other exchanges with John about interpreting the collapse models. Apparently he kept his position, and, in a letter I sent him, I gave voice once more to my difficulties with the wave function ontology and I asked him whether his satisfaction with it was related in one way or another to the fact that one might utilize it to ground an objective interpretation of things as we perceive them. His clear-cut answer is contained in a letter (Fig. 3.8) he sent to me on October 3, 1989, just one year before his premature death.

3.7 Bell at ICTP

The last time I met personally John was at the Abdus Salam ICTP, in the fall of 1989, on the occasion of the celebrations for the 25th anniversary of the establishment of this institution. All important speakers were Nobel Prize winners, but, fortunately, Abdus Salam was aware of the extreme importance of John's work, and he invited him to deliver a lecture which was chaired by Alain Aspect.

I think this was probably the last public general lecture he delivered, a wonderful speech by the title *First Class and Second Class Difficulties in Quantum Mechanics*. He went through all fundamental problems of quantum theory, he analyzed the (second class) difficulties connected with the divergences afflicting quantum field theories and the attempts to overcome them, describing the Glashow Salam and Weinberg unification and even commenting on string theories. The second half of his talk was entirely devoted to discussing the first class difficulties (those related to the foundations of quantum mechanics) and, within

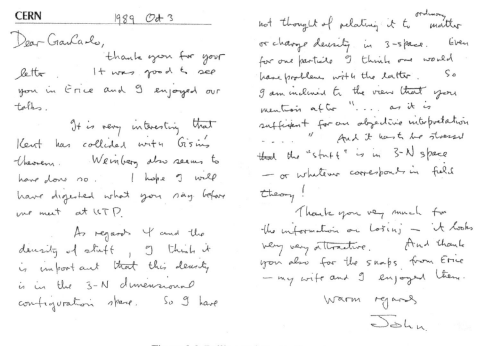

Figure 3.8 Bell's ontology confirmed.

this context, he discussed collapse models and the problems connected with their relativistic generalizations.

Just because this was one of the most stimulating of his talks, when some of my friends decided to organize a meeting for my 70th birthday (2005), I worked hard to get access to the video recording of his lecture, and I succeeded in producing a DVD version of it. However, due to the fact that the registration was not ideal (both from the visual and from the auditory point of view) and due to our desire to make the talk accessible to all scientists, A. Bassi and I decided to "decode" it and to publish it [26] in the special issue of *Journal of Physics A*: The Quantum Universe, collecting papers written by prestigious authors for this occasion.

For the interested reader I am reproducing here an (unfortunately low quality) image of John starting his lecture on November 2, 1989, as well as the beginning of the presentation we made for his talk (Fig. 3.9). I want to call the attention of readers on the fact that, even though various collections of his writings have been published recently, all of them, unfortunately, missed including the just-mentioned important contribution of his very last years, as well as the talk *Towards an Exact Quantum Mechanics* which he delivered on the occasion of the 70 birthday of Julian Schwinger, which has been published by World Scientific in 1989 in *Themes in Contemporary Physics II*, S. Deser and R.J. Finkelstein, eds.

As I have already mentioned, at that time, our attention, as well as that of all people interested in collapse models, was concentrated on the possibility of a relativistic generalization

IOP Publishing

JOURNAL OF PHYSICS A: MATHEMATICAL AND THEORETICAL

J. Phys. A: Math. Theor. 40 (2007) 2919–2933

doi:10.1088/1751-8113/40/12/002

The Trieste Lecture of John Stewart Bell

Delivered at Trieste on the occasion of the 25th Anniversary of the International Centre for Theoretical Physics, 2 November 1989

(A link to the video of this lecture is available on the electronic version of the article.)

General remarks by Angelo Bassi and GianCarlo Ghirardi

During the autumn of 1989 the International Centre for Theoretical Physics, Trieste, celebrated the 25th anniversary of its creation. Among the many prestigious speakers, who delivered extremely interesting lectures on that occasion, was the late John Stewart Bell. All lectures have been recorded on tape. We succeeded in getting a copy of John's lecture.

In the lecture, many of the arguments that John had lucidly stressed in his writings appear once more, but there are also extremely interesting new remarks which, to our knowledge, have not been presented elsewhere. In particular he decided, as pointed out by the very choice of the title of his lecture, to call attention to the fact that the theory presents two types of difficulties

Figure 3.9 John at the beginning of his talk and the introductory remarks to the published version.

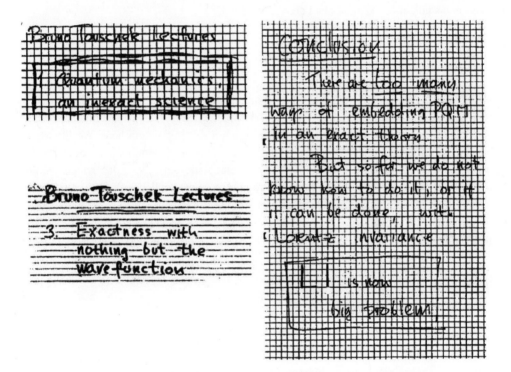

Figure 3.10 Extracts from Bell's transparencies for the B. Touschek lectures.

of such models. John had already stressed the fundamental importance of this problem. In fact, a few months before, when delivering in Rome a memorial Bruno Touschek series of lectures he concentrated, first of all, on the foundational problems of quantum theory, and then, in his second lecture, he discussed Bohmian mechanics and in the third one the collapse theories, as one can deduce by the images I have taken from his transparencies (Fig. 3.10).

I consider it particularly illuminating that he decided to close his "short course" by stressing the paramount importance of working out a consistent relativistic generalization of the theories he had analyzed in the previous lectures, by concluding that **L**(orentz) **I**(nvariance) is now the big problem. And I cannot avoid mentioning the specific attention he paid to this problem since the first time he discussed our collapse model, when, after having performed a quite smart analysis of this aspect by resorting to a two-times Scrödinger's equation, he concluded (see *Are there quantum jumps?*),

For myself, I see the GRW model as a very nice illustration of how, quantum mechanics , to become rational, requires only a change which is very small (on some measures). And I am particularly struck by the fact that the model is as Lorentz invariant as it could be in the nonrelativistic version. It takes away the grounds of my fear that any exact formulation of quantum mechanics must conflict with fundamental Lorentz invariance.

And in his talk at ICTP, mentioned above, he stated,

There is a whole line of relativistic research here which has been opened up as the Ghirardi-Rimini-Weber jump. And it remains to be seen whether it will work out well or ill. In any case there is a program where before Ghirardi, Rimini and Weber the fields were rather moribund.

Obviously, the problem of getting consistent relativistic generalizations of collapse models (as well as of Bohmian mechanics) has been investigated in many papers in recent years. We do not consider it appropriate to address such a subtle problem here. We limit ourselves to calling to the attention of the reader that while all deterministic hidden variable models, and thus Bohmian mechanics, admit relativistic generalizations which require the consideration of a (hidden) preferred reference frame, collapse models admit genuinely Lorentz invariant generalizations. The first one was introduced by Tumulka [24]; it relies on the so-called *flash ontology* and deals with many identical noninteracting fermions. Quite recently it has been proved [27] that, by adopting the appropriate method of attributing specific properties to physical systems in a relativistic context with reductions, it is possible to formulate a consistent relativistic model based on the *mass density ontology*.

3.8 An Important Celebration

Just one year after John's death, M. Bell, J. Ellis and D. Amati decided to devote a meeting at CERN to honoring this great scientist. The proceedings have been published by the Cambridge University Press under the title *Quantum Reflections*. The nine invited speakers were R. Penrose, H. Rauch, A. Aspect, G. C. Ghirardi, J. M. Leinaas, A. Shimony, K. Gottfried, N. D. Mermin and R. Jakiw. I really cannot avoid presenting an image, Fig. 3.11, of such an important event.

John Bell, even after his death, was celebrated in many other occasions on the following years, and particularly in the present year, which marks the 50th anniversary of the derivation of his revolutionary inequality. Just to mention an example, we (D. Dürr, S. Goldstein,

Figure 3.11 Celebration in honor of John at CERN, 1990.

N. Zanghì and myself) have devoted our sixth annual meeting on the foundations of quantum mechanics at Sexten precisely to him and to celebrating his inequality. R. Penrose was happy to take part in this event.[6]

3.9 A Synthetic Comment on Bell's Proof of Nonlocality

Here I intend to reconsider very briefly Bell's position concerning nonlocality. The main motivation for doing so is that Bell himself was always fully aware that there have been – and (let me state) there still are – basic misunderstandings concerning the extremely deep conceptual and philosophical implications of his work, even by part of great physicists.[7] Moreover, besides being aware of it, John was also quite upset by this fact.

To start with I will first of all mention an explicit sentence, which appears in *Bertlmann's socks and the nature of reality*, in which he has given clear voice to his disappointment with the way in which his work has been interpreted:

It is remarkably difficult to get this point across, that determinism is not a presupposition of the analysis ... My first paper on this subject starts with a summary of the EPR argument from locality to

---

[6] One can look at the registration of all lectures delivered there by going to the web site of the Sexten Centre for Astrophysics (www.sexten-cfa.eu/) and following the link to the meetings of 2014.

[7] Even the Editorial opening the special issue of *J. Phys. A* dedicated to J. Bell (*J. Phys. A*: 47, 420301, 2014) seems to suggest that in the derivation of Bell's inequality the problem of hidden variables has played a more important role than the (EPR) request of locality. Furthermore, the debate between T. Maudlin and R. F. Werner in the same special issue makes fully clear that, while Maudlin (in full agreement with Bell's views and, in my opinion, correctly) attributes the deserved value to the crucial issue of locality, other scientists have serious reservations about its relevance.

deterministic hidden variables. But the commentators have almost universally reported that it begins
with deterministic hidden variables.

Let me just mention that the term determinism (or equivalently the term realism) is used
to assert that the observables of a physical system have definite values (coinciding with
one of the eigenvalues of the associated self-adjoint operator) even before the measurement process is performed. On the other hand, and as is well known, the term *hidden variables* denotes mathematical entities which, either by themselves or in addition to, e.g., the
state vector, determine either the precise outcomes of the measurement of any observable
(deterministic hidden variable theories), or even only the probabilities of such outcomes
(stochastic hidden variable theories).

Given these premises, I can rephrase Bell's argument in complete generality. Let us
consider

- A completely general theory such that the maximal – in principle – specification of the
  state of a composite system with far apart (quantum-mechanically entangled) constituents
  determines uniquely the probabilities of all conceivable outcomes of single and correlated
  measurements.
- The state is formally specified by two types of variables μ and λ, which are, respectively,
  accessible and nonaccessible. The nonaccessible variables are distributed according to an
  appropriate non-negative distribution (over which averages have to be taken in order to
  get the quantum probabilities) $\rho(\lambda)$ such that $\int_\Lambda \rho(\lambda)d\lambda = 1$.
- The two measurement settings are chosen and the measurement processes are performed
  and completed in spacelike separated regions A* and B*.
- The specification of the "initial" state given by μ and λ refers to a spacelike surface which
  does not intersect the common region of the past light cones from A* and B* (so that each
  region is screened off from the other).
- The settings can be chosen freely by the experimenters (the free will assumption).
- The probabilities both of single events and of the correlations coincide with those of
  Q.M.

For simplicity let us make reference to an EPR-Bohm-like situation for a spin singlet with
settings a, b and outcomes A, B, and let us start with the standard relation for conditional
probabilities:

$$P(A, B|a, b; \mu, \lambda) = P(A|a, b; B; \mu, \lambda) \cdot P(B|a, b; \mu, \lambda). \tag{3.5}$$

Assuming

$$Outcome\ Independence \rightarrow P(A|a, b; B, \mu, \lambda) = P(A|a, b; \mu, \lambda)$$

and

$$Parameter\ Independence \rightarrow P(A|a, b; \mu, \lambda) = P(A|a; \mu, \lambda),$$

one immediately derives *the fundamental and unique request* of Bell, which we will call Bell's locality and denote as $\{B - Loc\}$:

$$B - Loc \longleftrightarrow P(A, B|a, b; \mu, \lambda) = P(A|a; \mu, \lambda) \cdot P(B|b; \mu, \lambda). \tag{3.6}$$

Finally, as already stated, averaging the probabilities over λ, one must get the quantum expectation values for the considered state.

The derivation is then straightforward. We know that, in the singlet state, one cannot get the same outcome in both spin measurements if they are performed along the same direction. Now,

$$\int_{\Lambda} \rho(\lambda)d\lambda P(A, A|a, a; \mu, \lambda) = 0 \rightarrow P(A, A|a, a; \mu, \lambda) = 0, a.e. \tag{3.7}$$

Use of $B - Loc$ implies that

$$[P(A|a, *; \mu, \lambda] \cdot [P(A|*, a; \mu, \lambda] = 0. \tag{3.8}$$

In the above equation we have used an asterisk to indicate that in the region B^* (A^*, respectively) no measurement (or, equivalently, any measurement whatsoever) is performed. From the above equation we have that one of the two factors of the product must vanish. On the other hand, $P(A|a, *; \mu, \lambda) = 0$, $\rightarrow P(-A|a, *; \mu, \lambda) = 1$, and, similarly, $P(A|*, a; \mu, \lambda) = 0$, $\rightarrow P(-A|*, a; \mu, \lambda) = 1$. In just the same way, taking into account also that $P(-A, -A|a, a; \mu, \lambda) = 0$, one proves that all the individual probabilities take either the value 1 or the value 0, and, as a consequence of $B - Loc$, the same holds for all the probabilities of correlated events. In conclusion, the request that $B - Loc$ hold implies that all probabilities take either the value 1 or zero, i.e., determinism (which we will denote as $\{Det\}$).

This is the precise sense of Bell's statement that, in his proof, determinism is *deduced* from the perfect EPR correlations and *not assumed*.

The rest of the story is known to everybody. Defining the quantities $E_\mu(a, b)$,

$$E_\mu(a, b) = \int_{\Lambda} \rho(\lambda)E(a, b; \mu, \lambda)d\lambda, \tag{3.9}$$

where

$$E(a, b; \mu, \lambda) = P(A = B|a, b; \mu, \lambda) - P(A \neq B|a, b; \mu, \lambda), \tag{3.10}$$

one trivially derives Bell's inequality in the Clauser-Horne form:

$$|E_\mu(a, b) + E_\mu(a, b') + E_\mu(a', b) - E_\mu(a', b')| \leq 2. \tag{3.11}$$

Identifying, as requested, $E_\mu(a, b)$ with the quantum expectation value $\langle \Psi | \sigma^{(1)} \cdot a \otimes \sigma^{(2)} \cdot b | \Psi \rangle$, one easily shows that, for appropriate directions (a, a', b, b') and for the singlet state, the considered combination violates the bound and can reach the value $2\sqrt{2}$.

At this point the logic of the argument should be clear:

$$\{Experimental\ Perfect\ Correlations\} \land \{B - Loc\} \supset \{Det\}$$

$$\{Det\} \land \{B - Loc\} \supset \{Bell's\ Inequality\}$$

$$\{General\ Quantum\ Correlations\} \supset \neg\{Bell's\ Inequality\}$$

$$\neg\{Bell's\ Inequality\} \supset \neg\{Det\} \lor \neg\{B - Loc\}$$

$$\neg\{Det\} \supset \neg\{Experimental\ Perfect\ Correlations\} \lor \neg\{B - Loc\}. \qquad (3.12)$$

Summarizing: $\{Natural\ Processes\} \supset \neg\{B - Loc\}$, i.e., *Nature is nonlocally causal.*

3.10 Conclusion: A Further Example of John's Humaneness

During a pause in the meeting at ICTP we were walking around the Miramare campus at Trieste and John told me, *GianCarlo, you know well how important I consider my interest in foundational problems. However I must state that I think that to devote oneself exclusively to this kind of studies is a luxury. One has also to do something more practical to get paid. This is why at CERN I am so involved in accelerator's physics.*

I replied immediately: *John, you are putting me in a very delicate position; in the last 20 years I have worked exclusively in the field of foundations of quantum mechanics.*

His answer was immediate and extremely comforting: *Oh, no, GianCarlo, you have completely ignored that in these years, as a lecturer, you have trained entire generations of young people in teaching them quantum theory. This fact fully justifies your salary!*

I sincerely believe that this is an appropriate anecdote to conclude this paper which is intended to honour the great John Bell.

References

[1] J.S. Bell, Time reversal in field theory, *Proc. Royal Soc. A*, **231**, 479 (1955).

[2] J.S. Bell, Current algebra and gauge variance, *Nuovo Cimento*, **50**, 129 (1967).

[3] J.S. Bell and R. Jakiw, A PCAC puzzle: $\pi^0 \to \gamma\gamma$ in the σ – Model, *Nuovo Cimento*, **60**, 47 (1969).

[4] A. Einstein, N. Rosen and B. Podolsky, Can quantum-mechanical description of physical reality be considered complete? *Phys. Rev.* **47**, 777 (1935).

[5] J.S. Bell, Einstein–Podolski–Rosen experiments, in *Proceedings of the Symposium on Frontiers Problems in High Energy Physics*, Pisa, June 1976, pp. 33–45.

[6] J.S. Bell, Beables for quantum field theory, 1984 Aug. 2, CERN-TH. 4035/84.

[7] J. von Neumann, *Mathematische Grundlagen der Quantenmechanik*, Springer, Berlin, 1932. English translation: *Mathematical Foundations of Quantum Mechanics*, Princeton University Press, Princeton, NJ, 1955.

[8] J.S. Bell, On the problem of hidden variables in quantum theory, *Rev. Mod. Phys.* **38**, 447 (1966).

[9] J.S. Bell, On the Einstein-Podolsky-Rosen Paradox, *Physics* **1**, 195 (1964).

[10] D. Bohm, A suggested interpretation of the quantum theory in terms of "hidden" variables. I, *Phys. Rev.*, **85**, 166 (1952), A suggested interpretation of the quantum theory in terms of "hidden" variables, II, *ibid.* 180 (1952).

[11] G.C. Ghirardi, A. Rimini and T. Weber, A model for a unified description of macroscopic and microscopic systems, in *Quantum Probabilities and Applications II*, Lecture Notes in Mathematics, Vol. 1136, Springer, Berlin (1985).

[12] G.C. Ghirardi, A. Rimini and T. Weber, Unified dynamics for microscopic and macroscopic systems, *Phys Rev.* D **34**, 470 (1986).

[13] J.S. Bell, Are there quantum jumps? in *Schrödinger. Centenary of a polymath*, Cambridge University Press, Cambridge, UK (1987).

[14] A. Einstein, in M. Born, *The Born Einstein Letters*, Glasgow, The Macmillan Press (1971).

[15] A.I.M. Rae, Can GRW theory be tested by experiments on SQUIDS? *J. Phys. A* **23**, L 57 (1990).

[16] A. Rimini, Spontaneous localization and superconductivity, in E. Beltrametti and J.M. Lévy Leblond (eds.), *Advances in Quantum Phenomena*, Plenum Press, New York (1995); M. Buffa, O. Nicrosini and A. Rimini, Dissipation and reduction effects of spontaneous localization on superconducting states, *Found. Phys. Lett.* **8**, 105 (1995).

[17] Q. Fu, Spontaneous radiation of free electrons in a nonrelativistic collapse model, *Phys. Rev. A* **56**, 1806 (1997)

[18] W. Marshall, C. Simon, R. Penrose and D. Bouwmeester, Towards quantum superpositions of a mirror, *Phys. Rev. Lett.* **91**, 130401 (2003).

[19] S.L. Adler, Lower and upper bounds on CSL parameters from latent image formation and IGM heating, *J. Phys. A* **40**, 2935 (2007); S.L. Adler and F. Ramazanoglu, Photon emission rate from atomic systems in the CSL model, *J. Phys. A* **40**, 13395 (2007); S.L. Adler and A. Bassi, Is quantum theory exact? *Science* **325**, 275 (2009); S.L. Adler, A. Bassi and S. Donadi, On spontaneous photon emission in collapse model, *J. Phys. A* **46** 245304 (2013).

[20] A. Bassi, D.A. Deckert and L. Ferialdi, Breaking quantum linearity: Constraints from human perception and cosmological implications, *Eur. Phys. Lett*, **92**, 50006 (2010); S. Donadi, A. Bassi, C. Curceanu, A. Di Domenico and B.C. Hiesmayr, Are collapse models testable via flavor oscillations? *Found. Phys* **43**, 813 (2013).

[21] R. Penrose, in A. Fokas, T.W.B. Kibble, A. Grigoriou, and B. Zegarlinski (eds.), *Mathematical Physics 2000*, Imperial College Press, London (2000).

[22] P. Pearle, Reduction of the state vector by a nonlinear Schroedinger equation, *Phys. Rev. D* **13**, 857 (1976).

[23] P. Pearle, Combining stochastic dynamical state-vector reduction with spontaneous localization, *Phys. Rev. A* **39**, 2277 (1989).

[24] R. Tumulka, A relativistic version of the Ghirardi-Rimini-Weber model, *J. Stat. Phys.* **125** 821 (2006).

[25] G.C. Ghirardi, R. Grassi and F. Benatti, Describing the macroscopic world: Closing the circle within the dynamical reduction program, *Found. Phys.* **25**, 5 (1995).

[26] A. Bassi and G. C. Ghirardi, The Trieste lecture of John Stewart Bell, *J. Phys. A* **40**, 2919 (2007).

[27] D. Bedingham, D. Dürr, G.C. Ghirardi, S. Goldstein, and N. Zanghì, Matter density and relativistic models of wave function collapse, *J. Stat. Phys.* **154**, 623 (2014).

Part II

Bell's Theorem

4

What Did Bell Really Prove?

JEAN BRICMONT

Abstract

The goal of this paper is to give a pedagogical introduction to Bell's theorem and its implications for our view of the physical world, in particular how it establishes the existence of nonlocal effects or of actions at a distance. We also discuss several misunderstandings of Bell's result and we explain how de Broglie-Bohm theory allows us to understand, to some extent, what nonlocality is.

4.1 Introduction

Although the goal of this paper is *not* to discuss the usual foundational issues of quantum mechanics, namely how to understand the quantum state[1] and its "collapse", it is useful to begin by recalling the problem. As is well known, when there are no measurements, the quantum state Ψ evolves according to the Schrödinger evolution,

$$i\partial_t \Psi = H\Psi = (H_0 + V)\Psi, \tag{4.1}$$

where H_0 is the free Hamiltonian and V the potential (in this paper, we set $\hbar = 1$). For a system of N particles, the function $\Psi = \Psi(x_1, x_2, \ldots, x_N, t)$ is defined on the space $\mathbb{R}^{3N}$ of all possible configurations of those particles and depends also on time. However, suppose that one measures an observable represented by an operator $\mathcal{A}$, having eigenvalues λ_i and eigenvectors Ψ_i, when the quantum state is Ψ. Then one writes Ψ in the basis of the Ψ_i's: $\Psi = \sum_i c_i \Psi_i$. The measurement yields the value λ_i with probability $|c_i|^2$ and the quantum state becomes, after the measurement, Ψ_i. This is called the collapse of the quantum state.

Thus, we have two different types of evolution for the quantum state: the Schrödinger evolution between measurements and the collapse during measurements.

How are we to understand this dual evolution? One can have at least two different sorts of answer to that question: either the quantum state represents the *information* that we have about a physical system, in which case its "collapse" simply means that we learn something about the system, or it represents something "physical", in which case the collapse may be seen as the result of a physical interaction between a measuring device and the system.

---

[1] We use this word to designate the wave function times the possible spin states.

Each of these views has its own difficulties, but we will not go into that. We simply note that these two views are at least suggested by the quantum formalism, which treats the "measurement process" as special. According to the first view, that process reveals something about the system; according to the second, it affects the system in some way and modifies its state.

In classical probability, if we throw a coin, it either lands heads or tails. If we do not look at the result, we may assign probability one-half to each outcome. This assignment is clearly related to our ignorance. Indeed, if we look at the state of the coin, after it has landed, our probability (if we still want to use this term) will become one for heads or for tails, depending on the result, and zero for the other possibility. In this case, we are in the first situation discussed above, and probabilities are related to information. When our information changes, our probabilities change also, but nothing physical happens to the system.

To contrast this with the situation in quantum mechanics, consider a spin measurement. Let us assume that one measures the spin of an electron, via a Stern-Gerlach apparatus, in a given direction and that the initial quantum state is in a superposed state: spin up + spin down (normalized). Of course, the probability of either result is again one-half. But what does that mean here? According to the view that the quantum state reflects our information, it means that the electron is either up or down *before* the measurement and that the measurement simply reveals which is the case. It would be like the coin that is either heads or tails, before one looks at it. Let us emphasize that this view does *not* mean that the quantum state would be up or down – the quantum state is what it is, namely a superposition – but that the spin of the electron itself would be up or down. In other words, the electron would have properties that are simply not encoded in the quantum state. These properties are sometimes called "hidden variables": "variables", because they characterize a given system, but vary from one system to another, even with the same quantum state, and "hidden" because they are not included in the description of the system by its quantum state. Many physicists are convinced that "hidden variable theories" were ruled out, on logical or empirical grounds, either by von Neumann in 1935 [1] or by others later, maybe even by Bell in 1964, but this is a misconception to which we will return later.

If, on the other hand, one thinks of the measurement as a physical process that affects the system, then the measurement would be analogous to throwing the coin rather than looking at it after it had fallen on one side. The measurement would, in that view, not be a measurement of anything, but rather a "random creation" of the result.

The reader may think that the discussion is pointless, or "metaphysical", because it is impossible to know which of these two views is the correct one. Maybe, but as we shall see, merely raising that question (is the measurement simply revealing preexisting properties or is it acting on the system?) has led to an extraordinary discovery.

Indeed, Einstein firmly held the first view, and that view is often expressed by saying that quantum mechanics is not complete, which is a natural terminology, since it means that each individual system has properties (such as having spin up or down) that are not reflected by its quantum state. This is like saying that describing a coin, once it has fallen on one of its faces, by the probability "one-half, one-half" is not a complete description of the state of the coin, since the latter is either heads or tails.

It is not entirely clear what people who maintained, in contradiction to Einstein, that quantum mechanics *is* a complete description of the system meant by that. Coming back to the analogy with the coin, one may imagine that the coin is inside a sealed box that nobody can open. In that case, the probabilistic description "one-half, one-half" could be called "complete" in the sense that we humans have no way to *know* more about the state of the coin. Nevertheless, one might also want to say that the word complete here means "relative to us", and that *in reality* the coin is either heads or tails.

It should be emphasized that the issue here is not whether one can manipulate or predict the value of those "hidden variables", but whether one can merely think of them as existing. It is possible that what the people who think that quantum mechanics is "complete" mean by this expression is simply that, given that those extra variables are impossible to predict or to manipulate, why bother with them? That may be a natural reaction, but it should not prevent us from trying to see what speculating about the existence of those variables may lead us to.

Einstein developed several ingenious arguments (one of them, in 1935, together with Podolsky and Rosen [2], to which we refer below by the letters EPR) to show indirectly that quantum mechanics *must* be incomplete, in the sense given above. These were based on thought experiments, but that were later performed, in a modified form. Contrary to another popular misconception, the arguments were not faulty, at least if they are stated as follows: if a certain assumption of locality or of no "action at a distance" is granted, *and nothing else, in particular no assumption whatsoever about "realism" or "determinism"* (more on that later), then quantum mechanics is incomplete, in the sense introduced here. Einstein, Podolsky and Rosen did not state their result like that, because, for them, the locality assumption was too obvious to be stated explicitly as a genuine assumption.[2] Moreover, for various reasons that we will discuss in Section 4.5, their argument was not generally understood, and certainly not in the form stated here.

Bell did show, almost 30 years later (in 1964), that the locality assumption *alone*, in the context of the EPR experiment, leads to a contradiction with quantum mechanical predictions that were later verified experimentally [3]. Unfortunately, Bell assumed, at the beginning of his own reasoning, that the argument of EPR was well known; and since the EPR argument was misunderstood, Bell's argument was also widely misunderstood.

We will start by a little-known, but very simple, thought experiment, that of Einstein's boxes. This example already allows us to raise the issue of locality. Then we will give a simple derivation of Bell's argument (due to [4]), *combined* with that of EPR; that is the simplest and clearest way to arrive at the conclusion, namely that the world is, in some

---

[2] However, in his 1949 "Reply to criticisms", Einstein did pose the question in the form of a dilemma; speaking of the EPR 'paradox', he wrote, "the paradox forces us to relinquish one of the following two assertions:

   (1)   the description by means of the ψ-function is complete.
   (2)   the real states of spatially separated objects are independent of each other" [15, p. 682].

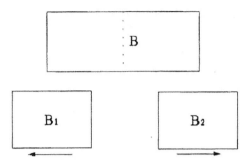

Figure 4.1 Einstein's boxes.

sense, nonlocal. Next, we will review and discuss some of the misunderstandings of both EPR and Bell among physicists. Finally, we will explain how the de Broglie-Bohm theory illustrates and explains, as far as one can, what nonlocality is.

4.2 Einstein's Boxes

Consider the following thought experiment.[3] A single particle is in a box B (see Figure 4.1), and its quantum state is $|$ state $>= |B>$, meaning that its quantum state is distributed over the box.[4] One then cuts the box into two half-boxes, B_1 and B_2, and the two half-boxes are then separated and sent as far apart as one wants.

According to ordinary quantum mechanics, the state becomes

$$\frac{1}{\sqrt{2}}(|B_1 > +|B_2 >),$$

where the state $|B_i >$ means that the particle "is" in box B_i, $i = 1, 2$ (and again, is distributed in that box). Here, we put quotation mark around the verb "is", because of the ambiguity inherent in the meaning of the quantum state: if it reflects our knowledge of the system, then the particle *is* in one of the boxes B_i, without quotation marks. But, if one thinks of the quantum state as being physical and of the position of the particle as being created or realized only when one measures it, then the quotation marks are necessary and "is" means: "would be found in box B_i after measurement".

Now, if one opens one of the boxes (say B_1) and one does *not* find the particle in it, one *knows* that it is in B_2. Therefore, the state "collapses" instantaneously: state $\longrightarrow |B_2 >$ (and, if one opens the box B_2, one will find the particle in it!).

Since B_1 and B_2 are as far apart as one wants, if we reject the notion of action at a distance, then it follows that acting on B_1, namely opening that box, cannot have any physical effect

---

[3] We base this Section on [6], where the description of the experiment is due to de Broglie [7, 8]. The original idea of Einstein was expressed in a letter to Schrödinger, written on June 19, 1935, soon after the EPR paper was published [9, p. 35]. Figure 4.1 is reproduced with permission from [6]; Copyright 2005 American Association of Physics Teachers.

[4] The precise distribution does not matter, provided it is spread over B_1 and B_2 as defined below; it could be distributed according to the square of the ground state wave function of a particle in the box B.

whatsoever on B_2. If opening the box B_1 leads to the collapse of the quantum state into one where the particle is necessarily in B_2, it must be that the particle was in B_2 all along. That is of course the common sense view and also the one that one would reach if the particle were replaced by any large enough object.

But then, one must admit that quantum mechanics is not complete, in the sense that Einstein gave to that word: there exists other variables than the quantum state that describe the system, since the quantum state does not tell us in which box the particle is and we just showed, assuming no action at a distance, that the particle *is* in one of the two boxes, before one opens either of them.

In any case, with his argument of the boxes, Einstein had at least proven the following dilemma: either there exists some sort of action at a distance in nature (opening the box B_1 changes the physical situation in B_2) or quantum mechanics is incomplete. Since action at a distance was anathema for him (and probably for everybody else at that time[5]), he thought that he had shown that quantum mechanics is incomplete.

There are many examples, at a macroscopic level, that would pose a similar dilemma and where one would side with Einstein in making assumptions, even very unnatural ones, that would preserve locality. Suppose that two people are located far away, each of whom tosses coins and the results are always either heads or tails, randomly, but are the same for both throwers. Or suppose that in two casinos, again far away from each other, the roulette wheel always ends up on the red or black color, again randomly but always the same in both casinos. Or imagine twins far apart that behave exactly in the same fashion. In all these examples (and in many others that are easy to imagine) one would naturally assume (even if it sounded very surprising) that the two coin throwers or the casino owners were able to manipulate their apparently random results and had coordinated them in advance or that genetic determinism was much stronger than one usually thinks. Who would suppose that one coin tosser immediately affects the result of the other one, far away, or that the spinning of the ball in one casino affects the motion of the other ball, or that the action of one twin affects the behavior of the other twin? In all these cases, one would assume a locality hypothesis; denying it would sound even more surprising than whatever one would have to assume to explain those odd correlations.

But one thing should be a truism, namely that those correlations pose a dilemma: either the results are coordinated in advance or there exists a nonlocal action.

Note that, compared to those examples, Einstein's assumption in the case of the boxes (incompleteness of quantum mechanics) was actually very natural.

As an aside, let us mention that the example of the boxes also raises a serious question about the quantum–classical transition. Indeed, if the quantum particle is replaced with a "classical" one, meaning a large enough object, nobody denies that the particle *is* in one of the boxes before one opens any of them. But where is the quantum/classical division? Usually the passage from quantum to classical physics is thought as some kind of limit; but a limit is something that one gets closer and closer to as a parameter varies. Here, we are

---

[5] Bohr's position was ambiguous on this issue; we will discuss it in Section 4.5.

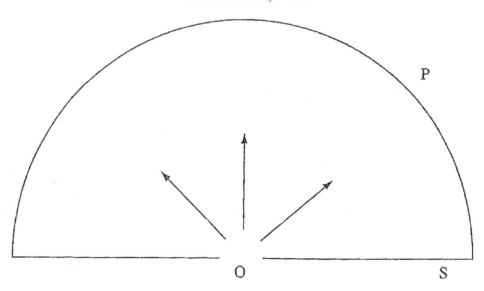

Figure 4.2 Einstein's objection at the 1927 Solvay Conference.

supposed to go from the statement "the particle is in neither of the boxes" to the statement "the particle is in one of them, but we do not know which one". This is an ontological jump and not the sort of continuous change that the notion of limit expresses.[6]

Actually, a somewhat similar argument was already put forward by Einstein in 1927, at the Solvay Conference.[7] Einstein considered a particle going through a hole, as in Figure 4.2.

In the situation described in the picture, the quantum state spreads itself in the half circle, but one always detects the particle at a given point. If the particle *is not* localized anywhere before its detection (think of it as a sort of "cloud" as spread out as the quantum state), then it must condense itself at a given point, again in a nonlocal fashion, since the part of the particle that is far away from the detection point must jump there instantaneously.

4.3 What Is Nonlocality

Let us consider what sort of nonlocality or action at a distance would be necessary to deny Einstein's conclusion about the incompleteness of quantum mechanics.

1. That action should be instantaneous, since the particle has to be entirely in box B_2 once we open box B_1. Of course, instantaneity is not a relativistic notion, so let us say, instantaneous in the reference frame where both boxes are at rest.

2. a. The action extends arbitrarily far, since the particle is entirely in box B_2 once we open box B_1; it is created there (since it was, by assumption, in neither box before B_1 was opened) and that fact does not change with the distance between the boxes.

---

[6] As we will see in Section 4.6, this problem does not arise in the de Broglie-Bohm theory.

[7] See [10, p. 486] or [11] for the "original" (published in French translation at the time of the Solvay Conference). In fact Einstein raised a similar issue as early as in 1909, at a meeting in Salzburg; see [10, p. 198]. Figure 4.2 is taken from [10, p. 440]. Originally in [11, p. 254].

b. The effect of the action does not decrease with the distance: indeed, the effect is the "creation" of the particle in box B_2 and that effect is the same irrespective of the distance between the boxes.

3. This effect is individuated: suppose we send out a thousand of pairs of half-boxes, each coming from the splitting in two of one box with a single particle in it. Then opening one half box will affect the state in the other half box, but not in any other box.

4. That action cannot be used to transmit messages: if we open one box, we learn what the state becomes in the other box, but we cannot use that to transmit a message from where one box is to where the other box is. Indeed, since we do not have any way, by acting on one box, to choose in which of the two boxes the particle will be, there is no way to use that experiment to send messages.

The impossibility of sending messages is sometimes taken to mean that there is nothing nonlocal going on. But nonlocality refers here to causal interactions as described (in principle) by physical theories. Messages are far more anthropocentric than that and require that humans be able to control these interactions to communicate. As remarked by Maudlin, the Big Bang and earthquakes cannot be used to send messages, but they have causal effects nevertheless [12, pp. 136–7].

Let us now compare that sort of nonlocality with that in Newton's gravity. The latter also allows actions at a distance: since the gravitational force depends on the distribution of matter in the universe, changing that distribution, say by moving my body, instantaneously affects all other bodies in the universe. That action at a distance has properties 1 and 2a, but not the others; of course that effect decreases with the distance, because of the inverse square law, and it affects all bodies at a given distance equally (it is not individuated). On the other hand, it can in principle be used to transmit messages: if I decide to choose, at every minute, to wave my arm or not to wave it, then one can use that choice of movements to encode a sequence of zeros and ones and, assuming that the gravitational effect can be detected, one can therefore transmit a message instantaneously and arbitrarily far (but the further away one tries to transmit it, the harder the detection). Of course, all this refers to Newton's *theory*. There were no experiments performed or suggested that could test whether gravitational forces really acted instantaneously (and, as we will see, this is a major difference from the situation in quantum mechanics).[8]

Post-Newtonian physics has tried to eliminate property 1, and classical electromagnetism and the general theory of relativity have kept only property 2a and the negation of 4. And, due to special relativity, the combination of 1 and the negation of 4 allows in principle the sending of messages into one's own past, so that, if 1 holds, 4 must hold also.

One may ask: does quantum mechanics prove that there are physical effects displaying properties 1–4? The example of the boxes does not allow that conclusion, because one can consistently think that the particle is always in one of the boxes. Indeed, in de Broglie

---

[8] It is well known that Newton did not like that aspect of his own theory. He said, "that one body may act upon another at a distance through a vacuum without the mediation of any thing else ... is to me so great an absurdity that I believe no man who has in philosophical matters any competent faculty of thinking can ever fall into it." [13, 14]. Newton thought that gravitation was mediated by particles moving at a finite speed, so that the effect of gravitation could not be instantaneous; see [15] for more details.

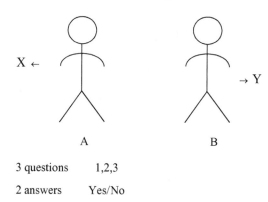

3 questions   1,2,3

2 answers   Yes/No

Figure 4.3 The anthropomorphic experiment.

and Bohm's theory (explained in Section 4.6), particles do have positions and trajectories and the predictions of the theory are the same as those of ordinary quantum mechanics. This is not the place to discuss the merits of de Broglie and Bohm's theory, but its mere existence and consistency imply that, in the example of the boxes, nothing forces us to accept nonlocality. It is true that de Broglie and Bohm's theory itself is nonlocal, but not when that theory is applied only to the situation of Einstein's boxes. In order to prove nonlocality in the sense introduced here, i.e., a phenomenon having properties 1–4 above, we have to turn to a more sophisticated scenario.

4.4 A Simple Proof of Nonlocality

4.4.1 An Anthropomorphic Thought Experiment

Let us start with an anthropomorphic thought experiment (which can be realized in principle by using quantum mechanics, as we will see in Subsection 4.4.2): two people, A (for Alice) and B (for Bob), are together in the middle of a room and go towards two different doors, located at X and Y. At the doors, each of them is given a number, 1, 2, 3 (let's call them "questions", although they do not have any particular meaning), and has to say "Yes" or "No" (the reason that we introduce three questions will be clear in the theorem below). That experience is repeated a large number of times, with A and B gathering together each time in the middle of the room, and the questions and the answers vary apparently at random. When A and B are together in the room, they can decide to follow whatever "strategy" they want in order to answer the questions, but the statistics of their answers, although they look random, have nevertheless to satisfy two striking properties.

The first property is that, when the same question is asked at X and Y, one always gets the same answer. How can that be realized? One obvious possibility is that A and B agree before moving towards the doors which answers they will give: they may decide, for example, to say both Yes if the question is 1, No if it is 2, and Yes if it is 3. They can choose different strategies at each repetition of the experiment and choose those strategies "at random" so that the answers will look random.

Another possibility is that, when *A* reaches door *X*, she calls *B* and tells him which question was asked and the answer she gave; then, of course, *B* can just give the same answer as *A* if he is asked the same question and any other answer if the question is different.

But let us assume that the answers are given simultaneously (in the reference frame in which the experience takes place), so that the second possibility is ruled out unless there exists some sort of action at a distance between *A* at *X* and *B* at *Y*. Maybe *A* and *B* communicate by telepathy . . . Of course, that is not to be taken seriously, but that is the sort of interaction that Einstein had in mind when he spoke of "spooky actions at a distance" [16, p. 158].

The question that the reader should ask herself at this point is whether there is *any other possibility*: either the answers are predetermined or (assuming simultaneity of the answers) there is a "spooky action at a distance", namely a communication of some sort takes place between *A* and *B after* they are asked the questions. This is similar to the dilemma about the boxes: either the particle is in one of the boxes or there is some sort of physical action between the two boxes.

Of course, this dilemma already arises if one asks only *one* question and the answers are perfectly correlated (the usefulness of having three questions will appear below).

A third possibility, which is sometimes suggested (probably out of desperation), is that the perfect correlations between the answers when the same questions are asked is simply a coincidence that does not need to be explained. Or, in the same vein, one sometimes claims that science limits itself to predictions, not to explanations. But the whole of science can be seen as an attempt to account for correlations or empirical regularities: the theory of gravitation, for example, accounts for the regularities in the motion of planets, satellites, etc. The atomic theory of matter accounts for the proportions of elements in chemical reactions. The effects of medicines account for the cure of diseases. To refuse to account for correlations, without giving any particular reason for doing so, is, in general, a very unscientific attitude.

In any case, refusing to face a question is not the same thing as answering it. So, let us proceed by accepting the dilemma, which is actually the EPR part of the argument, except that EPR did not formulate the issue with people but with quantum particles (we will come to that later) and not as a real dilemma since, for them, nonlocality was ruled out (see, however, footnote 2).

Now, there is a second peculiarity of the statistics of the answers: when the two questions asked of *A* and *B* are *different*, then the answers are the same only in one-fourth of the cases (the number $1/4$ being the result of a quantum mechanical calculation, as we shall see in the next subsection).

But that property, combined with the idea that the properties are predetermined, leads to a contradiction, which is a (very simple) version of *Bell's theorem*.[9]

---

[9] The argument given here is taken from [4]. In the original paper by Bell [17], the proof, although fundamentally the same, looked more complicated.

Theorem (Bell) We cannot have these two properties together:

- The answers are determined before the questions are asked and are the same on both sides.
- The frequency of having the same answers on both sides when the questions are different is $\frac{1}{4}$.

Proof There are 3 questions,   1 2 3,
and 2 answers,   Yes/No.
If the answers are given in advance, there exist $2^3 = 8$ possibilities:

1	2	3
Y	Y	Y
Y	Y	N
Y	N	Y
Y	N	N
N	Y	Y
N	Y	N
N	N	Y
N	N	N

In *each case* there are at least *two questions* with the same answer.
Therefore,

Frequency (answer to 1 = answer to 2)
$+$ Frequency (answer to 2 = answer to 3)
$+$ Frequency (answer to 3 = answer to 1) ≥ 1.
But if

Frequency (answer to 1 = answer to 2)
$=$ Frequency (answer to 2 = answer to 3)
$=$ Frequency (answer to 3 = answer to 1)
$= \dfrac{1}{4}$

we get $\Rightarrow \dfrac{3}{4} \geq 1$, which is a contradiction.   □

The inequality $\frac{3}{4} \geq 1$ is an example of *Bell's inequalities*.

Before drawing the conclusions of this theorem, let us see how the two people in the preceding section could realize these "impossible" statistics.

4.4.2 The Real Quantum Experiment

Let us first describe the situation in the previous section in a nonanthropomorphic manner. *A* and *B* are replaced by particles with spin $1/2$ (in reality, the experiments are made with polarized photons, but conceptually the two situations are similar; see [3] for some of the experiments).

At X and Y, there are Stern–Gerlach apparatus that "measure the spin" along some direction.

The "questions" 1, 2, 3 are three possible directions for that "measurement" (we put quotation marks here because, as we shall see below, there is no intrinsic property of the particles that is really measured in these experiments).

The answer Yes/No correspond to results Up/Down for the spin (we will define this correspondence more precisely below).

One sends the particles towards the apparatus and the initial state of the two particles is[10]

$$|\Psi> = \frac{1}{\sqrt{2}}(|A\ 1\ \uparrow> |B\ 1\ \downarrow> -|A\ 1\ \downarrow> |B\ 1\ \uparrow>)$$

$$= \frac{1}{\sqrt{2}}(|A\ 2\ \uparrow> |B\ 2\ \downarrow> -|A\ 2\ \downarrow> |B\ 2\ \uparrow>)$$

$$= \frac{1}{\sqrt{2}}(|A\ 3\ \uparrow> |B\ 3\ \downarrow> -|A\ 3\ \downarrow> |B\ 3\ \uparrow>), \tag{4.2}$$

where $|A\ 1\ \uparrow>$ is the state in which the particle A has its spin up in direction 1 (meaning that a particle in that state will have its spin up with certainty after a spin measurement in direction 1, or in other words, that state is the "up" eigenstate of the spin operator in direction 1), and the other symbols are defined analogously. We leave aside the issue of how to create such a state in practice and note only that it can be done (at least, in an analogous situation with photons and polarization instead of particles and spin). We also accept without proof the fact that this state has three similar representations in each of the directions 1, 2 or 3.

We also leave aside the "spatial" part of that quantum state: we assume implicitly that the state $|A\ 1\ \uparrow>$ is coupled to a wave function moving towards the door located at X, while the state $|B\ 1\ \downarrow>$ is coupled to a wave function moving towards the door located at Y, and similarly for the other parts of the state $|\Psi>$.

Consider now the standard quantum mechanical description of a measurement of the spin in direction 1 of A at X, without measuring anything on particle B, for the moment. If one sees $\uparrow$, the state becomes $|A\ 1\ \uparrow> |B\ 1\ \downarrow>$ (by the collapse rule). If one sees $\downarrow$, the state becomes $|A\ 1\ \downarrow> |B\ 1\ \uparrow>$. And, of course, we get similar results if one measures the spin in direction 2 or 3 of A.

But then the state changes *nonlocally* for particle B, because, if one sees $\uparrow$, the state "collapses", i.e., becomes $|A\ 1\ \uparrow> |B\ 1\ \downarrow>$, and the part $|A\ 1\ \downarrow> |B\ 1\ \uparrow>$ of the state has been suppressed by the collapse; in other words, after the measurement of A, any measurement of B at Y is guaranteed to yield the result opposite to what was found about A, while, before the measurement on A, the result of B was undetermined. This gives rise to the same dilemma as for Einstein's boxes: either the measurement on A affects the physical situation of B, or the particle B had its spin determined in advance, and anti-correlated with that of

---

[10] The fact that the three representations below are equal follows from the rotation invariance in spin space of that particular state $|\Psi>$.

A, which must also have been predetermined. Since we could of course measure the spin of B first and then of A, or do both simultaneously, one is led to the dilemma:

– either the spin values, up or down, were predetermined before the measurement, and in all three directions, because the same reasoning can be done for each one of them;
– or there is some form of action at a distance between X and Y (which can, in principle, be arbitrarily far from each other).

But the first assumption leads to a contradiction with observations made when the directions in which the spin is "measured" are *different* for A and B. To see this, denote by $v_A(\mathbf{a})$, $v_B(\mathbf{b})$ the pre-existing values of the results of measurements on A and B (assuming that they exist), where $\mathbf{a}$, $\mathbf{b}$ denote unit vectors in the directions 1, 2, 3, to be specified below, in which the spin is measured at X or Y. Let us make the following conventions at X and Y: $v_A(\mathbf{a}) = +1$ means that the answer is "Yes", $v_A(\mathbf{a}) = -1$ means that the answer is "No", but $v_B(\mathbf{b}) = +1$ means that the answer is "No", $v_A(\mathbf{b}) = -1$ means that the answer is "Yes". With that convention, we see that we always get the same answer when the same questions are asked on both sides. The contradiction comes from the fact that we get the same answer only $\frac{1}{4}$ of the time when one asks different questions at X and Y, and our Theorem shows that this is impossible.

Of course, one has to run the experiment a large number of times in order to get the "impossible" statistics (impossible without accepting the existence of actions at a distance).

Let us show how to obtain this factor $\frac{1}{4}$. This is an elementary quantum mechanical calculation. Compute first $\mathbf{E_{a,b}} \equiv < \Psi | \sigma_{\mathbf{a}}^A \otimes \sigma_{\mathbf{b}}^B | \Psi >$, where $\mathbf{a}$, $\mathbf{b}$ are the unit vectors in the directions (1, 2 or 3), and $\sigma_{\mathbf{a}}^A \otimes \sigma_{\mathbf{b}}^B$ is a tensor product of matrices, each one acting on the A or B part of the quantum state (with $\sigma_{\mathbf{a}} = a_1\sigma_1 + a_2\sigma_2 + a_3\sigma_3$, where, for $i = 1, 2, 3$, a_i are the components of $\mathbf{a}$ and σ_i the usual Pauli matrices). This quantity is bilinear in $\mathbf{a}$, $\mathbf{b}$ and rotation-invariant, so it must be of the form $\lambda \mathbf{a} \cdot \mathbf{b}$, for a certain $\lambda \in \mathbf{R}$.

For $\mathbf{a} = \mathbf{b}$, the result must be -1, because of the anti-correlations (if the spin is up at A, it must be down at B and vice versa). So $\lambda = -1$, and thus $\mathbf{E_{a,b}} = -\cos\theta$, where θ is the angle between the directions $\mathbf{a}$ and $\mathbf{b}$. We know that $v_A(\mathbf{a})$, $v_B(\mathbf{b}) = \pm 1$; thus,

$$\mathbf{E_{a,b}} = P(v_A(\mathbf{a}) = v_B(\mathbf{b})) - P(v_A(\mathbf{a}) = -v_B(\mathbf{b})) = 1 - 2P(v_A(\mathbf{a}) = -v_B(\mathbf{b})),$$

and $P(v_A(\mathbf{a}) = -v_B(\mathbf{b})) = \frac{1-\mathbf{E_{a,b}}}{2} = \frac{1+\cos\theta}{2}$.

One then chooses the directions:

$1 \Rightarrow \theta = 0$ degrees,
$2 \Rightarrow \theta = 120$ degrees,
$3 \Rightarrow \theta = 240$ degrees.

Since $\cos 120 = \cos 240 = -1/2$, we get $P(v_A(\mathbf{a}) = -v_B(\mathbf{b})) = \frac{1}{4}$. Thus we have perfect anti-correlations only $\frac{1}{4}$ of the time when $\mathbf{a}$ and $\mathbf{b}$ are different. This means, with our convention, that one gets the same answer when one asks different questions on both sides only $\frac{1}{4}$ of the time.

Finally, if one wants to reproduce the anthropomorphic experiment described in the previous subsection, one simply sends the particles towards the two doors once Alice (*A*) and Bob (*B*) have reached them. At each door, there is a Stern–Gerlach instrument, with which Alice and Bob "measure the spin" along the angles corresponding to the numbers given to them; then they answer the "questions" according to the results of their "measurements". In that way the "impossible" statistics mentioned in the Theorem will be reproduced.

4.4.3 Consequences

Let us first summarize what has been shown: assuming only locality, one derives from the perfect anti-correlations when the angles of detection are the same that the spin values preexist their measurement. Then the theorem in Section 4.4.1 shows that this fact, combined with the $\frac{1}{4}$ value for the frequency of unequal results when the measurement angles are different, leads to a contradiction.

Therefore the assumption of locality is false.[11]

To repeat: the EPR part of the argument shows that, if there are no pre-existing values, then the perfect correlations when the angles are the same imply some sort of action at a distance. The Bell part of the argument, i.e., the theorem of the previous subsection, shows that the mere assumption that there are pre-existing values leads to a contradiction when one takes into account the statistics of the results when the angles are different. That is why we used quotation marks in "measurement": there is no real property of the particle that is being "measured", since there are no spin values existing before the interaction with the "measuring" device (why this is so will become clearer in section 4.6).

But what does it mean? It means that some sort of action at a distance exists in Nature, but it does not tell us which sort. And we cannot answer that question without having a theory that goes beyond ordinary quantum mechanics. In ordinary quantum mechanics, what is nonlocal is the collapse of the quantum state, as we see in the transformation of (23.1) into $|A\ 1 \uparrow> |B\ 1 \downarrow>$ or $|A\ 1 \downarrow> |B\ 1 \uparrow>$, depending on the result of a measurement at A. This affects the state at B, since now the second part of $|\Psi>$ has been suppressed.

Since the meaning of the quantum state and of its collapse is ambiguous in ordinary quantum mechanics, it is not clear that this is a real physical effect. But, as we have emphasized, if there are no physical effects whatsoever, or if one interprets the collapse of the quantum state as a mere gain of information, then it means that we must have those predetermined values that lead to a contradiction.

It is important to notice that one cannot use this nonlocal effect to send messages. This is what contradicts all the pseudo-scientific uses of Bell's result: there is no telepathy of any

---

[11] As an aside, we note that an inequality similar to Bell's was derived later, in 1982, by Richard Feynman (with no reference made to Bell's work), who drew the following conclusion (what Feynman calls a local classical computer means more or less what we call locality here):

That's all. That's the difficulty. That's why quantum mechanics can't seem to be imitable by a local classical computer. I've entertained myself always by squeezing the difficulty of quantum mechanics into a smaller and smaller place, so as to get more and more worried about this particular item. It seems to be almost ridiculous that you can squeeze it to a numerical question that one thing is bigger than another.

Richard Feynman [18, p. 485]

sort that can be based on that result. And, if one could send messages, the theory of relativity implies that one could send messages into one's own past, which is certainly something that nobody is ready do accept.

The reason for the impossibility of sending messages is similar to what it was for Einstein's boxes. Each side sees a perfectly random sequence of results "spin up" or "spin down". Since there is no mechanism that allows, given the initial quantum state (23.1), to control or affect that result by acting on one side of the experiment, there is no way to send a message from one side to the other (see [19] for a more general proof of the impossibility of sending messages via EPR-Bell experiments).

But if each person tells the other which "measurements" have been made (1, 2 or 3), *without* telling which results are obtained, they both know which result has been obtained on the other side when the same measurement is made on both sides. Then, they both share a common sequence of Yes/No or up/down, which is a form of "information". This information was not transmitted directly, by hypothesis, and it cannot come from the source (the one that has emitted the two particles), because of the nonexistence of pre-existing spin values. Thus, some nonlocal transfer of information must have taken place[12]. This is one of the basis of the field of "quantum information", whose development will hopefully will lead to a better appreciation of the radical consequences of Bell's discovery.

But the "better appreciation", if it ever comes, will be in the future. Indeed, for the time being, it is mostly massive misunderstandings of Bell and of EPR that prevails and that we will now discuss[13].

4.5 Misunderstandings

4.5.1 Misunderstandings of Einstein–Podolsky–Rosen

The EPR article was not using the spin variables as above (this extension is due to Bohm [20]) but position and momentum. They considered two particles, starting from the same place, moving in opposite directions, so that their total momentum was conserved and equal to zero. Thus, by measuring the momentum of one particle, one could know the momentum of the other particle. But, if one measured instead the position of that one particle, one would know the position of the other particle (since they moved at the same speed in opposite directions). However, if the two particles are far apart and if locality holds, the choice that we make of the quantity to measure on one particle cannot affect the state of the other particle. Thus, that second particle must possess a well-defined momentum *and* a well-defined position, before any measurement, and this shows that quantum mechanics is incomplete, since the quantum state does not include those variables.

---

[12] See the book on relativity and non-locality by Maudlin [12] for a detailed discussion of the differences between messages and information and of what exactly is compatible or not with relativity.

[13] The following section has a lot in common with a paper by Maudlin [15]. In particular, Maudlin starts by quoting a video from *Physics World* [21] that nicely summarizes all the misunderstandings discussed below.

One should emphasize (as is done by Maudlin in [15]) that the concern of Einstein with "determinism" may not have been due to determinism itself but to his attachment to locality and to the fact that he understood (unlike most of his critics) that locality required determinism, at least in the sense that variables such as the spin values or positions and momenta *must* exist if measures on one side (on A) do not affect the physical situation on the other side (on B). Indeed, if the measurement on side A is truly "random", then it produces for states such as (23.1) a definite result at B, which is therefore no longer random (which it was also, of course, by symmetry, before the measurement at A); this means that, if we assume indeterminism of the results, we must accept that measurements on one side affect the physical situation on the other side nonlocally. Einstein's sentence "God does not play dice" is often quoted, but a more precise expression of his thought can be found in a 1942 letter:

It seems hard to sneak a look at God's cards. But that he plays dice and uses "telepathic" methods (as the present quantum theory requires of him) is something that I cannot believe for a moment.

Albert Einstein [22], quoted in [15]

So the problem that bothered Einstein was not simply determinism, but the fact that the latter implies "telepathy", i.e., action at a distance.

Coming back to the EPR article, one may think that their reasoning was not as transparent as one might have liked it to be.[14] But Einstein was completely clear when he wrote, in 1948:

If one asks what, irrespective of quantum mechanics, is characteristic of the world of ideas of physics, one is first of all struck by the following: the concepts of physics relate to a real outside world, that is, ideas are established relating to things such as bodies, fields, etc., which claim a 'real existence' that is independent of the perceiving subject – ideas which, on the other hand, have been brought into as secure a relationship as possible with the sense-data. It is further characteristic of these physical objects that they are thought of as arranged in a space time continuum. An essential aspect of this arrangement of things in physics is that they lay claim, at a certain time, to an existence independent of one another, provided these objects 'are situated in different parts of space'. . . .

The following idea characterizes the relative independence of objects far apart in space (A and B): external influence on A has no direct influence on B.

Albert Einstein [23] (reproduced in [16, pp. 170–71])

But Born, for example, missed Einstein's argument; he wrote, when he edited the Born–Einstein correspondence,

The root of the difference between Einstein and me was the axiom that events which happens in different places A and B are independent of one another, in the sense that an observation on the states of affairs at B cannot teach us anything about the state of affairs at A.

Max Born [16, p. 176]

---

[14] The article, according to Einstein, was written by Podolsky "for reasons of language", meaning that Einstein's English was far from perfect, and was not written with optimal clarity; see his June 19, 1935 letter to Schrödinger in [9, p. 35.]

As Bell says,

Misunderstanding could hardly be more complete. Einstein had no difficulty accepting that affairs in different places could be correlated. What he could not accept was that an intervention at one place could influence, immediately, affairs at the other.

John Bell [24, p. 144]

What Born said was that making an experiment at one place teaches us something about what is happening at another place, which is unsurprising. If, in the anthropomorphic example above, both people had agreed on a common strategy, one would learn what *B* would answer to question 1, 2 or 3, by asking that same question of *A*. In fact, in his comments about Einstein, Born gives the following example:

When a beam of light is split in two by reflection, double-refraction, etc., and these two beams take different paths, one can deduce the state of one of the beams at a remote point B from an observation at point A.

Max Born [16, p. 176]

But here Born is giving a classical example, where the polarization does pre-exist its measurement, which is exactly contrary to the idea that the quantum state is a complete description, since the latter does not, in general, specify any value of the position, momentum, spin, angular momentum, energy, etc., when it is a superposition of different eigenstates of the operator corresponding to the given physical quantity (as, for example, in the state (23.1)).

This shows that Born thought that, in our language, there are pre-existing answers, namely that the spins are up or down before the measurements, which means that, in fact, he agreed with Einstein that quantum mechanics is incomplete, but simply did not understand what Einstein meant by that.

There is also a rather widespread belief among physicists that Bohr adequately answered the EPR paper in [25]. But, unlike Born, whose misunderstanding of Einstein is clearly stated, Bohr's answer is hard to understand. EPR had written that

If, without in any way disturbing a system, we can predict with certainty (i.e. with probability equal to unity) the value of a physical quantity, then there exists an element of physical reality corresponding to this physical quantity,

Albert Einstein, Boris Podolsky, Nathan Rosen [2], reprinted in [26, pp. 138–41]

which again means that if, by doing something at A, we can learn something about the situation at B (for example, the value of the result of a spin measurement in a given direction[15]), then what we learn must exist before we learn it, since A and B can be far apart. Here of course, EPR assume locality.

---

[15] In the EPR paper, they considered position and momentum instead of spin, which can also be predicted at B once they are measured at A, due to conservation laws.

Bohr replied,

the wording of the above mentioned criterion . . . contains an ambiguity as regards the meaning of the expression "without in any way disturbing a system". Of course there is in a case like that just considered no question of a mechanical disturbance of the system under investigation during the last critical stage of the measuring procedure. But even at this stage there is essentially the question of *an influence on the very conditions which define the possible types of predictions regarding the future behavior of the system* . . . their argumentation does not justify their conclusion that quantum mechanical description is essentially incomplete . . . This description may be characterized as a rational utilization of all possibilities of unambiguous interpretation of measurements, compatible with the finite and uncontrollable interaction between the objects and the measuring instruments in the field of quantum theory.

Niels Bohr [25], quoted in [24, p. 155] (italics in the original)

Bell dissects that passage as follows:

Indeed I have very little idea what this means. I do not understand in what sense the word 'mechanical' is used, in characterizing the disturbances which Bohr does not contemplate, as distinct from those which he does. I do not know what the italicized passage means – 'an influence on the very conditions . . . '. Could it mean just that different experiments on the first system give different kinds of information about the second? But this was just one of the main points of EPR, who observed that one could learn *either* the position *or* the momentum of the second system. And then I do not understand the final reference to 'uncontrollable interactions between measuring instruments and objects', it seems just to ignore the essential point of EPR that in the absence of action at a distance, only the first system could be supposed disturbed by the first measurement and yet definite predictions become possible for the second system. Is Bohr just rejecting the premise – 'no action at a distance' – rather than refuting the argument?

John Bell [24, pp. 155–6]

One rather common misconception is to think that the goal of EPR was to beat the uncertainty principle and to show that one could measure, say, the spin in direction 1 at X and in direction 2 at Y and therefore know the values of the spin of both particles in directions associated with operators that do not commute (and, therefore, cannot be simultaneously measured, according to standard quantum mechanics).[16]

The main point of the EPR paper was not to claim that one could *measure* quantities that are impossible to measure simultaneously according to quantum mechanics, but rather that, if one can learn something about a physical system by making a measurement on a distant system, then, barring action at a distance, that "something" must already be there before the measurement is made on the distant system.

Einstein explicitly denied that his goal was to beat the uncertainty principle in his letter to Schrödinger of June 19, 1935. His argument, expressed in our language, was that, if one measures at A the spin in direction 1, when the state is given by (23.1), one may get, for

---

[16] Of course, since EPR were speaking of position and momentum instead of spin, it is those quantities that would have been simultaneously measured.

example, the state $|A\ 1\ \uparrow>\ |B\ 1\ \downarrow>$, and if one measures the spin in direction 2, one may get, say, the state $|A\ 2\ \downarrow>\ |B\ 2\ \uparrow>$. But then at B, one gets two different states, $|B\ 1\ \downarrow>$ and $|B\ 2\ \uparrow>$, which lead to different predictions for the future behavior of the system. So by choosing which quantity to measure at A, one changes the state at B in different ways: action at a distance! And that is what Einstein objected to. However, Einstein emphasized that "he couldn't care less" whether the collapsed states at A and B would be eigenstates of incompatible observables, such as the spin in directions 1 and 2 ([9, p. 38]).

In fact, the reason one cannot measure simultaneously the spin in direction 1 at X and in direction 2 at Y is that any one of those measurements affects the quantum state (by "collapsing" it) and affects it at both places, i.e., nonlocally, so that a measurement of the spin in direction 1 at X will change, in general, the probabilities of the results of a spin measurement in direction 2 at Y. In the EPR situation, it is because of the nonlocal character of the collapse that one cannot perform these simultaneous measurements. But, if things were perfectly local, then one could measure the spin in direction 1 at X (or the position of the particle at X) and the spin in direction 2 at Y (or the momentum of the particle at Y); since none of these measurements would, by the locality assumption, affect the state of the other particle, one would therefore know, because of the perfect correlations, the spin in both directions (or the position and the momentum) at X and at Y.

And that is why Bohr, in order to reply to EPR, had to "reject the premise – 'no action at a distance' – rather than refute the argument".[17]

4.5.2 Misunderstandings of Bell

The result of Bell, taken by itself and forgetting that of EPR, can be stated as a "no-hidden-variable theorem": the pre-existing values of the spins are by definition called hidden variables, because they are not part of the description provided by the quantum state. This expression is a misnomer, because, as we shall see in the next section, some "hidden variables" are not hidden at all. But, accepting this terminology for the moment, what Bell showed is that the mere supposition that the values of the spin pre-exist their "measurement", combined with the perfect anti-correlation when the axes along which measurements are made are the same and the $\frac{1}{4}$ result for measurements along different axes, leads to a contradiction. Since the last two claims are empirical, this means that these hidden variables or pre-existing values cannot exist. There have been other "no-hidden-variable theorems", most notably one due to John von Neumann [1], but the latter, unlike Bell's result, relied on some arbitrary assumptions, which we will not discuss (see [27, 28] for a discussion of this theorem).

But Bell, of course, always presented his result *in combination with* that of EPR, which shows that the mere assumption of locality, combined with the perfect correlation when the directions of measurement (or questions) are the same, implies the existence of those hidden

---

[17] This was also Einstein's analysis of Bohr's reply to the EPR paper; Einstein thought that Bohr rejected the premise of EPR that "the real situation of B could not be influenced (directly) by any measurement taken on A". See [5, pp. 681–2].

variables that are "impossible". So, for Bell, his result, combined with that of EPR, was not a "no-hidden-variable theorem", but a nonlocality theorem, the result on the impossibility of hidden variables being only one step in a two-step argument. Here is a quotation from Bell that states it clearly (here EPRB means EPR and Bohm, who reformulated the EPR argument in terms of spins [20]):

Let me summarize once again the logic that leads to the impasse. The EPRB correlations are such that the result of the experiment on one side immediately foretells that on the other, whenever the analyzers happen to be parallel. If we do not accept the intervention on one side as a causal influence on the other, we seem obliged to admit that the results on both sides are determined in advance anyway, independently of the intervention on the other side, by signals from the source and by the local magnet setting. But this has implications for non-parallel settings which conflict with those of quantum mechanics. So we *cannot* dismiss intervention on one side as a causal influence on the other.

John Bell [24, pp. 149–50]

He was also conscious of the misunderstandings of his results:

It is important to note that to the limited degree to which determinism[18] plays a role in the EPR argument, it is not assumed but inferred. What is held sacred is the principle of "local causality" – or "no action at a distance"... It is remarkably difficult to get this point across, that determinism is not a presupposition of the analysis.

John Bell [24, p. 143]

And he added, unfortunately only in a footnote,

My own first paper on this subject (*Physics* **1**, 195 (1965))[19] starts with a summary of the EPR argument *from locality to* deterministic hidden variables. But the commentators have almost universally reported that it begins with deterministic hidden variables.

John Bell [24, p. 157, footnote 10]

A famous physicist who is also such a commentator is Murray Gell-Mann, who wrote,

Some theoretical work of John Bell revealed that the EPRB experimental setup could be used to distinguish quantum mechanics from hypothetical hidden variable theories... After the publication of Bell's work, various teams of experimental physicists carried out the EPRB experiment. The result was eagerly awaited, although virtually all physicists were betting on the correctness of quantum mechanics, which was, in fact, vindicated by the outcome.

Murray Gell-Mann [29, p. 172]

So Gell-Mann opposes hidden variable theories to quantum mechanics, but the only hidden variables that Bell considered were precisely those that were needed, because of the EPR argument, to "save" locality. So, if there is a contradiction between the existence of those hidden variables and experiments, it is not just quantum mechanics that is vindicated, but locality that is refuted.

---

[18] Here, "determinism" refers the idea of pre-existing values (note of J.B.).
[19] Reprinted as Chapter 2 in [24] (note of J.B.).

Figure 4.4 Bertlmann's socks; taken from [24, p. 139].

In one of his most famous papers, "Bertlmann's socks and the nature of reality" [30], Bell gave the example of a person (Mr Bertlmann) who always wears socks of different colors (see Figure 4.4, taken from [30]). If we see that one sock is pink, we know automatically that the other sock is not pink (let's say it is green). That would be true even if the socks were arbitrarily far away. So, by looking at one sock, we learn something about the other sock, and there is nothing surprising about that, because socks *do have* a color whether we look at them or not. But what would we say if we were told that the socks have no color before we look at them? That would be surprising, of course, but the idea that quantum mechanics is complete means exactly that (if we replace the color of the socks by the values of the spin before measurement). But then looking at one sock would "create" non only the color of that sock but also the color of the other sock. And that would be even more surprising, because it implies the existence of actions at a distance, if the socks are far apart.

However, Murray Gell-Mann makes the following comment on the Bertlmann's socks paper:

The situation is like that of Bertlmann's socks, described by John Bell in one of his papers. Bertlmann is a mathematician who always wears one pink and one green sock. If you see just one of his feet and spot a green sock, you know immediately that his other foot sports a pink sock. Yet no signal is propagated from one foot to the other. Likewise no signal passes from one photon to the other in the experiment that confirms quantum mechanics. No action at a distance takes place.

Murray Gell-Mann [29, pp. 172–3]

This is not correct: it is true that, because of the random nature of the results, the experimental setup of EPRB cannot be used to send messages (or signals), as we saw in Subsection 4.4.3. But nevertheless, some action at a distance does take place. Of course, the goal of Gell-Mann in the passage quoted here is to dismiss pseudo-scientific exploitations of Bell's result, but his defense of science is misdirected: the behavior of quantum particles is *not* like that of Bertlmann's socks (which is indeed totally unsurprising), and that was the whole point of Bell's paper.

Eugene Wigner also saw Bell's result solely as a no-hidden-variables result:

The proof he [von Neumann] published...though it was made much more convincing later on by Kochen and Specker,[20] still uses assumptions which, in my opinion, can quite reasonably be questioned... In my opinion, the most convincing argument against the theory of hidden variables was presented by J. S. Bell.

Eugene Wigner [31, p. 291]

This is misleading, because Wigner considers only Bell's argument, which indeed shows that pre-existing spin values (or "hidden variables") cannot exist, but forgets the EPR part of the argument, which was the starting point for Bell, and which shows that these variables must exist if the world is local (see [32] for a further discussion of Wigner's views).

David Mermin summarized the situation described here in an amusing way:

Contemporary physicists come in two varieties.

Type 1 physicists are bothered by EPR and Bell's theorem.

Type 2 (the majority) are not, but one has to distinguish two subvarieties.

Type 2a physicists explain why they are not bothered. Their explanations tend either to miss the point entirely (like Born's to Einstein) or to contain physical assertions that can be shown to be false.

Type 2b are not bothered and refuse to explain why. Their position is unassailable. (There is a variant of type 2b who say that Bohr straightened out the whole business, but refuse to explain how.)

David Mermin [33]

Yet the same David Mermin also wrote (Bohm's theory, which we call de Broglie and Bohm's theory, will be explained in the next section),

Bell's theorem establishes that the value assigned to an observable must depend on the complete experimental arrangement under which it is measured, even when two arrangements differ only far from the region in which the value is ascertained - a fact that Bohm theory exemplifies, and that is now understood to be an unavoidable feature of any hidden-variables theory.

To those for whom nonlocality is anathema, Bell's Theorem finally spells the death of the hidden-variables program.

David Mermin [28, p. 814]

---

[20] Who proved another "no-hidden-variables" result [34]; see [28] for a discussion (note of J.B.).

But Bell's theorem shows that nonlocality, whether we consider it anathema or not, is an unavoidable feature of the world. What is even more surprising is that this comment comes at the end of a remarkably clear paper on Bell's theorem.

4.6 Why Is the Theory of de Broglie–Bohm Compatible with EPR–Bell?

To understand nonlocality, we need a theory that goes beyond ordinary quantum mechanics. In quantum mechanics, nonlocality is manifested by the collapse of the quantum state in situations such as that of EPR–Bell. But since the meaning of the quantum state is unclear, so is that of its collapse. By "going beyond ordinary quantum mechanics", we mean a theory that gives a clear meaning to the quantum state, and in particular decides whether the latter represents our information about the system or is rather something physical or both. One such theory is that of de Broglie and Bohm and, in that theory, nonlocality also acquires a clear meaning.[21]

Our goal here is not to discuss the de Broglie–Bohm theory in any detail, but only to explain enough of that theory to show how it allows us to understand nonlocality to some extent. In a nutshell, the de Broglie-Bohm theory is a "hidden variable theory" that eliminates the special role of "observations" in quantum mechanics and that is nonlocal, but that does not contain the hidden variables that are shown to be impossible by Bell (i.e., the spin values). Obviously, if certain variables are such that one cannot even conceive that they exist, they cannot be part of a consistent theory. And, by EPR + Bell, we know that any theory describing the world has to be nonlocal.

It cannot be emphasized enough that this is a quality of the de Broglie-Bohm theory: obviously if the theory was local, it could not be true! In fact, one of the motivations of Bell, when he derived his result, was to see if one could obtain an alternative to ordinary quantum mechanics that removed the central role of observations (as the de Broglie-Bohm theory does) while remaining local.

Moreover, as we will see, the "hidden variables" in that theory are not hidden at all.

All this sounds impossible, but here is how the theory goes:

In the de Broglie–Bohm theory, the state of a system is a pair (Ψ, X), where Ψ is the usual quantum state and X denotes the actual positions of all the particles in the system under consideration, $X = (X_1, \ldots, X_N)$. X are the hidden variables in this theory; this is obviously a misnomer, since particle positions are the only things that we ever directly observe (think of the double-slit experiment, for example). As we shall see below, this is also true when one "measures" the spin (and, in fact, any other quantum mechanical "observable").

A first remark about the de Broglie-Bohm theory is that, in the "Einstein boxes" experiment, the particle is always in one of the boxes, since it always has a position, so there is no paradox and no nonlocality.

---

[21] For the original papers of de Broglie see his contribution to the 1927 Solvay Conference in [10, 11]; the earlier work by de Broglie is also discussed in [10]. For the original papers of Bohm, see [35]. For a pedagogical introduction to that theory, see [36, 37]. For a more detailed exposition, see [24, 38–43].

The dynamics of the de Broglie-Bohm theory is as follows: both objects (Ψ, X) evolve in time; Ψ follows the usual Schrödinger equation (4.1):

$$i\partial_t \Psi = H\Psi = (H_0 + V)\Psi. \tag{4.3}$$

The evolution of the positions is guided by the quantum state: writing $\Psi = Re^{iS}$,

$$\dot{X}_k = \frac{1}{m_k} \frac{\text{Im}(\Psi^* \nabla_k \Psi)}{\Psi^* \Psi}(X_1, \ldots, X_N) = \frac{1}{m_k} \nabla_k S(X_1, \ldots, X_N) \tag{4.4}$$

for $k = 1, \ldots, N$, where $X_1, \ldots, X_N$ are the actual positions of the particles. For multicomponent quantum states, with spin, the products $\Psi^* \nabla_k \Psi$ and $\Psi^* \Psi$ are replaced by scalar products in spin space.

We are not going to discuss here in detail how the de Broglie-Bohm theory reproduces the results of ordinary quantum mechanics (see, e.g., [44] for a good discussion of the double slit experiment), but we will focus on one example: a spin measurement. Let us consider a particle whose initial state is $\Psi = \Phi(x, z)(|\uparrow> + |\downarrow>)$. $\Phi(x, z)$ denotes the spatial part of the quantum state and $(|\uparrow> + |\downarrow>)$ its (normalized) spin part (we consider here a two-dimensional motion, where z denotes the "vertical" direction in which the spin is "measured" by a Stern–Gerlach apparatus and x the transverse direction in which the particle moves towards the apparatus).[22] Assume that $\Phi(x, z)$ is a function whose support is concentrated around zero along both axes. As far as the free motion of the particle is concerned, we have rectilinear motion along the x axis. There is also some spreading of the wave function, but we will neglect that, assuming that the time of the experiment is short enough; we will furthermore neglect the motion along the x axis and replace $\Phi(x, z)$ with $\Phi(z)$. One can associate a measurement of spin with the introduction of an inhomogeneous magnetic field into the Hamiltonian, at some time, say 0 (see [41, Section 8.4] for a detailed discussion). The solution of Schrödinger's equation is then, with much simplification,

$$\Phi(z - t)|\uparrow> + \Phi(z + t)|\downarrow> . \tag{4.5}$$

This means that the particle has a state which is composed of two parts, one localized near t, the other near $-t$, along the z axis. If t is not too small, those two regions are far apart, and, by detecting the particle, we can see in which region it is. If it is near t, we say that the spin is up, and if it is near $-t$, we say that it is down.

The first thing to notice is that all we see in that "measurement" is the position of the particle. We never literally see the spin. In the de Broglie–Bohm theory, the particle has a position and therefore a trajectory; it will go either up or down.

One may ask how the de Broglie–Bohm theory reproduces the quantum statistics. Since the theory is deterministic, all "randomness" must be in the initial conditions. So one has to assume that, if we start with an ensemble of particles with the state $\Phi(x, z)(|\uparrow> + |\downarrow>)$, they will be distributed at the initial time according to the usual $|\Psi(x, z)|^2 = |\Phi(x, z)|^2$ distribution. The dynamics (4.3, 4.4) preserves that distribution, in the sense that the particles,

---

[22] See [37, pp. 145–160] for a more detailed discussion of these experiments within the theory of de Broglie and Bohm.

evolving according to (4.4) will be distributed at a later time t according to the $|\Psi(x, z, t)|^2$ distribution, where $\Psi(x, z, t)$ is the solution of Schrödinger's equation (4.3) at time t. Then obviously the usual quantum predictions are recovered. We will not discuss here or try to justify this statistical assumption on the initial conditions (see [39]), but we note that the de Broglie–Bohm theory answers the question about the status of the quantum state (does it represents our information about the system or is it rather something physical or both?) by "both": it has a clear physical role, through the guiding equation, (4.4), but also a statistical role (that is, it reflects our ignorance or our partial information) through the assumption on initial conditions.

Now, coming back to the spin measurement, suppose that Φ is symmetric in z: $\Phi(z) = \Phi(-z)$. Then the derivative $\partial_z \Phi(x, z)$ vanishes at $z = 0$ and, by (4.4), the particle velocity is zero for $z = 0$. Therefore, the particle never crosses the line $z = 0$. From this we conclude that the particle will go up if its initial condition satisfies $z > 0$ and down otherwise.

But, and here is the surprising point, suppose that we reverse the direction of the magnetic field. Then the solution of Schrödinger's equation becomes (again, with much simplification)

$$\Phi(z + t)| \uparrow> + \Phi(z - t)| \downarrow> . \tag{4.6}$$

So, now, if the particle is near $-t$, we say that the spin is up and if it is near t, we say that it is down.

But, if we do the experiment starting with the same quantum state and the same particle position, the particle will go in the upward direction again if its initial condition satisfies $z > 0$ and downward otherwise, since it cannot cross the line $z = 0$. But now, as we just saw, we declare that its spin is down if it goes in the positive z direction, and we declare that its spin is up if the particle goes in the negative z direction.

In other words, the value up or down of the spin is "contextual": that value does not depend only on the quantum state and the original particle position but also on the concrete arrangement of the "measuring" device. Here the quotation marks are introduced, because we see that there is no intrinsic property of the particle that is being "measured". Of course, since the system is deterministic, once we fix the initial state (quantum state and position) of the particle *and* the experimental device, the result of the experiment is predetermined. But that does not mean that the spin value that we "observe" is predetermined, because we can measure the spin along a given axis by orienting the magnetic field in one direction or the opposite one along that same axis. So the value of the spin depends on our conventions, which means that it does not exist as an intrinsic property of the particle. The same remark also holds for other quantum mechanical "observables" such as energy or momentum; see [40].

So not only is the word "hidden variable" misleading, but so are the words "observable" or "measurement", since, except when one measures positions, one does not observe or measure any intrinsic property of the system being observed or measured.

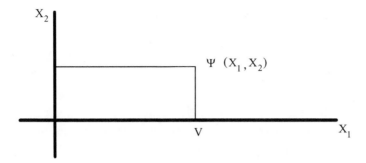

Figure 4.5 Nonlocality in de Broglie and Bohm's theory.

This vindicates in some sense Bohr's emphasis on

the impossibility of any sharp distinction between the behaviour of atomic objects and the interaction with the measuring instruments which serve to define the conditions under which the phenomena appear.

> *Niels Bohr [45, p. 210], quoted in [24, p. 2] (italics in original)*

But in the de Broglie-Bohm theory, this follows from the equations of the theory and not from some more or less a priori statement.

This is good news, since it explains how the de Broglie-Bohm theory can introduce hidden variables (the positions of the particles) without introducing the hidden variables (the spin values) that would lead to a contradiction, because of Bell's theorem.

Now, let us see why the de Broglie-Bohm theory is nonlocal. If we look at equation (4.4), we see that, if the quantum state factors as $\Psi = \prod_{i=1}^{N} \psi_i(x_i)$, then each particle is guided by its quantum state ψ_i and does not depend on the other ψ_j's, $j \neq i$. But if the state does not factor, which is the case for the state (23.1) (such states are called "entangled"), then the motion of each X_i is influenced by the value of the quantum state Ψ evaluated at the actual value $(X_1, \ldots, X_N)$ of the positions of all the particles of the system. So, in the example of the state (23.1), if we change the position of the particle A, we influence immediately the motion of particle B, through equation (4.4).

Of course, one might say that in a deterministic theory, changing the position of a particle would involve counterfactual reasoning, since all the motions are determined anyway. But a Schrödinger equation may involve "external" potentials (such as a magnetic field) and the latter can be viewed as representing an external action on the system described by that equation. In Figure 4.5, we consider two particles; we replace (to simplify matters) $\mathbb{R}^3 \times \mathbb{R}^3$ by $\mathbb{R} \times \mathbb{R}$ and denote by (X_1, X_2) the actual positions of the two particles.

Now suppose we act on the first particle, by introducing near the place where that particle is located a potential V (in equation (4.3)), for example, a magnetic field that is used to "measure" the spin. This of course does not affect the motion of either the first or the second particle directly, but it affects the behavior of the quantum state, through Schrödinger's equation (4.3), and, through the guiding equation (4.4), this affects the behavior of both particles simultaneously (when Ψ is not factorizable as a product of functions of X_1 and X_2, as is the case for the state [23.1]).

If we go back to the state (23.1), when one introduces a magnetic field as in the Hamiltonian to "measure" the spin in one of the three directions, let us say first at A, the particle at A follows an evolution as in (4.5) and, once the particle is coupled to a detector, that induces an effective collapse of the quantum state (see for example [41, Chapters 8, 9], for more details). In the de Broglie–Bohm theory, the quantum state never really collapses and always follows Schrödinger's equation. But, once the state of a microscopic system is coupled to that of a macroscopic system, interference effects become, in practice, impossible (this is also called decoherence). Then one can take the microscopic state coupled to the one that is seen for the macroscopic system as being the quantum state for all the future behavior of the microscopic system. Thus, there is an effective, but not fundamental, collapse (this is somewhat similar to the phenomenon of irreversibility in statistical mechanics). Suppose that the state collapses, if the magnetic field is along direction 1, on $|A\ 1\ \uparrow\rangle\ |B\ 1\ \downarrow\rangle$. This means that the state changes also at B (and instantaneously) and the future behavior of the particle at B (at which one may apply a field in direction 1, 2 or 3) will be guided by the state $|B\ 1\ \downarrow\rangle$. This of course allows the de Broglie–Bohm theory to recover the usual quantum predictions in the EPR–Bell situation.

A well-known problem is that whether the measurement is done first at A or at B is not a relativistically invariant notion – it will depend on our frame of reference. But this problem lies outside the scope of this paper. It is a problem caused by the existence of nonlocal effects and has nothing to do with the de Broglie–Bohm theory per se. Ordinary quantum mechanics also has this problem, but simply ignores it: for example, in relativistic quantum theories, including quantum field theories, the "collapse" operation is never treated in a relativistic way, and, because it is simultaneous, it poses the same problem as the one raised here. The only way out would be to say that the collapse simply reveals preexisting properties of the system, which is exactly the idea of preexisting spin values, which, as Bell showed, is inconsistent with quantum mechanical predictions. So the problem of making relativity and quantum mechanics truly compatible is deep and unsolved (irrespective of what one thinks of the de Broglie–Bohm theory).

4.7 Summary and Conclusions

During the debates over the "Copenhagen" interpretation of quantum mechanics, Einstein held a heterodox and minority view. He maintained that the quantum state is not a complete description of physical reality and that particles have properties beyond what is included in their quantum states. In other words, he thought that quantum mechanics provides a very accurate statistical description of quantum systems, but not a complete physical description of individual systems.

Einstein tried to give indirect arguments proving that quantum mechanics is incomplete and, in 1935, together with Podolsky and Rosen, he made the following reasoning: suppose that there is a quantity that I can measure on one system and that, because of conservation laws, immediately tells me the value of the corresponding quantity for another system which is far away. Then, if we assume that the measurement on the first system cannot affect the physical state of the second system, because of their spatial separation, the second system must have had that property all along.

The EPR argument was made unnecessarily complicated by considering two quantities (position and momentum) instead of one, each of which can be measured on the first system and then predicted for the second system. But, if all one wants to show is the incompleteness of quantum mechanics, then considering one quantity is enough, since the quantum state does not assign a definite value to that quantity for the second system, while the EPR argument shows that, barring nonlocal effects, there must be one.

The EPR argument was generally ignored (except mostly by Schrödinger [46]) and most physicists thought that it had been countered by Bohr (although Bohr's argument is far less clear than the EPR paper).

In 1964, almost thirty years after the EPR paper, and when their argument had been essentially forgotten, Bell showed that merely assuming that the quantities that EPR showed must exist if the world is local leads to a contradiction with quantum predictions that were later verified experimentally.

Thus, the question raised by EPR, whether the quantum mechanical description is complete, was answered in a way that Einstein would probably have liked least, as Bell said: by showing that the "obvious" assumption of locality made by EPR is actually false ([27, p. 11]).

But since most people had forgotten or misunderstood the EPR paper, Bell's argument was taken to be just one more proof that quantum mechanics is complete in the sense that no other variables that would complete the quantum description can be introduced without running into contradictions.

Of course, that was the opposite of what Bell showed and explicitly said: the conclusion of his argument, combined with that of EPR, is rather that there are nonlocal physical effects (and not just correlations) in Nature.

As for the completeness of quantum mechanics, while it is true that the EPR argument does not settle that issue, a more complete theory was proposed by de Broglie in 1927 and developed by Bohm in 1952. That theory of course is nonlocal, which is taken by people who misunderstand the EPR–Bell result as an argument against that theory, while, in reality, if that theory were local, it could not be empirically adequate.

The speculations of Einstein looked purely philosophical or even "metaphysical" to many physicists; for example, Pauli wrote,

As O. Stern said recently, one should no more rack one's brain about the problem of whether something one cannot know anything about exists all the same, than about the ancient question of how many angels are able to sit on the point of a needle. But it seems to me that Einstein's questions are ultimately always of this kind.

Wolfang Pauli [16, p. 223]

But those speculations have led to what is probably "the most profound discovery of science", to use Henri Stapp's phrase [47, p. 271]. They have also, because they lie at the basis of quantum information, led to various technological applications.

This should be a lesson for "pragmatists".

Acknowledgments

This paper is based on a talk given at the conference Quantum Theory Without Observers III ZIF, Bielefeld, April 23, 2013. I have benefitted from a great number of discussions about the meaning of Bell's theorem with many people, but especially with Detlef Dürr, Sheldon Goldstein and Tim Maudlin. I thank Dominique Lambert and Serge Dendas for useful remarks on this manuscript.

References

[1] J. von Neumann, *Mathematical Foundations of Quantum Mechanics*, Princeton University Press, Princeton, NJ, 1955.

[2] A. Einstein, B. Podolsky, and N. Rosen, Can quantum-mechanical description of reality be considered complete? *Physical Review* **47**, 777–80, 1935. Reprinted in [26, pp. 138–41].

[3] A. Aspect, J. Dalibard, and G. Roger, Experimental test of Bell's inequalities using time-varying analysers, *Physics Review Letters* **49**, 1804–7, 1982.

[4] D. Dürr, S. Goldstein, R. Tumulka, and N. Zanghì, John Bell and Bell's Theorem, in D.M. Borchert (ed.), *Encyclopedia of Philosophy*, Macmillan Reference, 2005.

[5] A. Einstein, Remarks concerning the essays brought together in this co-operative volume, in P.A. Schilpp (ed.), *Albert Einstein, Philosopher–Scientist*, Library of Living Philosophers, Evanston, IL, 1949, pp. 665–88.

[6] T. Norsen, Einstein's boxes, *American Journal of Physics* **73**, 164–76, 2005.

[7] L. de Broglie, L'interprétation de la mécanique ondulatoire, *Journal de Physique et le Radium*, **20**, 963–79, 1959.

[8] L. de Broglie, *The Current Interpretation of Wave Mechanics: A Critical Study*, Elsevier, Amsterdam, 1964.

[9] A. Fine, *The Shaky Game: Einstein Realism and the Quantum Theory*, University of Chicago Press, Chicago, 1986.

[10] G. Bacciagaluppi and A. Valentini, *Quantum Mechanics at the Crossroads. Reconsidering the 1927 Solvay Conference*, Cambridge University Press, Cambridge, 2009.

[11] Solvay Conference, *Electrons et photons: Rapports et discussions du 5e Conseil de Physique tenu á Bruxelles du 24 au 29 octobre 1927*, Instituts Solvay, Gauthier-Villars et cie, Paris, 1928, translated in [10].

[12] T. Maudlin, *Quantum Non-Locality and Relativity*, Blackwell, Cambridge, 1st ed., 1994, 3rd ed., 2011.

[13] E. McMullin, The explanation of distant action: Historical notes, in [16, pp. 272–302].

[14] I. Newton, Letter from Isaac Newton to Richard Bentley 25 February 1693, in H.W. Turnbull (ed.), *The Correspondence of Isaac Newton*, Vol. 3, pp. 253–4, Cambridge University Press, Cambridge, 1959.

[15] T. Maudlin, What Bell did, *Journal of Physics A: Mathematical and Theoretical*, **47**, 424010, 2014.

[16] M. Born (ed.), *The Born–Einstein Letters*, Macmillan, London, 1971.

[17] J.S. Bell, On the Einstein–Podolsky–Rosen paradox. *Physics* **1**, 195–200, 1964. Reprinted as Chapter 2 in [24].

[18] R. Feynman, Simulating physics with computers, *International Journal of Theoretical Physics*, **21**, 467–88, 1982.

[19] P.H. Eberhard, Bell's theorem and the different concepts of locality, *Nuovo Cimento B* **46**, 392–419, 1978.

[20] D. Bohm, *Quantum Theory*, Dover Publications, New York, new edition (1989); First edition: Prentice Hall, Englewood Cliffs, NJ, 1951.

[21] M. Archer, Why did Einstein say 'God doesn't play dice'? video available at physicsworld.com.

[22] A. Einstein, Letter to Cornelius Lanczos, March 21, 1942, Einstein Archive, 15–294. Translated in *Albert Einstein: The Human Side*, H. Dukas and B. Hoffmann, eds., Princeton University Press, Princeton, NJ, 1979, p. 68.

[23] A. Einstein, Quantum mechanics and reality, *Dialectica* **2**, 320–24, 1948.

[24] J.S. Bell, *Speakable and Unspeakable in Quantum Mechanics*, Cambridge University Press, Cambridge, 1993.

[25] N. Bohr, Can quantum-mechanical description of reality be considered complete? *Physics Review* **48**, 696–702, 1935. Reprinted in [26, pp. 145–51].

[26] J.A. Wheeler and W.H. Zurek (eds.), *Quantum Theory and Measurement*, Princeton University Press, Princeton, NJ, 1983.

[27] J.S. Bell, On the problem of hidden variables in quantum mechanics, *Reviews of Modern Physics* **38**, 447–52, 1966. Reprinted as Chapter 1 in [24].

[28] D. Mermin, Hidden variables and the two theorems of John Bell, *Review of Modern Physics*, **65**, 803–15, 1993.

[29] M. Gell-Mann, *The Quark and the Jaguar*, Little, Brown and Co., London, 1994.

[30] J.S. Bell, Bertlmann's socks and the nature of reality, *Journal de Physique*, **42**, C2 41–61, 1981. Reprinted as Chapter 16 in [24].

[31] E.P. Wigner, Interpretation of quantum mechanics, in [26, pp. 260–314].

[32] S. Goldstein, Quantum philosophy: The flight from reason in science, contribution to P. Gross, N. Levitt, and M.W. Lewis (eds.), *The Flight from Science and Reason*, *Annals of the New York Academy of Sciences* **775**, 119–25, 1996. Reprinted in [42, Chapter 4].

[33] D. Mermin, Is the moon there when nobody looks? Reality and the quantum theory, *Physics Today*, **38**, April 1985, 38–47.

[34] S. Kochen and E.P. Specker, The problem of hidden variables in quantum mechanics, *Journal of Mathematics and Mechanics*, **17**, 59–87, 1967.

[35] D. Bohm, A suggested interpretation of the quantum theory in terms of "hidden variables", Parts 1 and 2, *Physical Review* **89**, 166–93, 1952.

[36] D. Albert, Bohm's alternative to quantum mechanics, *Scientific American* **270** May 1994, 32–9.

[37] D. Albert, *Quantum Mechanics and Experience*, Harvard University Press, Cambridge, 1992.

[38] D. Bohm and B.J. Hiley, *The Undivided Universe*, Routledge, London, 1993.

[39] D. Dürr, S. Goldstein and N. Zanghì, Quantum equilibrium and the origin of absolute uncertainty, *Journal of Statistical Physics*, **67**, 843–907, 1992. Reprinted in [42, Chapter 2].

[40] D. Dürr, S. Goldstein and N. Zanghì, Naive realism about operators, *Erkenntnis*, **45**, 379–97, 1996.

[41] D. Dürr and S. Teufel, *Bohmian Mechanics. The Physics and Mathematics of Quantum Theory*, Springer, Berlin–Heidelberg, 2009.

[42] D. Dürr, S. Goldstein, and N. Zanghì, *Quantum Physics Without Quantum Philosophy*, Springer, Berlin–Heidelberg, 2012.

[43] P.R. Holland, *The Quantum Theory of Motion*, Cambridge University Press, Cambridge, 1995.

[44] J.P. Vigier, C. Dewdney, P.E. Holland and A. Kyprianidis, Causal particle trajectories and the interpretation of quantum mechanics, in B.J. Hiley and F.D. Peat (eds.), *Quantum Implications; Essays in Honour of David Bohm*, Routledge, London, 1987, 169–204.

[45] N. Bohr, Discussion with Einstein on epistemological problems in atomic physics, in P.A. Schilpp (ed.), *Albert Einstein, Philosopher–Scientist*, pp. 201–41, Library of Living Philosophers, Evanston, IL, 1949.

[46] E. Schrödinger, The present situation in quantum mechanics, *Proceedings of the American Philosophical Society*, **124**, 323–38. Translated from E. Schrödinger, Die gegenwärtige Situation in der Quantenmechanik, *Naturwissenschaften* **23**, 807–12; 823–8; 844–9, 1935. Reprinted in J.A. Wheeler and W.H. Zurek (eds.), *Quantum Theory and Measurement* Princeton University Press, Princeton, NJ, 1983, pp. 152–67.

[47] H.P. Stapp, Bell's theorem and world process, *Nuovo Cimento B* **29**, 270–76, 1975.

5

The Assumptions of Bell's Proof

RODERICH TUMULKA

Abstract

While it is widely agreed that Bell's theorem is an important result in the foundations of quantum physics, there is much disagreement about what exactly Bell's theorem shows. It is agreed that Bell derived a contradiction with experimental facts from some list of assumptions, thus showing that at least one of the assumptions must be wrong; but there is disagreement about what the assumptions were that went into the argument. In this paper, I make a few points in order to help clarify the situation.

5.1 Introduction

Different authors have expressed very different views about what Bell's theorem shows about physics. The disagreement concerns particularly the question of which assumptions go into the argument. Since we have to give up one of the assumptions leading to the empirically violated Bell inequality, knowing what the assumptions were is crucial. For example, if author X believes that the argument requires only one assumption, A_1, while author Y believes that it requires two, A_1 and A_2, then X will conclude that A_1 must be abandoned, while Y will conclude that either A_1 or A_2 must abandoned, so A_1 may well be true in our world if A_2 is false. In this paper, I consider several assumptions that have been mentioned in connection with Bell's theorem, and I look into their roles in the proof of Bell's theorem. While I do not say anything here that has never been said before, I hope that my remarks can nevertheless be helpful to some readers.

For the sake of definiteness, the version of the relevant experiment that I will consider involves two spin-$\frac{1}{2}$ particles, initially in the singlet spin state $\psi = 2^{-1/2}(|\uparrow\downarrow\rangle - |\downarrow\uparrow\rangle)$. Two widely separated experimenters, Alice and Bob, have one particle transported to each of them. At spacelike separation, they each choose a direction in space, represented by the unit vectors a and b, and carry out a quantum measurement of the spin observable in this direction on their particle, $(a \cdot \sigma) \otimes I$ and $I \otimes (b \cdot \sigma)$, with I the 2×2 unit matrix and σ the triple of Pauli matrices. The possible outcomes, A and B, are ± 1 on each side. Bell's theorem is the statement that under the assumptions we are about to discuss, Bell's inequality holds; quantum theory predicts violations of Bell's inequality, and experiments have confirmed these predictions.

5.2 Local Realism

The upshot of Bell's theorem, together with the empirical violation of Bell's inequality, is often described as refuting *local realism*; that is, Bell's theorem is described as requiring, as the two main assumptions, *locality* and *realism*. Let me focus first on realism. Sometimes different things are meant by this term, so let me formulate several options for what "realism" might mean here (without implying that "realism" is a good name for any of these conditions):

(R1) Every quantum observable (or at least $(a \cdot \sigma) \otimes I$ and $I \otimes (b \cdot \sigma)$ for every a and b) actually has a definite value even before any attempt to measure it; the measurement reveals that value.

(R2) The outcome of every experiment is predetermined by some ("hidden") variable λ.

(R3) There is some ("hidden") variable λ that influences the outcome in a probabilistic way, as represented by the probability $P(A, B|a, b, \lambda)$.

(R4) Every experiment has an unambiguous outcome, and records and memories of that outcome agree with what the outcome was at the space–time location of the experiment.

I comment on these hypotheses one by one, in reverse order.

(R4) is certainly an assumption required for the proof of Bell's theorem, and was clearly taken for granted by Bell. It would be false in the many-worlds view of quantum theory, as in that view every outcome is realized.[1] Bell himself acknowledged that his reasoning does not necessarily apply in a many-worlds framework [1]:

> The 'many world interpretation' ... may have something distinctive to say in connection with the 'Einstein Podolsky Rosen puzzle' [i.e., Bell's theorem], and it would be worthwhile, I think, to formulate some precise version of it to see if this is really so.

For an analysis of Bell's theorem in the many-worlds framework, I refer the reader to [2]. In the following, I will leave the many-worlds view aside. Thus, it seems that it would be *very* hard to abandon (R4). (R4) is perhaps the mildest form of "realism" one could think of: that macroscopic facts are unambiguous, do not change if we observe them, and can be remembered reliably. (R4) would be false if memories and records of outcomes changed after the fact, regardless of how many people observed the outcome and how the outcomes were recorded. That, of course, would be a veritable conspiracy theory. The upshot is that (R4) is an assumption that many, including myself, are not prepared to abandon. In the following, I will take (R4) for granted.

(R3) is really a vacuous assumption, because λ in $P(A, B|a, b, \lambda)$ could be just the wave function ψ of the particle pair, a variable which anyhow is well defined in this physical situation and does affect the probability distribution of the outcomes A, B. It is *also* allowed that λ involves more than just ψ, such as particle positions in Bohmian mechanics. But

---

[1] One of my readers thought I was implying that the many-worlds view is not a "realist" view. That is a misunderstanding, I was not implying that. The conditions (R1) through (R4) are considered here for their potential relevance to Bell's theorem, not for categorizing interpretations of quantum mechanics as realist ones and others.

$\lambda = \psi$ is a possibility that is included in the expression $P(A, B|a, b, \lambda)$, and thus, (R3) does not commit us to anything like Bohmian particles, that is, to the existence of any variables in addition to ψ (which is what is usually meant by "hidden variables"). The very notation $P(A, B|a, b, \lambda)$ in the literature on Bell's theorem has presumably suggested to many readers that Bell is making an assumption of hidden variables; after all, λ is the hidden variable! But in fact, an assumption of *no hidden variables* is equally allowed by this notation.

(R2) is a much stronger assumption than (R3); it is not vacuous, and λ could not be taken to be ψ here, as ψ does not determine the outcomes – it only provides probabilities. The assumption (R2) is why Bell's theorem has been called a no-hidden-variables theorem, the idea being that if we have to drop either locality or the assumption of deterministic hidden variables, it would be more attractive to drop the latter and keep locality.

The crucial point about (R2) is that Bell did not make this assumption. The proof of Bell's theorem does not require it. Before I explain why, let me quote what Bell wrote on this point later, in 1981 [3]:

It is important to note that to the limited degree to which *determinism* plays a role in the EPR argument [and thus the proof of Bell's theorem], it is not assumed but *inferred*. . . . It is remarkably difficult to get this point across, that determinism is not a *presupposition* of the analysis.

[emphasis in the original]

Now let us look at the role of (R2), i.e., of determinism, in the proof of Bell's theorem. Bell gave two proofs for Bell's theorem, the first in 1964 [4], the second in 1976 [5]. The second one (described also in [3]) is formulated in terms of the expression $P(A, B|a, b, \lambda)$; since it involves a variable λ that influences that outcomes A, B in a *probabilistic* way, it becomes clear that it is not assumed that λ influences A, B in a *deterministic* way. This illustrates that (R2) is not required for proving Bell's theorem.

The first proof, of 1964, did not assume (R2) either. And yet it can easily appear as if (R2) was being assumed, for the following reason. The proof has two parts; part one derives (R2) from the assumption of locality, and part two derives Bell's inequality from (R2) and locality. Any reader missing part one will conclude that Bell assumed (R2) *and* locality. And part one may be easy to miss, because it is short, and because all the mathematical work goes into part two, and because part one was, in fact, not developed by Bell but by Einstein, Podolsky, and Rosen (EPR) in their 1935 paper [6]. Bell complained later [3, footnote 10] that people missed part one of the proof:

My own first paper on this subject [i.e., [2]] starts with a summary of the EPR argument *from locality to* deterministic hidden variables. But the commentators have almost universally reported that it begins with deterministic hidden variables.

[emphasis in the original]

I will discuss EPR's argument for (R2) from locality in Section 5.7 below.

(R1) implies (R2), and like (R2), it need not be assumed for proving Bell's theorem. In fact, EPR's argument yields (R1) (in the version in parentheses, concerning only the observables $(a \cdot \sigma) \otimes I$ and $I \otimes (b \cdot \sigma)$ for all unit vectors a, b) if locality is assumed. Thus, the statement (R1) comes up within the 1964 proof of Bell's theorem, but not as an assumption. It does not even come up in Bell's second proof.

If the claims I have made are right, one should conclude that there is no *assumption of realism* that enters the proof of Bell's theorem next to the assumption of locality, and thus that we do not have the choice between the two options of abandoning realism and abandoning locality, but that we must abandon locality. One should conclude further that the widespread statement that "Bell's theorem refutes local realism" is misleading, and that Bell's theorem simply refutes locality. More detailed discussions of this point can be found in Refs. [7–10].

Bell himself contributed to the misunderstanding that his proof assumed (R2) when he wrote, as the first sentence of the "Conclusion" section of his 1964 paper [4]:

In a theory in which parameters are added to quantum mechanics to determine the results of individual measurements, without changing the statistical predictions, there must be a mechanism whereby the setting of one measuring device can influence the reading of another instrument, however remote.

This sentence, which would appear to be a summary of what is proved in the paper, clearly expresses an *assumption* of determinism as in (R2), and thus may give readers the wrong idea about Bell's theorem. As Norsen [11] has pointed out, Bell may have supposed, when writing this sentence, that his readers were aware that EPR had derived determinism from locality, that therefore it goes without saying that deterministic theories were the only remaining candidates for saving locality, and that therefore a description of the novel contribution of his paper needs to focus only on deterministic theories. Be that as it may, we should not let this sentence mislead us.

Another assumption of Bell's proof is the assumption of effective free will of the experimenters (or *no conspiracy*). This assumption, which can be expressed as $P(\lambda|a, b) = P(\lambda)$, means that the experimenter's choices a, b cannot be predicted by the two particles, or conversely, that the particles cannot force the experimenters to make a particular choice of a, b. If this were not true, then Bell's inequality could easily be violated by local mechanisms. In this paper, I will take this assumption for granted.

5.3 Kolmogorov's Axioms of Probability

It has been suggested that Bell's proof tacitly assumes the Kolmgorov axioms of probability, and that they can be questioned. Here are a few remarks about this suggestion. The Kolmogorov axioms summarize the usual notion of probability. They say that the events form a Boolean algebra (which means that for any two events E and F, the union $E \cup F$, the intersection $E \cap F$, and the complement E^c are also defined and obey the same rules as for sets), that for every event E its probability $P(E)$ is a real number with $0 \le P(E) \le 1$, and that P is additive,

$$P(E \cup F) = P(E) + P(F) \text{ when } E \cap F = \emptyset. \tag{5.1}$$

(Strictly speaking, P is "σ-additive," but the difference is not relevant here.)

Clearly, Bell's proof makes use of the Kolmogorov axioms for deriving the Bell inequality. It has been suggested that some of the Kolmogorov axioms might be violated in nature, so that locality may hold true even though the Bell inequality is violated. Specifically, it has been suggested that probabilities can be negative, or that additivity (5.1) can fail.

However, the λ in Bell's proof, which appears in expressions such as $P(A, B|a, b, \lambda)$, is a variable that has a definite value in every run of the experiment. Thus, as we repeat the experiment many times, any value of λ occurs with a certain frequency, and it is the limiting value of this frequency that is the relevant notion of probability in Bell's proof. Now, frequencies cannot be negative, and they are necessarily additive, because the number of occurrences of $E \cup F$ equals the number of occurrences of E plus that of F, if E and F are disjoint sets in the value space Λ of λ. In fact, since the events are subsets of the set Λ of all values of λ that occur, the events necessarily form a Boolean algebra. More generally, it lies in the nature of frequencies to obey the Kolmogorov axioms. That is why I see no room for the idea of saving locality by denying the Kolmogorov axioms.

5.4 Locality

At this point, it seems helpful to look more closely at the locality assumption, which I will call (L). It states

(L) If the space–time regions $\mathscr{A}$ and $\mathscr{B}$ are spacelike separated, then events in $\mathscr{A}$ cannot influence events in $\mathscr{B}$.

It has been argued about whether "locality" is a good name for this statement, and about whether it is an interesting question whether (L) is true. I will, however, leave these debates aside. After all, people can disagree about this and still agree about what Bell's theorem says; and my main concern in this paper is what Bell's theorem says.

In this section, I want to discuss the meaning of the hypothesis (L), or at least certain aspects of the meaning. The first question I raise is the following: In which way, if any, is (L) is different from the statement of *no signaling*? "No signaling" means no *superluminal* signaling, of course; that is,

(NS) If the space–time regions $\mathscr{A}$ and $\mathscr{B}$ are spacelike separated, then it is impossible to transmit a message, freely chosen by an agent (say, Alice) in $\mathscr{A}$, to another agent (say, Bob) in $\mathscr{B}$.

Clearly, (L) implies (NS), as signaling from $\mathscr{A}$ to $\mathscr{B}$ would constitute an instance of an influence from $\mathscr{A}$ to $\mathscr{B}$. And it is a natural thought that (NS) might also imply (L), because if some degree of freedom x in $\mathscr{A}$ could influence some degree of freedom y belonging to $\mathscr{B}$ then Alice might interact with x in such a way as to set it to a desired value encoding part of her message, and Bob might read off that part of the message from y. However, this reasoning may not go through if there are *limitations to control* or *limitations to knowledge*, which means that Alice cannot prepare x to have the value she desires, or that Bob cannot find out the value of y.

Limitations to control are quite familiar, as we have no control over the random out-come of a quantum experiment, once the initial quantum state of the experiment has been prepared. Limitations to knowledge, in contrast, appear to conflict with the principles of science: It may seem unscientific to believe in the physical existence of a variable that can-not be measured, perhaps like believing in angels. Put differently, if a variable y cannot be measured, it is natural to suspect that y is, in fact, not a well-defined variable – that it does not have a value. However, this conclusion is not necessarily correct, as the following example shows. Suppose that experimenter Carol chooses a direction c in 3-space, prepares a spin-$\frac{1}{2}$ particle with spin up in this direction (i.e., $c \cdot \sigma |\psi\rangle = |\psi\rangle$ for its quantum state $|\psi\rangle$), and hands it over to Donald, who does not know c. Donald cannot determine c by means of any experiment on the particle, even though there is a fact in nature about what $|\psi\rangle$ and c are: After all, Carol, who remembers $|\psi\rangle$ and c, could tell Donald and predict with certainty that a quantum measurement of $c \cdot \sigma$ will result in the outcome "up." The best Donald can do is to carry out a quantum measurement of $d \cdot \sigma$ in a direction d of his choice, which yields one bit of information, and allows a probabilistic statement about whether c is more likely to lie in the northern or southern hemisphere with respect to d, but no more than that. This example shows that there are facts in the world that cannot be revealed by any experiment – a limitation to knowledge about $y = c$.

So limitations to knowledge are a fact (see [12] for further discussion), but nevertheless they are rather unfamiliar, as we normally do not come across them when doing physics. The reason I elaborate on these limitations is that they are relevant to the statement of locality and to Bell's proof. Limitations to knowledge concern the distinction between real and observable; after all, such a limitation occurs when something is real and, at the same time, not observable. This distinction plays a role for reasoning about locality, as (L) refers to *real* events and influences, not to *observable* ones. Let me explain this by means of the "Einstein boxes example" (see [13] and references therein).

5.5 Einstein's Boxes and Reality

Split a 1-particle wavepacket into two wavepackets of equal size, transport one of them to Paris and the other to Tokyo, and keep each one in a box. At spacelike separation, two experimenters in Paris and Tokyo each apply a detector to their box; quantum mechanics predicts that one and only one of the two detectors clicks, each with probability 1/2. Was any violation of locality involved? Perhaps surprisingly, the answer is: That depends! That is, it depends on what the reality is like. Among the many possibilities for what the reality could be like in this experiment, let me describe two.

First, the possibility advocated by Einstein, in disagreement with the orthodox inter-pretation of quantum mechanics, was that the particle actually had a well-defined position before the detection; that is, that the particle actually was already in either Paris or Tokyo, and the detectors merely found out the location and made it known to the experimenters. If that scenario were right, (L) would hold true in this experiment, as the particle traveled slower than light, and the answer of each detector was determined simply by the presence or absence of the particle at the location of the detector, without any influence from the other location. The anti-correlation between the two detection results arises from a common cause in the common past, namely the particle's path toward either Paris or Tokyo.

Second, the possibility advocated by Bohr was that there is no fact about the particle position before we make a quantum measurement, and the outcome is generated randomly by nature at the moment of the quantum measurement. If this scenario were right, (L) would be violated in this experiment. After all, in a Lorentz frame in which the Paris experiment occurred first, nature made a random decision about its outcome with 50-50 chances, and this event had to influence, at spacelike separation, the outcome in Tokyo, which always is the opposite of that in Paris. Alternatively, in a Lorentz frame in which the Tokyo experiment occurred first, nature made a random decision about *its* outcome with 50-50 chances, and *this* event had to influence, at spacelike separation, the outcome in Paris.[2]

This example illustrates that (L) may hold in one scenario and not in another; that is, whether or not (L) holds depends on the scenario. Einstein's and Bohr's scenarios of the boxes example are observationally equivalent, but reality is quite different in the two scenarios, and the truth value of (L) depends on that reality. We may not know which, if either, of the two scenarios is correct, and we may never know. So we may never know whether (L) is violated in our world when we carry out Einstein's boxes experiment. In order to show that our world is nonlocal, we would have to show that *every* scenario, if it is compatible with the statistics of outcomes predicted by quantum mechanics and confirmed in experiment, has to violate locality. Bell's proof does exactly that.

For deciding whether (L) is valid or violated in a certain scenario, we need to consider the reality according to that scenario. Obviously, each scenario must commit itself to one particular picture of reality; that is, we cannot change our minds in the middle of the reasoning about whether electrons have definite positions or not. Considering reality includes considering reality independent of observation. That is something we often try to avoid in physics, but here we should not avoid it, here exactly that is appropriate and even required. If you think that, in experiments such as the one proposed by Bell, there is no reality independent of observation, then that is one possible scenario, just as the empty set is a set. But you need to consider other scenarios as well. If you are not willing to talk about reality then you cannot talk about locality. If you adopt the positivist attitude that in science one should make only operational statements (of the type "if we set up an experiment in such-and-such a way, then the possible outcomes are . . . and occur with probabilities . . . ") then you cannot distinguish between Einstein's and Bohr's scenarios, and cannot decide whether locality is valid or violated in them. And then you will not get the point of Bell's theorem.

As a historical remark, it was clear to Einstein that his scenario satisfied locality and Bohr's did not. He used his boxes example for an argument against Bohr's scenario:

---

[2] As a side remark, since relativity suggests that no frame is more "correct" than the other, we may fear that a conflict arises here: In a particular run of the experiment, did nature make the random decision in Paris, and did Paris influence Tokyo, or was it the other way around? I find it completely possible that there is no fact of the matter about *where* the random decision was taken, and in which direction the influence went. If we give up locality anyway, then the influence does not need to have a direction, and no conflict arises. I will come back to this point at the end of this section.

Einstein believed that locality holds in our world; when locality is taken for granted, it follows that Bohr's scenario is unacceptable, that electrons have definite positions at all times, and that quantum mechanics is incomplete because it leaves these positions out of the description.

One might be worried that the use of (what I called) *scenarios* amounts to adopting classical, rather than quantum, modes of reasoning. Perhaps Werner had in mind something like this when he recently claimed [14] that Bell made a tacit assumption by using classical, rather than quantum, probability calculus. Well, the rules of quantum mechanics apply to quantum observables, while classical rules apply to variables that have definite values. One may believe that electrons have definite positions at all times, or that they do not; and likewise with any other observable, one may believe that it has a definite value, or that it does not. In each of these scenarios, the locality condition refers to *events*, i.e., to the unambiguous *facts* about reality in this scenario. If a variable x belonging to space–time region $\mathscr{A}$ has a definite value, that is an event. If x does not have a definite value, then there is no event about the value of x. You can, for example, choose Einstein's or Bohr's scenario, in which reality is quite different, and in each scenario it is the reality of that scenario that is relevant to the locality condition. So you need to talk about the facts in this scenario, or, in other words, about those variables that *do* have definite values in this scenario. And to these variables, classical reasoning applies.

5.6 Nonlocality and Relativity

One motivation for doubting that Bell refuted locality arises from the impression that locality is part and parcel of relativity, so that, if we trust in relativity, it is not an option to abandon locality. However, that impression is not right, as it is actually possible to retain relativity and give up locality. This is best demonstrated by specifying a relativistic, nonlocal theory, and one such theory is the *GRW flash theory* [15, 16]; see also Refs. [17, 18] for discussions of this theory, as well as [19] for a different type of relativistic, nonlocal GRW theory. In the GRW flash theory, particles do not have world lines, but they have occasional world points; that is, the reality in this scenario consists of a discrete pattern of material space–time points called "flashes," along with a wave function associated with every spacelike hypersurface. The pattern of flashes is fundamentally random, and the theory prescribes the joint probability distribution of the space–time locations of all flashes as a function of the initial wave function.

This scenario is similar to the one of Bohr described above, in many ways: there are no definite particle positions (at most times); more generally, there are no further ("hidden") variables in addition to the wave function; the time evolution of a system is stochastic, so randomness is fundamental; the outcome of an experiment is generated randomly by nature at the moment of the experiment; locality is violated in Einstein's boxes experiment, as well as in Bell's experiment; there is no "mechanism" of transporting the influence from $\mathscr{A}$ to $\mathscr{B}$; instead, flashes simply occur randomly with a fundamental probability law that includes Bell's correlation, EPR's, and all others.

Another similar feature is the absence of direction of the nonlocal influence, mentioned in footnote 2. Let me explain. If locality is violated then there are some situations in which events in $\mathscr{A}$ *must have influenced* events in $\mathscr{B}$, or vice versa. Bell's theorem does not tell us which way the influence went. In the GRW flash theory, there are faster-than-light influences, but there is no fact in nature about which way the influence went – no fact about "who influenced whom," i.e., whether events in $\mathscr{A}$ influenced those in $\mathscr{B}$ or vice versa. In probability theory, two random variables can be correlated, and that is a symmetric relation, without any direction of influence. The situation in the GRW flash theory is exactly the same: the flashes in $\mathscr{A}$ and those in $\mathscr{B}$ are correlated in a nonlocal way, so there is an influence between them without a direction. If you feel that the word "influence" necessarily entails a direction, then perhaps this word is not ideal, and perhaps "interaction" [7] is a better word. (Of course, "interaction" usually means in physics "a term in the Hamiltonian," and here we would have to drop that meaning and return to the original meaning of "interaction" in English as a mutual effect.)

Bell called locality "local causality," a name with the disadvantage that it adds to the suggestion of direction. After all, a *causal influence* would seem to be one from a cause to an effect, and thus clearly have a direction. That is why I avoid the word "causal" in the context of locality.

5.7 The EPR Argument

I wish to return to my point of Section 5.2 that Bell's proof does not assume the hypotheses (R1) or (R2). As I said, this is easily visible in Bell's 1976 proof [5], while in his 1964 proof [2] it is owed to the EPR argument, which derives (R1) and thus (R2) from locality alone. Let us review this argument.

The EPR argument, for the experiment involving two spin-$\frac{1}{2}$ particles in the singlet state, can be put this way: Suppose that Alice and Bob always choose the z-direction, $a = b = (0, 0, 1)$. Quantum mechanics then predicts that the outcomes are perfectly anti correlated, $A = -B$ with probability 1. Assume locality. Alice's experiment takes place in a space–time region $\mathscr{A}$ and Bob's in $\mathscr{B}$ at spacelike separation. There is a Lorentz frame in which $\mathscr{A}$ is finished before $\mathscr{B}$ begins; thus, in this frame, there is a time at which Alice's experiment already has a definite outcome. She can therefore predict Bob's outcome with certainty, although she cannot transmit this information to Bob before Bob carries out his experiment. Anyway, Bob's outcome was already fixed on some spacelike hypersurface before his experiment. By locality, his outcome was not influenced by events in $\mathscr{A}$, in particular not by whether Alice did any experiment at all. Thus, the state of affairs inside the past light cone of $\mathscr{B}$, but before $\mathscr{B}$ itself, included a fact about the value B_z that Bob will obtain if he carries out a quantum measurement of $I \otimes \sigma_z$. In particular, B_z is a "hidden variable" in the sense that it cannot be read off from ψ. Since the argument works in the same way for any other direction b instead of z, there is a well-defined value B_b for every unit vector b, such that if Bob chooses b_0 then his outcome will be B_{b_0}. Since the argument works in the same way for Alice, her outcome also merely reveals the pre determined value A_a for the particular direction a she chose.

We see how locality enters the argument, and how (R1), and thus (R2), come out. EPR's reasoning is sometimes called a paradox, but the part of the reasoning that I just described is really not a paradox but an argument, showing that (L) implies (R1). The argument is correct, while the premise (L), I argue following Bell, is not. We also see how the argument refers to reality ("events," "state of affairs"), or to different possible scenarios of what reality might be like, in agreement with my point in Section 5.5 that locality cannot be formulated in purely operational terms.

It may be helpful to make explicit where exactly Bell makes use of the EPR argument in his 1964 paper. On the first page, he writes,

> the EPR argument is the following.... we make the hypothesis, and it seems one at least worth con-
> sidering, that if the two measurements are made at places remote from one another the orientation of
> one magnet does not influence the result obtained with the other. Since we can predict in advance the
> result of measuring any chosen component of σ_2, by previously measuring the same component of σ_1,
> it follows that the result of any such measurement must actually be predetermined. Since the initial
> quantum mechanical wave function does not determine the result of an individual measurement, this
> predetermination implies the possibility of a more complete specification of the state.

The "more complete specification of the state" means the specification of, in addition to ψ, the A_a and B_b for all unit vectors a and b, or of some variable λ of which all A_a and B_b are functions.

Actually, the way Bell described the argument in this paper is slightly inaccurate (a point also discussed in [11, Sect. III]), as his formulation of the hypothesis expresses not locality but another condition that we may call control locality (CL):

(CL) If the space–time regions $\mathscr{A}$ and $\mathscr{B}$ are spacelike separated, then Alice's actions in $\mathscr{A}$ cannot influence events in $\mathscr{B}$.

It differs from (L) in that the phrase "events in $\mathscr{A}$" has been replaced by "Alice's actions in $\mathscr{A}$." Since Alice's actions in $\mathscr{A}$ are particular events in $\mathscr{A}$, (L) implies (CL). However, while (L) alone implies (R1), (CL) does not, although (CL) and (R2) (i.e., determinism) together imply (R1). The reason (CL) without (R2) does not imply anything is that, for general stochastic theories, it is not even clear what (CL) should mean. In a deterministic theory, the meaning is clear, as there the outcome B is a function of λ and the parameters a and b, and (CL) means that this function does not depend on a. In a stochastic theory, however, it is not clear what it should mean that the parameter a does not influence the random outcome B, as it is not clear what the counterfactual statement "if a had been different, then B would have been the same" should mean; if we rerun the randomness, then B may come out differently even for the same a.[3] So the passage just quoted from Bell says that (CL) implies (R1), while the correct thing to say is that (L) implies (R1).

---

[3] The same difficulty arises with making sense of Conway and Kochen's MIN condition for stochastic theories [20].

To sum up, I have considered various statements that have been claimed to be assumed in the derivation of Bell's inequality and thus to be candidates for being refuted by experimental violations of Bell's inequality. I have given reasons for which some assumptions cannot be dropped, other statements are in fact not assumed in the derivation, and yet others are indeed refuted.

Acknowledgments

I wish to thank the many people I had discussions on this subject with. They are people with very different views, and I can only mention a few: Jürg Fröhlich, Shelly Goldstein, Richard Healey, Federico Holik, George Matsas, Tim Maudlin, Travis Norsen, Daniel Sudarsky, and Reinhard Werner. The author was supported in part by Grant no. 37433 from the John Templeton Foundation.

References

[1] J.S. Bell, Six possible worlds of quantum mechanics, In *Proceedings of the Nobel Symposium 65: Possible Worlds in Arts and Sciences*, Stockholm, August 11–15, 1986. Reprinted as chapter 20 of [21].

[2] V. Allori, S. Goldstein, R. Tumulka, and N. Zanghì, Many-worlds and Schrödinger's first quantum theory, *British Journal for the Philosophy of Science* **62**(1), 1–27 (2011) http://arxiv.org/abs/0903.2211.

[3] J.S. Bell, Bertlmann's socks and the nature of reality. *Journal de Physique* **42**, C2 41–61 (1981). Reprinted as chapter 16 of [21].

[4] J.S. Bell, On the Einstein–Podolsky–Rosen paradox, *Physics*, **1**, 195–200 (1964). Reprinted as chapter 2 of [21].

[5] J.S. Bell, The theory of local beables, *Epistemological Letters* **9**, 11 (1976). Reprinted as chapter 7 of [21].

[6] A. Einstein, B. Podolsky, and N. Rosen, Can quantum-mechanical description of physical reality be considered complete? *Physical Review* **47**, 777–80 (1935).

[7] S. Goldstein, T. Norsen, D.V. Tausk, and N. Zanghì, Bell's theorem, *Scholarpedia* **6(10)**, 8378 (2011), www.scholarpedia.org/article/Bell%27s_theorem.

[8] T. Maudlin, What Bell did, *Journal of Physics A: Mathematical and Theoretical* **47**, 424010 (2014).

[9] T. Maudlin, Reply to 'Comment on "What Bell did,"' *Journal of Physics A: Mathematical and Theoretical* **47**, 424012 (2014).

[10] T. Norsen, Against 'realism,' *Foundations of Physics* **37**(3), 311–40 (2007), http://arxiv.org/abs/quant-ph/0607057.

[11] T. Norsen, Are there really two different Bell's theorems? www.ijqf.org/groups-2/bells-theorem/forum/topic/are-there-really-two-different-bells-theorems/.

[12] C.W. Cowan and R. Tumulka, Epistemology of wave function collapse in quantum physics, *British Journal for the Philosophy of Science* **67**, 405–34 (2016), http://arxiv.org/abs/1307.0827.

[13] T. Norsen, Einstein's boxes, *American Journal of Physics* **73**(2), 164–76 (2005), http://arxiv.org/abs/quant-ph/0404016.

[14] R. Werner, Comment on 'What Bell did,' *Journal of Physics A: Mathematical and Theoretical* **47**, 424011 (2014),

[15] R. Tumulka, A relativistic version of the Ghirardi–Rimini–Weber model. *Journal of Statistical Physics* **125**, 821–40 (2006), http://arxiv.org/abs/quant-ph/0406094.

[16] R. Tumulka, Collapse and relativity, in A. Bassi, D. Dürr, T. Weber, and N. Zanghì (eds.), *Quantum Mechanics: Are There Quantum Jumps? and On the Present Status of Quantum Mechanics*, AIP Conference Proceedings **844**, 340–52, American Institute of Physics (2006), http://arxiv.org/abs/quant-ph/0602208.

[17] T. Maudlin, *Quantum Non-locality and Relativity: Metaphysical Intimations of Modern Physics* [third ed.], Blackwell, Oxford (2011).

[18] R. Tumulka, Comment on 'The free will theorem.' *Foundations of Physics* **37**, 186–97 (2007), http://arxiv.org/abs/quant-ph/0611283.

[19] D. Bedingham, D.Dürr, G.C. Ghirardi, S. Goldstein, and N. Zanghì, Matter density and relativistic models of wave function collapse, *Journal of Statistical Physics* **154**, 623–31 (2014) http://arxiv.org/abs/1111.1425.

[20] S. Goldstein, D.V. Tausk, R. Tumulka, and N. Zanghì, What does the free will theorem actually prove? *Notices of the American Mathematical Society* **57**(11), 1451–3 (2010), http://arxiv.org/abs/0905.4641.

[21] J.S. Bell, *Speakable and unspeakable in quantum mechanics*, Cambridge University Press, Cambridge (1987).

6

Bell on Bell's Theorem: The Changing Face of Nonlocality

HARVEY R. BROWN AND CHRISTOPHER G. TIMPSON

Abstract

Between 1964 and 1990, the notion of nonlocality in Bell's papers underwent a profound change as his nonlocality theorem gradually became detached from quantum mechanics, and referred to wider probabilistic theories involving correlations between separated beables. The proposition that standard quantum mechanics is itself nonlocal (more precisely, that it violates 'local causality') became divorced from the Bell theorem per se from 1976 on, although this important point is widely overlooked in the literature. In 1990, the year of his death, Bell would express serious misgivings about the mathematical form of the local causality condition and leave ill-defined the issue of the consistency between special relativity and violation of the Bell-type inequality. In our view, the significance of the Bell theorem, in both its deterministic and stochastic forms, can only be fully understood by taking into account the fact that a fully Lorentz covariant version of quantum theory, free of action at a distance, can be articulated in the Everett interpretation.

6.1 Introduction

John S. Bell's last word on his celebrated nonlocality theorem and its interpretation appeared in his 1990 paper 'La nouvelle cuisine', first published in the year of his untimely death. Bell was careful here to distinguish between the issue of 'no superluminal signalling' in quantum theory (both quantum field theory and quantum mechanics) and a principle he first introduced explicitly in 1976 and called 'local causality' [1]. In relation to the former, Bell expressed concerns that amplify doubts he had already expressed in 1976. These concerns touch on what is now widely known as the *no-signalling theorem* in quantum mechanics, and ultimately have to do with Bell's distaste for what he saw as an anthropocentric element in orthodox quantum thinking. In relation to local causality, Bell emphasised that his famous factorizability (no-correlations) condition is not to be seen 'as the *formulation* of local causality, but as a consequence thereof' and stressed how difficult he found it to articulate this consequence. He left the question of any strict inconsistency between violation of factorizability and special relativity theory unresolved, a not insignificant shift from his thinking up to the early 1980s. Bell felt strongly that correlations ought always to be apt for causal explanations – a view commonly attributed to the philosopher Hans Reichenbach (though perhaps in this particular context incorrectly – of which more below). But he recommended that factorizability 'be viewed with the utmost suspicion'.

Few if any commentators have remarked on this late ambivalence about the factorizability condition on Bell's part. Part of the problem facing him was, as he himself emphasised, that although the intuition behind local causality was based on the notion of *cause*, factorizability makes no reference to causes. Whatever else was bothering Bell, apart from the absence of a clear-cut relativistic motivation, is perhaps a matter of speculation. We suggest, at any rate, that there is an important difference between the notion of locality he introduced in his original 1964 theorem and that of local causality. The difference is essentially that in 1964 he was excluding a certain kind of instantaneous action at a distance, whereas the connection between local causality and such exclusion is not straightforward. More bluntly, and counterintuitively, violation of local causality does not necessarily imply action at a distance. The issue is model-dependent. There is a consistent Lorentz covariant model of quantum phenomena which violates local causality but is local in Bell's 1964 sense: the Everett picture. One of the themes of our paper is something that Bell himself emphasised, namely, that the consideration of detailed physical models is especially important in correcting wayward intuitions.

The importance of the distinction between Bell's 1964 and 1976 versions of his nonlocality theorem has been noted in the literature, particularly in the recent careful work of Wiseman [2] (see also Timpson and Brown [3]). But in the present paper we are more concerned with the notion of nonlocality than with the theorems per se; appreciation of the changing face of nonlocality in Bell's writings and its wider significance in the foundations of quantum mechanics is harder to find in the literature. In fact, there are categorical claims that Bell's understanding of nonlocality never underwent significant revision.[1] We think this unlikely, but more importantly, we wish to concentrate on how the 1964 and post-1976 notions of nonlocality *should* be understood.

The change in Bell's definition of locality was accompanied by a shift of thinking on his part and that of many other commentators from the 1970s onwards, in relation to the range of applicability of the Bell theorem. Originally confined to deterministic hidden-variables interpretations of quantum mechanics, the theorem was later correctly seen to apply to probabilistic theories generally. Thus it is a widely held view in the literature today that as a result of the predicted (and experimentally corroborated) violation of Bell-type inequalities, standard quantum mechanics itself is nonlocal.[2] And yet, as Bell himself acknowledged,

---

[1] See, e.g., Maudlin [4], Norsen [5], and Wiseman [2] for further discussion.

[2] Consider the following statements in the recent physics literature, which arguably are representative of a widespread view concerning the significance of the Bell nonlocality theorem. The first appears in a lengthy 2013 paper by A. Hobson on the nature of quantum reality in the *American Journal of Physics*: 'Violation of Bell's inequality shows that the [entangled state]…is, indeed, nonlocal in a way that cannot be interpreted classically' [6, p. 220]. In his reply to critics, Hobson writes: 'This violation [of the Bell inequality] means that the correlations are too tightly dependent on the non-local phase relationship between the two [entangled] systems to be explainable by purely local means. So such correlations do lead to nonlocality as a characteristic quantum phenomenon' [7, pp. 710–11].

The final quotation is the opening sentence in an extensive 2014 multiauthored paper on Bell nonlocality in *Reviews of Modern Physics*: 'In 1964, Bell proved that the predictions of quantum theory are incompatible with those of any physical

the proof that standard quantum mechanics violates local causality need not rest on the Bell theorem – or even entanglement. A proof is possible which is closely related to Einstein's pre-1935 argument for nonlocality in quantum mechanics.

We will attempt to spell these points out in more detail below by analysing the development of Bell's remarkable work from 1964 to 1990 and discussing the relevance of the modern Everettian stance on nonlocality.

6.2 The Original 1964 Theorem: Lessons of the de Broglie–Bohm Theory

The original version of Bell's nonlocality theorem was the fruit of a penetrating review of deterministic hidden variable theories published in 1966 [9], regrettably several years after it had been submitted for publication.[3] Bell achieved many things in this paper, not all of which have been duly recognised. In particular, there is still insufficient appreciation in the literature that in this paper, Bell was the first to prove the impossibility of noncontextual hidden variable theories, and that his interpretation of this result was the polar opposite to that of Simon B. Kochen and Ernst Specker, who independently and famously published a finitist, but much more complicated version of the proof in 1967 [10]. As Bell saw it, what he had shown was that a fragment of the contextualism built into the de Broglie–Bohm (pilot wave) hidden variable theory has to be found in any deterministic hidden variable theory consistent with the (at least state-independent) predictions of quantum mechanics, and that this is all well and good. In contrast, Kochen and Specker thought they had actually ruled out all hidden variable theories![4] Another interesting feature of Bell's thinking is this. Although he did not use the result in his simple version of the proof of the Bell–Kochen–Specker theorem, Bell realised that Gleason's theorem [11], and in particular its corollary concerning the continuity of frame functions on Hilbert spaces, ruled out noncontextualist hidden variable theories for most quantum systems. What Bell did not suspect in 1964, even so, was that the Gleason theorem could also be used to derive a nonlocality theorem for correlated spin-1 systems involving no inequalities.[5]

---

theory satisfying a natural notion of locality' [8, p. 420]. In fact, Bell did no such thing in 1964; his original version of the theorem refers only to deterministic hidden variable theories of quantum mechanics (see below). The definition of locality that the authors of the 2014 review paper actually give reflects that found in Bell's later papers, where determinism is no longer involved.
[3] For further details see Wiseman [2].   [4] See Brown [12] and Mermin [13].
[5] See Heywood and Redhead [15], Stairs [16], Brown and Svetlichny [17] and Elby [18]. The Heywood–Redhead and Stairs results predate Greenberger–Horne–Zeilinger [19] in terms of providing a Bell theorem with no inequalities, though this is widely overlooked in the literature; for example, Blaylock [20, footnote 35], and Brunner et al. [8] make no reference to these early spin-1 theorems. In the case of bipartite systems, the tight connection between contextualism and nonlocality means that the Bell–Kochen–Specker theorem raises a prima facie consistency problem for the very assumptions in the original 1964 Bell theorem as applied to quantum mechanics. To see how nonetheless consistency is assured, see Brown [21].

Nor did Bell suspect in 1964 that a weaker (state-dependent) version of the Bell–Kochen–Specker theorem for a single quantum system could be given in which a Bell-type inequality is shown to be both a consequence of noncontextualism and violated by quantum mechanics. This result is originally due to Home and Sengupta [22]. They considered the case of a single valence electron, say in an alkali atom, in the $^2P_{\frac{1}{2}}$ state, in which the electron wave function involves entanglement between spin and orbital angular momentum. A more manifestly self-consistent version of the proof was given in Foster and Brown [23]. For details of other state-dependent versions of the Bell–Kochen–Specker theorem see Brown [12, sect. 2(iv)], where it is argued that a result due to Belinfante [24] can be construed as such.

But for our present purposes, the relevant question that Bell raised in his review paper was whether the action at a distance built into the 1952 Bohm pilot wave theory [14] was characteristic of all hidden variable theories. As Bell noted, in the Bohm theory, in the case of measurements on a pair of separated entangled systems, 'the disposition of one piece of apparatus affects the results obtained with a distant piece'. Part of the answer to the question was provided in Bell's 1964 paper, amusingly published two years earlier. (Advocates of the retrocausal 'free will' loophole take heart.) Bell showed that any deterministic theory in which the hidden variable is associated with the quantum system alone (i.e., supplements the system's quantum state vector) must display action at a distance of a Bohmian nature, if it is to be consistent with the predictions of standard quantum mechanics. Specifically, what must be violated, in the case of spatially separated spin-1/2 systems in the singlet state, is that *The result of a Stern-Gerlach measurement of a spin component on either system does not depend on the setting of the magnet for the other system.* (The more general claim that in a *deterministic* hidden variable theory, the predicted outcome of a measurement of an observable in one of a pair of entangled systems does not depend on how a distant piece of equipment designed to measure any observable on the other system is set up, will be referred to as the *1964 locality assumption*.)

Bell's 1964 result was groundbreaking in itself. But it is fair to say that it was only part of the way to an answer to the specific question posed in the review paper, because the Bohm theory generally involves in its deterministic algorithm hidden variables associated with both the object system and the apparatus. Bell was to look at generalisations of this kind of theory in 1971, and unwittingly anticipated the later notion of local causality, as we shall see shortly.[6]

Bell started his 1964 paper by recalling that the 1935 Einstein–Podolsky–Rosen (EPR) argument demonstrates, using the strict, or perfect, correlations involved in entangled systems, that if action at a distance is to be avoided, the standard quantum mechanical state cannot be a complete description of the systems in question.[7] In particular, Bell seems to have accepted that the perfect anticorrelations for parallel spin components in the singlet state of spin-1/2 systems imply that a deterministic underpinning ('causality') is necessary if locality in this sense is to prevail, and that this is the reason that a local *deterministic* hidden variable theory is the focus of his 1964 paper.[8] Bell showed that a local deterministic hidden variable theory can easily be constructed to account for such correlations, so he was

---

[6] It is noteworthy that recently a third aspect of de Broglie–Bohm theory has been shown to be universal for hidden variable theories: that the quantum state cannot be entirely 'epistemic' in nature. See Pusey et al. [25] and Barrett et al. [26].

[7] We disagree with a recent claim that the EPR argument presupposed more than standard quantum mechanics (with measurement-induced collapse) and locality, the 'more' being counterfactual definiteness; see Zukowski and Brukner [27].

[8] See Maudlin [4] for making the case that Bell viewed determinism as a strict consequence of the EPR argument based on locality (no action at a distance). Certainly, after 1964, Bell was explicit: 'It is important to note that to the limited degree to which *determinism* played a role in the EPR argument, it is not assumed but *inferred*' (Bell [28, p. 143], original emphasis). However, disagreement has arisen in the literature as to the precise logic of Bell's 1964 theorem, with particular reference to the role of determinism. Some commentators see determinism as one of the assumptions of the theorem (view 1), while others see it as a consequence of the assumptions, which include the existence of perfect (anti)correlations. Clearly, if the former position is correct, then the empirical violation of the Bell inequality implies *either* indeterminism *or* nonlocality (in the 1964 sense of the term). A recent careful textual analysis of Bell's writings in the context of this debate is due to Wiseman [2], who provides grounds for thinking that in later life Bell's own reading of his 1964 logic – in line with view 2 – is questionable. Note that nothing in our paper hangs on this debate.

fatefully drawn to investigate the weaker correlations between nonparallel spin components on the distant systems: literally the EPR–Bohm scenario with a twist.

Insofar as Bell was to drop the deterministic requirement in his later papers, the connection with the EPR argument became less tangible. And this had to do with the fact that in its most general form, the derivation of Bell-type inequalities based on some version of the locality condition transcends quantum mechanics. It does not have to appeal to the existence of perfect EPR–Bohm (anti-)correlations between distant systems. We shall return to this point below, but in the meantime it is worth recalling the conclusion Bell draws from his 1964 theorem.

In a theory in which parameters are added to quantum mechanics to determine the results of individual measurements, without changing the statistical predictions, there must be a mechanism whereby the setting of one measurement device can influence the reading of another instrument, however remote. Moreover, the signal involved must propagate instantaneously, so that such a theory could not be Lorentz invariant. [29]

The conclusions in Bell's 1976 and 1990 papers are much more nuanced. There, as we shall see, there is no clear-cut claim of conflict with special relativity, largely because of explicit recognition of what is now widely known as the no-signalling theorem of quantum mechanics, which of course holds at the statistical level and hence is insensitive to the existence of any nonlocal hidden substratum.[9] It is interesting, though, that a degree of caution, if not skepticism, about the theorem emerges from Bell's remarks (see below).

6.3 The Legacy of Einstein–Podolsky–Rosen

We have seen that for Bell in 1964, the EPR argument was crucial in setting up the conditions that lead to an inequality which is violated by quantum mechanics. Einstein's own conviction that locality (absence of action at a distance) and the completeness of quantum mechanics are incompatible actually predate the 1935 collaboration with Podolsky and Rosen. In the 1927 Solvay conference, Einstein used the single-slit scenario to argue that detection of the particle at one point on the hemispherical measuring instrument means that all other points must instantaneously know not to detect, despite the wave function of the particle having a finite value at all such points prior to detection.[10] In fact, given the completeness assumption, measurement-induced collapse of the wave function involves action at a distance: *One does not need entanglement and EPR correlations to drive the nonlocality lesson home given such nonunitary processes.*

Unless one adopts something like Niels Bohr's philosophy. Here, the wave function/state vector is thought to be the complete description of the quantum system but somehow not in itself a physical object, and collapse (if one goes so far as to describe in quantum mechanics

---

[9] The no-signalling theorem in quantum mechanics only became prominent in the literature in the late 1970s, though its first enunciation effectively goes back to David Bohm's 1951 book *Quantum Theory* [30]. (Some historical details regarding the no-signalling theorem are found in Timpson and Brown [3, footnote 12]; Bell's 1976 discussion is regrettably overlooked.)

[10] Brown [31] and Norsen [32].

what goes on the measurement process, which Bohr did not) is thought not to represent a physical process.[11] So the EPR argument is best seen, like Newton's bucket thought experiment, as polemical in nature. Newton used the bucket to strike at the heart of Descartes' theory that for any body real motion is defined relative to the bodies immediately contiguous to it, themselves being taken to be at rest.[12] By cleverly exploiting *distant* correlations, Einstein hoped to bypass the feature of quantum mechanics that Bohr commonly used in defence of the claim that quantum mechanics is complete notwithstanding its statistical character, namely, the ineradicable *local* disturbance of the system caused by measurement. This aspect of the EPR argument was appreciated by Bell, who in 1981 devastatingly exposed the obscurity of Bohr's 1935 response to the argument.[13]

Be that as it may, it is well known that Einstein was unhappy with the way Podolsky had organised the argument, and in letters to Schrödinger had vented his frustration that the basic lesson had been 'smothered by the formalism'. Details concerning the form of the argument that Einstein preferred will not be rehearsed here,[14] yet it is worth mentioning the thought experiment known as 'Einstein's boxes'.[15] Writing to Schrödinger in 1935, Einstein imagined a box with a single classical particle inside. The box is then divided in two and the half-boxes spatially separated. The particle by chance ends up in one of them, and Einstein used this scenario to explain what he meant by an 'incomplete' description of the particle and what he meant by the 'separation principle' (locality), all with a view to better articulating the quantum EPR scenario involving entanglement. But it was de Broglie in 1964[16] who considered the analogue of the boxes scenario with a quantum particle, its wave function now in a superposition of components in each half-box prior to their opening. Given the strict anticorrelation between the outcomes of opening the boxes, this gedankenexperiment has all the features required for an EPR-type argument inferring nonlocality (action at a distance) from completeness of the wave functional description, but without entanglement. It is a variant of Einstein's 1927 single-slit diffraction argument, and both arguments will feature below.

6.4 Local Causality

Suppose now we step back from quantum mechanics and consider some hypothetical probabilistic theory involving microscopic systems. Suppose the probabilities referring to measurement outcomes in the theory are understood to be irreducible, the dynamics of the relevant processes being intrinsically stochastic.[17] Imagine further that a pure ensemble of spatially separated bipartite systems, along with measurement devices, can be prepared

---

[11] See, for example, Bell [33, p. 53].   [12] See Barbour [34, pp. 623–8].   [13] See Bell [28, Appendix 1].

[14] A useful account of Einstein's criticisms of the EPR paper is found in Fine [35, Chapters 3 and 5]; see also Maudlin [4] and Timpson and Brown [3] for further discussion of Einstein's overarching view on locality.

[15] This terminology is due to Fine [35, p. 37]; a fuller discussion of the history of the thought experiment is found in Norsen [32].

[16] See de Broglie [36] and Norsen [32].

[17] Technically what this means is that the theory allows no dispersion-free pure ensembles and yet is assumed to be 'statistically complete'. Recall that pure states are those which cannot be expressed as convex combinations of other distinct states. Statistical completeness entails that arbitrary ensembles of the joint system-apparatus which are pure – i.e., all joint systems therein share the same pure state – are homogeneous in the sense of von Neumann. A detailed analysis of this notion of statistical completeness, which is distinct from the EPR notion of completeness, is found in Elby et al. [38].

between which by hypothesis no interactions are taking place. Would we not expect the statistical outcomes of the measurement events on the separated systems to be independent, i.e., for there to be no correlations?

Something akin to this no-correlations condition was introduced into Bell's writings explicitly in 1976. We shall examine its formal representation, as well as its (dubious) connection with Hans Reichenbach's famous Common Cause Principle [37], in Section 6.8 below. In the meantime a little history may be useful.

In his second paper on nonlocality, published in 1971, Bell considered a class of hidden variable theories explicitly modelled on the 1952 Bohm theory [14], in which the predictions of measurement outcomes depend deterministically on hidden (hence uncontrollable) variables associated both with the object system and with the apparatus. Averaging over the latter, Bell noted that the theory now took the form of an *indeterministic* hidden variable theory for measurements occurring on the object system. In his derivation [34] of the Clauser–Horne–Shimony–Holt (CHSH) inequality [39] for this stochastic theory involving pairs of distant systems, Bell took as his locality condition that these average predictions on one system did not depend on the controllable setting of the distant measurement device. But another assumption was hidden in the derivation: that such averages for pairs of simultaneous measurements on the distant systems factorize.[18] It is true that this is a consequence of the locality of the background deterministic theory,[19] but it was a foretaste of what was to come in Bell's thinking about his theorem.[20]

Note that Bell himself was aware that if one added to his 1971 derivation the existence of strict EPR–Bohm anticorrelations in the spherically symmetric spin singlet state, then his original 1964 version of the inequality is obtained from that of CHSH. More to the point, Bell was conscious that allowing for the strict anticorrelations meant that the apparatus hidden variables can play no role in the predictions, so that even the apparent indeterminism of the averaged theory is spurious. But one sees in this 1971 paper that Bell was starting to follow CHSH in considering generalised local theories as the proper background to the Bell theorem, in which no constraints on correlations exist other than those imposed by locality.

The first clear articulation of this program for the case of genuinely stochastic (indeterministic) theories was due to John Clauser and Michael Horne in 1974 [41]. In defining locality in this context these authors explicitly distinguished between two predictive constraints: independence of the distant apparatus setting (sometimes called *parameter independence*) and the absence of correlations conditional on specification of the pure state of the pair of systems (*conditional outcome independence*, or *Jarrett completeness*[21]). The subtle question of how to motivate these constraints, and the connection, if any, with the principle of no action at a distance, will be discussed below in Section 6.8. In 1976, Suppes and Zanotti proved in a general way what Bell noted in his 1971 paper, namely that if

---

[18] See Brown and Svetlichny [17, p. 1385], and Brown [21, §II]). [19] Cf. Brown [21, §II].

[20] Bell noted that the 1971 locality condition 'Clearly...is appropriate also for *indeterminism* with a certain local character' ([40, fn. 10], original emphasis).

[21] The widespread appreciation that two distinct assumptions are at play in the early versions of the stochastic version of the Bell theorem is due in good part to the work of Jarrett [43].

one imposes these two 'locality' constraints for a stochastic theory of pairs of systems, and moreover postulates the existence of perfect (anti)correlations of the kind associated with the spin singlet state, then the theory is reduced to a deterministic one [42].[22]

Suppose only conditional outcome independence is assumed, and not parameter independence. Then the work of Suppes and Zanotti [41] implies in the case of the spin singlet state that the marginal probabilities for values of the spin components are zero or one in every case where the spin devices are parallel-oriented. So there will be elements of reality at one wing of the experiment which can be brought into and out of existence by varying the orientation of the device at the other wing. In this case, the joint locality condition reverts to the prohibition of action at a distance. However, it is important to bear in mind at this point that an implicit, but highly nontrivial assumption is being made in the Suppes and Zanotti argument, as well as in Bell's own application of perfect (anti)correlations in order to restore determinism: that in each measurement process, one and only one outcome is realised. We return to this issue in Section 6.9 below.

In 1976 there also appear for the first time in Bell's writings both the explicit notion and terminology of 'local causality' associated with stochastic theories of physical reality.[23] As he was famously to articulate in his later 1989 'Against measurement' [44], Bell was deeply suspicious of the cavalier way such common but nonfundamental notions as 'observable', 'system' and 'measurement' are often used in quantum mechanics. In his 1976 paper on 'local beables', Bell defines the factorizability condition not in terms of measurement outcomes on bipartite systems, but in terms of the beables in spacelike separated regions of space–time, conditional on the complete specification of all the beables belonging to the overlap of the backward light cones of these regions. There is no splitting of the local causality condition into anything like parameter independence and conditional outcome independence, because measurement processes do not explicitly figure in the analysis. Two inequalities are derived in Bell's 1976 paper. The first involves nothing but beables and their correlations; the second – a variation of the CHSH inequality – emerges from the first when 'in comparison with quantum mechanics', some beables are interpreted as controllable variables specifying the experimental setup, and some are 'either hidden or irrelevant' and averaged over as in the 1971 paper.

6.5 The 1976 Paper

There are several remarkable features of Bell's 1976 paper.

(1) More than ever before in Bell's writings, *his theorem stands apart from quantum mechanics.* In contradistinction to his 1971 paper, Bell starts by considering a loosely defined but *genuinely* indeterministic theory of nature, and the derivation of the (first)

---

[22] The same situation holds for the triplet state for spin-$\frac{1}{2}$ systems, despite its lack of spherical symmetry. This is because for any spin component on one subsystem, there is a spin component on the other with which it is perfectly correlated (though this is not in general the component antiparallel to the first, as in the case of the singlet state).

[23] Confusingly, at times Bell would use the term 'local causality' when he explicitly meant no action at a distance in the context of the EPR argument, as in Bell [28, p. 143].

inequality conspicuously holds without any reference to quantum features, and depends only on the requirement of local causality.

> We would like to form some [notion] of local causality in theories which are not deterministic, in which the correlations prescribed by the theory, for the beables, are weaker.
>
> *[1, p. 53]*

(2) For the first time, Bell states that 'ordinary quantum mechanics, even the relativistic quantum field theory, is not locally causal'. *The claim does not depend on his theorem. It does not depend on the violation of a Bell-type inequality: it does not even depend on entanglement in the usual sense.* The argument Bell gives concerns a single radioactive decay process and several spatially separated detectors; it is analogous to Einstein's 1927 argument involving single-slit diffraction, and to the Einstein–de Broglie boxes thought experiment, both referred to earlier. (This aspect of Bell's reasoning has been little appreciated in the literature, and it was partially lost sight of in his final 1990 paper.)

(3) Bell importantly qualifies the significance of local causality in the light of the no-signalling theorem, which makes its first appearance in his work, and which itself is qualified in terms of its 'human' origins. This hesitancy is repeated in 1990, as we see in the following section.

(4) When Bell does apply his theorem to quantum mechanics, it is in the context of a putative locally causal, indeterministic hidden variable 'completion' of the theory. But now there is no mention of the EPR-Bohm (anti)correlations: the legacy of EPR, so vital in the 1964 paper, has been put aside.

Let us ruminate a little on this last point. To ignore perfect EPR–Bohm (anti)correlations is not to deny their existence. But if they exist, then, as we have seen, in the usual setting, stochasticity collapses into determinism and the meaning of local causality reverts simply to the prohibition of action at a distance. It is understandable then that Maudlin in 2010 construes the legacy of Bell's work to have been the demonstration that standard quantum mechanics is nonlocal in the original 1964 sense, and that the proof does *not* presuppose the existence of hidden variables, or the related condition of 'counterfactual definiteness', as long as EPR–Bohm correlations are assumed and the possibility of Everettian branching is ruled out (see below). Rather, the existence of a deterministic substratum (and hence counterfactual definiteness) is *inferred* in the proof, and locality further constrains the phenomenological correlations in such a way as to satisfy a Bell inequality, which quantum mechanics violates in certain scenarios involving entangled states.[24]

This view of things does not seem to accord with Bell's thinking from 1976 on. As we have seen, the demonstration that quantum mechanics violates the new condition of local causality is fairly elementary and needs no appeal to the Bell theorem or even entanglement.

---

[24] Maudlin [4]. Maudlin is one of those defending view 2 mentioned in footnote 5 above, namely that determinism is not an assumption in Bell's original 1964 theorem. Again, we emphasise that we are interested here in a certain reading of the theorem, not in whether Bell himself adopted it in 1964.

The Bell theorem itself refers to locally causal stochastic theories generally, and when it is applied to the specific case of quantum mechanics, it is to some *stochastic* 'completion' of the theory which can have no justification in terms of the EPR argument. In 1981, in his famous paper 'Bertlmann's socks and the nature of reality', Bell made clear what his position on the EPR–Bohm anticorrelations came to be:

> Some residual imperfection of the set-up would spoil the perfect anticorrelations . . . So in the more sophisticated argument we will avoid any hypothesis of perfection.
>
> It was only in the context of perfect correlation (or anticorrelation) that *determinism* could be inferred for the relation of observation results to preexisting particle properties (for any indeterminism would have spoiled the correlation). Despite my insistence that determinism was inferred rather than assumed, you might still suspect somehow that it is a preoccupation with determinism that creates the problem. Note well then that the following argument [derivation of the CHSH inequality based on local causality] makes no mention whatever of determinism.
>
> *[28, emphasis in original]*

Granted, *perfect* (anti)correlations are operationally inaccessible; indeed, this was the motivation for the CHSH inequality. This inequality is testable in a way Bell's original 1964 inequality is not. But we should not lose sight of the fact that every experimental corroboration of the general correlation predictions in quantum mechanics for entangled states provides indirect evidence for the perfect (anti)correlations in the EPR–Bohm scenario, which in turn are connected with the fact that the spin singlet state has zero total spin angular momentum. It is noteworthy that the perfect (anti)correlations reappear in Bell's 1990 paper, as seen below.

We are thus led to the view that from 1976 on in Bell's writings, when he refers to a putative locally causal, stochastic, hidden variable 'completion' of quantum mechanics, this theory should be understood in the following sense: *Either it is a truncated theory restricted to certain nonparallel settings, or it is one incorporating an approximation to quantum mechanics whose correlation predictions concur with those of quantum mechanics for parallel settings to within experimental error (and so the theory may depend on the chosen experimental setup).* Otherwise, the theory is not stochastic at all, as we saw in Section 6.4.

6.6 Local Causality, No-Signalling and Relativity

The factorizability condition related to local causality first introduced by Bell in 1976 was itself to undergo a minor change in his later writings. He came to realise that it is unnecessary to conditionalize on complete specification of beables in an infinite space-time region, and a 'simpler' version of the condition involving beables in finite regions appeared informally in a footnote in [45].[25] A more systematic rendering of this 'simpler' version appeared

---

[25] Bell [45, fn. 7]. In his preface to the first edition of *Speakable*, Bell expressed regret that this slimmer version, which he had used in talks, had not been introduced earlier in his papers; see [33, p. xii]. There is also a reminder in the preface that 'If local causality in some theory is to be examined, then one must decide which of the many mathematical entities that appear are supposed to be real, and really here rather than there.' [*ibid*].

in Bell's final 1990 paper 'La nouvelle cuisine', which first appeared in the second edition of *Speakable and Unspeakable in Quantum Mechanics*.

The details need not detain us yet. But note that in the 1990 paper, when Bell discusses the reason for the violation of local causality by 'ordinary quantum mechanics', as in the 1976 paper, he makes no appeal to the violation of a Bell-type inequality and hence to the Bell theorem, but now he does refer to entanglement and in particular to an optical version of the EPR–Bohm scenario with perfect correlations. (Compare with point 2 in the previous section.)

Does the fact that quantum mechanics violates local causality mean it is inconsistent with special relativity? This is a question over which Bell wavered for a number of years, and in attempting to understand his position it is important to distinguish between 'ordinary' quantum mechanics and the hypothetical 'deeper' level of hidden variables.

Recall that in 1964, Bell categorically concluded that any deterministic hidden variable theory consistent with the standard quantum mechanical predictions could not be Lorentz covariant, though in 1976 he was more nuanced about the violation of local causality. In 1984 he wrote,

For me then this is the real problem with quantum theory: the apparently essential conflict between any sharp formulation and fundamental relativity. That is to say, we have an apparent incompatibility, at the deepest level, between the two fundamental pillars of contemporary theory ... [46]

Note the qualifying adjectives 'apparent' and 'fundamental'. In his influential essay of 1976, 'How to teach special relativity' [47], Bell advocated a dynamical Lorentzian peda-gogy in relation to the explanation of kinematic effects such as length contraction and time dilation. He stressed that one is not thereby committed to a Lorentzian 'philosophy' involv-ing a privileged inertial frame, as opposed to Einstein's austere philosophy which rejects the existence of such a frame. However, Bell also emphasised that 'The facts of physics do not oblige us to accept one philosophy rather than the other.' When he was asked in a recorded interview, first published in 1986, how he might respond to the possible existence of 'faster-than-light signalling' in the Aspect experiment, Bell returned to this theme:

I would say that the cheapest resolution is something like going back to relativity as it was before Einstein, when people like Lorentz and Poincaré thought that there was an aether – a preferred frame of reference – but that our measuring instruments were distorted by motion in such a way that we could not detect motion through the aether. Now, in that way you can imagine that there is a preferred frame of reference, and in this preferred frame of reference things do go faster than light. ... Behind the apparent Lorentz invariance of the phenomena, there is a deeper level which is not Lorentz invariant ... [This] pre-Einstein position of Lorentz and Poincaré, Larmor and Fitzgerald [*sic*], was perfectly coherent, and *is not inconsistent with relativity theory*.

[48, our emphasis]

This is a very different tone from the one at the end of Bell's 1964 paper, in which the failure of Lorentz covariance in any deterministic hidden variable theory is announced without any redemptive features. In 1986 he is adopting the perfectly defensible view that special relativity strictly only holds for physics that has phenomenological consequences, i.e., that is not 'hidden'. Note that Bell does not clarify here *what* goes faster than light at the 'deeper level' in quantum mechanics. Given the context of the discussion, it seems that something like the action at a distance in the Bohm theory is what Bell had in mind. But the question is more pressing if the issue is violation of local causality, when it is just 'ordinary' quantum mechanics or quantum field theory that is in question.

In his 'La nouvelle cuisine' paper, Bell harks back to his 1976 discussion of the non-locality theorem and provides a particularly careful demarcation between the condition of local causality and that of 'no-superluminal-signalling'. In connection with the latter, Bell investigates the case of external interventions in local relativistic quantum field theory and concludes, unsurprisingly, that the statistical predictions are insensitive to the introduction of external fields outside the backward light cone of the relevant 'observables'. But he is worried by what is truly meant by the notion of 'external' intervention. At the end of the paper he returns to the no-signalling theorem in standard quantum mechanics, and he similarly expresses concerns about its fundamentality, just as he had in 1976. He is worried, as always, about the vagueness or lack of conceptual sharpness in the theorem, which results from the anthropocentric element lurking behind the notions of measurement and preparation, and indeed this issue appears to be the lingering concern in the conclusion to the 1990 paper. *There is no categorical statement in the paper that violation of local causality is inconsistent with special relativity.*

6.7 Motivating Local Causality

In 'La nouvelle cuisine', Bell emphasised that even if a theory is well behaved relativistically in the strict sense of not allowing superluminal signalling, this is not enough to satisfy his causal intuitions. What then is the motivation for local causality? We need, at last, to look in detail at the definition and significance of the factorizability condition as given in this 1990 paper. Here is what Bell says:

A theory will be said to be locally causal if the probabilities attached to values of local beables in a space-time region 1 are unaltered by specification of values of local beables in a space-like separated region 2, when what happens in the backward light cone of 1 is already sufficiently specified, for example by a full specification of local beables in a space-time region 3 . . .

[49, pp. 239–40]

What is particularly relevant for our purposes is the intuition behind local causality that Bell provided in 1990:

The direct causes (and effects) are near by, and even the indirect causes (and effects) are no further away than permitted by the velocity of light.

[49, p. 239]

It is important for our subsequent discussion to highlight some key assumptions involved here. The first is the very existence of causal processes, which are normally taken to be time-asymmetric: causal influences propagate into the future, not the past. (Note the reference to backward light cones in the first quotation.) An awkwardness arises because there is nothing in the postulates of special relativity that picks out an arrow of time. (Whether the situation is any different in quantum mechanics is clearly an interpretation-dependent issue.) At any rate, operationally the notion of cause had to be cashed out by Bell in probabilistic terms; he conceded that the factorizability condition associated with local causality makes no explicit reference to causes. 'Note . . . that our definition of locally causal theories, although motivated by talk of "cause" and "effect", does not in the end explicitly involve these rather vague notions' [49, p. 240].

Note that the introduction of probabilities introduces further awkwardness into the picture. Special relativity is based on the claim that the fundamental equations governing the nongravitational interactions are Lorentz covariant. Such equations are normally assumed to be deterministic. How irreducible probabilities are supposed to inhabit a relativistic world is a difficult technical matter, which Bell, perhaps wisely, overlooks. And it is worth recognising at this point that unless probabilities in this context are understood themselves to be objective elements of reality, things go awry. There is an old view, by no means consensual of course, that probability in physics in general does not represent an objective element of reality.[26] If this were true, the representation of purportedly objective causal facts by constraints on probabilities becomes problematic and the status of factorizability somewhat questionable. In particular, the connection with the clear-cut condition of no action at a distance involved in Bell's 1964 paper becomes tenuous. This concern might seem misplaced in the context of an indeterministic theory, but what about a deterministic theory with intrinsic unpredictability, as in the Everett picture (see the next section)?

The second (widely held) assumption Bell is making is that the null cone structure of Minkowski space–time represents the boundaries of causal connectibility. The kinematics of special relativity implies that the speed of light is invariant across inertial frames, and the usual dynamical postulates (over and above Lorentz covariance) concerning the connection between energy, momentum and velocity imply that massive bodies cannot be accelerated up to the speed of light, let alone beyond. But whether tachyonic signals are inconsistent with special relativity is a more delicate matter, and Bell was fully aware of this. As he put it, 'What we have to do is add to the laws of relativity some responsible causal structure . . . we require [causal chains] . . . to go slower than light in any [inertial] frame of reference' [49]. Bell was effectively – and plausibly – assuming that tachyonic signals do not exist. But the justification Bell gave in Section 6.4 of the Nouvelle Cuisine paper is amusing. It involves considering the 'perfect tachyon crime' involving a gunman firing a tachyon gun. The movement of the murderer can be set up in such a way that according to the description of the deed relative to the rest frame of the victim, and the courts of justice, the trigger is pulled after the death of the victim! The ensuing 'relativity of morality' should be ruled out by the laws of nature, suggests Bell. This injunction is obviously high-handed – why should Nature

---

[26] A version of this view, in which probability is 'a numerical expression of human ignorance', has recently been defended by Tipler [50], for example. See also Caves et al. [51], Fuchs et al. [52] and Healey [53], and this volume, for various developments of related ideas.

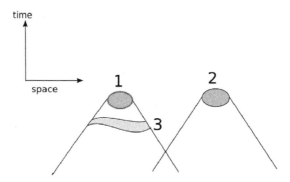

Figure 6.1 Space-time regions involved in Bell's 1990 statement of local causality.

care about our moral qualms? Playfully or otherwise, Bell is insisting on first principles that superluminal signalling be banished from Minkoswksi space-time, and as we mentioned in the last section, he goes on to show in the 1990 paper that quantum theory does not violate this injunction. But, to repeat, there is still a distance between no-signalling and local causality, particularly in the case where the candidate indeterministic theory is acting as a completion of ordinary quantum mechanics, as will be seen in the next section.

Bell emphasised that the factorizability condition is not to be seen 'as the *formulation* of local causality, but as a consequence thereof.' He was very concerned that in providing a clean mathematical consequence of local causality, one is 'likely to throw out the baby with the bathwater'. Indeed, careful analysis of the task that presents itself in the attempt to translate Bell's causal intuitions into well-defined probabilistic form shows how subtle the matter is. This has become especially clear through the recent work of Norsen [5] and Seevinck and Uffink [54]. However, we will stick to the specific form of factorizability Bell gave in the Nouvelle Cuisine paper as applicable to EPR–Bell experiments, where *the candidate indeterministic theory is now understood to be a completion of ordinary quantum mechanics* (subject to the qualification mentioned at the end of Section 6.5). Here, Bell extended the space–time region 3 in Figure 6.1 to cross the backward light cones of both regions 1 and 2 where they no longer overlap. In this extended region 3, Bell denoted by the symbol c the values of any number of variables describing the experimental setup as given in ordinary quantum mechanics, except for the final settings (of magnets, polarisers or the like) that are determined in regions 1 and 2. Presumably included in c is the entangled quantum state. Also defined in the extended region 3, and denoted by λ, are 'any number of additional complementary variables needed to complete quantum mechanics in the way envisaged by EPR'. Note that they too are not assumed to be spatially separable. Together c and λ give a 'complete specification of at least those parts of 3 blocking the two backward light cones'. Then Bell's specific 1990 factorizability condition is

$$P(A, B | a, b, c, \lambda) = P(A | a, c, \lambda) P(B | b, c, \lambda), \tag{6.1}$$

where $P(A, B | a, b, c, \lambda)$ is the joint probability in the candidate theory for outcome A associated with a binary random variable in an experiment in region 1 with setting (free variable) *a*, and outcome *B* in an experiment in region 2 with setting *b*, conditional on *c* and λ. The factors in the product on the RHS are the respective conditional marginal probabilities.

6.8 Unpacking Factorizability

As we have mentioned, ever since the 1974 work of Clauser and Horne, reinforced by that of Jarrett and also of Shimony (e.g., [55]), it has been regarded as conceptually helpful to separate any such factorizability condition into two components:

$$P(A|a, b, c, \lambda) = P(A|a, c, \lambda), \tag{6.2}$$

$$P(B|a, b, c, \lambda) = P(B|b, c, \lambda), \tag{6.3}$$

and

$$P(A|B, a, b, c, \lambda) = P(A|a, b, c, \lambda), \tag{6.4}$$

$$P(B|A, a, b, c, \lambda) = P(B|a, b, c, \lambda). \tag{6.5}$$

Equations (6.2) and (6.3) indicate that the marginal probabilities do not depend on the settings of the distant piece of apparatus. This 'parameter independence' assumption ('functional sufficiency', in the language of Seevinck and Uffink) is the probabilistic analogue of the locality assumption in Bell's 1964 paper. Equations (6.4) and (6.5), known, as we have said, as 'conditional outcome independence' or 'Jarrett completeness' ('statistical sufficiency' for Seevinck and Uffink), rule out correlations conditional on the combination of *c* and λ. A great deal of discussion has of course taken place since the 1970s about these conditions, and indeed their nonuniqueness, but let us note the claim by Seevinck and Uffink that each condition has the same motivation as the factorizability condition (6.1) itself: 'Both [conditions] are a consequence of local causality, and the appeal to notions of locality and causality used in implementing the functional and statistical sufficiency are just the same …" [54, p. 12].

But now that the candidate indeterministic theory is one into which quantum mechanics is embeddable, the specification of λ involves hidden variables. Would doubt not be thrown in this case on the plausibility of parameter independence (6.2) and (6.3), in particular, given Bell's own doubts whether what is 'hidden' can ever threaten relativity? Maybe, but as we have seen in the previous section, the issue is not just consistency with relativity, whether in the sense of requiring Lorentz covariance or in the sense of precluding tachyonic signalling. As regards outcome independence (6.4) and (6.5), ordinary quantum mechanics (in which λ is the empty set) violates it, but again no threat to a 'responsible causal structure' containing a ban on tachyonic signals automatically arises as a result.

Let us consider these conditions (6.4) and (6.5) in more detail. It is often regarded as a special case (within a relativistic space–time) of Reichenbach's 1956 Common Cause Principle,[27] which is based on the general notion that if events *X* and *Y* are correlated, then either *X*(*Y*) is the cause of *Y*(*X*) or there is a common cause in the past [37, Sect. 19].

---

[27] See for example van Fraassen [56], Butterfield [57–59], Henson [60].

Conditional on the common cause, the correlations between X and Y must disappear (the so-called 'screening-off' condition). But although this principle is a useful heuristic in daily life, Reichenbach's own writings on the matter should make us pause before applying it to microphenomena (whether classical or quantum) involving a small number of systems. What is amply clear in Reichenbach's 1956 discussion of the principle is that it holds for 'macrostatistics', that is to say, systems sufficiently complex to display entropic behaviour. For Reichenbach, the notions of cause and effect only make sense in macro-phenomena displaying quasi-irreversible behaviour: 'The distinction between cause and effect is revealed to be a matter of entropy and to coincide with the difference between past and future' [37, p. 155]. There is of course irreversibility associated with the measurement processes themselves in an EPR–Bell experiment, but it may not be enough to resolve the conundrum in Reichenbach's own terms, although he was not entirely consistent on the whole matter.[28]

To the extent that Reichenbach's common cause principle *is* cogently applicable to Bell-type experiments, there is arguably a missing element in Reichenbach's analysis, which was largely motivated by common-place correlations involving familiar day-to-day phenomena. In the case of a strictly indeterministic microphysics of the kind Bell considered from 1976 on, the motivation for conditional outcome independence is arguably stronger than anything Reichenbach had in mind. Bell was envisaging outcomes of measurements in regions 1 and 2 which are *intrinsically* random, associated with *irreducible* probabilities. *What kind of explanation can there be for correlations between separate, strictly random processes?* Note that the puzzle does not require reference to space-like separations defined relative to the light cone structure; it would presumably arise even in Galilean space–time for simultaneous events as long as the dynamics in the candidate indeterministic theory did not allow instantaneous interactions between separated systems. The profound oddity of such correlations – let us call it the *randomness problem* – goes hand in hand with the natural view that probabilities in strictly indeterministic theories are, *pace* Tipler, objective. (We shall return to the randomness problem in Section 6.9.1.)

Bell himself did not put things quite this way. A noteworthy element of his surprisingly tentative, but always honest, thinking in the Nouvelle Cuisine paper comes in the Conclusion:

Do we then have to fall back on 'no signalling faster than light' as the expression of the fundamental causal structure of contemporary theoretical physics? That is hard for me to accept. For one thing *we have lost the idea that correlations can be explained*, or at least this idea awaits reformulation.

[Emphasis added]

Precisely what Bell meant here is perhaps not entirely clear. We take him to be saying that if we stop short of demanding of a causally well-behaved theory that it satisfy the

---

[28] At the end of his 1956 book, Reichenbach asserted that quantum mechanics – *even at the statistical level* – no more introduces a fundamental arrow of time than does classical mechanics (p. 211). As for probabilities themselves, Reichenbach defined them in terms of relative frequencies, so they are time-symmetric in his book. It is then hard to see how he could apply his common cause principle to quantum phenomena involving small numbers of systems, and yet he seems to have done so! Reichenbach states in the book that the statistics of pairs of identical bosonic particles engenders 'causal anomalies' – i.e., violate his principles – unless the assignment of physical identity to the particles is given up (p. 234).

factorizability condition (local causality) and impose only the no-signalling constraint (a recent suggestion of such an approach is Zukowski and Brukner [27], for example), then we are no longer in a position to explain, or to expect to explain, the quantum correlations. Such a view would compromise the motivation for the search for a locally causal stochastic completion of quantum mechanics. It would diminish the importance of the conclusion of post-1976 version of the Bell theorem, which is that locally causal, stochastic explanations of quantum correlations are, in the end, unavailable.

How bad is that? Such a blunt question is partly motivated by Bell's own doubts: earlier in the 1990 paper he recommended that the factorizability condition 'be viewed with the utmost suspicion'. Could it be that his misgivings had to do with the fact that all the motivation for the condition provided in his papers is couched in abstract terms, with no guidance from concrete models? Bell himself stressed the importance of the role of models in checking our physical intuitions; in particular, he laid stress on study of the details of the de Broglie–Bohm theory in the context of theorising about the nature of deterministic hidden variable theories. The same kind of lesson was stressed by Dickson [61]. In this spirit, we think that the Everett picture of quantum mechanics can play an illuminating role in understanding the significance of factorizability. It is a model in which, remarkably, *there is intrinsic unpredictability, but no strict indeterminism*, thus opening the door to a nuanced notion of randomness, and more importantly, probability. The branching structure of the universe gives rise to a subtle reinterpretation of the notion of correlations between entangled systems. Most significantly, the Everett version of quantum mechanics provides a Lorentz covariant picture of the world in which factorizability fails but there is no action at a distance.

6.9 Locality in the Everett Picture

Amongst those who have taken Everett's approach to quantum theory at all seriously as an option, it is a commonplace that – given an Everettian interpretation – quantum theory is (dynamically) local – there is no action at a distance. Indeed, this is often taken as one of the main selling points of an Everettian approach (cf. Vaidman [62], for example). Everett himself noted that his approach obviated the 'fictitious paradox' of EPR (Everett [63, Sect. 5.3]), though he did not go into detail, beyond noting the crucial role of the collapse of the wave function in the EPR argument – collapse which is of course absent in purely unitary Everettian quantum mechanics. Page [64] filled in the details of how the absence of collapse in Everett circumvents the EPR dilemma. Other discussions of locality in Everett, covering also the extension to Bell's argument as well as EPR's, include amongst others Vaidman [65], Tipler [66], Bacciagaluppi [67], and Timpson and Brown [3].[29]

---

[29] See also Wallace [68, Chapter 8]. Another important stream in the locality-in-Everett literature begins with Deutsch and Hayden [69] and their emphasis on locality in Everettian Heisenberg-picture quantum mechanics. (See also [70–73].) However, Deutsch and Hayden's claims are best understood not as addressing the question of locality in the sense of no action at a distance, but rather as addressing the distinct question of the *separability of the states* of one's fundamental theory. See [74–77]. Another example of an essentially Everettian treatment is the recent Brassard and Raymon-Robichaud [78].

A number of interrelated factors are involved in Everett's theory being local.[30] First is the point already noted that by remaining within purely unitary quantum mechanics – thus eschewing collapse – one moves out of range of the EPR argument, even if one believes (contra Bohr et al.) that the wave function should be understood realistically. Second, the point that the unitary dynamics itself derives from local Hamiltonians (in the relativistic context, moreover, Lorentz covariant Hamiltonians) which include only point like interactions. Third, the fact that the fundamental states of the theory are *nonseparable*, due to entanglement: it is not the case that the properties of a joint system are determined by the properties of the individual parts taken in isolation. And fourth, the crucial fact that following a measurement, it is generally not the case that there is one unique outcome of the measurement. It is this last, of course, which lies behind the 'many worlds' conception of the theory: following a measurement interaction between a measuring apparatus and a system not in an eigenstate of the quantity being measured, the measuring apparatus will be left in a superposition of its indicator states. Unitary processes of decoherence lead to effective noninteraction between the different terms in this superposition and lead to effective classicality for their future evolution over time. As further items interact with the apparatus – whether environmental (noise) degrees of freedom, or perhaps observers in the lab – they too become drawn linearly into the superposition, becoming part of the overall entangled state. An effective branching structure of macroscopically determinate goings-on emerges (at least at a suitably coarse-grained level) and we can reasonably call these emergent branches of effectively isolated, quasi-classical, macroscopically determinate goings-on *worlds*. Within particular branches in the structure will be found measuring apparatuses each giving definite readings for the outcome of experiments performed, and correlated with these can be found observers, each of whom will see their measuring device indicating a single particular result of measurement. But zooming out (in a God's-eye view) from a particular branch will be seen all the other branches, each with a different result of measurement being recorded and observed, all coexisting equally; and all underpinned by (supervenient on) the deterministically, unitarily, evolving universal wave function.

Now why is this fourth factor – the nonuniqueness in the complete state of the world of the measurement outcomes for a given measurement – so significant when it comes to considering locality? Some commentators point out – plausibly – that it is simply an implicit premise in Bell's discussions that specified measurements, when performed, *do* have unique outcomes: that the beables that obtain in a given region of space–time where a measurement has taken place do fix a unique value for the outcome of that measurement (do fix, that is, a unique value for the quantity measured). When this implicit premise is given up, Bell's reasoning just doesn't apply, they note (Wallace [68, p. 310], Blaylock [20] and Maudlin [4]).[31] In that case, we can infer nothing as to the presence or absence of locality in a theory when it violates a Bell inequality, or equivalently, when it violates factorizability.

---

[30] Whilst we speak of *Everett's theory* or of *Everettian quantum mechanics* or of *the Everett interpretation*, it is of course true that Everett's original ideas have been developed in rather different directions by different authors. Nothing much in our present discussion of locality hangs on this, in our view, but for the record, we take the current canonical form of Everretianism to be that articulated in [68, 79–83].

[31] In Wiseman [2], that measurements yield unique outcomes is stipulated in Axiom 1, p. 4; see also the first paragraph of Section 7 therein.

This response is correct, so far as it goes, but (i) it is perhaps less obvious how 'unique outcomes' functions as an assumption in Bell's reasoning once he moved to consider stochastic theories and local causality as formulated from 1976 on, rather than the deterministic theories of 1964, and (ii) one can hope to say a little more to illuminate the situation than just that this implicit premise fails. Indeed, regarding (ii), it is in our view rather helpful to go into some of the details of how in fact the Everett interpretation deals with the generation of correlations in EPR and Bell settings in a local manner, when this proves so difficult or impossible to do for other approaches. For convenience we will recapitulate here the analysis we gave in Timpson and Brown [3] before going on to make some further points. We will return to address (i) briefly in Section 6.9.2.

6.9.1 EPR and Bell Correlations in the Everettian Setting

It is a straightforward matter to apply Everettian measurement theory in the context of EPR–Bell scenarios. (Very similar analyses can also be found elsewhere, for example in [50, 66].) We will discuss once again, by way of concrete example, the Bohm version of the EPR experiment, involving spin. We begin with two spin-1/2 systems (labelled 1 and 2) prepared in a singlet state. The relevant degrees of freedom (the 'pointer variables') of measuring apparatus m_A, m_B in widely separated regions A and B of space we can model as two-state systems with basis states $\{|\uparrow\rangle_A, |\downarrow\rangle_A\}$, $\{|\uparrow\rangle_B, |\downarrow\rangle_B\}$ respectively. Measurement by apparatus A of the component of spin at an angle θ from the z-axis on system 1 would have the effect

$$U(\theta) \begin{cases} |\uparrow_\theta\rangle_1 |\uparrow_\theta\rangle_A & \mapsto & |\uparrow_\theta\rangle_1 |\uparrow_\theta\rangle_A \\ |\downarrow_\theta\rangle_1 |\uparrow_\theta\rangle_A & \mapsto & |\downarrow_\theta\rangle_1 |\downarrow_\theta\rangle_A, \end{cases} \tag{6.6}$$

where $|\uparrow_\theta\rangle$, $|\downarrow_\theta\rangle$ are eigenstates of spin in the rotated direction; similarly for measurement of system 2 by apparatus B.[32]

Case 1: Perfect Correlations (EPR)

Now consider the case in which our measuring apparatuses are perfectly aligned with one another, as in the EPR example. The initial state of the whole system is

$$|\uparrow\rangle_A \frac{1}{\sqrt{2}}(|\uparrow\rangle_1|\downarrow\rangle_2 - |\downarrow\rangle_1|\uparrow\rangle_2)|\uparrow\rangle_B. \tag{6.7}$$

Note that the states of the measuring apparatuses factorise: at this stage, they are independent of the states of the spin-1/2 particles. For the measurements to be made, systems 1 and 2 are taken to regions A and B, respectively, and measurement interactions of the form (6.6) occur (the time order of these interactions is immaterial); we finish with the state

$$\frac{1}{\sqrt{2}}(|\uparrow\rangle_A|\uparrow\rangle_1|\downarrow\rangle_2|\downarrow\rangle_B - |\downarrow\rangle_A|\downarrow\rangle_1|\uparrow\rangle_2|\uparrow\rangle_B). \tag{6.8}$$

---

[32] Of course, for a measurement *truly* to have taken place, the indicator states of the apparatus would also need to have been irreversibly decohered by their environment, and themselves to be robust against decoherence. We can put these complications to one side for the purposes of our schematic model.

Now definite states of the measuring device m_A are correlated with definite spin states of particle 1, as are definite states of the measuring device m_B with definite spin states of particle 2. But since all of the outcomes of the measurements are realised, there is no question of the obtaining of one definite value of spin at one side forcing the anticorrelated value of spin to be obtained at the far side. Both sets of values become realised, relative to different states of the apparatus (and relative to subsequent observers if they are introduced into the model) – as we put it before,

There is, as it were, no dash to ensure agreement between the two sides to be a source of non-locality and potentially give rise to problems with Lorentz covariance. [3]

However, it is important to note that following the measurements at A and B, not only does each measured system have a definite spin state relative to the indicator state of the device that has measured it, but the systems and measuring apparatuses in each region (e.g., system 1 and apparatus m_A in A) have definite spin and indicator states relative to definite spin and indicator states of the system and apparatus in the *other* region (e.g., system 2 and apparatus m_B in B). That is, following the two local measurements, from the point of view of the systems in one region, the states of the systems in the far region correspond to a definite, in fact perfectly anticorrelated, measurement outcome. This is in contrast to the general case of nonparallel spin measurements at A and B, as we shall see in a moment.

In this parallel-settings EPR case we have a deterministic explanation of how the perfect (anti-) correlations come about. Given the initial state that was prepared, and given the measurements that were going to be performed, it was always going to be the case that a spin-up outcome for system 1 would be correlated with a spin-down outcome for system 2 and vice versa, once both sets of local measurements were completed. The perfect (anti-) correlations unfold deterministically from the initial entangled state given the local measurement interactions in regions A and B respectively. There is no puzzle here about how independent, intrinsically stochastic processes taking place in spacelike separated regions could nonetheless end up being correlated, since the evolution, and the explanation of how the correlations come about, is purely deterministic: *up* for system 1 was always going to be correlated with *down* for system 2, as was *down* for system 1 with *up* for system 2, both cases (of course) being superposed in the final state.[33]

Case 2: Nonaligned Measurements (Bell)

As we have already noted, one of the plethora of good points in Bell [29] is of course that the parallel-settings case is in an important sense not very interesting. For this restricted class of measurement settings, Bell readily provided a local (deterministic) hidden variable model. Moreover, from a practical point of view, cases in which one has managed exactly to

---

[33] Note that given that both possible outcomes for each spin measurement obtain simultaneously in the final state, the Bell–Suppes–Zanotti argument that any local theory which predicts perfect (anti-) correlations must collapse into the kind of deterministic theory Bell considered in [29], does not obtain. Everett is a deterministic theory, but it does not belong to the class of deterministic completions of quantum mechanics considered in Bell [29], since it is *also* a probabilistic (stochastic) theory – at the emergent level at which measurement outcomes are part of the story.

align one's Stern–Gerlach magnets with one's colleagues' way over yonder will form a set of measure zero. For both these reasons, then, it is important to consider what happens in the general case – as in the setup required to derive Bell's inequalities – where the measurement angles are not aligned.

In this case we may write the initial state as

$$|\uparrow\rangle_A \frac{1}{\sqrt{2}} \left(|\uparrow\rangle_1 |\downarrow\rangle_2 - |\downarrow\rangle_1 |\uparrow\rangle_2 \right) |\uparrow_\theta\rangle_B, \tag{6.9}$$

where we have assumed a relative angle θ between the directions of measurement. If we write $|\uparrow\rangle = \alpha|\uparrow_\theta\rangle + \beta|\downarrow_\theta\rangle$, $|\downarrow\rangle = \alpha'|\uparrow_\theta\rangle + \beta'|\downarrow_\theta\rangle$, we can express this joint state as

$$|\uparrow\rangle_A \frac{1}{\sqrt{2}} \left[|\uparrow\rangle_1 \left(\alpha|\uparrow_\theta\rangle_2 + \beta|\downarrow_\theta\rangle_2 \right) - |\downarrow\rangle_1 \left(\alpha'|\uparrow_\theta\rangle_2 + \beta'|\downarrow_\theta\rangle_2 \right) \right] |\uparrow_\theta\rangle_B. \tag{6.10}$$

We can then see that following the measurements at A and B (the time order of these two measurements is again immaterial, of course), we will have

$$\frac{1}{\sqrt{2}} \left[|\uparrow\rangle_A |\uparrow\rangle_1 \left(\alpha|\uparrow_\theta\rangle_2 |\uparrow_\theta\rangle_B + \beta|\downarrow_\theta\rangle_2 |\downarrow_\theta\rangle_B \right) \right.$$
$$\left. - |\downarrow\rangle_A |\downarrow\rangle_1 \left(\alpha'|\uparrow_\theta\rangle_2 |\uparrow_\theta\rangle_B + \beta'|\downarrow_\theta\rangle_2 |\downarrow_\theta\rangle_B \right) \right]. \tag{6.11}$$

Here, relative to states representing a definite outcome of measurement in region A, there is no definite outcome in B; system 2 and apparatus m_B are just entangled, with no definite spin and indicator states. Similarly, from the point of view of definite spin and indicator states of 2 and m_B (a definite outcome in region B), there is no definite outcome of measurement in region A.

For nonparallel spin measurements, then, unlike the parallel case, there needs to be a third measurement (or measurement like interaction), comparing (effectively) the outcomes from A and B, in order to make definite spin and indicator states from one side definite relative to definite spin and indicator states from the other. Thus systems from A and B have to be brought (or come) together and a joint measurement be performed (or a measurement like interaction take place), leading to a state such as

$$\frac{1}{\sqrt{2}} \left[\alpha|\uparrow\rangle_A |\uparrow\rangle_1 |\uparrow_\theta\rangle_2 |\uparrow_\theta\rangle_B |\uparrow\uparrow\rangle_C + \beta|\uparrow\rangle_A |\uparrow\rangle_1 |\downarrow_\theta\rangle_2 |\downarrow_\theta\rangle_B |\uparrow\downarrow\rangle_C \right.$$
$$\left. - \alpha'|\downarrow\rangle_A |\downarrow\rangle_1 |\uparrow_\theta\rangle_2 |\uparrow_\theta\rangle_B |\downarrow\uparrow\rangle_C - \beta'|\downarrow\rangle_A |\downarrow\rangle_1 |\downarrow_\theta\rangle_2 |\downarrow_\theta\rangle_B |\downarrow\downarrow\rangle_C \right], \tag{6.12}$$

where the states $|\uparrow\uparrow\rangle_C$, etc., are the indicator states of the comparing apparatus. Following this third measurement-interaction, which can only take place in the overlap of the future light cones of the measurements at A and B, a definite outcome for the spin measurement in one region finally obtains, relative to a definite outcome for the measurement in the other. That is, we can only think of the *correlations* between measurement outcomes on the two sides of the experiment actually obtaining in the overlap of the future light-cones of the measurement events – they do not obtain before then and, *a fortiori*, they do not obtain instantaneously. On each side *locally* there are definite measurement outcomes (superposed with one another) as soon as each local measurement is complete, but there is no correlation between the measurement outcomes on the two sides until later on, when a suitable entangling operation between systems from the two regions *A* and *B* can take place.

It is important to note what this means for how probability statements derived from the Born rule for joint measurements on spacelike separated systems should be understood in this context. Given the initial entangled singlet state, we can make formal statements about what the probability distribution over joint measurement outcomes for spacelike separated measurements is. In general these joint probabilities will be nontrivial and in fact Bell-inequality-violating. But physically, there is nothing for such joint probabilities to be joint probabilities *of* until one reaches the overlap region of the future light-cones of the measurement events, since it is only in this overlap region that measurement outcomes for the individual measurements on either side can become definite with respect to one another, in general. Before then, the joint probabilities are only formal statements, regarding what one would expect to see, were one to compare the results of measurements on the two sides. This state of affairs does much to take the sting out of the randomness problem mentioned above in Section 6.8: the original separated measurement events are neither fundamentally random, nor correlated in the straightforward sense of classical physics.

6.9.2 The Role of Nonseparability

Is there anything left to be explained? Arguably yes: it is the nontrivial fact that the correlations *will* be found to obtain in the future, given a future comparison measurement.

So far we have primarily emphasised the role of the failure of uniqueness for measurement outcomes in permitting Everettian quantum mechanics to produce EPR and Bell correlations without any action at a distance, and we have seen in some detail how in fact these correlations can be explained as arising following a local dynamics from the initial entangled state. Earlier, however, we noted that nonseparability was also an important part of the explanation of how Everett could provide a local story for EPR and Bell correlations. We shall expand on this now, and in so doing return to the point raised above that it may not be entirely obvious whether or how uniqueness of outcomes features as an assumption in Bell's reasoning from 1976 on, once the idea of local causality – or rather its formulation in terms of factorizability – was introduced.

One obvious point to make straightaway is that if the initial pure state of the two spins was not entangled, i.e., was separable, then there could not be *any* correlation between the spin measurements on the two sides of the experiment, let alone inequality-violating ones. (This point holds in more general theories than quantum mechanics also.) But more deeply, what is going on is that nonseparability allows there to be facts about the relations between spatially separated systems which go above and beyond – are not determined by – the intrinsic (i.e., locally defined) properties of those systems individually. In particular, there can be facts about how things in spatial region *A* will be correlated with (related to) things in spatial region B, without its being the case that how things are in A and how things are in B fix these relations.[34]

In the Bell experiment with spins we have just discussed, if one only had failure of uniqueness of outcomes for the measurements on each side, in an otherwise separable theory, then no nontrivial correlations could obtain between the outcomes of the measurements. Everettian quantum mechanics exploits both nonuniqueness of outcomes and nonseparability in accounting for EPR and Bell correlations without action at a distance. In fact, it is the particular way that nonseparability features in the theory which entails nonuniqueness for the measurement outcomes.[35]

Now: the Everett interpretation shows that a theory can be local in the sense of satisfying no-action-at-a-distance, whilst failing to be locally causal: it violates the factorizability condition, or equivalently, the condition in terms of the probabilities attached to local beables in a given space-time region being independent of goings-on in spacelike regions, once the state of the past light-cone of that region is sufficiently specified. What, then, went wrong with Bell's formulation of local causality, as an expression of locality? It is not perspicuous, at least to us, why the mere failure of uniqueness of measurement outcomes should make local causality – as Bell formulated it mathematically – inapposite as an expression of a principle of locality. So there is something more to be said here. In our view what needs to be said is that, at root, *it is moving to the context of nonseparable theories* which makes Bell's mathematically formulated conditions fail properly to capture his intuitive notions of locality.[36]

To substantiate this thought, we need to return to Reichenbach's principle of the common cause. Recall that in order to move from Bell's informal statement of local causality – that the proximate causes of events should be near them, and causal chains leading up to these events should lie on or within their past light-cones – to its mathematical formulation in terms of factorizability or equivalently in terms of screening off, something like Reichenbach's principle needs to be appealed to: statistical correlations between events must be explained either by direct causal links between them, or in terms of a Reichenbachian common cause in the past. With such a principle in place, we can connect mathematical statements in terms of probabilities to physical claims about causal links, and thereby to claims about action at a distance. However, once we move to a situation where our theory can be nonseparable, the common cause principle is unnatural and unmotivated.[37] This is for a very simple reason: in a nonseparable theory there is a *further* way in which correlations can be explained which Reichenbach's stipulations miss: correlations between systems (e.g., the

---

[34] Some early explorations of this idea in the context of Bell's theorem include Howard [84, 85], Teller [86, 87], French [88], Healey [89, 90] and Mermin [91–93].

[35] We leave it as an open question whether or not in any nonseparable theory which is dynamically local, but violates local causality, uniqueness of measurement outcomes fails.

[36] In making this claim, we are therefore taking issue with some of the conclusions of the otherwise splendid discussion of Henson [94]. We will expand on our disagreement with Henson in full on another occasion.

[37] Seevinck and Uffink [54] argue that we need not see Bell's mathematical conditions as resting on Reichenbach's principle, but on the notions of locality, causality and statistical sufficiency. In our view, their analysis in terms of statistical sufficiency would also be uncompelling in the context of nonseparable theories, for much the same reasons as Reichenbach's principle is. Again, we leave a detailed development of this claim to another forum.

fact that certain correlations between measurement outcomes *will* be found to obtain in the future) can be explained directly by irreducible relational properties holding between the systems, relational properties which themselves can be further explained in dynamical terms as arising under local dynamics from a previous nonseparable state for the total system. Which is precisely what happens in the Everettian context, for example.

In sum, we can see Bell's mathematical formulation(s) of the intuitive idea of local causality as instantiating Reichenbach's principle of the common cause[38] as applied in the case of measurements on systems in spacelike separated regions, where we are assuming the standard (or naive) conception of what constraints relativity imposes on causal processes. It may be that Reichenbach's principle is plausible enough when one considers separable theories, but it is unacceptable when one considers nonseparable theories. It is therefore, at the most straightforward level, simply because Reichenbach's principle is not apt for worlds which may be nonseparable that Bell's formal statements of local causality go wrong, and that there can be theories such as Everett's which are not locally causal, but which are local, in the sense that they involve no action at a distance.

We noted at the end of Section 6.8 that Bell appeared to sound a note of despair at the end of Nouvelle Cuisine when contemplating the prospect that local causality might not turn out to be an adequate statement of locality in physics. In particular, he worried that should his formulations of local causality not be apt and that we had instead to settle for some other statement of locality (such as no-signalling) which would allow the existence of Bell-inequality-violating correlations whilst one's theory counted as fully local, then we would have 'lost the idea that correlations can be explained', to quote him again. And this would indeed seem a worrying thing. But it seems to us, at least, that Bell need not have cause to despair in the circumstances which we have sketched. In our view, correlations which violate factorizability, whilst yet arising from a dynamically local theory, need not be condemned to be unexplainable: we just need to free ourselves from a Reichenbachian-common-cause straitjacket of what suitable explanation could be. Put another way, we can all actually fully *agree* with the most basic sentiment which commentators draw from Reichenbach, namely that correlations should be explainable, whilst disagreeing with his specific formulation of what causal explanation (or maybe just *explanation*) in terms of factors in the past must be like.[39] Specifically, it need not be the case that the factor in the past should be some classical random variable which screens off the correlations. A perfectly acceptable, nonequivalent alternative form of explanation of correlations in terms of factors in the past would seem to be in terms of the evolution of a later nonseparable state from an earlier nonseparable state. We saw an instance of this, of course, in the Everettian case earlier. The *general* story as to how a nonseparable theory can locally explain Bell-inequality-violating correlations would be that the correlations are entailed by some suitable

---

[38] Modulo the historical subtleties noted earlier about whether in Reichenbach's own setting the principle should really be thought to apply at the micro-level at all.

[39] Accordingly, we commend the approach of Cavalcanti and Lal [95], who explicitly separate Reichenbach's principle into two components: first that correlations should be causally explicable either by direct interaction or common cause in the past; second that explanation by common cause in the past takes the particular form that Reichenbach imposed. One can maintain the first idea whilst denying the second.

nonseparable joint state. And if one has a detailed story of the contents of one's local and nonseparable theory (i.e., a detailed specification of its kinematics and dynamics – including measurement theory) then one will have a perfectly good explanation of how this comes about. One's explanation will be in terms of how that particular nonseparable joint state evolved out of some previous (generally) nonseparable state.[40]

6.9.3 Maudlin's Challenge

We have sought to explain how the Everett interpretation provides one concrete example illustrating that Bell's mathematical, probabilistic formulation of local causality does, as he feared it might, fail adequately to capture the notion of locality. An important challenge to a conclusion of this kind is presented in characteristically trenchant and pithy form by Maudlin [4], however, to which we must now turn.

Maudlin argues that

Because the many worlds interpretation fails to make The Predictions [a subset of the predictions of standard quantum theory], Bell's theorem has nothing to say about it.... And reciprocally, the existence of the many worlds interpretation can in principle shed no light on Bell's reasoning because it falls outside the scope of his concerns.

[4, Sect. III]

We will consider these claims in turn.

The first thing Maudlin includes in 'The Predictions' is the claim that measurements (of spin, for the EPR–Bohm experiment we have been considering) have unique outcomes, where 'Born's rule provides the means of calculating the probability of each of the ... outcomes' [4]. The further elements are the prediction of perfect EPR correlations for parallel measurements and the prediction of Bell-inequality-violating correlations for certain specific choices of nonaligned measurements. It is because he sees uniqueness of measurement outcomes as a necessary condition for the further predictions of EPR and Bell correlations to have any content that Maudlin believes the Everett interpretation gets into trouble:

If every experiment carried out on particle 1 yields both [possible outcomes] ... and every experiment carried out on particle 2 yields both [possible outcomes], what can it mean to say that the outcomes on the two sides are always correlated or always anticorrelated [as in the EPR–Bohm scenario] or agree only [some percentage] ... of the time [as in the EPR–Bell scenario]? For such claims to have any content, particular results on one side must be associated with particular results on the other so that the terms 'agree' and 'disagree' make sense.

[4, Sect. III]

The first thing to note is that from the point of view which we have been entertaining, it is simply question-begging to include 'uniqueness of measurement outcomes' amongst

---

[40] Thus we offer a picture in terms of the fundamental states of a theory and their dynamical – law-governed – evolution. Is this a *causal* form of explanation? If it is thought not, to our minds it is not clear that that matters. It is certainly *physical explanation* by one gold standard.

the predictions of *standard quantum theory*. For what, after all, even *is* standard quantum theory? Arguably there is no such thing – there is a standard quantum *algorithm*, which experimentalists know how to apply to get a good fit to experiment, and where in particular they find that the Born rule gives an excellent fit to the results they see. But anything beyond this is up for grabs, and may be understood differently in different interpretations (or different theories) offered to underpin the success of the standard quantum algorithm. In particular, from an Everettian point of view, the *observation* of unique outcomes of measurements does not at all entail *uniqueness of the outcomes in the complete state of the world*. For the Everettian, the success of the standard quantum algorithm can be guaranteed without requiring the latter form of uniqueness.

But even if uniqueness of measurement outcomes can not be thought to be part of the *predictions* of standard quantum theory (as opposed to being part of the presuppositions in various standard approaches to the theory) it might even so – as Maudlin alleges – be a necessary condition for making sense of the prediction of correlations, for the reasons Maudlin states above. We disagree, however.

First, return to the Everettian treatment of perfect EPR correlations as we described above (Section 6.9.1). Here we saw how the standard, local, unitary dynamics would lead deterministically to the case in which *up* on one side of the experiment was correlated with *down* on the other, and vice versa. Relative to definite results on one side of the experiment, there are definite results on the other. It is quite clear what it means for the measurement results always to be correlated in this case, notwithstanding the fact that both options for the correlated outcomes are superposed together in the overall final entangled state. Note for future reference that in this case we do not have to appeal to the Born rule to understand the prediction that the results on the two sides of the experiment will be perfectly correlated with one another.

Second, we return to our treatment of the nonaligned Bell case (Section 6.9.1). Here one will need to appeal to the Born rule in order to predict that there are correlations, but importantly, as we described, it is only in the overlap of the future light-cones of the measurement events, given a suitable comparison interaction between systems from the two sides, that measurement outcomes from one side will become definite relative to measurement outcomes from the other. Thus what is required to make sense of the prediction of Bell correlations is that a measurement in the overlap of the future light-cones of the initial measurement events, one which compares locally, at a point, records of the outcomes from the two regions *A* and *B* where the initial measurements took place, should give a suitable probability distribution for what the results of that local comparison will be. In the nonaligned case, that 'particular results on one side [are] associated with particular results on the other' comes about subsequent to the initial measurements being made, and is brought about by there being a suitable later comparison interaction. It is after this further interaction that 'the terms "agree" and "disagree" make sense' as applied to the results of the earlier measurements.

It is possible that one might remain unsatisfied by this. After all, one might say that – at least for the Bell-correlations case, if not the EPR – we have had to appeal to the Born rule, and perhaps the real thrust of the worry coming from nonuniqueness of measurement outcomes is that sense can not be made of how the Born rule could apply to govern the probabilities of outcomes of measurements, if all the outcomes occur. This, of course, is a longstanding and respectable objection to the Everett interpretation, but notice *that there is nothing specific about EPR or Bell correlations in this*. Given our analysis, the alleged difficulty in predicting Bell correlations has been reduced to the difficulty of making sense of probability in Everett for local measurements. And in our view, this amounts to reduction to – plausibly – a previously solved case (see particularly the work of Saunders, Deutsch, Wallace and Greaves).[41] Even if one remains agnostic about whether or not Everett does give an adequate account of probability for local measurements, our point is simply that *insofar* as Everett is a player at all as a viable interpretation of quantum theory, it provides a concrete counterexample to Bell's probabilistic formulation of local causality. (In our view, of course, Everett *is* a very significant player.)[42]

In effect, we have already stated our response to Maudlin's reciprocal claim that 'the existence of the many worlds interpretation can shed no light on Bell's reasoning because it falls outside the scope of his concerns'. On the contrary. Once it is recognised that the uniqueness of measurement outcomes is question-begging if assumed to be a requirement on there being well-formed theories which make probabilistic predictions for the results of measurements in various space–time regions, Everett plainly falls within the scope of Bell's post-1976 reasoning and is entirely germane to the adequacy of his probabilistic formulation of local causality. In 1986, Bell noted,

The 'many world interpretation' seems to me an extravagant . . . hypothesis . . . And yet . . . It may have something distinctive to say in connection with the 'Einstein Podolsky Rosen puzzle'. [101]

We think he was dead right on this last point.

6.10 Conclusions

We have seen that there was a very significant shift in Bell's notion of nonlocality between 1964 and 1976, a shift that occurred in parallel with his thinking moving away from focusing on completions of quantum mechanics in the spirit of EPR. Crucially, violating the 1964' locality condition gives rise to action at a distance, whereas violating local causality of 1976 need not. We have noted, moreover, that as Bell's thinking developed, he came more and more to recognise that there need be no straightforward conflict between violation of either of his locality conditions and the demands of relativity.

In our discussion of locality in the Everett interpretation we have sought to provide a constructive example illustrating precisely how a theory can be dynamically local, whilst violating local causality; and we emphasised the interconnected roles of the failure of

---

[41] Saunders [82], Deutsch [96], Greaves [97] and Wallace [68, Part II; 98–100].

[42] There is a final worry which might be motivating Maudlin: He may find it obscure how the Everettian story about the emergence of determinate – but nonetheless superposed – measurement outcomes and experiences of the everyday world is supposed to work. But again this is just a general objection to Everett, and one on which there has been a great deal of persuasive work; see again Saunders et al. [83] and Wallace [68] for the state of the art.

uniqueness of measurement outcomes and of nonseparability in achieving this. We think that Bell was right to have had doubts in 1990 regarding whether he had managed, in his mathematical, probabilistic, statements of local causality adequately to capture the concept of locality. But we have suggested that even if local causality is rejected as the expression of locality, it need not follow that one is doomed to having to put up with unexplained correlations, as Bell feared one might be. For as we have explained, nonseparable theories allow additional ways in which correlations can be causally explained without action at a distance.

Acknowledgements

We would like to thank Mary Bell and Shan Gao, for the invitation to contribute to this auspicious volume celebrating the 50th anniversary of Bell's seminal 1964 paper. We would also like to thank Ray Lal, Owen Maroney, Rob Spekkens, and David Wallace for helpful discussions. CGT's work on this paper was partially supported by a grant from the Templeton World Charity Foundation.

References

[1] Bell, John S. 1976, The theory of local beables, *Epistemological Letters*, March. Repr. in [33, Chap. 7].

[2] Wiseman, Howard 2014, The two Bell's theorems of John Bell, *Journal of Physics A* **47**, 063028.

[3] Timpson, Christopher G, and Brown, Harvey R. 2002, Entanglement and relativity, in Rosella Lupacchini and Vincenzo Fano (eds.), *Understanding Physical Knowledge* (University of Bologna, CLUEB), arXiv:quant-ph/0212140.

[4] Maudlin, Tim 2010, What Bell proved: A reply to Blaylock, *American Journal of Physics* **78**(1), 121–5.

[5] Norsen, Travis 2011, J.S. Bell's concept of local causality, *American Journal of Physics* **79**, 1261.

[6] Hobson, Art 2013, There are no particles, there are only fields, *American Journal of Physics* **81**, 211.

[7] Hobson, Art 2013, Response to M.S. de Bianchi and M. Nauenberg, *American Journal of Physics* **81**, 709.

[8] Brunner, Nicolas, Cavalcanti, Daniel, Pironio, Stefano, Scarani, Valerio, and Wehner, Stephanie. 2014, Bell nonlocality, *Reviews of Modern Physics* **86**, 419–78.

[9] Bell, John S. 1966, On the problem of hidden variables in quantum mechanics, *Reviews of Modern Physics* **38**, 447–52. Repr. in Bell [33, Chap. 1].

[10] Kochen, Simon and Specker, Ernst 1967, The problem of hidden variables in quantum mechanics, *Journal of Mathematics and Mechanics* **17**, 59–87.

[11] Gleason, A. 1957, Measures on the closed subspaces of Hilbert space, *Journal of Mathematics and Mechanics* **6**(6).

[12] Brown, Harvey R. 1992, Bell's other theorem and its connection with nonlocality. Part I, in A. van der Merwe, F. Selleri and G. Tarozzi (eds.), *Bell's Theorem and the Foundations of Modern Physics* (World Scientific), pp. 104–16.

[13] Mermin, N. David 1993, Hidden variables and the two theorems of John Bell, *Reviews of Modern Physics* **65**, 803–15.

[14] Bohm, David 1952, A suggested interpretation of the quantum theory in terms of hidden variables, I and II, *Physical Review* **85**, 166–79, 180–93.

[15] Heywood, Peter and Redhead, Michael 1983, Nonlocality and the Kochen–Specker paradox, *Foundations of Physics* **13**(5), 481–99.

[16] Stairs, Alan 1983, Quantum logic, realism, and value definiteness, *Philosophy of Science* **50**(4), 578–602.

[17] Brown, Harvey R. and Svetlichny, George 1990, Nonlocality and Gleason's lemma: Part I. Deterministic theories, *Foundations of Physics* **20**(11), 1379–87.

[18] Elby, Andrew 1990, Nonlocality and Gleason's lemma. Part 2. Stochastic theories, *Foundations of Physics* **20**(11), 1389–97.

[19] Greenberger, Daniel, Horne, Michael and Zeilinger, Anton 1989, Going beyond Bell's theorem, in M. Kafatos (ed.), *Bell's Theorem, Quantum Theory, and Conceptions of the Universe* (Kluwer), pp. 73–6.

[20] Blaylock, Guy 2010, The EPR paradox, Bell's inequality, and the question of locality, *American Journal of Physics* **78**(1), 111–20.

[21] Brown, Harvey R. 1991, Nonlocality in quantum mechanics, *Aristotelian Society Supplementary Volume* **65**, 141–59.

[22] Home, Dipankar and Sengupta, S. 1984, Bell's inequality and non-contextual dispersion-free states, *Physics Letters A* **102**(4), 159–62.

[23] Foster, Sarah and Brown, Harvey R. 1987, A reformulation and generalisation of a recent Bell-type theorem for a valence electron, unpublished manuscript.

[24] Belinfante, F.J. 1973, *A Survey of Hidden-Variable Theories* (Pergamon Press).

[25] Pusey, Matthew, Barrett, Jonathan and Rudolph, Terry 2012, On the reality of the quantum state, *Nature Physics* **8**, 475–8.

[26] Barrett, Jonathan, Cavalcanti, Eric G., Lal, Raymond, and Maroney, Owen J.E. 2014, No ψ-epistemic model can fully explain the indistinguishability of quantum states, *Physical Review Letters* **112**, 250403.

[27] Zukowski, Marek and Brukner, Caslav 2014, Quantum non-locality – It ain't necessarily so . . . , *Journal of Physics A* **47**(42), 424009.

[28] Bell, John S. 1981, Bertlmann's socks and the nature of reality, *Journal de Physique* **42**(3C2), C241–61. Repr. in [33, Chap. 16].

[29] Bell, John S. 1964, On the Einstein–Podolsky–Rosen Paradox, *Physics* **1**, 195–200. Repr. in Bell [33, Chap. 2].

[30] Bohm, David 1951, *Quantum Mechanics* (Prentice-Hall).

[31] Brown, Harvey R. 1981, O debate Einstein–Bohr na mecânica quântica, *Cadernos de História e Filosofia da Ciência (Brazil)* **2**, 59–81.

[32] Norsen, Travis 2005, Einstein's boxes, *American Journal of Physics* **73**(2), 164–76.

[33] Bell, John S. 1987, *Speakable and Unspeakable in Quantum Mechanics* (Cambridge University Press). Second edition 2004.

[34] Barbour, Julian 1989, *Absolute or Relative Motion: Volume 1. The Discovery of Dynamics* (Oxford University Press).

[35] Fine, Arthur 1986, *The Shaky Game* (Chicago University Press).

[36] de Broglie, Louis 1964, *The Current Interpretation of Wave Mechanics: A Critical Study* (Elsevier).

[37] Reichenbach, Hans 1956, *The Direction of Time* (University of California Press). Repr. Dover 1999.

[38] Elby, Andrew, Brown, Harvey R. and Foster, Sarah 1993, What makes a theory physically 'complete'? *Foundations of Physics* **23**(7), 971–85.

[39] Clauser, John F., Horne, Michael A., Shimony, Abner and Holt, Richard A. 1969, Proposed experiment to test local hidden-variable theories, *Physical Review Letters* **20**, 880.

[40] Bell, John S. 1971, Introduction to the hidden variable question, in *Foundations of Quantum Mechanics* (Academic Press, New York). Repr. in [33, Chap. 4].

[41] Clauser, John F. and Horne, Michael A. 1974, Experimental consequences of objective local theories, *Physical Review D* **10**, 526–35.

[42] Suppes, Patrick and Zanotti, M. 1976, On the determinism of hidden variables theories with strict correlation and conditional statistical independence, in Patrick Suppes (ed.), *Logic and Probability in Quantum Mechanics* (Springer), pp. 445–55.

[43] Jarrett, Jon P. 1984, On the physical significance of the locality conditions in the Bell arguments, *Noûs* **18**(4), 569–89.

[44] Bell, John S. 1989, Against 'measurement', in *62 Years of Uncertainty* (Plenum). Repr. in [33, Chap. 23].

[45] Bell, John S. 1986, EPR correlations and EPW distributions, in *New Techniques and Ideas in Quantum Measurement Theory* (New York Academy of Sciences). Repr. in [33, Chap. 20].

[46] Bell, John S. 1984, Speakable and unspeakable in quantum mechanics. Repr. in [33, Chap. 18].

[47] Bell, John S. 1976, How to teach special relativity, *Progress in Scientific Culture* **1**(2). Repr. in [33, Chap. 9].

[48] Bell, John S. 1986, Chapter 3 in Paul Davies and Julian Brown (eds.), *The Ghost in the Atom* (Cambridge University Press).

[49] Bell, John S. 1990, La nouvelle cuisine, in A. Sarlemijn and P. Kroes (eds.), *Between Science and Technology*, Amsterdam: Elsevier, Chap. 6, p. 97. Repr. in [33, Chap. 24].

[50] Tipler, Frank J. 2014, Quantum nonlocality does not exist, *Proceedings of the National Academy of Sciences* **111**(31), 11281–6.

[51] Caves, Carlton, Fuchs, Christopher A. and Shack, Rüdiger 2007, Subjective probability and quantum certainty, *Studies in History and Philosophy of Modern Physics* **38**, 255–74.

[52] Fuchs, Christopher A., Mermin, N. David and Schack, Rüdiger 2014, An introduction to QBism with an application to the locality of quantum mechanics, *American Journal of Physics* **82**(8), 749–54.

[53] Healey, Richard 2013, How to use quantum theory locally to explain EPR–Bell correlations, in Vassilios Karakostas and Dennis Dieks (eds.), *EPSA11: Perspectives and Foundational Problems in Philosophy of Science*, The European Philosophy of Science Association Proceedings, Vol. 2 (Springer), pp. 195–205.

[54] Seevinck, Michael P. and Uffink, Jos 2011, Not throwing out the baby with the bathwater: Bell's condition of local causality mathematically 'sharp and clean', in D. Dieks, W.J. Gonzalez, S. Hartmann, T. Uebel and M. Weber (eds.), *Explanation, Prediction and Confirmation* (Springer).

[55] Shimony, Abner 1984, Controllable and uncontrollable non-locality, in S. Kamefuchi (ed.), *Proceedings of the International Symposium: Foundations of Quantum Mechanics in the Light of New Technology* (Physical Society of Japan), pp. 225–30.

[56] van Fraassen, Bas 1982, The Charybdis of realism: Epistemological implications of Bell's inequality, in [102, pp. 97–113].

[57] Butterfield, Jeremy 1989, A space-time approach to the Bell inequality, in [102, pp. 114–44].

[58] Butterfield, Jeremy 1992, Bell's theorem: What it takes. *British Journal for the Philosophy of Science* **43**(1), 41–83.

[59] Butterfield, Jeremy 2007, Stochastic Einstein locality revisited, *The British Journal for the Philosophy of Science* **58**(4), 805–67.

[60] Henson, Joe 2010, Causality, Bell's theorem, and ontic definiteness. arXiv:quant-ph/1102.2855.

[61] Dickson, Michael 1998, *Quantum Chance and Non-locality* (Cambridge University Press).

[62] Vaidman, Lev 2014, *Many-Worlds Interpretation of Quantum Mechanics*, in Edward N. Zalta (ed.), the Stanford Encyclopedia of Philosophy (Spring 2014 ed.) URL=http://plato.stanford.edu/archives/spr2014/entries/qm-manyworlds.

[63] Everett, Hugh 1957, "Relative state" formulation of quantum mechanics *Reviews of Modern Physics* **29**, 454–62.

[64] Page, Don 1982, The Einstein–Podolsky–Rosen physical reality is completely described by quantum mechanics, *Physics Letters A* **91**, 57–60.

[65] Vaidman, Lev 1994, On the paradoxical aspects of new quantum experiments, in D. Hull, M. Forbes and R.M. Burian (eds.), *PSA 1994* (Philosophy of Science Association), Vol. 1, pp. 211–7.

[66] Tipler, Frank J. 2000, *Does quantum nonlocality exist? Bell's theorem and the many-worlds interpretation*, arXiv:quant-ph/0003146.

[67] Bacciagaluppi, Guido 2002, Remarks on space–time and locality in Everett's interpretation, in Jeremy Butterfield, and Tomas Placek (eds.), *Nonlocality and Modality*, NATO Science Series II (Kluwer Academic), Pitt-Phil-Sci 00000504.

[68] Wallace, David 2012, *The Emergent Multiverse: Quantum Mechanics according to the Everett Interpretation* (Oxford University Press).

[69] Deutsch, David and Hayden, Patrick 2000, Information flow in entangled quantum systems, *Proceedings of the Royal Society of London A* **456**, 1759–74.

[70] Hewitt-Horsman, Clare and Vedral, Vlatko 2007, Entanglement without nonlocality, *Physical Review A* **76**, 062319–1–8.

[71] Rubin, Mark A. 2001, Locality in the Everett interpretation of Heisenberg-picture quantum mechanics, *Foundations of Physics Letters* **14**, 301–22.

[72] Rubin, Mark A. 2002, Locality in the Everett interpretation of quantum field theory, *Foundations of Physics* **32**, 1495–523.

[73] Rubin, Mark A. 2011, Observers and locality in Everett quantum field theory, *Foundations of Physics* **41**, 1236–62.

[74] Arntzenius, Frank 2012, *Space, Time and Stuff* (Oxford University Press), Chap. 4.

[75] Timpson, Christopher G. 2005, Nonlocality and information flow: The approach of Deutsch and Hayden, *Foundations of Physics* **35**(2), 313–43.

[76] Wallace, David and Timpson, Christopher G. 2007, Non-locality and gauge freedom in Deutsch and Hayden's formulation of quantum mechanics, *Foundations of Physics* **37**(6), 951–5.

[77] Wallace, David and Timpson, Christopher G. 2014, *Quantum Mechanics on Space–Time II*. [Forthcoming].

[78] Brassard, Gilles and Raymon-Robichaud, Paul 2013, Can free will emerge from determinism in quantum theory, in *Is Science Compatible with Free Will?:*

Exploring Free Will and Consciousness in the Light of Quantum Physics and Neuroscience (Springer), pp. 41–61.

[79] Saunders, Simon 1995, Time, quantum mechanics, and decoherence, *Synthese* **102**, 235–66.

[80] Saunders, Simon 1996, Relativism, in R. Clifton (ed.), *Perspectives on Quantum Reality* (Kluwer Academic Publishers), pp. 125–42.

[81] Saunders, Simon 1996, Time, quantum mechanics, and tense, *Synthese* **107**, 19–53.

[82] Saunders, Simon 1998, Time, quantum mechanics, and probability, *Synthese* **114**, 373–404.

[83] Saunders, Simon Barrett, Jonathan, Kent, Adrian and Wallace, David (eds.) 2010, *Many Worlds? Everett, Realism and Quantum Mechanics* (Oxford University Press).

[84] Howard, Don 1985, Einstein on locality and separability, *Studies in History and Philosophy of Science Part A* **16**(3), 171–201.

[85] Howard, Don 1989, Holism, separability, and the metaphysical implications of the Bell experiments, in [102, pp. 224–53].

[86] Teller, Paul 1986, Relational holism and quantum mechanics, *The British Journal for the Philosophy of Science* **37**(1), 71–81.

[87] Teller, Paul 1989, Relativity, relational holism, and the Bell inequalities, in [102, pp. 208–23].

[88] French, Steven 1989, Individuality, supervenience and Bell's theorem, *Philosophical Studies* **55**(1), 1–22.

[89] Healey, Richard 1991, Holism and nonseparability, *The Journal of Philosophy* **88**(8), 393–421.

[90] Healey, Richard 1994, Nonseparable processes and causal explanation, *Studies in History and Philosophy of Science* **25**, 337–74.

[91] Mermin, N. David 1998, The Ithaca interpretation of quantum mechanics, *Pramana* **51**, 549–65.

[92] Mermin, N. David 1998, What is quantum mechanics trying to tell us? *American Journal of Physics* **66**, 753.

[93] Mermin, N. David 1999, What do these correlations know about reality? Nonlocality and the absurd, *Foundations of Physics* **29**, 571–87.

[94] Henson, Joe 2013, Non-separability does not relieve the problem of Bell's theorem, *Foundations of Physics* **43**(8), 1008–38.

[95] Cavalcanti, Eric G. and Lal, Raymond 2014, On modifications of Reichenbach's principle of common cause in light of Bell's theorem, *Journal of Physics A* **47**, 424018.

[96] Deutsch, David 1999, Quantum theory of probability and decisions, *Proceedings of the Royal Society of London A* **455**(1988), 3129–37.

[97] Greaves, Hilary 2004, Understanding Deutsch's probability in a deterministic multiverse, *Studies in History and Philosophy of Modern Physics* **35**(3), 423–56.

[98] Wallace, David 2003, Everettian rationality: Defending Deutsch's approach to probability in the Everett interpretation, *Studies in History and Philosophy of Modern Physics* **34**(3), 415–39.

[99] Wallace, David 2007, Quantum probability from subjective likelihood: Improving on Deutsch's proof of the probability rule, *Studies in History and Philosophy of Modern Physics* **38**(2), 311–32.

[100] Wallace, David 2010, How to prove the Born rule, in [83, Chap. 8].

[101] Bell, John S. 1986, Six possible worlds of quantum mechanics, in Sture Allen (ed.), *Proceedings of the Nobel Symposium 65: Possible Worlds in Arts and Sciences* (The Nobel Foundation). Repr. in [33, Chap. 20].

[102] Cushing, James T. and McMullin, Ernan 1989, *Philosophical Consequences of Quantum Theory: Reflections on Bell's Theorem* (University of Notre Dame Press).

7

Experimental Tests of Bell Inequalities

MARCO GENOVESE

7.1 Introduction

7.1.1 Annus mirabilis 1964

1964 was a wonderful year for physics. In this annus mirabilis, summarising and simplifying a little, the quark model was proposed by Gell-Mann and Zweig and charm and colour properties were introduced, Higgs published his work on the scalar boson, the Ω boson and CP violation were discovered, cosmic background radiation was identified and, finally, Bell inequalities were suggested.

Most of these works were later subjects of a Nobel prize. Unluckily John Bell left us too early for this achievement, but, at least in my opinion, his work [1] is probably the most fruitful among all these huge advances, since now we can recognise that it was the root of a fundamental change in our vision of the physical world, providing, on one hand, the possibility of a test of quantum mechanics against a whole class of theories (substantially all the theories that we could define as "classical") and, on the other hand, introducing a clear notion of quantum nonlocality that represents not only a challenge in understanding quantum mechanics, but also one of the most important resources paving the way to quantum technologies that were developed in the last decades based on the quantum correlations discussed in the Bell paper.

In the following we will present the experimental progress in testing Bell inequalities. Rather amazingly even 50 years after Bell's paper, a conclusive test of his proposal is still missing [2, 3].

Nevertheless, huge progress has been achieved with respect to the first experiments. We will discuss these developments and the remaining problems for a conclusive test of Bell inequalities in detail.

7.1.2 The Bell Inequalities, Their Hypotheses and the Experimental Loopholes

In order to understand the experimental difficulties in achieving a final test of Bell inequalities, let us analyse a little how these inequalities are derived and the explicit and implicit hypotheses they contain.

The proof of Bell inequalities usually starts by considering two separated subsystems of an entangled state, which are addressed to two independent measurement apparatuses measuring the expectation values of two dichotomic observables, A_a and B_b. a, b are two parameters describing the settings of the measuring apparatus A and B, respectively.

The two measurements must be independent (which at the end means they should be spacelike separated events).

Bell demonstrated [4] that joint expectation values of the two observables satisfy some inequality for every local realistic theory (LRT), the so-called Bell inequalities (BI). On the other hand, this inequality can be violated in standard quantum mechanics (SQM) for specific choices of parameters.

As discussed in the theoretical chapters, several different equivalent (at least in the two-dimensional case, that is, the one of experimental interest) "Bell inequalities" exist. In the following we consider the one proposed by Clauser, Horne, Shimony, and Holt (CHSH), which has found wide experimental use.

The proof is relatively simple. It supposes that the results of measurements are determined by a hidden variable x (or a set of hidden variables) distributed with a certain probability distribution $\rho(x)$.

After having introduced the expected value for joint measurements $C(a, b)$, let us consider the inequality

$$
\begin{aligned}
|C(a, b) - C(a, c)| &\leq \int_X |A_a(x)B_b(x) - A_a(x)B_c(x)|\rho(x)dx \\
&= \int_X |A_a(x)B_b(x)|[1 - B_b(x)B_c(x)]\rho(x)dx \\
&= \int_X [1 - B_b(x)B_c(x)]\rho(x)dx
\end{aligned}
\tag{7.1}
$$

Introducing $0 \leq \delta \leq 1$ such that for some values b, b' we have $C(b', b) = 1 - \delta$, by splitting the set X into two regions $X_\pm = \{x | A_{b'}(x) = \pm B_b(x)\}$ we have

$$
\int_{X_-} dx\rho(x) = \delta/2,
\tag{7.2}
$$

where $C(b', b) = 1 - \delta = \int_X B_b^2(x)\rho(x)dx - 2\int_{X_-} B_b^2(x)\rho(x)dx$. Hence,

$$
\begin{aligned}
\int_X B_b(x)B_c(x)\rho(x)dx &\geq \int_X A_{b'}(x)B_c(x)\rho(x)dx - 2\int_{X_-} |A_{b'}(x)B_c(x)|\rho(x)dx \\
&= C(b', c) - \delta.
\end{aligned}
\tag{7.3}
$$

From this result and Eq. (7.1) follows the CHSH inequality [5],

$$
S = |C(a, b) - C(a, c)| + C(b', b) + C(b', c) \leq 2.
\tag{7.4}
$$

Let us now consider the hypotheses needed to derive this result:

- There exist one (or more) "hidden variable" predetermining the values of observables (realism).
- There is no nonlocal effect, i.e., $C(a, b) = \int_X A_a(x)B_b(x)\rho(x)dx$ (Bell definition of locality).

– The measurements can be chosen independently and randomly by the two observers (freedom of choice).

These hypotheses condition how an eventual experiment must be realised in order to perform a conclusive test of Bell inequality.

A first point is that the two measurements must be set independently. Apart from some "conspiracy" (absence of free will, absence of the possibility of having independent random numbers, etc.), this requires that the settings of the two measuring apparatuses be spacelike separated events, i.e., no communication is possible, even in principle, that could influence one setting on the basis of the other. If this is not the case, the experiment will suffer from the so-called *spacelike loophole*.

A second relevant point is that one should be able to detect all the pairs involved in the experiment or, at least, a sufficiently large fraction of them. If only a subsample of the total number of produced entangled systems are really detected, one needs to introduce an additional assumption, that the measured sample is a faithful representation of the whole. Indeed, a subsample could contain a distribution of the hidden variables different from the total one, since the hidden variable values can also be related to the probability of the state to be observed. This means that when the observed sample is not a sufficiently large fraction of the total set of pairs, then the additional hypothesis of having an unbiased measured subsample is introduced [6–9]. If this is the case, the experiment will suffer from the so-called *detection loophole*.

By considering the effect of losses (including all causes, from losses on the paths to detection efficiency), one can deduce that for maximally entangled states, a detection-loophole-free test of LRT requires observing at least 82.84% of the total sample. This value can eventually be reduced to 66.7% for non-maximally-entangled systems [10].[1]

Additional requirements are also

– that the number of emitted particles is independent of measurement settings, the *production rate loophole*;
– that the presence of a coincidence window does not allow a situation in a hidden variable scheme where a local setting may change the time at which a local event happens (*coincidence loophole*) [11].

7.2 Testing Bell Inequalities

7.2.1 Starting to Test BI

1970s Experiments

Immediately after the publication of Bell's work, the quest for a suitable system for an experimental test started.

Several different systems were analysed (such as $K\bar{K}$, $\Lambda\bar{\Lambda}$, and entangled pairs of ions); however, entangled photons appeared from the beginning to be the most suitable candidate.

---

[1] I.e., for a pure bipartite state when the two components do not have an equal weight; see Subsection 7.2.1.

The first experiments were performed in the 1970s using polarisation-entangled photon pairs as

$$|\Phi^+\rangle = \frac{|H\rangle|H\rangle + |V\rangle|V\rangle}{\sqrt{2}}, \qquad (7.5)$$

where H and V denote the horizontal and vertical polarisation, respectively.

These states were produced using either a cascade atom decay or positronium decay.

Nevertheless, all these experiments suffered from the spacelike loophole and presented other additional hypotheses.

Experiments with positronium produced in atomic decays were realised by [12–15] with results in agreement with SQM, i.e., showing a BI violation (with the exception of Ref. [13]).

However, the difficulty of selecting polarisation of high-energy (gamma) photons produced in positronium decay substantially limited the performance of this source (the polarisation was estimated through the scattering distribution by means of the Klein–Nishina formula, introducing the hypothesis that this result can be correctly related to the one that would have been obtained by using linear polarisers, using quantum mechanics). For this reason this source was later abandoned.

An alternative was offered by entangled photons produced in atomic cascade decay, which are in the visible region of the spectrum, so that polarisation can easily be selected. However, in this case, the atom carries away part of the momentum and thus photon directions are not well correlated. This led to a severe detection loophole, detection efficiencies typically being under 1%.

The first BI test was performed in 1972 by Freedman and Clauser [16] using polarisation-entangled photon pairs at 551 and 423 nm produced in a $4p^2\,{}^1S_0 \rightarrow 4p4s\,{}^1P_1 \rightarrow 4s^2\,{}^1S_0$ cascade in calcium. Subsequent experiments were realised by Clauser [17] and by Fry and Thompson [18] (photons at 436 and 254 nm produced in $7^3S_1 \rightarrow 6^3P_1 \rightarrow 6^1S_0\,{}^{200}$Hg decay).

These seminal experiments demonstrated a clear violation of Bell inequalities, with the only exception of [19], where a systematic error was later identified in the form of stresses in the walls of the bulb containing the electron gun and mercury vapour.

A First Attempt to Eliminate Spacelike Loopholes

The experiment realised in 1982 in Orsay [20] was the culminating effort of this series of experiments, addressing for the first time the spacelike separation between the two observers.

Here [20] entangled photons had wave lengths of 422.7 and 551.3 nm, respectively. They were generated by cascade decay in calcium-40, $(J = 0) \rightarrow (J = 1) \rightarrow (J = 0)$, particularly suited for coincidence experiments, since the lifetime of the intermediate level is rather short (about 5 ns).

The two entangled photons were then sent to two photomultipliers, preceded by a polarisation measurement set, 6 m from the source. The possibility of achieving a spacelike separation between the two detections was due to the use of rapid acousto-optical switches operating at 50 MHz, addressing on two different paths photons with different selected polarisation.

It must be noted that some concerns about the real elimination of spacelike loopholes still remained, due to the fact that the switching among polarisation measurement bases was not random, but periodic.

The observed violation of BI was by five standard deviations. However, only a very small fraction of generated pairs were really observed, leading to a very severe detection loophole.

7.2.2 A New Source: PDC Experiments

Early Experiments with PDC

After the Orsay experiment, no new experimental development in testing BI followed for a few years. Essentially, photon cascade experiments had reached their limit and new sources were necessary.

The breakthrough was represented by the development of sources of entangled photons based on parametric down-conversion (PDC) at the end of the 1980s, together with the availability of silicon avalanche photodiode detectors (SPAD), with a much higher detection efficiency than earlier phototubes.

PDC is an exclusively quantum effect arising from vacuum fluctuation that occurs in the interaction between a high-energy electromagnetic field and the atoms of a nonlinear dielectric birefringent crystal, such as lithium iodate ($LiIO_3$) or β-barium borate (BBO). This spontaneous emission consists in a very low-probability ($\approx 10^{-9}$) decay of a photon with higher frequency into twin conjugated photons such that the sums of their frequencies and wavevectors correspond to the frequency and wavevector of the decaying photon (i.e., energy and momentum are conserved).

Two types of PDC exist. In type I, PDC photons with the same polarisation are emitted in circumcentric cones. In type II, PDC photons with orthogonal polarisation are emitted in cones shifted due to birefringence.

The type I PDC two-photons state presents a phase and momentum entanglement that can be eventually applied to Bell inequality measurement using interferometers.

On the other hand, polarisation-entangled states can be obtained by recombining two correlated photons emitted from a type I PDC in a beam splitter after rotating the polarisation of one of the two photons.

This scheme allowed a first application of PDC to test BI in 1988 [21, 22]. However, this kind of setup required postselection of the cases when both the photons had exited different ports of the beam splitter, limiting the usefulness in overcoming the detection loophole.

Therefore, further developments of PDC BI tests required either the use of schemes based on interferometry or the development of bright sources of polarisation-entangled photons. These two lines of research will be the subject of the next two subsections.

Interferometric Experiments

The first scheme for testing BI through an interferometer was proposed by Franson [23] and consisted in placing two Mach–Zehnder interferometers (MZI) on the path of the two correlated photons. If the long arms of the interferometers add a tunable phase ϕ_i ($i = 1, 2$ corresponding to interferometer 1,2, respectively) to the short ones, the final state is

$$\Psi_{fr} = \frac{1}{2} \left[|s_1\rangle|s_2\rangle + |l_1\rangle|l_2\rangle e^{i(\phi_1+\phi_2)} + e^{i(\phi_1)}|l_1\rangle|s_2\rangle + e^{i(\phi_2)}|s_1\rangle|l_2\rangle \right], \qquad (7.6)$$

where s, l denote the short and long paths, respectively.

If both photons follow either the short or the long path, they are registered as a coincidence count by the detectors placed at the exits of the MZIs. If not, they are not registered as coincidences (i.e., only 50% of the pairs are selected).

The coincidence rate of photons arriving simultaneously is (a_i being the destruction operator of a photon in mode i)

$$R_c \propto \eta_1\eta_2\langle\Psi_{fr}|a_1^\dagger a_2^\dagger a_1 a_2|\Psi_{fr}\rangle = \frac{1}{4}\eta_1\eta_2[1 + cos(\phi_1 + \phi_2)], \qquad (7.7)$$

where $\eta_1\eta_2$ are quantum efficiencies of the detectors following the MZI on paths 1 and 2, respectively. The striking fact about this equation for the coincidence rate is that it can be modulated perfectly using either of the widely separated phase plates. This "nonlocal" effect can be used for testing Bell inequalities (e.g., as in 7.4) using the phases ϕ_1, ϕ_2.

Franson's scheme was realised for the first time in [24], where a BBO crystal was pumped by an argon ion laser beam in a collinear regime producing type I PDC photon pairs at 916 nm. A beam splitter separated the two photons, which were then addressed to the two MZIs, leading to a seven-standard-deviation BI violation. Later 16-standard-deviations (only inferred from visibility) were achieved in [25, 26].

It is also worth mentioning that a modified scheme for exploiting momentum–phase entanglement was used in a previous experiment [27] where a 10-standard-deviation violation was observed.

As mentioned, all these schemes suffered from 50% pairs postselection.

A setup overcoming this problem, i.e., without the problem of eliminating the long–short terms, was then realised [28] by using type II PDC and polarising beam splitters in the interferometers, reaching $(95.0 \pm 1.4)\%$ visibility (even if no real test of Bell inequalities was made). In this configuration either the horizontally polarised photon follows the long path, while its vertical twin travels on the long one, or both follow short paths.

An important advantage of interferometric schemes is that they can be easily implemented in fiber. This represents an important issue when one wants to propagate photons for long distances either to achieve clear spacelike separation of measurements or for quantum communication application.

A first Franson scheme experiment on long distance entanglement transmission was presented in [29], achieving 86.9% visibility. Of the two entangled photons, the one at 820 nm was addressed to a single-mode fiber interferometer; the other, with a 1.3 μm wavelength,

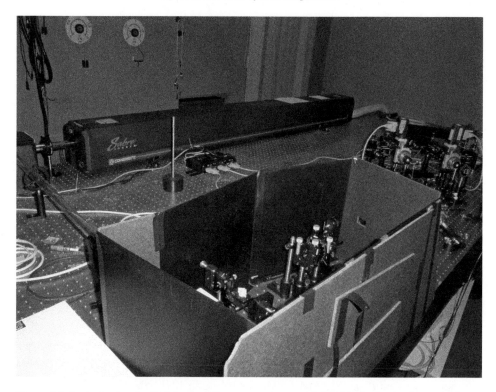

Figure 7.1 Typical setup for Bell inequalities measurement with PDC light. In the foreground one can observe nonlinear crystals where entangled photons are generated by the pump beam emitted by a laser (in the background). On the right-hand side are the polarisation selection apparatus and photodetectors.

was propagated through a 4.3 km single-mode telecom communication fiber before reaching the interferometer.

In a following experiment, a separation longer than 10 km [30, 31] allowed definitively closing the locality loophole (assuming that the passive coupler randomly selects which interferometer analyzes the photon) at the same time with a polarisation entanglement scheme [32] (where real random selection was done). In this case a CHSH inequality violation of $S = 2.92 \pm 0.18$ was achieved. Finally, a following upgrade [33] allowed realising measurements whose temporal order was invertible by changing the reference frame.

Bright Sources of Polarisation-Entangled Photons

The development of bright sources of polarisation-entangled photons was the following significant step in the experimental path toward a final test of BI (see Fig. 7.1).

In the middle of the 1990s two ideas circulated. One, proposed by Hardy [34], consisted in superimposing the emissions, with orthogonal polarisation, of two type I crystals;

the other in selecting the intersections of two degenerate (i.e., with the same wavelength) emission cones (of orthogonal polarisation) in type II PDC [35].

For type II sources, the first example was realised in a collinear regime. Here the two degenerate photons are emitted in two tangent cones then by selecting their intersection point the two orthogonally polarized correlated photons exit in the same direction and can be separated by a beam splitter, generating an entangled state when one postselects events in which they have exited the beam splitter in different directions

This setup allowed a 10-standard-deviation violation of BI [4].

However, in the collinear regime, due to postselection after the beam splitter, only 50% of original pairs are selected. Therefore, it is advantageous to use noncollinear schemes [36], selecting the two intersections of orthogonally polarised cones. It must be noticed that the state deriving from this superposition is not yet entangled, since, due to birefringence in the nonlinear crystal, ordinary and extraordinary photons propagate with different velocities and different directions inside the medium. Thus, longitudinal and transverse walk-offs (i.e., optical beam displacement due to birefringence) must be compensated for, restoring indistinguishability between the two polarisations and producing an entangled state. This can be achieved by inserting some birefringent medium along the optical path of the photons.

Ref. [37] realised the first noncollinear type II PDC set up by pumping a 3-mm-long BBO crystal with a 351 argon laser beam. The longitudinal (385 fs) walkoff (the transverse one being negligible) was compensated for by an additional BBO crystal. All four Bell states, were generated leading to a 102-standard-deviation violation of the CHSH inequality ($S = -2.6489 \pm 0.0064$).

In the following years, several very bright sources were realised [38–41], up to dozens of coincidence counts per second for mW pump power ($77\,s^{-1}$ in [38]) in a traditional crystal and even up to a measured coincidence flux of hundreds per second for mW pump power using periodically poled crystals (i.e., crystals whose susceptibility is periodically modulated, producing a constructive interference in the emission), e.g., $300\,s^{-1}$ in [39].

These bright sources found application both in closing the spacelike loophole without any doubt [32] and in long-distance open air transmission and testing of BI (more than 600 m with $S = 2.41 \pm 0.10$ in [41], 13 km in [42] with $S = 2.45 \pm 0.09$, and finally 144 km in [43]).

As hinted, in developing bright sources of polarised-photon-entangled states, an alternative to type II PDC stems from superimposing the emissions, with orthogonal polarisations, of two type I PDC crystals whose optical axes are orthogonal [34], generating the state (eventually nonmaximally-entangled when $f \neq 1$ [39])

$$|\psi_{NME}\rangle = \frac{|H\rangle|H\rangle + f|V\rangle|V\rangle}{\sqrt{(1 + |f|^2)}}, \qquad (7.8)$$

where the value of the parameter f can be tuned by tuning the pump beam properties.

In 1999 two different setups realising this scheme and demonstrating large BI violation were produced. In the first, the emission of two thin adjacent 0.59-mm-long type I BBO crystals was superimposed [44], achieving $S = 2.7007 \pm 0.0029$. The scheme is very simple and thus found widespread use. Nevertheless, the short length of the crystals is dictated by the need to have good superposition of the subsequent emissions.

The second was based on a more complicated set where the emissions of two 1-cm-long LiIO$_3$ crystals were superimposed by an optical condenser [45].

These two setups were realized using a continuous wave pump laser. However, for timing reasons, the pulsed regime is preferable. When the pump pulses are very short (femtosecond regime), the photon pair production amplitudes corresponding to different regions inside the crystal do not sum coherently [46]. This initially required either using thin ($\approx$100 μm) nonlinear crystals [47] or narrowband spectral filters (to increase the coherence length) [48–50]. However, these solutions significantly reduce the available flux of entangled photon pairs.

More recently, bright sources in the pulsed regime were realised by inserting nonlinear crystals into interferometers [51–54] (such as a Sagnac interferometer).

Altogether, efficient bright sources of bipartite polarisation-entangled photons became available in the second part of the 1990s. Very-high-visibility experiments were realised, measuring BI violation close to the theoretical bound (Cirel'son bound); see Fig. 7.2.

The spacelike loophole was clearly eliminated and other loopholes concerning the statistics of the samples [55] did not appear to be a major concern, since clear elimination of the detection loophole would largely solve these issues as well.

Thus, the last step toward a conclusive test of BI remained the elimination of the detection loophole (at the same time as the spacelike one), at most 30% total detection efficiency having been approached in these latest experiments: thus some space remains for specific LRT [2, 9, 56] exploiting this loophole (and some of them were falsified by specific experimental tests [2, 57]).

Two possible directions were considered (the possibility of using higher dimensional systems requiring lower detection efficiencies being substantially limited by the difficulty of producing these states): either to abandon photons for other systems or to look for high-detection-efficiency photo detectors.

7.2.3 Towards a Conclusive Test of Bell Inequalities

Test of Local Realism with Atoms

Even if most tests of local realism performed up to now have been realised with photons, several other systems have been considered. Some of them have mainly historical interest, since the problems connected with their use stopped the research in this sense. Among them we can mention polarisation correlation of $^1S_0$ proton pairs produced in nuclear reactions [58] or K mesons [59–77] (where the detection loophole reappears due to decay channel selection [78–80]). In other cases they are still very far from a conclusive test, as with neutron interferometry [81], where large losses in interferometers make it impossible to

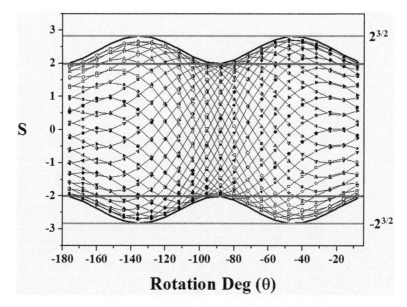

Rotation Deg (θ)

Figure 7.2 Measured values of the CHSH parameter S (F. Bovino, I.P. Degiovanni concession) are plotted versus θ and the various curves are associated with different values of the parameter ξ (used to parameteize the entangled states $|\varphi(\xi)\rangle = \cos(\xi)|\phi^+\rangle + \sin(\xi)|\psi^-\rangle$). The thicker curves correspond to the theoretically predicted S bounds [94]. Plotted points are the measured values [95] of S for any pair $\{\xi, \theta\}$. Good qualitative and quantitative agreement is seen between theoretical and experimental bounds, even if the experimental upper (lower) bounds stands slightly below (above) the theoretical predictions. These effects are, as usual, imputable to noise and imperfections associated with the polarization preservation and measurement of same-setup components, namely HWPs, PBSs, and fibers.

eliminate the detection loophole, or atoms entangled in a superposition involving two circular Rydberg states produced by single-photon exchange in a high-Q cavity [82] where it is difficult to have high purity of states.

Also, entangled superconducting systems, even if they allowed the detection loophole to be closed [83], are probably not the best choice for eliminating the spacelike loophole at the same time.

The use of entangled atoms or photon–atom entangled states is of greater interest [84] (which should nonetheless solve the detection efficiency problem for the photon).

In principle, atomic states can be detected with very high efficiency, and, indeed, in an experiment with Be$^+$ ions [85], a clear violation of BI was observed with a detection efficiency of $\approx 98\%$. However, in this case the measurement of the two ions was not disjoined. More recently, Yb$^+$ in two traps separated by 1 m [86] and 20 m [87] as used. Nevertheless, to eliminate the spacelike loophole, at least 300 m would be required.

In principle, these schemes could lead to a conclusive test of BI, in particular in configurations where distant atoms are entangled by exchanging entangled photons.

7.2.4 The Elimination of the Detection Loophole in Photon Experiments

As mentioned, the most important problem of BI tests with photons is the detection efficiency of SPADs, which, although far larger than that of phototubes, is substantially limited to be smaller than 70%, preventing, also in principle, a final test of BI without a fair sampling assumption.

A possible way out of this problem is represented by cryogenic detectors, which in principle can have a detection efficiency close to 1.

After a decade without specific progress in testing BI with photons, recently two experiments have claimed to have overcome the detection loophole by exploiting transition-edge sensors (TESs), i.e., microcalorimeters based on a superconducting thin film operating next to transition and working as a very sensitive thermometer. TESs can reach very high detection efficiency, have discrimination in number of impinging photons, and are substantially free from dark counts.

In the first experiment [88], degenerate pairs of photons were produced in a Sagnac type II PDC source where a periodically poled KTP was pumped by a laser beam at 405 nm. Nonmaximally-entangled states (with a weight $f \approx 0.3$) were used to reduce the detection efficiency as needed to eliminate the detection loophole. After a polarisation measurement, performed with a half-wave plate followed by a calcite polariser, both photons were addressed through fibres to two TESs (with a total arm efficiency of 73.8% and 78.6%, respectively).

Possible coincidence and production rate loopholes, which were suggested for these experiments, were later shown not to be a real problem [89].

In the second experiment [90], nonmaximally-entangled pairs of photons (with a weight $f \approx 0.26$) were produced by pumping two thin type I BBO crystals with a 355 nm laser beam pulsed (avoiding in this way problems with coincidence loopholes) at a 25 kHz repetition rate through a Pockels cell. Again, entangled photons were detected by TES detectors.

Of course, the task of eliminating at the same time detection and spacelike loopholes is formidable, since propagating the photons at a sufficient distance to eliminate the spacelike loophole unavoidably compromises the overall channel efficiency.

Nevertheless, both groups seem to proceed in this direction. Thus, there are chances that a conclusive experiment on BI will arrive from photon experiments in the next few years.

7.3 Conclusions

In the fifty years since Bell's paper, continuous progress has led from seminal experimental tests of Bell inequalities to several experiments with a high statistical violation of them.

This huge experimental work leaves very few doubts on the fact the world described by quantum mechanics is really nonlocal, and nowadays this "spooky action at a distance" has become a common notion not only among physicists, but diffused in the larger scientific community, being now a resource for quantum technologies.

Even if a very small space is left for a different result, a completely loophole-free violation of the Bell inequality is still missing. Nonetheless, as described in this contribution, a conclusive result could be realised within a few years.

In conclusion, Bell inequalities and their experimental tests have represented a fundamental contribution to our understanding of nature, confirming the most paradoxical aspect of quantum mechanics, nonlocality, and making it possible to exclude any possible deterministic extension of quantum mechanics.

Furthermore, quantum correlations not only have played a fundamental role in the discussion on foundations of quantum mechanics [91], but also now represent a fundamental tool for developing quantum technologies, and in this field BI have often become a tool for testing quantum resources in ordinary laboratory work.

If Bell inequalities have played an exponentially growing role in the last 50 years in physics, everything suggests that their role will keep on growing in the next 50, both in application to quantum technologies and as an instrument for understanding the secrets of nature.

Felix qui potuit rerum cognoscere causas.

Appendix

After this work was submitted, two experiments realized a loophole-free test of Bell inequalities [92], overcoming some statistical limitations of a slightly earlier work [93].

In a little more detail, we have seen that a few conditions emerged that must be fulfilled in order to perform a conclusive test of the Bell inequality.

The main ones are as follows:

- The two measurements must be set independently (locality loophole).
- The choice of the setting must be truly random (freedom-of-choice loophole).
- One should be able to detect all the pairs involved in the experiment or, at least, a sufficiently large fraction of them (detection loophole).

Furthermore,

- The number of emitted particles must be independent by measurement settings (production rate loophole).
- The presence of a coincidence window in a hidden variable scheme must not allow a situation where a local setting may change the time at which the local event happens (coincidence loophole).
- An eventual memory of previous measurements must be considered in the statistical analysis, since the data can be nonindependent and identically distributed (memory loophole).

When all these conditions are satisfied, no room is left for local realistic hidden variable theories. Realistic theories not excluded by such an experiment would require very unusual "conspiracies," such as superdeterminism (i.e., all the evolution of physical systems are predetermined, including, e.g., generation of random numbers).

The two mentioned works realized all these conditions:

- The two measuring apparatus were placed at several meters distance. A careful analysis of the spacelike separation (taking into account delays in transmission, etc.) of setting choices and measurements was done. Thus, the locality loophole is overcome.
- The use of high-detection-efficiency superconducting detectors together with non-maximally-entangled states (as suggested by Eberhard) allowed a detection-loophole-free experiment.
- Independent random number generators based on laser phase diffusion guaranteed the elimination of the freedom-of-choice loophole (except in the presence of super determinism or other hypotheses that, by definition, do not allow a test through Bell inequalities).
- A perfect random choice of settings, as realized, did not permit a production rate loophole.
- The use of a pulsed source eliminated the coincidence loophole.
- An involved statistical analysis did not leave room for a memory loophole.

In summary, these experiments unequivocally tested Bell inequalities without any additional hypotheses.

References

[1] J.S. Bell, On the Einstein–Podolsky–Rosen paradox, *Physics* (1964), 195.
[2] M. Genovese, Research on hidden variable theories: A review of recent progresses, *Phys. Rep.* **413** (2005), 319.
[3] N. Brunner, D. Cavalcanti, S. Pironio, V. Scarani, and S. Wehner, Bell nonlocality, *Rev. Mod. Phys.* **86** (2014), 419.
[4] T.E. Kiess, Y.H. Shih, A.V. Sergienko, and C.O. Alley Einstein–Podolsky–Rosen–Bohm experiment using pairs of light quanta produced by type II parametric down-conversion, *Phys. Rev. Lett.* **71** (1993), 3893.
[5] J. Clauser et al., Proposed experiment to test local hidden-variable theories, *Phys. Rev. Lett.* (1969), 880.
[6] E. Santos, Unreliability of performed tests of Bell's inequality using parametric down-converted photons, *Phys. Lett. A* **212** (1996), 10.
[7] L. De Caro and A. Garuccio, Bell's inequality, trichotomic observables, and supplementary assumptions, *Phys. Rev. A* **54** (1996), 174.
[8] J.F. Clauser and M.A. Horne, Experimental consequences of objective local theories, *Phys. Rev. D* **10** (1974), 526.
[9] P.M. Pearle, Hidden-variable example based upon data rejection, *Phys. Rev. D* **2** (1970), 1418.
[10] P.H. Eberhard, Background level and counter efficiencies required for a loophole-free Einstein–Podolsky–Rosen experiment, *Phys. Rev. A* **47** (1993), R747.
[11] J. Larson and R. Gill, Bell's inequality and the coincidence–time loophole, *Eur. Phys. Lett.* **67** (2004), 707.

[12] L.R. Kasday, J.D. Ullman and C.S. Wu, *Nuovo Cim. B* **25** (1975), 633.

[13] G. Faraci, D. Gutkowski, S. Notarrigo, and A.R. Pennisi, *Lett. Nuov. Cim.* **9** (1974), 607.

[14] A.R. Wilson et al., Measurement of the relative planes of polarization of annihilation quanta as a function of separation distance, *J. Phys. G.* **2** (1976), 613.

[15] M. Bruno et al., *Nuovo Cim. B* **40** (1977), 142.

[16] S.J. Freedman and J.F. Clauser, Experimental test of local hidden-variable theories, *Phys. Rev. Lett.* (1972), 938.

[17] J.F. Clauser, Experimental tests of realistic local theories via Bell's theorem, *Phys. Rev. Lett.* **36** (1976), 1223.

[18] E.S. Fry and R.C. Thompson, Experimental test of local hidden-variable theories, *Phys. Rev. Lett.* **37** (1976), 465.

[19] R.A. Holt and F.M. Pipkin, unpublished (1973).

[20] A. Aspect, J. Dalibard, and G. Roger, Experimental test of Bell's inequalities using time-varying analyzers, *Phys. Rev. Lett.* **49** (1982), 1804.

[21] Y.H. Shih, A.V. Sergienko, and M.H. Rubin, Einstein–Podolsky–Rosen state for space–time variables in a two-photon interference experiment, *Phys. Rev. A* **47** (1993) 1288.

[22] Z.J. Ou and L. Mandel, Violation of Bell's inequality and classical probability in a two-photon correlation experiment, *Phys. Rev. Lett.* **61** (1988), 50.

[23] J.P. Franson, Bell inequality for position and time, *Phys. Rev. Lett.* **62** (1989), 2205.

[24] J. Brendel, E. Mohler, and W. Martienssen, Experimental test of Bell's inequality for energy and time, *Eur. Phys. Lett.* **20** (1992), 275.

[25] P.G. Kwiat, A.M. Steinberg, and R.Y. Chiao, High-visibility interference in a Bell-inequality experiment for energy and time, *Phys. Rev. A* **47** (1993), R2472.

[26] R.Y. Chiao, P.G. Kwiat, and A.M. Steinberg, Quantum non-locality in two-photon experiments at Berkeley, *Quantum Semiclass. Opt.* **7** (1995), 259.

[27] J.G. Rarity and P.R. Tapster, Experimental violation of Bell's inequality based on phase and momentum, *Phys. Rev. Lett.* **64** (1990), 2495.

[28] D.V. Strekalov et al., Postselection-free energy-time entanglement, *Phys. Rev. A* **54** (1996), R1.

[29] P.R. Tapster, J.G. Rarity, and P.C.M. Owens, Violation of Bell's inequality over 4 km of optical fiber, *Phys. Rev. Lett.* **73** (1994), 1923.

[30] W. Tittel, J. Brendel, H. Zbinden, and N. Gisin, Violation of Bell inequalities by photons more than 10 km apart, *Phys. Rev. Lett.* **81** (1998), 3563.

[31] W. Tittel, J. Brendel, N. Gisin, and H. Zbinden, Long-distance Bell-type tests using energy-time entangled photons, *Phys. Rev. A* **59** (1999), 4150.

[32] G. Weihs et al., Violation of Bell's inequality under strict Einstein locality conditions, *Phys. Rev. Lett.* **81** (1998), 5039.

[33] H. Zbinden et al., Experimental test of relativistic quantum state collapse with moving reference frames, *J. Phys. A Math. Gen.* **34** (2001), 7103.

[34] L. Hardy, Nonlocality for two particles without inequalities for almost all entangled states, *Phys. Lett. A* **161** (1992), 326.

[35] A. Garuccio, in D. Greenberger (ed.), "Fundamental Problems in Quantum Theory" (New York Academy of Sciences, 1995).

[36] P. Kwiat, P.H. Eberhard, A.M. Steinberg, and R.Y. Chiao, New high-intensity source of polarization-entangled photon pairs, *Phys. Rev. A* **49** (1994), 3209.

[37] P.G. Kwiat et al., New high-intensity source of polarization-entangled photon pairs, *Phys. Rev. Lett.* **75** (1995), 4337.

[38] C. Kurtsiefer, M. Oberparleiter and H. Weinfurter, High-efficiency entangled photon pair collection in type-II parametric fluorescence, *Phys. Rev. A* **64** (2001), 023802.

[39] C.E. Kuklewicz et al., High-flux source of polarization-entangled photons from a periodically poled KTiOPO4 parametric down-converter, *Phys. Rev. A* **69** (2004), 013807.

[40] G. Di Giuseppe et al., Entangled-photon generation from parametric down-conversion in media with inhomogeneous nonlinearity, *Phys. Rev. A* **66** (2002), 013801.

[41] M. Aspelmeyer et al., Long-distance free-space distribution of quantum entanglement, *Science* **301** (2003), 621.

[42] C.-Z. Peng et al., Experimental free-space distribution of entangled photon pairs over 13 km: Towards satellite-based global quantum communication, *Phys. Rev. Lett.* **94** (2005), 150501.

[43] R. Ursin et al., quant-ph 0607182; *Nat. Phys.* 3 (2007), 481.

[44] A.G. White et al., Nonmaximally entangled states: Production, characterization, and utilization, *Phys. Rev. Lett.* **83** (1999), 3103.

[45] G. Brida et al., A first test of Wigner function local realistic model, *Phys. Lett. A* **299** (2002), 121.

[46] T.E. Keller and M.H. Rubin, Theory of two-photon entanglement for spontaneous parametric down-conversion driven by a narrow pump pulse, *Phys. Rev. A* **56** (1997), 1534.

[47] A.V. Sergienko et al., Quantum cryptography using femtosecond-pulsed parametric down-conversion, *Phys. Rev. A* **60** (1999), R2622.

[48] W.P. Grice and I.A. Walmsley, Spectral information and distinguishability in type-II down-conversion with a broadband pump, *Phys. Rev. A* **56** (1997), 1627.

[49] W.P. Grice et al., Spectral distinguishability in ultrafast parametric down-conversion, *Phys. Rev. A* **57** (1998), R2289.

[50] G. Di Giuseppe et al., Quantum interference and indistinguishability with femtosecond pulses, *Phys. Rev. A* **56** (1997), R21.

[51] Y. Kim, S.P. Kulik, and Y. Shih, Generation of polarisation-entangled photon pairs at 1550 nm using two PPLN waveguides, *Phys. Rev. A* **63** (2001), 060301(R).

[52] H. Poh et al., Eliminating spectral distinguishability in ultrafast spontaneous parametric down-conversion, *Phys. Rev. A* **80** (2009), 043815.

[53] M. Barbieri, F. De Martini, G. Di Nepi, and P. Mataloni, Violation of Bell inequalities and quantum tomography with pure-states, Werner-states and maximally entangled mixed states created by a universal quantum entangler, quant-ph 0303018.

[54] C. Cinelli, M. Barbieri, F. De Martini, and P. Mataloni, Experimental realization of hyper-entangled two-photon states, quant-ph/ 0406148.

[55] A. Khrennikov, What does probability theory tell us about Bell's inequality? *Adv. Sci. Lett.* **2** (2009), 488.
L. Accardi, *Foundations of Probability and Physics-5*, L. Accardi et al. (eds.), AIP Conference Proceedings 1101 (2009, New York), p. 3.
C. Garola and S. Sozzo, The ESR model: A proposal for a noncontextual and local Hilbert space extension of QM, *Eur. Phys. Lett.*, **86** (2009), 20009.

[56] N. Gisin and B. Gisin, A local hidden variable model of quantum correlation exploiting the detection loophole, *Phys. Lett. A* **260** (1999), 323.
T.W. Marshal and E. Santos, Stochastic optics: A reaffirmation of the wave nature of light, *Found. Phys.* **18** (1988), 185. A. Casado et al., Dependence on crystal parameters of the correlation time between signal and idler beams in parametric down conversion calculated in the Wigner representation, *Eur. Phys. J. D* **11** (2000), 465.

A. Casado et al., Spectrum of the parametric down converted radiation calculated in the Wigner function formalism, *Eur. Phys. J. D* **13** (2001) 109.

[57] G. Brida, M. Genovese, M. Gramegna, and E. Predazzi, A conclusive experiment to throw more light on 'light', *Phys. Lett. A* **328** (2004), 313; G. Brida, M. Genovese and F. Piacentini G. Brida, M. Genovese, M. Gramegna, and E. Predazzi, Experimental local realism tests without fair sampling assumption, *Eur. Phys. J. D* **44** (2007), 577.

[58] M. Lamehi-Rachti and W. Mittig, Quantum mechanics and hidden variables: A test of Bell's inequality by the measurement of the spin correlation in low-energy proton-proton scattering. *Phys. Rev. D* **14** (1976), 2543.

[59] F. Uchiyama, Generalized Bell inequality in two neutral kaon systems, *Phys. Lett. A* **231** (1997), 295.

[60] R.A. Bertlmann, W. Grimus and B.C. Hiesmayr, Bell inequality and CP violation in the neutral kaon system, *Phys. Lett. A* **289** (2001), 21.

[61] F. Benatti and R. Floreanini, Bell's locality and ϵ'/ϵ, *Phys. Rev. D* **57** (1998), R1332.

[62] F. Benatti and R. Floreanini, Direct CP-violation as a test of quantum mechanics, *Eur. Phys. J. C* **13** (2000), 267.

[63] R.A. Bertlmann and W. Grimus, Quantum-mechanical interference over macroscopic distances in the B0B0 system, *Phys. Lett. B* **392**, (1997), 426; R.A. Bertlmann and W. Grimus, How devious are deviations from quantum mechanics: The case of the B0B0 system, *Phys. Rev. D* **58** (1998), 034014.

[64] G.C. Ghirardi et al., The DAΦNE Physical Handbook, L. Maiani, G. Pancheri, and N. Paver (eds.) (INFN, Frascati, 1992), Vol. I.

[65] J. Six, Test of the nonseparability of the $K^0 K^0$ system, *Phys. Lett. B* **114**, (1982) 200.

[66] P.H. Eberhard, Testing the non-locality of quantum theory in two-kaon systems, *Nucl. Phys. B* **398** (1993), 155.

[67] A. Di Domenico, Testing quantum mechanics in the neutral kaon system at a ϕ−factory, *Nucl. Phys. B* **450** (1995), 293.

[68] B. Ancochea et al., Bell inequalities for K0K0 pairs from ϕ-resonance decays, *Phys. Rev. D* **60** (1999), 094008.

[69] A. Bramon and M. Nowakowski, Bell inequalities for entangled pairs of neutral kaons, *Phys. Rev. Lett.* **83** (1999), 1.

[70] N. Gisin and A. Go, EPR test with photons and kaons: Analogies, *Am. J. Phys.* **69** (2001), 264.

[71] B.C. Hiesmayr, A generalized Bell inequality and decoherence for the $K^o K^o$, *Found. Phys. Lett.* **14** (2001), 231.

[72] R.A. Bertlmann, A. Bramon, G. Garbarino, and B.C. Hiermayr, Violation of a Bell inequality in particle physics experimentally verified? *Phys. Lett. A* **332** (2004), 355. A. Bramon et al., Bell's inequality tests: From photons to B, mesons. quant-ph0410122.

[73] R.H. Dalitz and G. Garbarino, Local realistic theories and quantum mechanics for the two-neutral-kaon system, quant-ph 0011108.

[74] P. Privitera and F. Selleri, Quantum mechanics versus local realism for neutral kaon pairs, *Phys. Lett. B* **296** (1992), 261.

[75] F. Selleri, Incompatibility between local realism and quantum mechanics for pairs of neutral kaons, *Phys. Rev. A* **56** (1997), 3493.

[76] A. Pompili and F. Selleri, On a possible EPR experiment with $B_d^0 B_d^0$ pairs, *Eur. Phys. J. C* **14** (2000), 469.

[77] A. Bramon and G. Garbarino, Novel Bell's inequalities for entangled $K0\ \bar{K}0$ pairs, *Phys. Rev. Lett.* **88** (2002), 040403.

[78] M. Genovese, C. Novero, and E. Predazzi, Can experimental tests of Bell inequalities performed with pseudoscalar mesons be definitive? *Phys. Lett. B* **513** (2001), 401.

[79] M. Genovese, C. Novero, and E. Predazzi, On the conclusive tests of local realism and pseudoscalar mesons, *Found. Phys.* **32** (2002), 589.

[80] M. Genovese, Entanglement properties of kaons and tests of hidden-variable models, *Phys. Rev. A* **69** (2004), 022103.

[81] Y. Hasegawa et al., Violation of a Bell-like inequality in single-neutron interferometry, *Nature* **425** (2003), 45.

[82] E. Hagley et al., Generation of Einstein-Podolsky-Rosen pairs of atoms, *Phys. Rev. Lett.* **79**, 1 (1997).

[83] M. Ansmann et al., Violation of Bell's inequality in Josephson phase qubits, *Nature* **461** (2009), 504.

[84] D.L. Moehring, M.J. Madsen, B.B. Blinov, and C. Monroe, Experimental Bell inequality violation with an atom and a photon, *Phys. Rev. Lett.* **93** (2004), 090410.

[85] M.A. Rowe et al., Experimental violation of a Bell's inequality with efficient detection, *Nature* **409** (2001), 791.

[86] D.N. Matsukevich, P. Maunz, D.L. Moehring, S. Olmschenk, and C. Monroe, Bell inequality violation with two remote atomic qubits, *Phys. Rev. Lett.* **100** (2008), 140404.

[87] J. Hofmann et al., Heralded entanglement between widely separated atoms, *Science* **337** (2012), 72.

[88] M. Giustina et al., Bell violation using entangled photons without the fair-sampling assumption, *Nature* **497** (2013), 227.

[89] J. Larsson et al., Bell-inequality violation with entangled photons, free of the coincidence-time loophole, *Phys. Rev. A* **90** (2014), 032107.

[90] B.G. Christensen et al., Detection-loophole-free test of quantum nonlocality, and applications, *Phys. Rev. Lett.* **111** (2013), 130406.

[91] G. Auletta (ed.), *Foundations and Interpretations of Quantum Mechanics* (World Scientific, Singapore, 2000).

[92] M. Giustina et al., *Phys. Rev. Lett.* **115**, 250401 (2015); Lynden K. Shalm et al., Phys. Rev. Lett. **115**, 250402 (2015).

[93] B. Hensen et al., *Nature* **526**, 682 (2015).

[94] A. Cabello, Proposed experiment to test the bounds of quantum correlations, *Phys. Rev. Lett.* **92** (2004), 060403.

[95] F.A. Bovino, G. Castagnoli, I.P. Degiovanni, and S. Castelletto, Experimental evidence for bounds on quantum correlations, *Phys. Rev. Lett.* **92** (2004), 060404.

8

Bell's Theorem without Inequalities: On the Inception and Scope of the GHZ Theorem

OLIVAL FREIRE JR. AND OSVALDO PESSOA JR.

8.1 Introduction

Since its inception, fifty years ago, Bell's theorem has had a long history not only of experimental tests but also of theoretical developments. Studying pairs of correlated quantum-mechanical particles separated in space, in a composite "entangled" state, Bell [1] showed that the joint ascription of hidden variables and locality to the system led to an inequality that is violated by the predictions of quantum mechanics. Fifteen years later, experiments confirmed the predictions of quantum mechanics, ruling out a large class of local realist theories.

One of the most meaningful theoretical developments that followed Bell's work was the Greenberger–Horne–Zeilinger (GHZ) theorem, also known as Bell's theorem without inequalities. In 1989, the American physicists Daniel Greenberger and Michael Horne, who had been working on Bell's theorem since the late 1960s, together with the Austrian physicist Anton Zeilinger, introduced a novelty into the testing of entanglement, extending Bell's theorem in a different and interesting direction. According to Franck Laloë [2],

For many years, everyone thought that Bell had basically exhausted the subject by considering all really interesting situations, and that two-spin systems provided the most spectacular quantum violations of local realism. It therefore came as a surprise to many when in 1989 Greenberger, Horne, and Zeilinger (GHZ) showed that systems containing more than two correlated particles may actually exhibit even more dramatic violations of local realism.

The trio analyzed the Einstein, Podolsky, and Rosen 1935 argument once again and were able to write what are now called Greenberger–Horne–Zeilinger (GHZ) entangled states, involving three or four correlated spin-$1/2$ particles, leading to conflicts between local realist theories and quantum mechanics. However, unlike Bell's theorem, the conflict now was not of a statistical nature, as was the case with Bell's theorem, insofar as measurements on a single GHZ state could lead to conflicting predictions with local realistic models [3–5].[1] In this paper we present the history of the creation of this theorem and analyze its scope.

---

[1] The first paper is the original presentation of the GHZ theorem. The second paper, co-authored with Abner Shimony, contains a more detailed presentation of the theorem and its proof and suggests possible experiments, including momentum and energy correlations among three and more photons produced through parametric down-conversion. The third paper presents Greenberger's recollections of the background of the original paper. On the background of the GHZ theorem, see also [7, pp. 236–44].

8.2 The Men behind the GHZ Theorem

In the early 1980s, Greeenberger, Horne, and Zeilinger began to collaborate around the subject of neutron interferometry at the Massachusetts Institute of Technology (MIT), but they came from different backgrounds concerning the research on the foundations of quantum mechanics. As early as 1969, while doing his PhD dissertation at Boston University under the supervision of Abner Shimony, Horne began to work on foundations. His challenge was to carry Bell's theorem to a laboratory test, in which he succeeded, but not alone. What is now called the CHSH paper [6], which is the adaptation of Bell's original theorem to a real-life experiment, resulted from the collaboration of Shimony and Horne with John Clauser, an experimental physicist who was independently working on the same subject at the University of California at Berkeley, and Richard Holt, who was doing his PhD at Harvard under Francis Pipkin on the same issue. The CHSH paper triggered a string of experimental tests, which continue to date, leading ultimately to the wide acknowledgment of entanglement as a new physical effect. In the early 1970s, however, Horne thought this subject was dead and turned to work with neutrons while teaching at Stonehill College, near Boston.

Zeilinger's interests in foundational issues flourished while he studied at the University of Vienna, and were favored by the flexible curriculum at this university at that time and by the intellectual climate of physics in Vienna – with its mix of science and philosophy – a legacy coming from the late nineteenth century. In addition, working under the supervision of Helmut Rauch on neutron interferometry, he benefited from Rauch's support for research on the foundations of quantum mechanics.[2] In 1976, as a consequence of this interest shared with Rauch, Zeilinger went to a conference in Erice, Italy, organized by John Bell, fully dedicated to the foundations of quantum mechanics. Over there, while the hottest topic was Bell's experiments, Zeilinger talked about neutron interferometry. He presented a report on experiments held by Rauch's team confirming a counterintuitive quantum prediction. This prediction states that a neutron quantum state changes its sign after a 2π rotation, only recovering the original sign after a 4π rotation [8]. Zeilinger went to Erice unaware of entanglement but came back fascinated by the subject.

Greenberger was a high-energy theorist working at the City College of New York when he decided to move to a domain where connections between quantum mechanics and gravity could be revealed. He chose neutron experiments precisely to look for these connections and met Zeilinger and Horne at a neutron conference in Grenoble [5]. In the early 1970s, Clifford Shull's laboratory at MIT became the meeting point for the GHZ trio. Shull was working on neutron scattering, which earned him the 1994 Nobel Prize. Zeilinger was there as a postdoctoral student and later as a visiting professor. Horne taught at Stonehill College and conducted research at MIT. Greenberger was always around the lab. Basic experiments with neutron interferometry were the bread and butter of the trio.

For Zeilinger, the GHZ theorem was a reward for a risky professional change. In the mid-1980s he had decided to leave neutron interferometry to build a research program in

---

[2] Anton Zeilinger, interviewed by Olival Freire, 30 June 2014, American Institute of Physics (AIP). For Rauch's research on neutron interferometry and its relation to foundational issues, see the review [9].

quantum and atomic optics from scratch. Zeilinger was supported by the Austrian Science Foundation and was looking for the basics in the new field. Sometimes he did this through interaction with other teams, such as Leonard Mandel's in Rochester. Reasons for this choice were related to his understanding that these fields offered more opportunities than neutron interferometry. He was increasingly attracted by the foundations of quantum physics and particularly by the features of entanglement.

The collaboration among the trio started with joint work involving only Zeilinger and Horne and concerned quantum two-particle interference. Indeed, this topic, nowadays a hallmark of quantum light behavior, was independently exploited by two teams and arose in physics through two different and independent paths: on one hand, via quantum optics, and on the other, via scientists who were working on neutron interferometry, such as Anton Zeilinger and Michael Horne. In 1985, unaware of the achievements in the quantum optics community, Zeilinger and Horne tried to combine the interferometry experiments they were doing with Bell's theorem. Then they suggested a new experiment with Bell's theorem using light, but instead of using correlation among polarizations (internal variables) they used linear momenta. They concluded that the quantum description of two-particle interferometry was completely analogous to the description of singlet spin states used by Bell [10]. However, they did not know how to produce such states in laboratories, because they "didn't know where to get a source that would emit pairs of particles in opposite directions." When Horne read the Ghosh–Mandel paper [11], they sent their paper to Mandel, who reacted by saying, "This is so much simpler than the way we describe it, you should publish it." They called Horne's former supervisor, Abner Shimony, and wrote the "two-particle interferometry" paper explaining "the fundamental ideas of the recently opened field of two-particle interferometry, which employs spatially separated, quantum mechanically entangled two-particle states" [12].[3]

While neutron interferometry had been the common ground of collaboration among Zeilinger, Greenberger, and Horne, the GHZ theorem was a result that had implications far beyond their original interest. The first intuition related to the GHZ theorem came from Greenberger, who asked Zeilinger and Horne, "Do you think there would be something interesting with three particles that are entangled? Would there be any difference, something new to learn with a three-particle entanglement?" The work matured while he spent a sabbatical in Vienna working with Zeilinger. "I have a Bell's theorem without inequalities," was the manner in which he reported his results to Horne.[4]

8.3 Reception of the GHZ Theorem

The GHZ theorem was ready in 1986, but the result remained unpublished for three years. In 1988 Greenberger presented the proof at a conference at George Mason University (which led to publication the following year), and in 1989 it was presented at a conference in Sicily

---

[3] The 1985 Horne and Zeilinger paper was prepared for a conference in Joensuu, Finland, dedicated to the 50th anniversary of the EPR paper. This was the first of Zeilinger's papers to deal with Bell's theorem. Interview with Michael Horne by Joan Bromberg, 12 September 2002, AIP. Interview with Anton Zeilinger by Olival Freire Jr., 30 June 2014, AIP.

[4] Interview with Michael Horne by Joan Bromberg, 12 September 2002, AIP. Interview with Anton Zeilinger by Olival Freire Jr., 30 June and 2 July 2014, AIP.

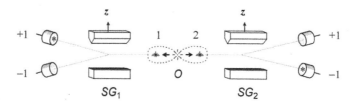

Figure 8.1 Two spin-$\frac{1}{2}$ particles emitted in an entangled state with perfect anticorrelation.

in honor of Werner Heisenberg, attended by N. David Mermin, from the United States, and Michael Redhead, from England, who began to publicize it [13, 14].[5] Zeilinger recalls when he met Bell at a conference in Amherst, in 1990, how enthusiastic Bell was about the result. Unfortunately, Bell's untimely death in that same year prevented him from following the subject. After this initial favorable reception, Greenberger, Horne, and Zeilinger realized the subject deserved a better presentation than simply a conference paper.[6] They called Abner Shimony to join them in the writing of a more complete paper explaining the theorem and envisioning possible experiments. As we have seen, it was the second time Shimony was called on by Horne and Zeilinger to further exploit and present their original ideas.

8.4 The Bell Inequality

To understand the GHZ theorem, one might begin with the celebrated two-particle spin singlet state, with zero total angular momentum, introduced by David Bohm in his 1951 textbook, *Quantum Theory*, and depicted in Fig. 8.1.

Stern Gerlach (SG) magnets separate the beams in such a way that, for each particle, the probability of measuring each of the two possible outcomes is $\frac{1}{2}$. The observable being measured is the z-component of spin, with the two possible eigenvalues (outcomes) conventionally chosen as $+1$ and -1, as shown in the figure. The singlet state is such that in ideal experimental situations, in which both SG magnets point in the same angle (the "superclassical" case, according to [3]), one measures perfect anticorrelation, meaning that if particle 1 is measured with the value of the spin component $I = +1$, particle 2 will necessarily be measured with the value $II = -1$, and vice versa.

The experiment shown in Fig. 8.1 could be readily explained by a local realist theory: one could suppose that the spin particles are emitted with definite and opposite spin components in the z direction. What such a local realist theory cannot do is explain the fact that perfect anticorrelation would occur *for any angle* of the SG magnets, assuming that both magnets and pairs of detectors are quickly rotated at the same angle, right after the emission of the pair of particles.

---

[5] Gilder [15].

[6] In fact, there already was some competition over the most general proof of the theorem between Greenberger, Horne, and Zeilinger, on one hand, and Clifton, Redhead, and Butterfield, on the other. This competition is recorded in the paper by Clifton *et al.* [14] in a "note added in proof" on p. 182.

This feature of rotational symmetry in the experimental outcomes implies that the quantum mechanical representation of the singlet state must have rotational symmetry. This appears in the following expression:

$$|\Psi_s\rangle = \frac{1}{\sqrt{2}} \cdot |+_z\rangle_1 |-_z\rangle_2 - \frac{1}{\sqrt{2}} \cdot |-_z\rangle_1 |+_z\rangle_2. \tag{8.1}$$

The notation $|+_z\rangle_2$ indicates the eigenstate of the z component of spin in the positive direction, associated with eigenvalue $+1$, for particle 2. That this expression is rotationally invariant may be verified by replacing each individual state by its representation in another basis (corresponding to the eigenstates for another angle of the SG magnets).

The proof that local realist theories cannot account for the predictions of quantum mechanics for pairs of entangled particles was derived by Bell in 1964 by means of an inequality that limits the predictions of any local realist theory, but may be violated by the quantum mechanical predictions. The version derived by Clauser, Horne, Shimony, and Holt [6] is

$$|c(a, b) + c(a', b) + c(a, b') - c(a', b')| \leq 2. \tag{8.2}$$

If $I(a)$ and $II(b)$ represent the results of measurements for each pair of particles performed with SG magnets adjusted respectively at angles a and b, then the correlation coefficient $c(a, b)$ is the mean of the products $I(a) \cdot II(b)$ averaged over all the pairs of particles.

Bell [1] postulated the existence of a set of hidden variables λ that uniquely determine the values of the measurement outcomes. The general expressions for values (either 1 or -1) predicted by the realist hidden variables theory are $I(a, b, \lambda)$ and $II(a, b, \lambda)$. In order to impose the property of *locality* in these models, Bell dropped the dependence of a measurement outcome on the far-away SG magnet; as a result of this factorizability condition, the values of outcomes are written simply as $I(a, \lambda)$ and $II(b, \lambda)$. With this, Bell was able to derive his inequality.

Quantum mechanics predicts that the correlation function for the singlet state is given by

$$c_{\Psi_s}(a, b) = -\cos(a - b). \tag{8.3}$$

One may find values for a, a', b, and b' for which the values obtained in (8.3) violate inequality (8.2).

One notes in this inequality that its experimental verification must involve four experimental runs, with four pairs of angles of the magnets. This introduces additional hypotheses of fair sampling in the experimental test of local realist theories.

If an experimental test could be done with a single setup, this would reduce the additional assumptions used to rule out local realist theories. The first derivation of a version of Bell's theorem without inequalities was done by Heywood and Redhead [16], based on simplified versions of the Kochen–Specker theorem. A simpler and more testable approach was that of Greenberger, Horne, and Zeilinger [3], who derived a proof involving four spin-$^1/_2$ particles.

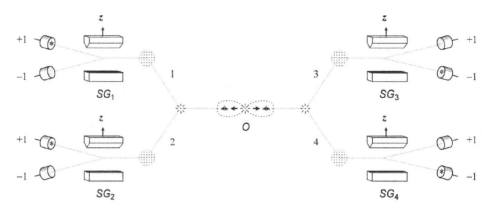

Figure 8.2 Experimental setup for four entangled spin-$\frac{1}{2}$ particles, proposed by GHZ, with every SG magnet pointing at the same angle z.

8.5 Scope of the GHZ Theorem

The four-particle entangled state proposed by Greenberger *et al.* [3] is written as follows:

$$|\Psi_{ss}\rangle = \frac{1}{\sqrt{2}} \cdot |+_z\rangle_1 |+_z\rangle_2 |-_z\rangle_3 |-_z\rangle_4 - \frac{1}{\sqrt{2}} \cdot |-_z\rangle_1 |-_z\rangle_2 |+_z\rangle_3 |+_z\rangle_4. \qquad (8.4)$$

This state may be prepared in a way similar to the two-particle system of Fig. 8.1, assuming that each of the two entangled particles has spin 1, and that each of them decays into two spin-$\frac{1}{2}$ particles. This is represented in Fig. 8.2.

One notices again strict anticorrelation between measurements on the left side of the picture and on the right, as in Fig. 8.1. The situation imagined by GHZ to test local realist theories involves magnets at different angles, respectively a, b, c, d, so that the strict anticorrelation of the superclassical case will not occur.

The quantum mechanical correlation function associated with these measurements is similar to the two-particle case, and gives

$$c_{\Psi_{ss}}(a, b, c, d) = -\cos(a + b - c - d). \qquad (8.5)$$

One can now consider a case of strict anticorrelation ($c_{\Psi_{ss}} = -1$), of which the setup of Fig. 8.2 is a special case, and also a case of strict correlation ($c_{\Psi_{ss}} = 1$):

$$\begin{aligned} &\text{If } a + b - c - d = 0, &&\text{then } c_{\Psi_{ss}} = -1; \\ &\text{If } a + b - c - d = \pi, &&\text{then } c_{\Psi_{ss}} = 1. \end{aligned} \qquad (8.6)$$

These situations have the property that if one measures the spin components of three particles, one knows for sure (in an ideal experimental situation) the value of the fourth one (in the special case depicted in Fig. 8.2, a single measurement is sufficient, as indicated in Eq. (8.4)). This is analogous, as pointed out by Greenberger *et al.* [3], to what happens in the Einstein, Podolsky, and Rosen argument, which is based on counterfactual strict anticorrelation measurements.

Let us now postulate the existence of hidden variables λ that uniquely determine the measurement outcomes $I(a, \lambda), II(b, \lambda), III(c, \lambda)$, and $IV(d, \lambda)$ of the respective components of spin. As in the derivation of Bell's theorem, the factorizability condition has been imposed, which expresses the locality of the model. The value of each $I(a, \lambda), \ldots$, is either 1 or -1.

Expressions (8.6) may be rewritten taking into account that, in the superclassical cases of strict correlation or anticorrelation, the product $I(a) \cdot II(b) \cdot III(c) \cdot IV(d)$ is equal to the correlation coefficient $c_{\psi ss}$:

$$
\begin{aligned}
&\text{If } a+b-c-d = 0, \quad \text{then } I(a, \lambda) \cdot II(b, \lambda) \cdot III(c, \lambda) \cdot IV(d, \lambda) = -1; \\
&\text{If } a+b-c-d = \pi, \quad \text{then } I(a, \lambda) \cdot II(b, \lambda) \cdot III(c, \lambda) \cdot IV(d, \lambda) = 1.
\end{aligned} \tag{8.7}
$$

GHZ were able to show that these two expressions lead to a contradiction. Starting with the first expression, one may derive four special cases, where ϕ is an arbitrary angle:

$$
\begin{aligned}
I(0, \lambda) \cdot II(0, \lambda) \cdot III(0, \lambda) \cdot IV(0, \lambda) &= -1, \\
I(\phi, \lambda) \cdot II(0, \lambda) \cdot III(\phi, \lambda) \cdot IV(0, \lambda) &= -1 \\
I(\phi, \lambda) \cdot II(0, \lambda) \cdot III(0, \lambda) \cdot IV(\phi, \lambda) &= -1, \\
I(2\phi, \lambda) \cdot II(0, \lambda) \cdot III(\phi, \lambda) \cdot IV(\phi, \lambda) &= -1.
\end{aligned} \tag{8.8}
$$

Multiplying the first three equations of (8.8), and recalling that the values I, II, III, and IV are either 1 or -1, one obtains

$$
I(0, \lambda) \cdot II(0, \lambda) \cdot III(\phi, \lambda) \cdot IV(\phi, \lambda) = -1. \tag{8.9}
$$

Comparing this with the last equation of (8.8), one concludes that $I(2\phi, \lambda) = I(0, \lambda)$, i.e., $I(\psi, \lambda)$ has a constant value for all ψ. That would be very strange, for if $\psi = \pi$, one expects that the value of I will have a negative sign in relation to the value for $\psi = 0$.

For this to constitute a mathematical contradiction, one may take the second expression of (8.7) and derive the following special case:

$$
I(\phi + \pi, \lambda) \cdot II(0, \lambda) \cdot III(\phi, \lambda) \cdot IV(0, \lambda) = 1. \tag{8.10}
$$

Together with the second of equations (8.8), one concludes that $I(\phi + \pi, \lambda) = -I(\phi, \lambda)$, which contradicts the previous assertion that $I(\psi, \lambda)$ will have a constant value for all ψ.

Therefore, the assumption that quantum mechanics can be formulated as a realist hidden variable theory leads to a contradiction, without the use of inequalities!

8.6 Epilogue: GHZ Experiments

The production of GHZ states in laboratories was neither an easy job nor a quick achievement. It took a decade. In fact, from 1990 on, Bell's reaction, in particular, motivated Zeilinger to think immediately about taking this theorem to the laboratory benches. The road to experiments, however, was dependent on both conceptual and experimental advances. Zeilinger considers it the most challenging experiment he has ever carried out. The quest for the required expertise, however, brought important preliminary results and spinoffs, such as the concept of "entanglement swapping" and the experiment on teleportation [17, 18].[7] The twentieth century closed with Zeilinger successfully obtaining the

---

[7] Anton Zeilinger, interviewed by Olival Freire Jr., ibid.

experimental production of GHZ states that are in agreement with quantum theory and in disagreement with local realistic theories [19].

References

[1] Bell, J.S. 1964, On the Einstein–Podolsky–Rosen paradox, *Physics* **1**, 195–200.

[2] Laloë, F. 2012, *Do We Really Understand Quantum Mechanics?* (Cambridge University Press, Cambridge).

[3] Greenberger, D., Horne, M., and Zeilinger, A. 1989, Going beyond Bell's theorem. In: Kafatos, M.C. (eds.), *Bell's Theorem, Quantum Theory and Conceptions of the Universe* (Kluwer, Dordrecht), pp. 69–72.

[4] Greenberger, D.M., Horne, M.A., Shimony, A., and Zeilinger, A. 1990, Bell's theorem without inequalities, *American Journal of Physics* **58**(12), 1131–43.

[5] Greenberger, D. 2002, The history of the GHZ paper, in R.A. Bertlmann and A. Zeilinger (eds.). *Quantum (Un)speakables: From Bell to Quantum Information.* Berlin: Springer, pp. 281–6.

[6] Clauser, J.F., Horne, M.A., Shimony, A., and Holt, R.A. 1969, Proposed experiment to test local hidden-variable theories, *Physical Review Letters* **23**(15), 880–84.

[7] Whitaker, A. 2012, *The New Quantum Age* (Oxford University Press, Oxford).

[8] Rauch, H., Zeilinger, A., Badurek, G., Wilfing, A., Bauspiess, W., and Bonse, U. 1975, Verification of coherent spinor rotation of fermions, *Physics Letters A* **54**(6), 425–7.

[9] Rauch, H. 2012, Quantum physics with neutrons: From spinor symmetry to Kochen–Specker phenomena, *Foundations of Physics* **42**(1), 153–72.

[10] Horne, M.A. and Zeilinger, A. 1985, A Bell-type EPR experiment using linear momenta, in P. Lahti and P. Mittelstaedt (eds.), *Symposium on the Foundations of Modern Physics: 50 Years of the Einstein–Podolsky–Rosen Gedanken Experiment* (World Scientific, Singapore), pp. 435–9.

[11] Ghosh, R. and Mandel, L. 1987, Observation of nonclassical effects in the interference of two photons, *Physical Review Letters* **59**(17), 1903–5.

[12] Horne, M.A., Shimony, A., and Zeilinger, A. 1989, Two-particle interferometry, *Physical Review Letters* **62**(19), 2209–12.

[13] Mermin, N.D. 1990, What's wrong with these elements of reality? *Physics Today* **43**(6), 9–11.

[14] Clifton, R.K., Redhead, M.L.G., and Butterfield, J.N. 1991, Generalization of the Greenberger–Horne–Zeilinger algebraic proof of nonlocality, *Foundations of Physics* **21**(2), 149–84.

[15] Gilder, L. 2008, *The Age of Entanglement – When Quantum Physics Was Reborn* (Knopf, New York).

[16] Heywood, P. and Redhead, M.L.G. 1983, Nonlocality and the Kochen–Specker paradox, *Foundations of Physics* **13**: 481–99.

[17] Żukowski, M., Zeilinger, A., Horne, M.A., and Ekert, A.K. 1993, "Event-ready-detectors" Bell experiment via entanglement swapping, *Physical Review Letters* **71**(26), 4287–90.

[18] Bouwmeester, D., Pan, J.W., Mattle, K., Eibl, M., Weinfurter, H., and Zeilinger, A. 1997, Experimental quantum teleportation, *Nature* **39**, 575–9.

[19] Bouwmeester, D., Pan, J.W., Daniell, M., Weinfurter, H., and Zeilinger, A. 1999, Observation of three-photon Greenberger–Horne–Zeilinger entanglement, *Physical Review Letters* **82**(7), 1345–9.

Part III
Nonlocality: Illusion or Reality?

9

Strengthening Bell's Theorem: Removing the Hidden-Variable Assumption

HENRY P. STAPP

9.1 Spooky Action at a Distance

In the context of correlation experiments involving pairs of experiments performed at very nearly the same time in very far-apart experimental regions, Einstein famously said [1],

But on one supposition we should in my opinion hold absolutely fast: "The real factual situation of the system S2 is independent of what is done with system S1 which is spatially separated from the former."

This demand is incompatible with the basic ideas of standard (Copenhagen/orthodox) quantum mechanics, which makes two relevant claims:

(1) Experimenters in the two labs make "local free choices" that determine which experiments will be performed in their respective labs. These choices are free in the sense of not being predetermined by the prior history of the physically described aspects of the universe, and they are localized in the sense that the physical effects of these free choices are inserted into the physically described aspects of the universe only within the laboratory in which the associated experiment is being performed.

(2) These choices of what is done with the system being measured in one lab can (because of a measurement-induced global collapse of the quantum state) influence the outcome of the experiment performed at very close to the same time in the very faraway lab.

This influence of *what is done with* the system being measured in one region upon the outcome appearing at very nearly the same time in a very faraway lab was called "spooky action at a distance" by Einstein, and was rejected by him as a possible feature of "reality."

9.2 John Bell's Quasi-classical Statistical Theory

Responding to the seeming existence in the quantum world of "spooky actions," John Bell [2] proposed a possible alternative to the standard approach that might conceivably be able to reconcile quantum spookiness with "reality." This alternative approach rests on the fact that quantum mechanics is a statistical theory. We already have in physics a statistical theory called "classical statistical mechanics." In that theory the statistical state of a system is expressed as a *sum* of terms, each of which is a possible *real physical state* λ of the system multiplied by a probability factor.

Bell conjectured that quantum mechanics, being a statistical theory, might have the same kind of structure. Such a structure would satisfy the desired properties of "locality" and "reality" (local realism) if, for each real physical state λ in this sum, the relationships between the chosen measurements in the two regions and the appearing outcomes are expressed as products of two factors, with each factor depending upon the measurement and outcome in just one of the two regions. The question is then whether the statistical properties of such a statistical ensemble can be consistent with the statistical predictions of quantum mechanics.

Bell and his associates proved that the answer is No! They considered, for example, the empirical situation that physicists describe by saying that two spin-$\frac{1}{2}$ particles are created in the so-called spin-singlet state, and then travel to two far-apart but nearly simultaneous experimental regions. The experimenter in each region freely chooses and performs one of the two alternative possible experiments available to him. Bell et al. then prove that the predictions of quantum mechanics cannot be satisfied if the base states λ satisfy the "factorization property" demand of "local realism." A theory satisfying this demand is called a "local hidden-variable theory" because the asserted underlying reality is described by variables that cannot be directly apprehended.

9.3 Two Problems with Bell's Theorems

Bell-type theorems, if *considered as proofs of the logical need for spooky actions at a distance in a theory that entails the predictions of quantum mechanics*, have two problems. The first is that the theorems postulate a reality structure basically identical to that of classical statistical mechanics. Bell's theorems then show that imposing locality (factorizability for each fixed λ) *within this classical-type reality structure* is incompatible with some predictions of quantum mechanics. But that result can be regarded as merely added confirmation of the fact that quantum mechanics is logically incompatible with the conceptual structure of classical mechanics. Simply shifting to a classically conceived statistical level does not eliminate the essential conceptual dependence on the known-to-be-false concepts of classical physics.

The second problem is that the condition of local realism is implemented by a factorization property, described above, that goes far beyond Einstein's demand for no spookiness. In addition to the nondependence of outcomes in a region upon what is done in the faraway region-local realism entails also what Shimony calls "outcome independence." That condition goes significantly beyond what Einstein demanded, which is merely the nondependence of the factual reality (occurring outcome) in one region on *the choice of experiment performed in the faraway region*. Outcome independence demands that the outcome in each region be independent also of the *outcome* in the other region.

That property, outcome independence, is not something that one wants to *postulate* if a resulting incompatibility with predictions of quantum mechanics is supposed to entail the existence of spooky actions at a distance!

The unwanted independence assumption is not just a minor fine point. Consider the simple example of two billiard balls, one black, one white, shot out in opposite directions to two far-apart labs. This physical example allows – given the initial symmetrical physical state – the outcome in one region to be correlated with the "outcome" appearing in the other region, without any hint of any spooky action at a distance: a black ball in one region entails a white ball in the other, and vice versa, without any spooky action. Hence Bell's theorems do not address – or claim to address – the key question of the compatibility of Einstein's demand for no spookiness with the predictions of (relativistic) quantum mechanics. Bell's theorems are based on the stronger assumption of local hidden variables.

Bell's theorems (regarded as proofs of the need for spooky actions) are thus deficient in two ways: they bring in from classical (statistical) mechanics an alien-to-quantum-mechanics idea of reality; and they assume, in the process of proving a contradiction, a certain property of outcome independence that can lead to a violation of quantum predictions without entailing the lack of spookiness that Einstein demanded.

The question thus arises of whether the need for spooky interactions can be proved simply from the validity of some empirically well validated predictions of standard quantum mechanics, without introducing Bell's essentially classical hidden variables? The answer is "Yes"!

9.4 The Proof

The following proof of the need for spooky actions places no conditions at all on any underlying process or reality, beyond the macroscopic predictions of quantum mechanics: it deals exclusively with connections between macroscopic measurable properties. This change is achieved by taking Bell's parameter λ to label, now, the different experiments in a very large set of simultaneously performed similar experiments, rather than the different possible basic microscopic states λ of the statistical ensembles. The ontology thereby becomes essentially different, though the mathematics is similar. The macroscopic experimental arrangements are the ones already described.

In the design of this experiment the physicists are imagining that a certain initial macroscopic preparation procedure will produce a pair of tiny invisible (spin-$1/2$) particles in what is called the singlet state. These two particles are imagined to fly out in opposite directions to two faraway experimental regions. Each of these experimental regions contains a Stern–Gerlach device that has a directed preferred axis that is perpendicular to the incoming beam. Two detection devices are placed to detect particles deflected along this preferred axis or in opposite directions. Each of these two devices will produce a visible signal (or an auditory click) if the imagined invisible particle reaches it.

The location of the individual detector is specified by the angle Φ of the directed preferred axis such that a displacement along that particular direction locates the detector. Clearly, the two detectors in the same experimental region will then be specified by two angles Φ that differ by $180°$. For example, if one detector is displaced "up" ($\Phi = 90°$) then the other is displaced "down" ($\Phi = -90°$). The angle $\Phi = 0°$ labels a common deflection to the right in both regions: e.g., along the positive x axis in the usual x–y plane.

Under these macroscopic experimental conditions, quantum theory predicts that, if the detectors are 100% efficient, and if, moreover, the geometry is perfectly arranged, then for each created pair of particles – which are moving in opposite directions to the two different regions – exactly one of the two detectors in each region will produce a signal (i.e., "fire"). The key prediction of quantum theory for this experimental setup is that the fraction F of the particle pairs for which the detectors that fire in the first and second regions are located at angles $\Phi 1$ and $\Phi 2$, respectively, is given by the formula $F = (1 - \text{Cosine}[\Phi 1 - \Phi 2])/4$.

In the experiment under consideration, there are two alternative possible experiments in the left-hand lab and two alternative possible experiments in the right-hand lab, making $2 \times 2 = 4$ alternative possible pairs of experiments. For each single experiment (on one side) there are two detectors, and hence two angles Φ. Thus there are altogether $4 \times 2 \times 2 = 16$ F's.

I take the large set of similar experiment to have 1,000 experiments. Then the fractions F of 1,000 are entered into the 16 associated boxes of Diagram 9.1.

Diagram 1

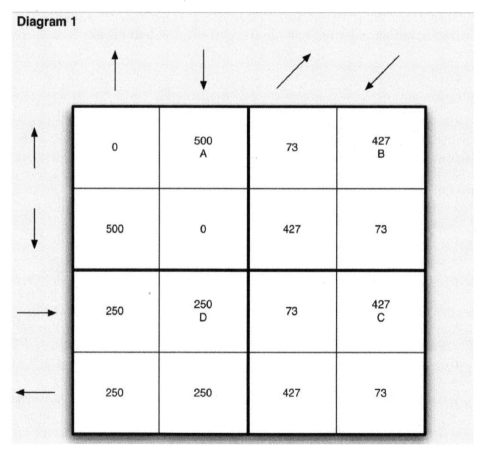

Diagram 9.1

In Diagram 9.1, the first and second *rows* correspond to the two detectors in the *first* possible set-up in the left-hand region. The third and fourth rows correspond to the two detectors in the *second* possible set-up in the left-hand region. The four *columns* correspond in the analogous way to the detectors in the right-hand region. The arrows on the periphery show the directions of the displacements of the detectors associated with the corresponding row or column.

For example, in the top left 2-by-2 box, if the locations of the two detectors (one in each region) that fire together are both specified by the same angle, $\Phi1 = \Phi2$, then, because Cosine $0 = 1$, each specified pair of detectors will *never* fire together: if one of these two specified detectors fires, then the other will not fire. If $\Phi1$ is some fixed angle and $\Phi2$ differs from it by $180°$ then, because Cosine $180° = -1$, these two specified detectors will, under the ideal measurement conditions, fire together for *half* of the created pairs. If $\Phi1$ is some fixed angle and $\Phi2$ differs from it by $90°$, then these two specified detectors will fire together for *one-fourth* of the pairs. If $\Phi1$ is some fixed angle and $\Phi2$ differs from it by $45°$, then these two specified detectors will fire together, in the long run, for *close to 7.3%* of the pairs. If $\Phi1$ is some fixed angle and $\Phi2$ differs from it by $135°$, then these two specified detectors will fire together, in the long run, for *close to 42.7%* of the created pairs.

I have listed these particular predictions because they are assumed to be valid in the following proof of the need for near-instantaneous transfer of information between the two far-apart, but nearly simultaneous, experimental space–time regions. These particular predictions have been massively confirmed empirically.

The second assumption is localized free choices. The point here is that physical theories make predictions about experiments performed by experimenters with devices that detect or measure properties of the systems whose properties are being probed by these devices. The theory entails that the various settings of the devices will correspond to probe-associated properties of the system being probed.

Of course, in an actual situation, these specified parts of the experimental setup are all parts of a universe that also includes the experimenter and whatever the experimenter uses to actually fix the experimental settings. Such a "choosing" part of the universe *could*, however, *conceivably* causally affect not only to the setting of the associated measuring device but, say, via the distant past, also other aspects of the experiment. Those unsuspected linkages via the past could then be responsible for systematic correlations between the empirical conditions in the two regions – correlations that are empirically dependent on which experiments are chosen and performed but are empirically independent of *how* the experimental setups are chosen.

In view of the limitless number of ways one could arrange to have the experimental setup specified, and the empirically verified fact that the predictions are found to be valid independent of *how* the setup is chosen, it is reasonable to assume that the choices of the experimental setups can be arranged so that they are not systematically connected to the specified empirical aspects of the experiment except via these choices of the experimental setup. This is the assumption of *localized free choices*. It is needed to rule out the (remote) possibility that the choice of the setup is significantly and systematically entering the dynamics in some way other than as just the localized fixing of the experimental setup.

Suppose, then, that we have the two far-apart experimental regions, and in each region an experimenter who can freely choose one or the other of two alternative possible experimental setups. Suppose we have, in a certain region called the source region, a certain mechanical procedure to which we give the name "creation of N individual experimental instances," where N is a large number, say a thousand. At an appropriate later time the experimenters in the two regions make and implement their localized free choices pertaining to which of the two alternative possible experiments will be set up in their respective experimental regions. At a slightly later time each of the two experimenters looks at and sees, in each of the N individual instances, which one of his two detection devices has fired, and then records the angle Φ that labels that detector, thereby recording the outcome that occurs in that individual instance.

There are altogether two times two, or four, alternative possible experimental setups. Diagram 9.1 gives, for each of these four alternative possible setups, the number of individual instances, from the full set of 1,000, that produce firings in the pair of detectors located at the pair of angles Φ specified along the left-hand and top boundaries of the full diagram. For example, the four little boxes in the first two rows and the first two columns correspond to the case in which the experimenter in the left-hand region sets his two detectors at "up" ($\Phi 1 = 90°$) and "down" ($\Phi 1 = -90°$s), while the experimenter in the right-hand region sets his two detectors also at "up" ($\Phi 2 = 90°$) and "down" ($\Phi 2 = -90°$). In this case the expected distribution (modulo fluctuations) of the thousand instances is 500 in the box in which $\Phi 1 = 90°$ and $\Phi 2 = -90°$ and the other 500 in the box in which $\Phi 1 = -90°$ and $\Phi 2 = 90°$.

The fluctuations become smaller and smaller as N get larger and larger. So I will, for simplicity, ignore them in this discussion and treat the predictions as exact already for $N = 1,000$.

The two experimental regions are arranged to be essentially simultaneous, *very* far apart, and very tiny relative to their separation. These two regions will be called the left and right regions.

The "no essentially instantaneous transfer of information about localized free choices" assumption made here is that, no matter which experiment is performed in a region, the outcome appearing there is independent of which experiment is freely chosen and performed in the faraway region. This means, for example, that if the experiment on the right is changed from the case represented by the left-hand two columns to the case represented by the right-hand two columns, then the *particular set* of 500 instances – from the full set of 1,000 – that are represented by the 500 in the top row second column get shifted into the two boxes of the top row in the second two columns.

More generally, a change in the experiment performed on the right shifts the individual instances – in the set of 1,000 individual instanced – horizontally in the same row, whereas a change in the experiment performed on the left shifts the individual instance vertically.

Diagram 9.1 then shows how, by a double application of the "no FTL" condition, a subset of the set of 500 instances occupying box A gets shifted via box B to box C, which must then contain at least $427 - 73 = 354$ of the original 500 instances in A. However, the application of the two changes in the other order, via D, demands that the subset of instances in A that can be in C can be no greater than 250. That is a contradiction. Thus one cannot maintain simultaneously both the general rule of no FTL transfers of information and four very basic and empirically confirmed predictions of quantum mechanics.

In more detail, the argument then goes as follows. Let the pairs (individual instances) in the ordered sequence of the 1,000 created pairs be numbered from 1 to 1,000. Suppose that the actually chosen pair of measurements corresponds to the first two rows and the first two columns in the diagram. This is the experiment in which, in each region, the displacements of the two detectors are "up" and "down." Under this condition, quantum theory predicts that for some particular 500-member subset of the full set of 1,000 individual instances (created pairs) the outcomes conform to the specifications associated with the little box labeled A. The corresponding 500-member subset of the full set of 1,000 positive integers is called Set A. This Set A is a particular subset of 500 integers from set $\{1, 2, \ldots, 1,000\}$. The first four elements in Set A might be, for example, $\{1, 3, 4, 7\}$.

If the local free choice in the right-hand region had gone the other way, then the prediction of quantum mechanics is that the thousand integers would be distributed in the indicated way among the four little boxes that lie in one of the first two rows and also in one of the *second* two columns, with the integer in each of these four little boxes specifying the number of instances in the subset of the original set of 1,000 individual instances that lead to that specified outcome. Each such outcome consists, of course, of a pair of outcomes, one in the left-hand experimental region and specified by the row, the other in the right-hand experimental region and specified by the column.

If we now add the locality condition, then the demand that the macroscopic situation in the left-hand region be undisturbed by the reversal of the localized free choice made by the experimenter in the (faraway) right-hand region means that the set of 500 integers in Set A must be distributed between the two little boxes standing directly to the right of the little box A. Thus the Set B, consisting of the 427 integers in box B, would be a 427-member subset of the 500 integers in Set A.

These conclusions were based on the supposition that the actual choice of experiment on the left was the option represented by the top two rows and the leftmost two columns in diagram 9.1. However, having changed the choice in the right-hand region to the one that is represented by the *rightmost* two columns – the possibility of which is which is entailed by Einstein's reference to a dependence on "what is done with" the faraway system – we next apply the locality hypothesis to conclude that changing the choice on the left must leave the outcomes on the right undisturbed. That means changing the top two rows to the bottom two rows, leaving the integers that label the particular experiment in the set of 1,000 experiments in the same column. This means that the 427 elements in box B must be distributed among the two boxes that lie directly beneath it. Thus box C must include at least $427 - 73 = 354$ of the 500 integers in Set A.

Repeating the argument, but reversing the order in which the two reversals are made, we conclude, from exactly the same line of reasoning, that box C can contain no more than 250 of the 500 integers in box A. Thus the conditions on Set C that arise from the two different possible orderings of the two reversals are contradictory!

A contradiction is thus established between the consequences of the two alternative possible ways of ordering these two reversals of localized free choices. Because, due to the locality hypothesis being examined, no information about the choice made in either region is present in the other region, no information pertaining to the order in which the two experiments are performed is available in either region. Hence nothing pertaining to outcomes can depend upon the relative ordering of these two spacelike separated reversals of the two choices.

This argument uses only macroscopic predictions of quantum mechanics – without any conditions on, or mention of, any microstructure from whence these macroscopic properties come – to demonstrate the logical inconsistency of combining a certain 16 (empirically validated) predictions of quantum mechanics with the locality hypothesis that for each of the two experimental regions there is no faster-than-light transfer to the second region of information about macroscopically localized free choices made in the first.

The proofs of Bell's theorem are rightly identified as proofs of the incompatibility of local realism with the predictions of quantum mechanics. But local realism brings in both alien-to-quantum-theory classical concepts and also an outcome-independence condition whose inclusion nullifies those theorems as possible proofs of the need for spooky actions at a distance. Both of these features are avoided in the present proof.

As regards Einstein's reality condition, namely that the no-spooky-action condition pertains to the "real factual situation," one must, of course, use the quantum conception of the real factual situation, not an invalid classical concept. In ontologically construed orthodox quantum mechanics (in the contemporary relativistic quantum field theory version that I use) the real factual situation evolves in a way that depends upon the experimenter's free choices and nature's responses to those choices. The no-spooky-action condition is a condition on these choice-dependent real factual situations – namely outcomes observed under the chosen conditions – that is inconsistent with certain basic predictions of the theory. That is what has just been proved. In classical mechanics there are no analogous free choices: the physical past uniquely determines the physical present and future.

The Einstein idea of no spooky actions involves comparing two or more situations, only one of which can actually occur. This is the kind of condition that occurs in modal logic considerations involving "counterfactuals." But here this modal aspect does not bring in any of the subtleties or uncertainties that plague general modal logic. For in our case the specified condition is a completely well-defined and unambiguous (trial) mathematical assumption of the nondependence of a nearby outcome upon a faraway free choice between two alternative possible probing actions. The proof does not get entangled with the subtle issues that arise in general modal logic. Everything is just as well defined as in ordinary logic.

In this proof there is no assumption of a hidden variable of an essentially classical kind lying behind the ontologically construed orthodox quantum theory. The phenomena are rationally understandable, rather, in terms of an evolving quantum state of the universe that represents potentialities for experiences that evolve via a Schroedinger-like equation punctuated by an ordered sequence of psychophysical events each of which is an observer's personal experience accompanied by a collapse of the quantum state of the universe that brings that evolving state into conformity with a self-initiated experience of that observer, and also with the associated experiences of all members in his community of communicating observers.

The bottom line is that if scientists want to understand, within a science-based framework, the structure of human experience then they must move beyond the classical conception of reality.

References

[1] Einstein, A., in P.A. Schilpp (ed.), *Albert Einstein: Philosopher-Scientist* (Tudor, New York, 1949), p. 85.
[2] Bell, J.S., *Speakable and Unspeakable in Quantum Mechanics* (Cambridge University Press, Cambridge, 1987), pp. 1, 15.

10

Is Any Theory Compatible with the Quantum Predictions Necessarily Nonlocal?

BERNARD D'ESPAGNAT

10.1 Introduction

Important advances in scientific knowledge may, not infrequently, be looked at from several angles. Here, in investigating the question the article title defines, part of our attention will have to be turned on the relationships of Bell's theorem to the old, but still not fully cleared up, question of the nature of causality.

The word "nonlocality" means the violation of local causality, the initial meaning of which was that no influence, that is, no causal *action*, can be propagated faster than light. For a long time it was a received view that actions somewhat akin to the basically time-directed ones that living beings perform also take place within inanimate nature. This conception of causality may be called the *causality–action theory*. But later, having in view Hume's serious reservations concerning it, renowned philosophers of science claimed it might and should be dropped, since the effects it was meant to account for could be explained just as well by considering that what is at work is simply a conjunction of physical laws expressed by means of differential equations – a view that later Hempel and Oppenheim generalized and called the D-N (deductive–nomological) model of explanation (reprinted in [1]). Here the (widely accepted, as it seems) corresponding conception of causality will, for simplicity, be called the *law-centered theory of causality*.

But does the latter theory subsume, at least within physics, all the meanings of the word 'cause' that true understanding calls for? Among philosophers the matter is not settled. Hempel himself granted that in several respects the D-N model is somehow inadequate for analyzing the cause–effect relationship [2]. And it seems indeed that a significant argument favoring a 'no' answer to the question might have been put forward as early as 1907 (although, to our knowledge, it was not), when Einstein pointed out that according to his then two years old special relativity (SR) theory, if we assumed an effect could follow its cause sooner than light velocity permits, then in some other frames of reference it would occur before its cause. For, concerning this consequence of the assumption, he wrote, "in my opinion, regarded as pure logic . . . it contains no contradiction; however it absolutely clashes with the character of our total experience, and in this way is proved the impossibility of the hypothesis" [3]. Now, through the words "it contains no contradiction" Einstein granted that the premise he invoked for inferring the limitation (local causality) inherent in

this impossibility could logically follow neither from the basic laws of prerelativistic classical physics (be it only because it is time-directed, while the said laws are not) nor, clearly,
from the two basic postulates than induce the whole of SR (relativity and independence
of the speed of light from the speed of its source). In other words, he then supplemented
SR with a new element, and the type of causality this element calls for is not reducible
to the law-centered theory of causality. It seems, therefore, that it is only by referring to
something akin to the older causality–action theory of causality, interpreting causality as a
time-directed action of the cause on the effect, that the meaning of a notion of local causality
can be made fully explicit. To discriminate the resulting conception of local causality from
other definitions of it that have been or could be proposed, it is called here 'local causality
in the first sense'.

Now, while the fact that Einstein thus indirectly invoked the causality–action theory
is not of course a sufficient reason for taking it to be scientifically valid, still it should
prevent us from branding it as unscientific just on the basis of controversial philosophical
arguments. And anyhow, to disprove local causality, Bell clearly had to take that theory into
consideration, which in, for example, his last paper on the subject [7] he did through the
words "The direct causes (and effects) are near by", which set in relief the notion of *direct*
causes operating between purely physical events.

Two problems, however, arise. One of them bears on the exact nature of the premises
Bell had to assume. The other one, expressed by the very title of this article, is to infer
from this study whether or not some remarkable general premise such as realism must be
assumed for a watertight proof of nonlocality. They are investigated in Sections 10.2 and
10.3, respectively. Section 10.4 draws conclusions.

10.2 Bell's Premises

In the course of years Bell developed several versions of his proof, described in the articles
[4–7], respectively. The premises of the first one still have close links to the EPR paper. In
it the EPR criterion of reality is implicitly made use of, leading, through the consideration
of a special case (an EPRB type experiment with parallel orientations of instruments), to
a proof of determinism. Now, this criterion and its implications – this one in particular –
are somewhat generally held in suspicion within the physics community. Moreover, the
article leads to an inequality that, currently at least, is not experimentally testable. For these
reasons and for brevity, its premises are not analyzed here. We focus on those of the three
other papers, the general purpose of which is (and merely is) to look for an answer to a
question that, in Section 10.3 of [5], Bell, in substance, stated as follows: Could it not be
that quantum theory is a fragment of a more complete hidden variable theory that, contrary
to standard quantum theory, *has* local causality?[1]

---

[1] All the Bell articles here referred to are reprinted in Bell's *Speakable and Unspeakable in Quantum Mechanics*, 2nd ed., Cambridge: Cambridge University Press, 2004.

10.2.1 The 1976 Premises

In nondeterministic theories the fact that 'local causality in the first sense' is grounded on the causality–action notion raises special difficulties. This presumably is why in the 1976 paper Bell propounded a meaning of the words "local causality" making the notion compatible with the law-centered theory of causality and apt therefore to be subjected to mathematical analysis, while maximally preserving, of course, the main features of our intuitive notion of what a cause is. It was in this spirit that, in that article, he defined locally causal theories as being those in which, R and R' being two finite, spatially separated space–time regions, the probability of an event A happening in R cannot depend on any event B happening in R' when all the beables (roughly, 'elements of reality') within the overlap N of the backward light cones of R and R' are specified. This condition of completeness is indeed necessary for the fact that events in N may well constitute causes common to both A and B to be taken into account, and for information about B to be prevented from affecting the probability $P(A)$ of A without B being in any way a direct cause of A in the causality–action sense, with the consequence that it is only when all the beables in N are specified and hence assumed fixed, that the absence of any direct causal action of B on A implies that the probability of A is independent from B. In other terms, it is only when the just stated 'condition of completeness' is fulfilled that from 'local causality in the first sense' it is possible to infer consequences by applying standard probability rules, and that in particular it is possible to make use of the standard rule

$$P(A, B) = P(A|B)P(B) \tag{10.1}$$

(where $P(A, B)$ is the "probability of both A and B" and $P(A|B)$ the "probability of A if B") and infer from it and the just inferred independence of $P(A|B)$ from B the factorization

$$P(A, B) = P(A)P(B), \tag{10.2}$$

that is, the mutual independence of the probabilities of A and B.

The rest of Bell's reasoning is well known. From this independence and with the help of straightforward mathematical calculations, Bell inferred an inequality (called 'Bell's inequality') and pointed out that, at least with some choices of the orientations of the instruments, it is at variance with both the quantum mechanical predictions and the experimental outcomes, from which he could conclude that 'local causality in the first sense' is disproved for all theories in which all the beables in the overlap N of the two backward light cones are specified. And it is only to such theories that, in this article, Bell gave the name 'locally causal theories'. Only they, therefore, are disproved by the article in question – not all of those, if there are any more, merely satisfying 'local causality in the first sense'.

In the paper, Bell stated his hope that a gain in precision might be possible by concentrating on the notion the word 'beable' expresses rather than on the wooly one of 'observables', because by definition beables, he wrote, 'are there'. These two last terms, 'are' and 'there' neatly summarize in fact Bell's thinking as to what may be called realism and its necessity in physics. The need for the first one, a mood of 'to be', springs, in his view, from the fact that a theory can truly be precise only if it describes 'a bit of what exists [is], thought of independently of its modification through observation', as Einstein wrote [8], a conception that may be called 'Einsteinian realism'. And as for the second, the word 'there', apparently it is just a shortening for 'embedded in space (or spacetime)'. In that sense, however, the notion is still quite extensive (total energy, for example, is a beable). Most of the observables we deal with (instrument settings in particular) can be assigned to some bounded space–time region, and the corresponding beables Bell calls local.

10.2.2 The 1981 Premises

Thus, in the 1976 paper, realism is unquestionably a premise, explicitly stated and considered necessary for the conclusion reached to be valid. Moreover, in it the class of locally causal (and finally refuted) theories is, as shown above, subject to a definite limitation due to recourse to the condition of completeness mentioned above. Most remarkably, in the 1981 paper, which aimed at maximal generality, Bell got rid of both conditions. There, neither the word 'beable' nor any reference to the standard rule (10.1) (which caused the limitation) appears. And still, just as did the foregoing one, this article concluded in favor of local causality violation. A question thus arises: were these limitations truly necessary for proving the latter?

The matter is worth detailed examination. Since, in this domain, only correlations are experimentally reachable, in [9] Bell first considered a simple example of correlations and proposed next to transpose it, mutatis mutandis, to the experiment under study. To begin with, he noted that although heart attacks are typically stochastic, still, between the ones that take place in two distant cities not interacting in this field – Antwerp and Brussels, say – statistics show significant correlations. But they are easily explainable, he pointed out, by attributing them to the fact that the probability of an attack depends on several causal factors, such as especially hot days or eating too much on Sundays, that are the same in the two cities and that, taken together, constitute the state λ of the outside world at a given time.

To investigate the pertinence of the planned transposition, let us take a close look at this example. We observe first that when the causal factors do not change – let it be granted that in one minute time this is the case – the probabilities $P(A)$ of one attack in Antwerp and $P(B)$ of one attack in Brussels are independent. Thus between them and the probability $P(A, B)$ that one attack in Antwerp and one in Brussels take place simultaneously within one minute time, equality (10.2) above holds good. And clearly this is true during ordinary days (world state λ_1) as well as during hot days (world state λ_2), even though the involved probability values are not the same in the two cases. On the other hand, an elementary calculation shows that when they are not, the overall mean probability calculated, say, within a one-year period that within any given minute an attack takes place in Antwerp and another one in Brussels is not a product of the similarly evaluated mean probabilities concerning

Antwerp and Brussels separately. A correlation then exists, obviously due to the fact that within a one-year time the involved causal factor – here temperature – varied, so that we had to consider a combination of the two states λ_1 and λ_2 we assumed it could be in. To recover factorization we have to separate, by thought, the set of minutes composing a whole year into two subsets, E_1 and E_2, composed respectively of all the minutes of ordinary days ($\lambda = \lambda_1$) and all the minutes of hot days ($\lambda = \lambda_2$). At all the minutes composing one of them we assume the considered causal factor (temperature) is the same, so that in each of them $P(A, B)$ factors, which may be written

$$P(A, B|a, b, \lambda_i) = P(A|a, \lambda_i)P(B|b, \lambda_i); \quad i = 1, 2, \tag{10.3}$$

where a and b are stable data specific to Antwerp and Brussels, respectively, such as, say, the average sanitary level in the two cities. In each of the two sub-ensembles, independence of the probabilities is thus recovered. (Note in passing that in [6] the sentence introducing (10.3) (Eq. (10) of the paper) is not entirely clear. Since this formula is an equality, it can be true only if *all* causal factors are kept fixed, and then what Bell calls 'the residual fluctuations' are just the probabilities of attacks, assumed intrinsic and independent as a premise.)

This example, of course, is highly schematic. In reality, many causal factors such as temperature intervene simultaneously. In the general case, symbol λ_i stands for many numbers, each of which is the value one of these causal factors takes in subensemble E_i. We assume here for simplicity that the number of causal factors is finite and that each one can take only a finite number of values, and we say that λ_i is complete if it specifies the value that every one of the existing causal factors has within E_i. Factorization (10.3) then applies for the λ_i. The number of such λ_i may of course be very large, and is unknown in general, even though the above assumptions make it finite.

The idea that feasible measurements bearing exclusively on one particular E_i could actually be performed is unbelievable. It may in fact be considered that actual measurements can only bear on the above mentioned mean overall probability, which is the mean value $P(A, B|a, b)$, over all λ_i, of $P(A, B|a, b, \lambda_i)$. If $\rho(\lambda_i)$ is the (unknown) probability of λ_i this yields, because of (10.3),

$$P(A, B|a, b) = \sum_i \rho(\lambda_i)P(A|a, \lambda_i)P(B|b, \lambda_i), \tag{10.4}$$

and, as we know, deriving from (10.4) the Bell or CHSH inequalities is just a matter of straightforward mathematical calculations. Were we to fancy (just for the sake of argument, since within this purely classical problem there is no reason to expect they should) that statistical results are at variance with these inequalities, we would have to grant that our premises were incomplete. One possibility is that parents of attack victims in Antwerp phoned the information to relatives in Brussels, who were so moved that they themselves suffered attacks, for it is easily seen that under such an assumption the inequalities in question cannot be derived from (10.4). But note that this is assuming a typical phenomenon of direct action by a cause on its effect.

Bell's transposition of this example to the study of experiments of the EPRB type simply consisted in the conjecture – hereafter called 'the Conjecture' – that in these experiments observed correlations are explainable in the same way as those of the example. More precisely, what is assumed in the Conjecture is that when, on a given pair, Alice's instrument, oriented along *a*, registers outcome *A* and, on the same pair, Bob's instrument, oriented along *b*, registers outcome *B*, the joint probability $P(A, B|a, b)$ must be given by formula (10.4) [(12) of [6]], from which the Bell and CHSH inequalities follow. Their experimental violation in the EPRB experiments therefore reveals, just as above, incompleteness of the premises. And for adequately supplementing it, since the Conjecture was endorsed, no other way appears than to assume, as in the attack example above, that some direct causal influence of Bob's registered outcome *B* on the one, *A*, that Alice registered indeed took place, or vice versa, which, when the two measurements are spacelike separated, implies a violation of 'local causality in the first sense'.

Compared with the way of justifying nonlocality Bell had put forward in [5], this one is advantageous in several respects. First, it is more general: as we saw, it has no need of the "beable" notion (in it Bell's mention of causal factors being "kept fixed" concerns but an intermediate stage of the argument), nor, as Bell himself pointed out, of any localization of the λ_i, the causal factors inside which might even involve wave functions. On these grounds, it may seem that it does not call for realism as a necessary premise (a question to be investigated further in Section 10.3). Moreover, it would seem that it justifies rejecting *any* theory postulating 'local causality in the first sense', whereas, as we also saw, the theories the 1976 approach disproved were those called there 'locally causal' in a somewhat restricted meaning of the phrase.

On the other hand, however, this way of proceeding also has the serious inconvenience that it implies endorsing the Conjecture and that the arguments that might justify this move are not crystal clear. For indeed, the heart attack model is grounded on the ideas, taken as premises, that heart attacks are basically random, that in the absence of variable causal factors (such as temperature) the probabilities of an attack occurring in Antwerp and of one occurring in Brussels are independent, and that the need for considering subsets E_i and the corresponding labels λ_i essentially comes from the need, by thought, to group into sets and subsets the events in which every causal factor keeps the same value. Now while, in the model, all these various ingredients (essential for the reasoning) are quite clear, not every one of them has a clear parallel in an EPRB experiment on quantum systems. True, there are similarities: In the model causal factors really exist, and in a quantum theory admitting of hidden variables, it is assumed that they also exist. But on the other hand, in the model, Eq. (10.4) could be derived from Eq. (10.3) because (a) both involve probabilities and (b) for any λ_i these probabilities are mutually independent by assumption. In the case of an EPRB quantum experiment admitting of hidden variables, the theory also involves probabilities, because a definite wave function is part of the premises. But then independence, for any given λ_i, of the probabilities appearing in (10.3) is, to say the least, problematic. So it seems that in such an EPRB experiment Eq. (10.4) cannot be properly derived. Or, more precisely, it appears that the mode of derivation of (10.4) made use of in [6] cannot be meaningfully carried over to such experiments. This implies that, in dealing with the latter ones, the Conjecture, a vital element of the 1981 paper, can merely be considered a hypothesis or, more appropriately, just a definition of what it has been decided to call 'local causality'. And, apparently, it is indeed in this sense that it was understood by physicists who, not as convinced as Einstein and Bell were of the necessity of coming back to a realism of some sort, aimed at interpreting the latter's work on the subject as proving nonlocality quite independent of realism. But, as it seems, the proof thus constructed, interesting and valuable as it is, still is one of a violation of a notion whose relationship with 'local causality in the first sense' is not crystal clear.

10.2.3 The 1990 Premises

While, apparently, Bell did not point out in his writings the hardly binding character of the Conjecture, he surely was aware of it, and this is probably why, in his 1990 paper, he preferred strict inference to maximal generality and substituted for his 1981 approach a procedure very much akin to the 1976 one. Indeed, both explicitly invoke realism and make use of the word 'beable', which confirms the continuity of Bell's strong belief in realism, considered necessary, not for obscure metaphysical reasons but simply because he held it to be one of the main conditions a physical theory must satisfy in order to achieve sufficient precision.

Another strong analogy between these two modes of proof is that both make use of the standard probability rule (10.1). In Subsection 10.2.1 we saw how from this rule and the local causality hypothesis Bell could derive the Bell–CHSH inequalities by preventing information about B from affecting the probability $P(A)$ of A without B being in any way a direct cause of A in the causality–action sense. To that end, *it is necessary that every beable a variation of which might cause events to happen jointly in both R and R′ should be kept fixed*, a requisite that, for future reference, will be called here '*Condition C*'. It is easily seen that assuming that all the parameters (beables) in region N are specified, hence fixed, indeed fulfills Condition C. But it so happens that this region is not the only one in this respect, and in [7] Bell made the same assumption concerning another one he called 'Region *3*' which is suitable as well. It is a spacelike one that cuts both backward lightcones in the future of region N.

Let it be recalled that in any theory fulfilling Condition C, the experimentally observed violation of the CHSH inequality cannot be explained otherwise than by assuming some direct causal action of B on A (or of A on B). Since these events are spacelike separated, such a causal action constitutes a violation of 'local causality in the first sense', which is indeed what Bell actually meant to prove. On the other hand, here as in [5], Condition C restricts the types of theories for which nonlocality is thus proven; and, admittedly, it might be feared that it restricts them to an unsatisfactory degree. In this respect, however, it should be noted, as T. Norsen appropriately pointed out [9], that Condition C may well be taken to concern only the particular hypothetical theory that is being considered as a possible candidate. It then does not mean we should 'know everything' concerning a specified space–time region but merely that, given a candidate theory, the space–time region in question should include all of the beables (called λ here) whose existence is assumed by this theory.

10.3 Is Realism a Necessary Premise for Proving Nonlocality?

For very powerful reasons he often explained, John Bell, as we already stressed, was a strong supporter of Einsteinian realism. His papers were written in this spirit, and it was noted above that indeed realism has a significant role in most of his proofs of nonlocality and in particular in the one to which, finally, he seems to have attached the greatest weight. To a great extent this justifies the often-appearing statement that Bell disproved 'local realism'. On the other hand, such a statement suggests the possibility of a choice. It seems it implies not only that realism and locality cannot both be true but also that if realism is not assumed locality might conceivably be preserved. Now it has recently been claimed that this, in fact, is not the case; that what is really proved goes beyond this, in that it refutes locality independently of whether any additional premise such as realism is assumed. This section is an attempt at clarifying this question – which of course forces us to, provisionally at least, set aside the Bell's aforementioned reasons for advocating realism.

One of the arguments (see, e.g., [10]) by means of which the just-mentioned thesis was defended went as follows. It started from the observation that the correlation between Alice's and Bob's measurement outcomes 'may actually arise out of a statistical mixture of different situations' traditionally labelled λ. It then proceeded in four steps, namely

1. writing down the standard probability rule (10.1) as

$$P(A, B|a, b, \lambda) = P(A|a, b, B, \lambda)P(B|a, b, \lambda); \tag{10.5}$$

2. defining the locality assumption to be that for any λ, 'what happens on Alice's side does not depend on what happens on Bob's side, and vice versa';
3. inferring from this that

$$P(A|a, b, B, \lambda) = P(A|a, \lambda) \tag{10.6}$$

and

$$P(B|a, b, \lambda) = P(B|b, \lambda); \tag{10.7}$$

4. inserting (10.6) and (10.7) into (10.5), which yields, for each λ, the factorization

$$P(A, B|a, b, \lambda) = P(A|a, \lambda)P(B|b, \lambda), \tag{10.8}$$

from which the Bell–CHSH inequalities are derived in the standard way recalled above; local causality is said to be thereby refuted.

It was then pointed out (as Bell himself had done in ([6]) that the λ's may specify any state of affairs in the outside world, including even quantum states, the notion of which is generally not considered to particularly rely on realism; and that this also holds true concerning the instrument orientations a and b, which may be considered mere observables. In view of this all, it was finally claimed that proving nonlocality necessitates no additional premise and in particular no reference to realism.

Now, this is a conclusion that might conceivably have been drawn from Bell's 1981 approach, but it was shown above that the latter was not really convincing. The point here is that the reasoning above is at variance with the stand Bell finally took in his 1990 article, in which, as we saw, he stressed the necessity, for the argument to go through, of fulfilling Condition C above. And indeed, bearing this in mind, it must be observed that, in Step 2 above, the notion 'dependence' is ambiguous. One meaning of the word is that B is one of the direct causes of A in the traditional causality–action sense; however, in probability calculus, it has the wider meaning that A depends on B whenever the probability $P(A)$ of A depends on B, the dependence being due either to B being the direct cause of A or to the fact that A and B had a common cause in their past. If what is required is that the investigated notion should somehow be connected with local causality in the causality–action sense (i.e., in what we called the 'first sense', the one Bell clearly adopted), the meaning to be retained is clearly the first one. But the fact that A does not depend on B in this sense or, in other terms, that B is not a direct cause of A, does not imply that the probabilities $P(A)$ and $P(B)$ are independent – in other words, it does not imply factorization of $P(A, B)$ – since, to repeat, A and B may have shared a common direct cause in their past. Hence the reasoning fails.

On the other hand, a remark by T. Norsen [11] may dissuade us from being categorical on this point, for, while he agreed on the necessity of taking Condition C into account, he pointed out that the latter 'commits us, really, to nothing', presumably meaning by this that theories are produced by us and that therefore *we* decide what premises we impart to them; that, for example, we may freely decide that in region *3* of [7] just one causal factor exists, which is the pair wave function, and that it is specified, which just means it is kept the same in all of the measurements on individual pairs that compose a given experiment. This is true, but still it does not rule out the possibility that the not-yet-discovered 'truly correct' theory, the one describing nature 'as it really is', is not the one we have in mind. So, as we see, the conclusion of the just-reported reasoning may still be retained, but only provided the view (deemed obsolete by many) that nature obeys fixed eternal laws that are not yet known (and may never be) is definitely rejected.

10.4 Back to the Question Raised by the Article's Title and Conclusion

As we just saw, the validity of the statement that nonlocality is true quite independent of whether or not Einsteinian realism is true cannot be completely salvaged in the mind of any thinker, even taking Norsen's remark into account. But to answer the question this section is meant to deal with, the safest procedure could well be another one, namely searching for a conception compatible both with local causality and with all the verifiable predictions from quantum mechanics. If, by any chance, one can be found, the question will automatically be answered in the negative. Now, we know already that no realist theory (in the sense of Einsteinian realism) can satisfy both requisites. But conceivably some nonrealist one might.

Admittedly the conception we think of lies rather far from beaten tracks. It is suggested by significant aspects of Rovelli's *Relational Quantum Mechanics* [12], a theory quite a number of physicists deem acceptable. From the latter, it borrows the idea that the notion 'relativity' should be extended to the perceptions individuals have of the properties (attributes) of systems. Just as the speed of a vehicle is not the same relative to someone standing on the sidewalk and relative to a driver overtaking it, this conception assumes that the properties of any object must not be considered attached to that object but rather to its observers, and are therefore relative to each observer observing it. As we see, it is clearly centered on observation and information. It is incompatible with Einsteinian realism, and since realism has here been identified with the latter, we should indeed consider it to be a variety of antirealism.

Let one of the most salient features of this theory be briefly recalled. As we know, in the eighty-years-old EPR reasoning the there defined notion 'elements of reality' enjoys an absolute meaning, in the sense that once the existence of one such element has been duly derived, it is valid for any observer whatsoever (in its name the word 'reality', akin to Bell's 'beable', is meant to stress precisely that). In an EPRB type experiment once, via reasoning making use of local causality, the value of some quantity has been shown to be an element of reality, it is this absoluteness that entails that the value in question remains the same also when, instead of the measurement that induced it, the measurement of some other quantity is performed; and that also entails it must be the same for any observer. This is what made it possible for EPR to infer that quantum mechanics is not a complete theory. But within the nonrealist conception considered here, such a derivation is impossible, for if we tried to define in EPR's way, some kind of 'element of reality' notion, this could not enjoy absoluteness, since it would be necessary to distinguish within it the 'elements of reality' relative to the various observers; those, for example, relative to Alice and to Bob respectively in an EPRB experiment. From this it follows that the well-known reasoning by means of which EPR proved the incompatibility between standard quantum mechanics (without hidden variables) and local causality does not go through. In other words, this conception appears compatible with local causality.

Of course the question then arises of whether or not its compatibility with the quantum mechanical observational predictions can be refuted in Bell's manner, by showing that it entails an inequality such as those of Bell and CHSH. But it is easily seen that it cannot. For indeed, according to the conception in question, the set of notions that cannot be defined to be observer-independent includes not only elements of reality but also the probabilities that some definite observation will be made. In it, for example, in the second member of equation (10.4), the probability $P(A|a, \lambda_i)$ has a meaning only for Alice. For Bob it does not have any, which implies of course that, for Bob, its product with any given number has no sense either. For Alice the reverse is true: neither $P(B|b, \lambda_i)$ nor its product with any number has a meaning. This implies that the product $P(A|a, \lambda_i)P(B|b, \lambda_i)$ has a meaning neither for Bob nor for Alice (nor for any other observer, of course). Consequently, a joint probability such as $P(A, B|a, b)$, as yielded by (10.4), simply has no meaning at all. And since the significant terms in both the Bell and the CHSH inequalities are ultimately composed of such expressions, it follows that the inequalities are meaningless as well. Their possible incompatibility with the quantum mechanical predictions and with experiment is therefore void of significance.

A nonrealist conception borrowed from an interpretation of the quantum mechanical formalism considered acceptable and interesting by a number of physicists has been described, whose incompatibility with local causality cannot be established by any presently known method. Admittedly this result is *not* sufficient for proving that, to derive nonlocality from the validity of the quantum observational predictions, realism must be assumed. The reason is that, in this respect, what we saw is merely a counterexample to the view that such a derivation is possible in complete generality, without anything else being postulated, which only means that some other premise must be assumed, not that the latter *must* be realism. On the other hand, currently, no other premise is available to play that role efficiently, so result in question reinforces an impression the reading of Bell's articles already imparts, the one that assuming realism is by far the safest way to establish nonlocality on truly firm grounds. In any case, to the question that forms the title of this article it gives with certainty (except for people bluntly rejecting anything approaching the Rovelli approach!) a definite answer: the answer 'no'.

References

[1] Hempel, C.G. 1965, *Aspects of Scientific Explanation*, New York: The Free Press.
[2] Hempel, C.G. 1988, A problem concerning the inferential function of scientific theories, in A. Grünbaum and W.C. Salmon (eds.), *The Limitations of Deductivism*, Berkeley, CA: University of California Press, pp. 19–36, quoted by M. Kisler, *La causalité dans la philosophie contemporaine*, Paris: *Intellectica* **1** (38), 139–85.
[3] Einstein, A. 1907, *Ann. Phys.* **23**, 371, quoted in [7].
[4] Bell, J.S. 1964, On the Einstein–Podolsky–Rosen paradox, *Physics*, **1**, 195–200.
[5] Bell, J.S. 1976, The theory of local beables, *Epistemological Letters*, March.
[6] Bell, J.S. 1981, Bertlmann's socks and the nature of reality, Colloque C2, suppl. au numéro 3, *Journal de Physique* **42**, 41–61.
[7] Bell, J.S. 1990, La nouvelle cuisine, in A. Sarlemijn and P. Kroes (eds.), *Between Science and Technology*, New York: Elsevier, North-Holland, pp. 97–115.
[8] Einstein, A. 1949, *Reply to Criticism*. Library of Living Philosophers, Lasalle (IL), quoted by M. Paty, Einstein philosophe, Paris: P.U.F. (1993).
[9] Norsen, T. 2011, J.S.Bell's concept of local causality. *Am. J. Phys.* **79**, 1261, and ArXiv:quant-ph/0707.0401v3.

[10] Gisin, N. 2012, Non-realism: Deep thought or a soft option? *Found. Phys.* **42**, 80–85, and ArXiv:quant-ph/0901.4255.

[11] Norsen, T. 2006, Counter-factual meaningfulness and the Bell and CHSH inequalities. ArXiv:quant-ph/0606084v1.

[12] Rovelli, C. 1996. Relational quantum mechanics, *Int. J. Theor. Phys.* **35**, 1637–78,

11

Local Causality, Probability and Explanation

RICHARD A. HEALEY

Abstract

In papers published in the 25 years following his famous 1964 proof, John Bell refined and reformulated his views on locality and causality. Although his formulations of local causality were in terms of probability, he had little to say about that notion. But assumptions about probability are implicit in his arguments and conclusions. Probability does not conform to these assumptions when quantum mechanics is applied to account for the particular correlations Bell argues are locally inexplicable. This account involves no superluminal action and there is even a sense in which it is local, but it is in tension with the requirement that the direct causes and effects of events be nearby.

11.1 Introduction

I never met John Bell, but his writings have supplied me with a continual source of new insights as I read and reread them over 40 years. As I worked toward a rather different understanding of quantum mechanics he was foremost in my mind as a severe but honest critic of such attempts. We all would love to know what Einstein would have made of Bell's theorem. I confess that the deep regret I feel that Bell cannot respond to this paper is sometimes assuaged by a sense of relief.

11.2 Locality and Local Causality

In his seminal 1964 paper [1], John Bell expressed locality as the requirement

that the result of a measurement on one system be unaffected by operations on a distant system with which it has interacted in the past.

[2, p. 14]

This seems to require that the result of a measurement would have been the same, no matter what operations had been performed on such a distant system. But suppose the result of a measurement were the outcome of an indeterministic process. Then the result of the measurement might have been different even if exactly the same operations (if any) had been performed on that distant system. So can no indeterministic theory satisfy the locality requirement? Bell felt no need to address that awkward question in his 1964 paper [1], since he took the EPR argument to establish that any additional variables needed to restore locality and causality would have to determine a unique result of a measurement. Indeterminism was not an option:

Since we can predict in advance the result of measuring any chosen component of σ_2 [in the Bohm–EPR scenario], by previously measuring the same component of σ_1, it follows that the result of any such measurement must actually be predetermined.

[2, p. 15]

Afterwards he repeatedly stressed that any theory proposed as an attempt to complete quantum theory while restoring locality and causality need not be *assumed* to be deterministic: to recover such perfect (anti)correlations it would *have* to be deterministic. This argument warrants closer examination, and I will come back to it. But in later work Bell offered formulations of locality conditions tailored to theories that were not deterministic.

An initial motivation may have been to facilitate experimental tests of attempted local, causal completions of quantum mechanics by suitable measurements of spin or polarization components on pairs of separated systems represented by entangled quantum states. Inevitable apparatus imperfections would make it impossible to confirm a quantum prediction of perfect (anti)correlation for matched components, so experiment alone could not require a local, causal theory to reproduce them. Einstein himself thought some theory might come to underlie quantum mechanics much as statistical mechanics underlies thermodynamics. In each case there would be circumstances in which the more basic theory (correctly) predicted deviations from behavior the less basic theory leads one to expect.

But by 1975 a second motivation had become apparent – the hope that by revising or reformulating quantum mechanics as a theory of local beables one might remove ambiguity and arrive at increased precision. It is in this context that Bell now introduces a requirement of local causality. This differs in two ways from his earlier locality requirement. It is not a requirement on the world, but on theories of local beables: and it applies to theories that are probabilistic, with deterministic theories treated as a special case in which all probabilities are 0 or 1 (and densities are delta functions).[1] Bell [3, 4] designs his requirement of local causality as a generalization of a requirement of local determinism met by Maxwell's electromagnetic theory. In source-free Maxwellian electromagnetism, the local beables are the values of the electric and magnetic fields at each point $(\mathbf{x}, t)$. This theory is locally deterministic because the field values in a space–time region are uniquely determined by their values at an earlier moment in a finite volume of space that fully closes the backward light cone of that region.

---

[1] While Bell did not make this explicit in 1975, his 1990 paper also notes the analogy with source-free Maxwellian electromagnetism, and there he does say:

The deterministic case is a limit of the probabilistic case, the probabilities becoming delta functions.

[2, p. 240]

Local causality arises by generalizing to theories in which the assignment of values to some beables Λ implies, not necessarily a particular value, but a probability distribution $\Pr(A|\lambda)$, for another beable A. Here is how Bell [2, p. 54] defines it (in my notation):

Let N denote a specification of *all* the beables, of some theory, belonging to the overlap of the backward light cones of space-like separated regions 1 and 2. Let Λ be a specification of some beables from the remainder of the backward light cone of 1, and B of some beables in the region 2. Then in a *locally causal theory* $\Pr(A|\Lambda, N, B) = \Pr(A|\Lambda, N)$ whenever both probabilities are given by the theory.

If M is a specification of some beables from the backward light cone of 2 but not of 1, then (assuming the joint probability distribution $\Pr(A, B|\Lambda, M, N)$ exists)

$$\Pr(A, B|\Lambda, M, N) = \Pr(A|\Lambda, M, N, B) \cdot \Pr(B|\Lambda, M, N) \qquad (11.1)$$

$$= \Pr(A|\Lambda, N) \cdot \Pr(B|\Lambda, M), \qquad (11.2)$$

where (11.1) follows from the definition of conditional probability, and (11.2) follows for any locally causal theory. This means that any theory of local beables that is locally causal satisfies the condition

$$\Pr(A, B|\Lambda, M, N) = \Pr(A|\Lambda, N) \cdot \Pr(B|\Lambda, M). \qquad (11.3)$$

In his 1990 presentation Bell modified his formulation of local causality, in part in response to constructive criticisms. He also defended his revised formulation by appeal to an *Intuitive Principle* of local causality (IP), namely

The direct causes (and effects) of events are near by, and even the indirect causes (and effects) are no further away than permitted by the velocity of light.

[2, p. 239]

Here is Bell's revised formulation of *Local Causality* (LC):

A theory will be said to be locally causal if the probabilities attached to values of local beables in a space-time region 1 are unaltered by specification of values of local beables in a space-like separated region 2, when what happens in the backward light cone of 1 is already sufficiently specified, for example by a full specification of local beables in a space-time region 3 [a thick "slice" that fully closes the backward light cone of region 1 wholly outside the backward light cone of 2].

[2, p. 240]

Bell [5] then applies this condition to a schematic experimental scenario involving a linear polarization measurement on each photon in an entangled pair in which the polarizer setting a and outcome recording A for one photon occur in region 1, while those (b, B) for the other photon occur in region 2. He derives a condition analogous to (11.3) and uses it to prove a CHSH inequality whose violation is predicted by quantum mechanics for the

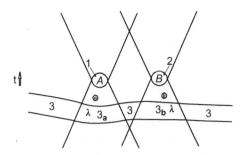

Figure 11.1 Space–time diagram for *Local Causality*.

chosen entangled state for certain sets of choices of a, b:[2]

$$\Pr(A, B|a, b, c, \lambda) = \Pr(A|a, c, \lambda) \cdot \Pr(B|b, c, \lambda). \qquad \text{(Factorizability)}$$

Not only did this proof not assume the theory of local beables was deterministic, but also even this (Factorizability) condition was not assumed but derived from the reformulated local causality requirement.

Bell's formulation of local causality (LC) has been carefully analyzed by Norsen [6], whose analysis has been further improved by Seevinck and Uffink [7]. They have focused in their analyses on what exactly is involved in a sufficient specification of what happens in the backward light cone of 1. This specification could fail to be sufficient through failing to mention local beables in 3 correlated with local beables in 2 through a joint correlation with local beables in the overlap of the backward light cones of 1 and 2. A violation of a local causality condition that did not require such a sufficient specification would pose no threat to the intuitive principle of local causality: specification of beables in 2 could alter the probabilities of beables in 1 if *unspecified* beables in 3 were correlated with both through a (factorizable) common cause in the overlap of the backward light cones of 1 and 2. On the other hand, requiring a specification of *all* local beables in 3 may render the condition (LC) inapplicable in attempting to show how theories meeting it predict correlations different from those successfully predicted by quantum theory.

To see the problem, consider the setup for the intended application depicted in Figure 11.1. A, B describe macroscopic events,[3] each usually referred to as the detection of a photon linearly polarized either vertically or horizontally relative to an a- or b-axis, respectively: a, b label events at which an axis is selected by rotating through an angle $a°$, $b°$, respectively, from some fixed direction in a plane. The region previously labeled 3

---

[2] Both λ and c are assumed to be confined to region 3 (now symmetrically extended to close the backward light cone of 2 also): c stands for the values of magnitudes characterizing the experimental setup in terms admitted by ordinary quantum mechanics, while λ specifies the values of magnitudes introduced by the theory assumed to complete quantum mechanics. It will not be necessary to mention c in what follows.

[3] I use each of 'A','B' to denote a random variable with values $\{V_A, H_A\}$, $\{V_B, H_B\}$, respectively, e_A ($\bar{e}_A$) denotes the event in region 1 of Alice's photon being registered as vertically (horizontally) polarized. e_B ($\bar{e}_B$) denotes the event in region 2 of Bob's photon being registered as vertically (horizontally) polarized.

has been relabeled as 3a, a matching region 3b has been added in the backward light cone of 2, and '3' now labels the entire continuous "stack" of space-like hypersurfaces right across the backward light cones of 1 and 2, shielding off these light cones' overlap from 1, 2 themselves. Note that each of 1, a is space like separated from each of 2, b.

In some theories, a complete specification of local beables in 3 would constrain (or even determine) the selection events a, b. But in the intended application a, b must be treated as free variables in the following sense: in applying a theory to a scenario of the relevant kind, each of a, b is to be specifiable independently in a theoretical model, and both are taken to be specifiable independent of a specification of local beables in region 3. Since this may exclude some *complete* theoretical specifications of beables in region 3, it is best not to require such completeness. Instead, one should say exactly what it is for a specification to be sufficient.

Seevinck and Uffink [7] clarify this notion of sufficiency as a combination of functional and statistical sufficiency, rendering the label b and random variable B (respectively) redundant for predicting $\Pr_{a,b}(A|B, \lambda)$, the probability a theory specifies for beable A representing the outcome recorded in region 1 given beables a, b representing the free choices of what the apparatus settings are in subregions of 1, 2, respectively, conditional on outcome B in region 2 and beable specification λ in region 3. This implies that

$$\Pr_{a,b}(A|B, \lambda) = \Pr_a(A|\lambda). \tag{11.4a}$$

Notice that a, b are no longer treated as random variables, as befits their status as the locus of free choice. It would be unreasonable to require a theory of local beables to predict the probability that the experimenters make one free choice rather than another: but treating a, b as random variables (as in Bell's formulation of Factorizability) would imply the existence of probabilities of the form $\Pr(a|\lambda)$, $\Pr(b|\lambda)$.

By symmetry, interchanging '1' with '2', 'A' with 'B' and 'a' with 'b' implies that

$$\Pr_{a,b}(B|A, \lambda) = \Pr_b(B|\lambda). \tag{11.4b}$$

Seevinck and Uffink [7] offer equations (11.4a) and (11.4b) as their mathematically sharp and clean (re)formulation of the condition of local causality. Together, these equations imply the condition

$$\Pr_{a,b}(A, B|\lambda) = \Pr_a(A|\lambda) \times \Pr_b(B|\lambda) \qquad \text{(Factorizability}_{\text{SU}})$$

used to derive CHSH inequalities. Experimental evidence that these inequalities are violated by the observed correlations in just the way quantum theory leads one to expect may then be taken to disconfirm Bell's intuitive causality principle.

In more detail, Seevinck and Uffink [7] claim that orthodox quantum mechanics violates the statistical sufficiency conditions (commonly known as Outcome Independence, following Shimony)

$$\Pr_{a,b}(A|B, \lambda) = \Pr_{a,b}(A|\lambda), \tag{11.5a}$$

$$\Pr_{a,b}(B|A, \lambda) = \Pr_{a,b}(B|\lambda), \tag{11.5b}$$

while conforming to the functional sufficiency conditions (commonly known as Parameter Independence, following Shimony)

$$\mathrm{Pr}_{a,b}(A|\lambda) = \mathrm{Pr}_a(A|\lambda), \tag{11.6a}$$

$$\mathrm{Pr}_{a,b}(B|\lambda) = \mathrm{Pr}_b(B|\lambda). \tag{11.6b}$$

Statistical sufficiency is a condition employed by statisticians in situations where considerations of locality and causality simply do not arise. But in this application the failure of quantum theory to provide a specification of beables in region 3 such that the outcome B is always redundant for determining the probability of outcome A (and similarly with 'A', 'B' interchanged) has clear connections to local causality, as Seevinck and Uffink's analysis [7] has shown.

In the light of Seevinck and Uffink's analysis, perhaps Bell's local causality condition (LC) should be reformulated as follows:

(LC$_{SU}$)   A theory is said to be locally causal$_{SU}$ if it acknowledges a class R_λ of beables λ in space–time region 3 whose values may be attached independent of the choice of a, b and are then sufficient to render b functionally redundant and B statistically redundant for the task of specifying the probability of A in region 1.

The notions of statistical and functional redundancy appealed to here are as follows:

For $\lambda \varepsilon R_\lambda$, λ renders B statistically redundant for the task of specifying the probability of A iff $\mathrm{Pr}_{a,b}(A|B, \lambda) = \mathrm{Pr}_{a,b}(A|\lambda)$.
For $\lambda \varepsilon R_\lambda$, λ renders b functionally redundant for the task of specifying the probability of A iff $\mathrm{Pr}_{a,b}(A|\lambda) = \mathrm{Pr}_a(A|\lambda)$.

Though admittedly less general than (LC), (LC)$_{SU}$ seems less problematic but just as well motivated by (IP), as applied to the scenario depicted in Figure 11.1. If correlations in violation of the CHSH inequality are locally inexplicable insofar as no theory of local beables can explain them in a way consistent with (LC), then they surely also count as locally inexplicable in so far as no theory of local beables can explain them in a way consistent with (LC)$_{SU}$. But Bell himself said we should regard his step from (IP) to (LC) with the utmost suspicion, and that is what I shall do. My grounds for suspicion are my belief that quantum mechanics *itself* helps us to explain the particular correlations violating CHSH inequalities that Bell [2, pp. 151–2] claimed to be locally inexplicable without action at a distance. Moreover, that explanation involves no superluminal action, and there is even a sense in which it is local.

To assess the status of (LC) (or (LC)$_{SU}$) in quantum mechanics, one needs to say first how it is applied to yield probabilities attached to values of local beables in a space–time region 1 and then what it would be for these to be altered by specification of values of local beables in a space like separated region 2. This is not a straightforward matter. The Born rule may be correctly applied to yield more than one chance for the same event in region 1, and there is more than one way to understand the requirement that these chances be unaltered by what happens in region 2. As we will see, the upshot is that while Born rule probabilities do violate (Factorizability) (or [Factorizability$_{SU}$]) here, this counts as a violation of (LC) (or (LC)$_{SU}$) only if that condition is applied in a way that is not motivated by (IP)'s prohibition of space like causal influences.

11.3 Probability and Chance

Bell credited his formulation of local causality (LC) with avoiding the 'rather vague notions' of cause and effect by replacing them with a condition of probabilistic independence. The connection to (IP)'s motivating talk of 'cause' and 'effect' is provided by the thought that a cause alters (and typically raises) the chance of its effect. But this connection can be made only by using the general probabilities supplied by a theory to supply chances of particular events.

By chance I mean the definite, single-case probability of an individual event such as rain tomorrow in Tucson. As in this example, its chance depends on *when* the event occurs – afterwards, it is always 0 or 1: and it may vary up until that time as history unfolds. Chance is important because of its conceptual connections to belief and action. The chance of e provides an agent's best guide to how strongly to believe that e occurs, when not in a position to be certain that it does.[4] And the comparison between e's chances according as (s)he does or does not do D are critical in the agent's decision about whether to do D. These connections explain why the chance of an event defaults to 0 or 1 when the agent is in a position to be certain about it – typically, after it does or does not occur.

Probabilistic theories may be useful guides to the chances of events, but what they directly yield are not chances but general probabilities of the form $\mathrm{Pr}_C(E)$ for an event of type E relative to reference class C. To apply such a general probability to yield the chance of e, you need to specify the type E of e and also the reference class C. A probabilistic theory may offer alternative specifications when applied to determine the chance of e, in which case it becomes necessary to choose the appropriate specifications. Actuarial tables may be helpful when estimating the chance that you will live to be 100, but you differ in all kinds of ways from every individual whose death figures in those tables. What you want is the most complete available specification of *your* situation: this may include much irrelevant information, but it is not necessary to exclude this since it will not affect the chance anyway. In Minkowski space–time, the conceptual connection between chance (Ch) and the degree of belief (Cr) it prescribes is captured in this version of David Lewis's Principal Principle that implicitly defines chance:[5]

The chance of e at p, conditional on any information I_p about the contents of p's past light cone satisfies: $\mathrm{Cr}_p(e/I_p) =_{\mathrm{df}} \mathrm{Ch}_p(e)$.

Now consider an agent who accepts quantum theory and wishes to determine the chance of the event e_A that the next photon detected by Alice registers as vertically polarized (V_A).

---

[4] I use the "tenseless present" rather than the more idiomatic future tense here for reasons that will soon become clear.

[5] See Ismael [8]. I have slightly altered her notation to avoid conflict with my own. Here 'e' ambiguously denotes both an event and the proposition that it occurs. Cr stands for credence: an agent's degree of belief in a proposition, represented on a scale from 0 to 1 and required to conform to the standard axioms of probability.

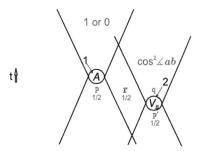

Figure 11.2 "The" chance of V_A.

Assuming that the state $\left|\Phi^+\right\rangle = \frac{1}{\sqrt{2}}\left(|H\rangle\,|H\rangle + |V\rangle\,|V\rangle\right)$ was prepared and the settings a, b chosen long before, the agent is also in a position to be certain what these were. The agent is then in a position to use the Born rule to determine the chance of e_A. But that chance must be relativized not just to a time, but (relativistically) to a space–time point. So consider the diagram in Figure 11.2.

As this shows, if the outcome in region **2** is of type V_B then $\mathrm{Ch}_p(e_A) = \mathrm{Ch}_{p'}(e_A) = \mathrm{Ch}_r(e_A) = \frac{1}{2}$, but $\mathrm{Ch}_q(e_A) = \cos^2 \angle ab$. These are the chances that follow from application of the Born rule to state $\left|\Phi^+\right\rangle$, given settings a, b. In each case the event e_A of the next photon detected in **1**'s registering as vertically polarized has been specified as of type V_A, and the specification of the reference class at least includes the state and settings. Specifically,

$$\mathrm{Ch}_p(e_A) = \mathrm{Pr}_{a,b}^{\Phi^+}(V_A) = ||\hat{P}^A(V)\Phi^+||^2 = \frac{1}{2},$$

$$\mathrm{Ch}_q(e_A) = \mathrm{Pr}_{a,b}^{\Phi^+}(V_A|V_B) \equiv \frac{\mathrm{Pr}_{a,b}^{\Phi^+}(V_A, V_B)}{\mathrm{Pr}_{b}^{\Phi^+}(V_B)} = \frac{\frac{1}{2}\cos^2 \angle ab}{\frac{1}{2}} = \cos^2 \angle ab.$$

Note that the reference class used in calculating $\mathrm{Ch}_q(e_A)$ is narrower: it is further restricted by specification of the outcome as of type V_B in region **2**.

Any agent who accepts quantum theory and is (momentarily) located at space–time point x should match credence in e_A to $\mathrm{Ch}_x(e_A)$ because it is precisely the role of chance to reflect the epistemic bearing of all information accessible at x on facts not so accessible, and to accept quantum theory is to treat it as an expert when assessing the chances. This is so whether or not an agent is *actually* located at x – fortunately, since it is obviously a gross idealization to locate the epistemic deliberations of a physically situated agent at a space–time point! A hypothetical agent located at q in the forward light cone of region **2** (but not **1**) has access to the additional information that the outcome in **2** is of type V_B: so the reference class used to infer the chance of e_A at q from the Born rule should include that information. That is why $\mathrm{Ch}_q(e_A)$ is determined by the conditional Born probability $\mathrm{Pr}_{a,b}^{\Phi^+}(V_A|V_B)$, but $\mathrm{Ch}_p(e_A)$ is determined by the unconditional Born probability $\mathrm{Pr}_{a,b}^{\Phi^+}(V_A)$.

In the special case where the settings a, b coincide (the polarizers are perfectly aligned), application of the Born rule yields the chances depicted in Figure 11.3.

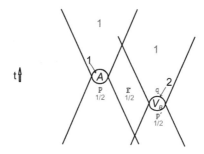

Figure 11.3 EPR–Bohm chances of V_A.

Bell [2, pp. 240–41] took this example as a simple demonstration that ordinary quantum mechanics is not locally causal, crediting the argument of EPR [9]. He begins

Each of the counters considered separately has on each repetition of the experiment a 50% chance of saying 'yes'.

Each of the chances $\mathrm{Ch}_p(e_A)$, $\mathrm{Ch}_{p'}(e_B)$ is $\frac{1}{2}$, as Bell says: but $\mathrm{Ch}_q(e_A) = 1$. After noting that quantum theory here requires a perfect correlation between the outcomes in **1**, **2**, he continues,

So specification of the result on one side permits a 100% confident prediction of the previously totally uncertain result on the other side. Now in ordinary quantum mechanics there just *is* nothing but the wavefunction for calculating probabilities. There is then no question of making the result on one side redundant on the other by more fully specifying events in some space-time region **3**. We have a violation of local causality.

It is true that (Factorizability) and (Factorizability$_{\mathrm{SU}}$) fail here, since $\mathrm{Pr}_{a,a}^{\Phi^+}(V_A|V_B) = 1$, $\mathrm{Pr}_{a,a}^{\Phi^+}(H_A|V_B) = 0$, while $\mathrm{Pr}_{a,a}^{\Phi^+}(A) = \mathrm{Pr}_{a,a}^{\Phi^+}(B) = \frac{1}{2}$. But does that constitute a violation of local causality? (LC) and (LC$_{\mathrm{SU}}$) are both conditions straightforwardly applicable to a theory whose general probabilities yield a *unique* chance for each possible outcome in **1** prior to its occurrence. In the case of quantum theory, however, the general Born rule probabilities yield *multiple* chances for each possible outcome in **1**, each at the same time (in the laboratory frame): $\mathrm{Ch}_p(e_A) = \frac{1}{2}$, but $\mathrm{Ch}_q(e_A) = 1$ (assuming the outcome in **2** is of type V_B). When (LC) speaks of '*the* probabilities attached to $[e_A, \bar{e}_A]$ in a space–time region **1** being unaltered by specification of $[V_B]$ in a space-like separated region **2**' (my italics), which probabilities are these?

Since the connection to (IP)'s motivating talk of 'cause' and 'effect' is provided by the thought that a cause alters the chance of its effect, (LC) is motivated only if applied to the *chances* of e_A, $\bar{e}_A$ in region **1**. But $\mathrm{Ch}_p(e_A)$ is *not* altered by specification of V_B in the space like separated region **2**: its value depends only on what happens in the backward light cone of **1**, in conformity to its role in prescribing $\mathrm{Cr}_p(e_A)$. Of course $\mathrm{Ch}_q(e_A)$ does depend on the outcome in **2**. If it did not, it could not fulfill its constitutive role of prescribing $\mathrm{Cr}_q(e_A/I_q)$

no matter what information I_q provided about the contents of q's past light cone. It follows that $\text{Ch}_q(e_A)$ is not altered but *specified* by specification of the result in **2**.

Only for a hypothetical agent whose world line has entered the future light cone of **2** at q is it true that specification of the result in **2** permits a 100% confident prediction of the previously totally uncertain result on the other side. A hypothetical agent at p is not in a position to make a 100% confident prediction. For such an agent the result in **1** remains totally uncertain: what happens in **2** makes no difference to what (s)he should believe, since region **2** is outside the backward light cone of p. That is why it is $\text{Ch}_p(e_A)$, not $\text{Ch}_q(e_A)$, that says what is certain at p. Newtonian absolute time fostered the illusion of the occurrence of future events becoming certain for everyone at the same time – when they occurred, if not sooner. Relativity requires certainty, like chance, to be relativized to space–time points – idealized locations of hypothetical knowers.

So does ordinary quantum mechanics violate local causality? If 'the probabilities' (LC) speaks of are $\text{Pr}_{a,b}^{\Phi^+}(V_A)$, $\text{Pr}_{a,b}^{\Phi^+}(H_A)$, and the condition that these be unaltered is understood to be that $\text{Pr}_{a,b}^{\Phi^+}(V_A) = \text{Pr}_{a,a}^{\Phi^+}(V_A|B)$, $\text{Pr}_{a,b}^{\Phi^+}(H_A) = \text{Pr}_{a,a}^{\Phi^+}(H_A|B)$, then ordinary quantum mechanics violates (LC). But if this is all (LC) means, then it is not motivated by (IP) and its violation does not imply that the quantum world is nonlocal in that there are superluminal causal relations between distant events. For (LC) to be motivated by causal considerations such as (IP), the probabilities (LC) speaks of must be understood to be chances, including $\text{Ch}_p(e_A)$ and $\text{Ch}_q(e_A)$. But neither of these would be altered by the specification of the outcome $[V_B]$ in a space like separated region, so local causality would not be violated. Although it remains unclear exactly how (LC) (or (LC$_{\text{SU}}$)) is supposed to be applied to quantum mechanics, one way of applying it is unmotivated by (IP), while if it is applied in another way quantum mechanics does *not* violate this local causality condition.

11.4 Chance and Causation

Suppose in the EPR–Bohm scenario that the outcome in region **2** had been of type H_B instead of V_B: then $\text{Ch}_q(e_A)$ would have been 0 instead of 1. Suppose that the polarization axis for the measurement in region **2** had been rotated through $60°$: then $\text{Ch}_q(e_A)$ would have been $\frac{1}{4}$ or $\frac{3}{4}$, depending on the outcome in region **2**. Or suppose that no polarization measurement had been performed in region **2**: then $\text{Ch}_q(e_A)$ would have been $\frac{1}{2}$. This shows that $\text{Ch}_q(e_A)$ depends counterfactually on the polarization measurement in **2** and also on its outcome. Another way to understand talk of "alteration" of "the" probability of an event of type V_A is as the difference between the actual value of $\text{Ch}_q(e_A)$ and what its value would have been had the polarization measurement in **2** or its outcome been different. Do such counterfactual "alterations" in $\text{Ch}_q(e_A)$ not amount to *causal* dependence between space like separated events, in violation of (IP)? There are several reasons why they do not.

The first reason is that while $\text{Ch}_q(e_A)$ would be different in each of these counterfactual scenarios, in none of them would $\text{Ch}_p(e_A)$ differ from $\frac{1}{2}$, so the "local" chance of e_A is insensitive to all such counterfactual variations in what happens in **2**. If one wishes to infer causal from counterfactual dependence of "the" chance of a result in **1** on what happens in **2**, then only one of two relevant candidates for "the" chance displays such counterfactual dependence. For those who think of chance as itself a kind of indeterministic cause – a localized physical propensity whose actualization may produce an effect – $\text{Ch}_p(e_A)$ seems better qualified for the title of "the" chance of e_A than $\text{Ch}_q(e_A)$.

The role of chance in decisions provides the second reason. Just as the chance of e tells you everything you need to know to figure out how strongly to believe e, the causal dependence of e on d tells you everything you need to know about e and d when deciding whether to do d (assuming you are not indifferent about e). As Huw Price [10] put it, 'causal dependence should be regarded as an analyst-expert about the conditional credences required by an evidential decision maker'.

Consider the situation of a hypothetical agent Bob at p' deciding whether to act by affecting what happens in **2** to try to get outcome e_A in **1**. Bob can choose not to measure anything, or he can choose to measure polarization with respect to any axis b. If he were to measure nothing, $\text{Ch}_q(e_A)$ would be $\frac{1}{2}$. If he were to measure polarization with respect to the same axis as Alice, then $\text{Ch}_q(e_A)$ would be either 0 or 1, with an equal chance (at his momentary location p') of either outcome. Since he can neither know nor affect which of these chances it will be, he must base his decision on his best estimate of $\text{Ch}_q(e_A)$ in accordance with Ismael's [8] Ignorance Principle:

Where you're not sure about the chances, form a mixture of the chances assigned by different theories of chance with weights determined by your relative confidence in those theories.

Following this principle, Bob should assign $\text{Ch}_q(e_A)$ the estimated value $\frac{1}{2} \cdot 0 + \frac{1}{2} \cdot 1 = \frac{1}{2}$, and base his decision on that. Since measuring polarization with respect to the same axis as Alice would not raise his estimated chance of securing outcome e_A in **1**, he should eliminate this option *whether or not he could execute it*. His estimated value of $\text{Ch}_q(e_A)$ were he to measure polarization with respect to an axis rotated $60°$ from Alice's is also $\frac{1}{2}$ ($\frac{1}{2} \cdot \frac{1}{4} + \frac{1}{2} \cdot \frac{3}{4} = \frac{1}{2}$). Similarly for any other angle. This essentially recapitulates part of the content of the no-signalling theorems, going back to Eberhard [11]. Bell [2, pp. 237–8] shows why manipulation of external fields at p' or in **2** would also fail to alter Bob's estimated value of $\text{Ch}_q(e_A)$.

But what if Bob had simply arranged for the measurement in **2** to have had the different *outcome* $\bar{e}_B$? Then $\text{Ch}_q(e_A)$ would have been 0 instead of 1. No one who accepts quantum mechanics can countenance this counterfactual scenario. The Born rule implies that $\text{Pr}_b^{\Phi^+}(H_B) = \frac{1}{2}$, and anyone who accepts quantum mechanics accepts the implication that $\text{Ch}_{p'}(\bar{e}_B) = \frac{1}{2}$. So anyone who accepts quantum mechanics will have credence $\text{Cr}_{p'}(\bar{e}_B/I_{p'}) = \frac{1}{2}$ no matter what he takes to happen in the backward light cone of p' (as specified by $I_{p'}$).[6] If he accepts quantum mechanics, Bob will conclude that there is nothing it makes sense to contemplate doing to alter his estimate of $\text{Ch}_{p'}(\bar{e}_B)$, and so there is no conceivable counterfactual scenario in which one in Bob's position arranges for the

---

[6] A unitary evolution $\Phi^+ \Rightarrow \Xi^+$ corresponding to a *local* interaction there would still yield $\text{Pr}_b^{\Xi^+}(H_B) = \frac{1}{2}$.

measurement in **2** to have had the different outcome $\bar{e}_B$. In general, there is causal dependence between events in **1** and **2** only if it makes sense to speak of an intervention in one of these regions that would affect a hypothetical agent's estimated chance of what happens in the other. Anyone who accepts quantum mechanics should deny that makes sense.

Perhaps the most basic reason that counterfactual dependencies between happenings in region **2** and the chance(s) of e_A are no sign of causal dependence is that chances are not beables, and are incapable of entering into causal relations. That Bell thought they behaved like beables is suggested by the paper [3] in which he introduced local causality as a natural generalization of local determinism:

In Maxwell's theory, the fields in any space–time region **1** are determined by those in any space region V, at some time t, which fully closes the backward light cone of **1**. Because the region V is limited, localized, we will say the theory exhibits *local determinism*. We would like to form some no[ta]tion of *local causality* in theories which are not deterministic, in which the correlations prescribed by the theory, for the beables, are weaker.

[2, p. 53]

It seems that Bell thought the chances prescribed by a theory that is not deterministic were analogous to the beables of Maxwell's electromagnetism, so that while local determinism (locally) specified the local[ized] beables (e.g., fields), local causality should (locally) specify the local[ized] *chances* of beables, where those chances (like local beables) are themselves localized physical magnitudes.

Others have joined Bell in this view of chances as localized physical magnitudes. But quantum mechanics teaches us that chances are *not* localized physical propensities whose actualization may produce an effect. Maudlin says what he means by calling probabilities objective:

there could be probabilities that arise from fundamental physics, probabilities that attach to actual or possible events in virtue solely of their physical description and independent of the existence of cognizers. These are what I mean by *objective probabilities*.

[12, p. 294]

Although quantum chances do attach to actual or possible events, they are not objective in this sense. As we saw, the chance of outcome e_A does not attach to it in virtue solely of its physical description: the *chances* of e_A attach also in virtue of its space–time relations to different space–time locations. Each such location offers the epistemic perspective of a situated agent, even in a world with no such agents. The existence of these chances is independent of the existence of cognizers. But it is only because we are not merely cognizers but physically situated agents that we have needed to develop a concept of chance tailored to our needs as informationally deprived agents. Quantum chance admirably meets those needs: an omniscient God could describe and understand the physical world without it.

While they are neither physical entities nor physical magnitudes, quantum chances are objective in a different sense. They supply an objective prescription for the credences of an agent in any physical situation. Anyone who accepts quantum mechanics is committed to following that prescription.

11.5 A view of quantum mechanics

As I see it [13], it is not the function of quantum states, observables, probabilities or the Schrödinger equation to represent or describe the condition or behavior of a physical system with which they are associated. These elements function in other ways when a quantum model is applied in predicting or explaining physical phenomena such as non-localized correlations. Assignment of a quantum state may be viewed as merely the first step in a procedure that licenses a user of quantum mechanics to express claims about physical systems in descriptive language and then warrants that user in adopting appropriate epistemic attitudes toward some of these claims. The language in which such claims are expressed is not the language of quantum states or operators, and the claims are not about probabilities or measurement results: they are about the values of physical magnitudes, and I will refer to them as *magnitude claims*. Magnitude claims were made by physicists and others before the development of quantum mechanics and continue to be made, some in the same terms, others in terms newly introduced as part of some scientific advance. But even though quantum mechanics represents an enormous scientific advance, claims about quantum states, operators and probability distributions are not magnitude claims.

The quantum state has two roles. One is in the algorithm provided by the Born Rule for assigning probabilities to significant claims of the form $M_\Delta(s)$: The value of M on s lies in Δ, where M is a physical magnitude, s is a physical system and Δ is a Borel set of real numbers. In what follows, I will call a descriptive claim of the form $M_\Delta(s)$ a *canonical magnitude claim*. For two such claims the formal algorithm may be stated as follows:

$$\Pr(M_\Delta(s), N_\Gamma(s)) = \mathrm{Tr}(\rho \hat{P}^M[\Delta] \cdot \hat{P}^N[\Gamma]). \qquad \text{(Born Rule)}$$

Here ρ represents a quantum state as a density operator on a Hilbert space $\mathcal{H}_s$ and $\hat{P}^M[\Delta]$ is the value for Δ of the projection-valued measure defined by the unique self-adjoint operator on $\mathcal{H}_s$ corresponding to M.

But the significance of a claim such as $M_\Delta(s)$ varies with the circumstances to which it relates. Accordingly, a quantum state plays a second role by modulating the content of $M_\Delta(s)$ or any other magnitude claim by modifying its inferential relations to other claims. Because I believe the nature of this modulation of content renders inappropriate the metaphor of magnitudes corresponding to elements of reality, I recommend against thinking of magnitudes that figure in canonical or other magnitude claims as beables, even though many such magnitude claims are true. But if one insists on calling magnitudes that figure in magnitude claims beables, these magnitudes are not beables introduced by quantum mechanics – they are at most beables recognized in its applications.[7]

---

[7] Compare Bell [2, p. 55].

The quantum state is not a beable in this view. Indeed, since none of the distinctively quantum elements of a quantum model qualifies as a beable introduced by the theory, quantum mechanics has no beables of its own. Viewed this way, a quantum state does not describe or represent some new element of physical reality.[8] But neither is it the quantum state's role to describe or represent the epistemic state of any actual agent. A quantum state assignment is objectively true (or false): in that deflationary sense a quantum state is objectively real. But its function is not to say what the world is like but to help an agent applying quantum mechanics to predict and explain what happens in it. It is physical conditions in the world that make a quantum state assignment true (or false). True quantum state assignments are backed by true magnitude claims, though some of these are typically about physical systems other than that to which the state is assigned.

Any application of quantum mechanics involves claims describing a physical situation. While it is considered appropriate to make claims about where individual particles are detected contributing to the interference pattern in a contemporary interference experiment, claims about through which slit each particle went are frequently alleged to be 'meaningless'. In its second role the quantum state offers guidance on the inferential powers, and hence the content, of canonical magnitude claims.

The key idea here is that even assuming unitary evolution of a joint quantum state of system and environment, delocalization of system state coherence into the environment will typically render descriptive claims about experimental outcomes and the condition of apparatus and other macroscopic objects appropriate by endowing these claims with enough content to license an agent to adopt epistemic attitudes toward them, and in particular to apply the Born Rule. But an application of quantum mechanics to determine whether this is so will not require referring to any system as "macroscopic", as an "apparatus" or as an "environment". All that counts is how a quantum state of a supersystem evolves in a model, given a Hamiltonian associated with an interaction between the system of interest and the rest of that supersystem.

It is important to note that since the formulation of the Born Rule now involves no explicit or implicit reference to "measurement", Bell's strictures [2, pp. 213–31] against the presence of the term 'measurement' in a precise formulation of quantum mechanics are met. None of the other proscribed terms 'classical', 'macroscopic', 'irreversible' or 'information' appears in its stead.

Since an agent's assignment of a quantum state does not serve to represent a system's properties, her reassignment of a "collapsed" state on gaining new information represents no change in that system's properties. That is why collapse is not a physical process, in this view of quantum mechanics. Nor does the Schrödinger equation express a fundamental physical law: to assign a quantum state to a system is not to represent its dynamical properties. A formulation of quantum mechanics has no need to include a statement distinguishing the circumstances in which physical processes of Schrödinger evolution and "collapse" occur. An agent can use quantum mechanics to track changes of the dynamical

---

[8] Compare Bell [2, p. 53]: 'this does not bother us if we do not grant beable status to the wave-function.'

properties of a system by noting what magnitude claims are significant and true of it at various times. But quantum mechanics itself does not imply any such claim, even when an agent would be correct to assign a system a quantum state, appropriately apply the Born Rule, and conclude that the claim has probability 1.

Quantum states are relational, in this interpretation. When agents (actually or merely hypothetically) occupy relevantly different physical situations, they should assign different quantum states to one and the same system, even though these different quantum state assignments are equally correct. The primary function of Born probabilities is to offer a physically situated agent authoritative advice on how to apportion degrees of belief concerning contentful canonical magnitude claims that the agent is not currently in a position to check. That is why the Born Rule should be applied by differently situated agents to assign different chances to a single canonical magnitude claim $M_\Delta(s)$ about a system s in a given situation. These different chance assignments will then be equally objective and equally correct.

The physical situation of a (hypothetical or actual) agent will change with (local) time. The agent may come to be in a position to check the truth-values of previously inaccessible magnitude claims, some of which may be taken truly to describe outcomes of measurements. If a quantum state is to continue to provide the agent with good guidance concerning still inaccessible magnitude claims, it must be updated to reflect these newly accessible truths. The required alteration in the quantum state is not a physical process involving the system, in conflict with Schrödinger evolution. What has changed is just the physical relation of the agent to events whose occurrence is described by true magnitude claims. This is not represented by a discontinuous change in the quantum state of some model: it corresponds to adoption of a *new* quantum model that incorporates additional information, newly accessible to the user of quantum mechanics.

The preceding paragraphs contained a lot of talk of agents. To forestall misunderstandings, I emphasize that quantum mechanics is not about agents or their states of knowledge or belief: A precise formulation of quantum mechanics will not speak of such things in its models any more than it will speak of agents' measuring, observing or preparing activities. If quantum mechanics is about anything it is about the quantum systems, states, observables and probability measures that figure in its models. Quantum mechanics, like all scientific theories, was developed by (human) agents for the use of agents (not necessarily human: while insisting that any agent be physically situated, I direct further inquiry on the constitution of agents to cognitive scientists). Trivially, only an agent can apply a theory for whatever purpose. So any account of a predictive, explanatory or other application of quantum mechanics naturally involves talk of agents.

11.6 How to use quantum mechanics to explain nonlocalized correlations

In [14] Bell argued that

certain particular correlations, realizable according to quantum mechanics, are locally inexplicable. They cannot be explained, that is to say, without action at a distance.

[2, pp. 151–2]

The particular correlations to which Bell refers arise, for example, in the EPR–Bohm scenario in which pairs of spin-$\frac{1}{2}$ particles are prepared in a singlet spin state, and then at widely separated locations each element of a pair is passed through a Stern–Gerlach magnet and detected either in the upper or in the lower part of a screen. By calling them realizable rather than realized, he acknowledged the experimental difficulties associated with actually producing statistics supporting them in the laboratory (or elsewhere). Enormous improvements in experimental technique since 1981 have overcome most of the difficulties associated with performing a loophole-free test of CHSH or other so-called Bell inequalities and at the same time provided very strong statistical evidence for quantum mechanical predictions in analogous experiments. Since the improvements have been most dramatic for experiments involving polarization measurements on entangled photons, it is appropriate to refer back to the experimental scenario discussed by Bell in 1990 [2, pp. 232–48].

Suppose photon pairs are prepared at a central source in the entangled polarization state $\Phi^+ = 1/\sqrt{2}(|HH\rangle + |VV\rangle)$, and the photons in a pair are both subsequently detected in coincidence at two widely separated locations after each has passed through a polarizing beam splitter (PBS) with axis set at a, b, respectively. If a photon is detected with polarization parallel to this axis, a macroscopic record signifies 'yes': if it is detected with polarization perpendicular to this axis, the record signifies 'no'. Let the record 'yes' at one location be the event of a magnitude A taking on value $+1$, 'no' the event of A taking on value -1, and similarly for B at the other location. Let a be a locally generated signal that quickly sets the axis of the PBS on the A side to an angle a° from some standard direction, and similarly for b on the B side. Assume that this is done so that each of a and A's taking on a value is space like separated from each of b and B's taking on a value. In this scenario quantum mechanics predicts that, for $a^\circ = 0^\circ$, $a'^\circ = 45^\circ$, $b^\circ = 22\frac{1}{2}^\circ$, $b'^\circ = -22\frac{1}{2}^\circ$,

$$E(a, b) + E(a, b') + E(a', b) - E(a', b') = 2\sqrt{2}, \tag{11.7}$$

where, for example, $E(a, b) \equiv \mathrm{Pr}_{a,b}(+1, +1) + \mathrm{Pr}_{a,b}(-1, -1) - \mathrm{Pr}_{a,b}(+1, -1) - \mathrm{Pr}_{a,b}(-1, +1)$. This is in violation of the CHSH inequality

$$|E(a, b) + E(a, b') + E(a', b) - E(a', b')| \leq 2 \tag{CHSH}$$

that follows from (Factorizability$_{\mathrm{SU}}$). Bell claims that these correlations are realizable according to quantum mechanics but that they cannot be explained without action at a distance. While it is generally acknowledged that quantum mechanics successfully predicts Bell's particular correlations, demonstrating this will illustrate the present view of quantum mechanics. To decide whether it also explains them, we need to ask what more is required of an explanation.

What we take to be a satisfactory explanation has changed during the development of physics, and we may confidently expect such change to continue. One who accepts quantum mechanics is able to offer a novel kind of explanation. Nevertheless, explanations of phenomena using quantum mechanics may be seen to meet two very general conditions met by many, if not all, good explanations in physics:

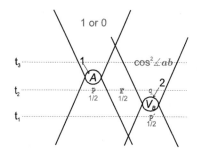

Figure 11.4 Explaining Bell's correlations.

 (i) They show that the phenomenon to be explained was to be expected, and
(ii) They say what it depends on.

Quantum mechanics enables us to give explanations meeting both conditions.

Meeting the first condition is straightforward. Anyone accepting quantum mechanics can use the Born Rule applied to state Φ^+ to calculate joint probabilities such as $\Pr_{a,b}(A, B)$ and go on to derive (11.7). So for anyone who accepts quantum mechanics, violation of the CHSH inequalities is to be expected. But it is worth showing in more detail how quantum mechanics can be applied to derive (11.7), because this will help to exhibit the relational nature of quantum states and probabilities while making it clear that a precise formulation of quantum mechanics need not use the word 'measurement' or any other term on Bell's list of proscribed words [2, p. 215].

Figure 11.4 is a space–time diagram depicting space like separated polarization measurements by Alice and Bob in regions 1, 2 respectively on a photon pair. Time is represented in the laboratory frame. At t_1 each takes the polarization state of the $L - R$ photon pair to be $\Phi^+ = 1/\sqrt{2}(|HH\rangle + |VV\rangle)$. What justifies this quantum state assignment is their knowledge of the conditions under which the photon pair was produced – perhaps by parametric down-conversion of laser light by passage through a nonlinear crystal. Such knowledge depends on information about the physical systems involved in producing the pair. This state assignment is backed by significant magnitude claims about such systems – if anything counts as a claim about beables recognized by quantum mechanics, these do. Then Alice measures polarization of photon L along axis $a°$ and Bob measures polarization of photon R along axis $b°$. Decoherence at the photon detectors licenses both of them to treat the Born Rule measure corresponding to state assignment Φ^+ as a probability distribution over significant canonical magnitude claims about the values of A, B.

At t_1, Alice and Bob should both assign state Φ^+ and apply the Born Rule to calculate the joint probability $\Pr_{a,b}^{\Phi^+}(A, B)$ assigned to claims about the values of magnitudes A, B – claims that we may, but need not, choose to describe as records of polarization measurements along both $a°, b°$ axes – and hence the (well-defined) conditional probability $\Pr_{a,b}^{\Phi^+}(A, B)/\Pr_{a,b}^{\Phi^+}(B) = |\langle A|B\rangle|^2$. Each will then expect the observed nonlocalized correlations between the outcomes of polarization measurements in regions 1, 2 when the detectors are set along the a°, b° axes. They will expect analogous correlations as these axes are varied, and so they will expect (11.7) (violating the CHSH inequality) in such a scenario.

At t_2, after recording polarization V_B for R, Bob should assign pure state $|V_B\rangle$ to L and use the Born Rule to calculate the probabilities $\Pr_a^{|V_B>}(A) = |\langle A|V_B\rangle|^2$ for Alice to record polarization of L with respect to the a°-axis. At t_2, Alice should assign state $\hat{\rho} = \frac{1}{2}\hat{1}$ to L, and use the Born Rule to calculate probability $\Pr_a^\rho(A) = \frac{1}{2}$ that she will record either polarization of L with respect to the a°-axis. In this way each forms expectations as to the outcome of Alice's measurement on the best information available to him or her at t_2. Alice's statistics for her outcomes in many repetitions of the experiment are just what her quantum state $\frac{1}{2}\hat{1}$ for L led her to expect, thereby helping to explain her results. Bob's statistics for Alice's outcomes (in many repetitions in which his outcome is V_B) are just what his quantum state $|V_B\rangle$ for L led him to expect, thereby helping him to explain Alice's results.

There is no question as to which, if either, of the quantum states $|V_B\rangle$, $\frac{1}{2}\hat{1}$ was the *real* state of Alice's photon at t_2. Neither of the different probabilities $|\langle A|V_B\rangle|^2$ or $\frac{1}{2}$ represents a unique physical propensity at t_2 of Alice's outcome – even though neither its chance at p nor its chance at q is subjective. This discussion applies independent of the time-order in Alice's frame of the regions 1, 2 in Figure 11.1: had she been moving away from Bob fast enough, she would have represented 1 as earlier than 2.

It is widely acknowledged that one cannot explain a phenomenon merely by showing that it was to be expected in the circumstances. To repeat a hackneyed counterexample, the falling barometer does not explain the coming storm even though it gives one reason to expect a storm in the circumstances. As a joint effect of a common cause, a symptom does not explain its other independent effects. But a system's being in a quantum state at a time is not a symptom of causes specified by the true magnitude claims that back it, since it is not an event distinct from the conditions those claims describe. Each event figuring in Bell's particular correlations is truly described by a canonical magnitude claim. We may choose to describe some, but not all, such events as an outcome of a quantum measurement on a system: the probabilities of many of those events depend counterfactually on the particular entangled state assigned at t_1 – if that state had been different, so would these probabilities have been. But this dependence is not causal. In quantum mechanics, neither states nor probabilities are the sorts of things that can bear causal relations: in Bell's terminology, they are not beables.

When relativized to the physical situation of an actual or hypothetical agent, a quantum state assignment is objectively true or false – which one depends on the state of the world. More specifically, a quantum state assignment is made true by the true magnitude claims that back it. One true magnitude claim backing the assignment of $|V_B\rangle$ to L at q reports the outcome of Bob's polarization measurement in region 2 of Figure 11.1; but there are others, since this would not have been the correct assignment had the correct state assignment at p' been $|H_A\rangle|V_B\rangle$. We also need to ask for the backing of the entangled state Φ^+.

There are many ways of preparing state Φ^+, and this might also be the right state to assign to some naturally occurring photon pairs that needed no preparation. In each case there is a characterization in terms of some set of true magnitude claims describing the systems and events involved; these back the state assignment Φ^+. It may be difficult or even impossible to give this characterization in a particular case, but that is just an epistemic problem that need not be solved even by experimenters skilled in preparing or otherwise assigning this state. Φ^+ will be correctly assigned at p' only if some set of true magnitude claims backing that assignment is accessible from p'; events making them true must lie in the backward light-cone of p'.

A quantum state counterfactually depends on the true magnitude claims that back it in somewhat the same way that a dispositional property depends on its categorical basis. The state Φ^+ may be backed by alternative sets of true magnitude claims, just as a person may owe his immunity to smallpox to any of a variety of categorical properties. If Walt owes his smallpox immunity to antibodies, his possession of antibodies does not cause his immunity; it is what his immunity consists in. No more is the state Φ^+ caused by its backing magnitude claims; a statement assigning state Φ^+ is true only if backed by some true magnitude claims of the right kind. A quantum state is counterfactually dependent on whatever magnitude claims back it because backing is a kind of determination or constitution relation, not because it is a causal relation.

In this view, a quantum state causally depends neither on the physical situation of the (hypothetical or actual) agent assigning it nor on any of its backing magnitude claims. The correct state $|V_B\rangle$ to be assigned to L at q is not causally dependent on anything about Bob's physical situation even if he happens to be located at q; it is not causally dependent on the outcome of Bob's polarization measurement in region 2; and it is not causally dependent on how Bob sets his polarizer in region 2. But a quantum state assignment is not just a function of the subjective epistemic state of any agent: If Bob or anyone else were to assign a state other than $|V_B\rangle$ to L at q he or she would be making a mistake.

The quantum derivation of (11.7) shows not only that Bell's particular correlations were to be expected, but also what they depend on. They depend counterfactually but not causally on the quantum state Φ^+, and they also depend counterfactually on that state's backing conditions, as described by true magnitude claims. The status of the quantum state disqualifies it from participation in causal relations, but true magnitude claims may be taken to describe beables recognized by quantum mechanics. To decide which conditions backing any of the states involved in their explanation describe causes of Bell's particular correlations or the events they correlate, we need to return to the connection between causation and chance.

The intuition that, other things being equal, a cause raises (or at least alters) the chance of its effect is best cashed out in terms of an interventionist counterfactual: c is a cause of e just in case c, e are distinct actual events and there is some conceivable intervention on c whose occurrence would have altered the chance of e. Such an intervention need not be the act of an agent; it could involve any modification in c of the right kind. Woodward [15, p. 98] is one influential attempt to say what kind of external influence this would involve. Note that Einstein's formulation of a principle of local action also appeals to intervention:

The following idea characterizes the relative independence of objects far apart in space (A and B): external influence on A has no *immediate* ('unmittelbar') influence on B; this is known as the 'principle of local action'. Einstein

[16, pp. 321–2]

I used the idea of intervention to argue against any causal dependence between events in **1** and **2**: anyone who accepts quantum mechanics accepts that it makes no sense to speak of an intervention in one of these regions that would affect a hypothetical agent's estimated chance of what happens in the other. So even though the outcome e_B in **2** backs the assignment $|V_B\rangle$ to L at q, the outcome in **1** does not depend causally on e_B. For similar reasons, neither does the outcome in **2** depend on that in **1**. The same idea can now be used to show that both these outcomes *do* depend causally on whatever event o in the overlap of the backward light cones of **1** and **2** warranted assignment of state Φ^+ – an event truly described by magnitude claims that backed this assignment.

Assume first that the events a, b at which the polarizers are set on a particular occasion occur in the overlap of the backward light cones of **1** and **2**. This assumption will later be dropped. Let r be a point outside the future light cones of e_A, e_B but within the future light cone of the event o. Let $e_A \uplus e_B$ be the event of the joint occurrence of e_A, e_B. This is an event of a type to which the Born rule is applicable; the application yields its chance $\mathrm{Ch}_r(e_A \uplus e_B) = \mathrm{Pr}^{\Phi^+}_{a,b}(V_A, V_B) = \frac{1}{2}\cos^2 \angle ab$. We already saw that $\mathrm{Ch}_r(e_A) = \mathrm{Pr}^{\Phi^+}_{a,b}(V_A) = \frac{1}{2} = \mathrm{Pr}^{\Phi^+}_{a,b}(V_B) = \mathrm{Ch}_r(e_B)$. The event o affects all these chances: Had a different event o' occurred backing the assignment of a different state (e.g., $|H_A\rangle|V_B\rangle$), or no event backing any state assignment, then any or all of these chances could have been different. Since it makes sense to speak of an agent altering the chance of event o at s in its past light cone, we have

$$\mathrm{Ch}_r(e_A \uplus e_B | do - o) \neq \mathrm{Ch}_r(e_A \uplus e_B),$$
$$\mathrm{Ch}_r(e_A | do - o) \neq \mathrm{Ch}_r(e_A),$$
$$\mathrm{Ch}_r(e_B | do - o) \neq \mathrm{Ch}_r(e_B),$$

where $do - o$ means o is the result of an intervention without which o would not have occurred. It follows that e_A, e_B, $e_A \uplus e_B$ are each causally dependent on o: o is a common cause of e_A, e_B even though the probabilities of events of these types do not factorize. The same reasoning applies to each registered photon pair on any occasion at any settings a, b. So the second requirement on explanation is met: The separate recording events, as well as the event of their joint occurrence, depend *causally* on the event o that serves to back assignment of state Φ^+ to the photon pairs involved in this scenario.

By rejecting any possibility of an intervention expressed by $do - e_B$ or $do - \bar{e}_B$, anyone accepting quantum mechanics should deny that $\mathrm{Ch}_{p(q)}(e_A | do - e_B) \neq \mathrm{Ch}_{p(q)}(e_A | do - \bar{e}_B)$ is true or even meaningful. Nevertheless, $\mathrm{Ch}_q(e_A | e_B) \neq \mathrm{Ch}_q(e_A | \bar{e}_B)$: in this sense e_A depends counterfactually but not causally on e_B. Does such counterfactual dependence provide reason enough to conclude that e_A is part of the explanation of e_B? An obvious objection is that because of the symmetry of the situation with **1** and **2** space like separated, there is an equally strong reason to conclude that e_B is part of the explanation of e_A, contrary to the fundamentally asymmetric nature of the explanation relation. But one can see that this objection is not decisive by paying attention to the contrasting epistemic perspectives associated with the different physical situations of hypothetical agents Alice* and Bob* with world lines confined to interiors of the light cones of **1, 2** respectively.

As his world line enters the future light cone of **2**, Bob* comes into position to know the outcome at **2** while still physically unable to observe the outcome at **1**. His epistemic situation is then analogous to that of a hypothetical agent Chris in a world with Newtonian absolute time, in a position to know the outcome of past events but physically unable to observe any future event. Many have been tempted to elevate the epistemic asymmetry of Chris's situation into a global metaphysical asymmetry in which the future is open while the past is fixed and settled. It is then a short step to a metaphysical view of explanation as a productive relation in which the fixed past gives rise to the (otherwise) open future, either deterministically or stochastically.[9]

Such a move from epistemology to metaphysics should always be treated with deep suspicion. But in this case it is clearly inappropriate in a relativistic space–time, since the "open futures" of agents such as Alice* and Bob* cannot be unified into *the* open future. This prompts a retreat to a metaphysically drained view of explanation as rooted in cognitive concerns of a physically situated agent, motivated by the need to unify, extend and efficiently deploy the limited information to which it has access.

For many purposes it is appropriate to regard the entire scientific community as a (spatially) distributed agent, and to think of the provision of scientific explanations as aiding *our* collective epistemic and practical goals. This is appropriate insofar as localized agents share an epistemic perspective, with access to the same information about what has happened. But Alice* and Bob* do not have access to the same information at time t_2 or t_3, since they are then space like separated. So it is entirely appropriate for Bob* to use e_B to explain e_A and for Alice* to use e_A to explain e_B. This does not make explanation a subjective matter, for two reasons. There is an objective physical difference between the situations of Alice* and Bob* underlying the asymmetry of their epistemic perspectives: and by adopting either perspective in thought (as I have encouraged the reader to do), anyone can come to appreciate how each explanation can help make Bell's correlations seem less puzzling. Admittedly, neither explanation is very deep, and I will end by noting one puzzle that remains.

By meeting both minimal requirements on explanation, the application of quantum theory enables us to explain Bell's correlations. But is this explanation local? Several senses of locality are relevant here. The explanation involves no superluminal causal dependence. As stated, the condition of Local Causality is not applicable to the quantum mechanical explanation, since it presupposes the uniqueness of the probability to which it refers. (Factorizability$_{SU}$) (and presumably also (Factorizability)) are violated, but Bell [2, p. 243] preferred to see (Factorizability) as not a formulation but a consequence of 'local causality'. I have argued that it is not. To retain its connection to (IP), a version of Local Causality

---

[9] See, for example, Maudlin [17, pp. 173–8].

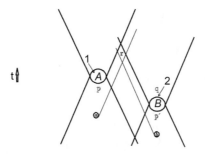

Figure 11.5 A late chance.

should speak of chances rather than general probabilities. A version that equates the unconditional chance of e_A to its chance conditional on e_B holds, no matter how these chances are relativized to the same space–time point. But a version that is clearly motivated by the intuitive principle (IP) would rather equate the unconditional chance of e_A to its chance conditional on *an intervention that produces* e_B. However, this version is inapplicable, since acceptance of quantum mechanics renders senseless talk of interventions producing e_B.

The explanation one can give by applying quantum mechanics appeals to *chances* that are localized, insofar as they are assigned at space–time points that may be thought to offer the momentary perspective of a hypothetical idealized agent whose credences they would guide. But these chances are not quantum beables and they are not physical propensities capable of manifestation at those locations (or anywhere else). The only causes figuring in the explanation are localized where the physical systems are whose magnitudes back the assignment of state Φ^+. That chances are not propensities becomes clear when one drops the assumption that a, b occur in the overlap of the backward light cones of **1** and **2**, as depicted in Figure 11.5.

If a, b are set at the last moment, the chance of $e_A \uplus e_B$ that figures in its explanation may be located *later* in the laboratory frame than e_A, e_B. If chance were a physical propensity, it should act *before* its manifestation. But chances are not propensities – proximate causes of localized events. They are a localized agent's objective guide to credence about epistemically inaccessible events.

I will conclude by noting one sense in which the explanation one can give using quantum mechanics is not local as it stands. Though it is a (nonfactorizable) cause of events of types A, B in regions **1**, **2**, respectively, the event o is not connected to its effects by any spatiotemporally continuous causal process described by quantum mechanics. This puts the explanation in tension with the *first* conjunct of Bell's [2, p. 239] intuitive locality principle (IP):

The direct causes (and effects) of events are near by, and even the indirect causes (and effects) are no further away than permitted by the velocity of light.

o is separated from both recording events in regions **1**, **2** in time, and from at least one in space. If o is not merely a cause but a *direct* cause of these events, then it violates the first conjunct of (IP), because it is *not* nearby. But if one adopts the present view of quantum mechanics, the theory has no resources to describe any causes mediating between *o* and these recording events. So while their quantum explanation is not explicitly inconsistent with the first conjunct of (IP), mediating causes could be found only by constructing a new theory. Bell's work has clearly delineated the obstacles that would have to be overcome on that path.

References

[1] Bell, J.S. [1964] On the Einstein–Podolsky–Rosen paradox, *Physics* **1**, 195–200.

[2] Bell, J.S. [2004] *Speakable and Unspeakable in Quantum Mechanics* (2nd rev. ed.), Cambridge: Cambridge University Press.

[3] Bell, J.S. [1975] The Theory of Local Beables. TH-2053-CERN, July 28.

[4] Bell, J.S. [1985] The theory of local beables, *Dialectica* **39**, 86–96.

[5] Bell, J.S. [1990] La nouvelle cuisine. In A. Sarlemijn and P. Krose (eds.), *Between Science and Technology*, pp. 97–115.

[6] Norsen, T. [2011] John S. Bell's concept of local causality, *American Journal of Physics* **79**, 1261–75.

[7] Seevinck, M.P. and Uffink, J. [2011] Not throwing out the baby with the bathwater: Bell's condition of local causality mathematically 'sharp and clean', in D. Dieks et al. (eds.), *Explanation, Prediction and Confirmation*, Berlin: Springer, pp. 425–50.

[8] Ismael, J. [2008] *Raid!* Dissolving the big, bad bug, *Nous* **42**, 292–307.

[9] Einstein, A., Podolsky, B., and Rosen, N. [1935] Can quantum-mechanical description of physical reality be considered complete, *Physical Review* **47**, 777–80.

[10] Price, H. [2012] Causation, chance and the rational significance of supernatural evidence, *Philosophical Review* **121**, 483–538.

[11] Eberhard, P. [1978] Bell's theorem and the different concepts of locality, *Nuovo Cimento* **B46**, 392–419.

[12] Maudlin, T. [2011] Three roads to objective probability, In Beisbart, C. and Hartmann, S. (eds.), *Probabilities in Physics*, Oxford: Oxford University Press, pp. 293–319.

[13] Healey, R. [2012] Quantum theory: A pragmatist approach, *British Journal for the Philosophy of Science* **63**, 729–71.

[14] Bell, J.S. [1981] Bertlmann's socks and the nature of reality, *Journal de Physique*, Colloque 2, suppl. au numero 3, Tome 42, C2 41–61.

[15] Woodward, J. [2003] *Making Things Happen*, Oxford: Oxford University Press.

[16] Einstein, A. [1948] Quanten-mechanik und Wirklichkeit, *Dialectica* **2**, 320–23.

[17] Maudlin, T. [2007] *The Metaphysics within Physics*, Oxford: Oxford University Press.

12

The Bell Inequality and the Many-Worlds Interpretation

LEV VAIDMAN

It is argued that the lesson we should learn from Bell's inequalities is not that quantum mechanics requires some kind of action at a distance, but that it leads us to believe in parallel worlds.

12.1 Introduction

Bell's work [1] led to a revolution in our understanding of nature. I remember attending my first physics conference, on "Microphysical Reality and Quantum Formalism," in Urbino, 1985. Most of the talks were about Aspect's experiment [2] confirming the nonlocality of quantum mechanics based on experimental violations of Bell's inequalities. Although I did not share the skepticism of many speakers regarding the results of Aspect, I was not ready to accept that a local action in one place can instantaneously change anything at another place. So, while for the majority the lesson from Bell was that quantum mechanics requires some "spooky action at a distance", I was led by Bell's result to an alternative revolutionary change in our view of nature. I saw no other way than accepting the many-worlds interpretation (MWI) of quantum mechanics [3, 4].

I shall start by presenting the Einstein–Podolsky–Rosen (EPR) [5] argument. Then I will present Bell's presented using the Greenberger–Horne–Zeilinger setup [6] in the form proposed by Mermin [7, 8]. The discussion of nonlocality will suggest that Bell's inequalities are the only manifestation of action at a distance in nature. The demonstration of the necessity of action at a distance will be done through a detailed analysis of the GHZ experiment. Then I shall show how multiple worlds resolve the problem of action at a distance. After discussing the issue of nonlocality in the MWI, I shall conclude by citing Bell's views on the MWI.

12.2 EPR–Bell–GHZ

The story of Bell cannot be told without first describing the EPR argument. Instead of following the historical route, I shall use the GHZ setup, which, in my view, is the clearest way to explain the EPR and Bell's discovery.

There are three separate sites with Alice, Bob and Charley that share an entangled state of three spin-$\frac{1}{2}$ particles, the GHZ state:

$$|GHZ\rangle = \frac{1}{\sqrt{2}}(|\uparrow_z\rangle_A|\uparrow_z\rangle_B|\uparrow_z\rangle_C - |\downarrow_z\rangle_A|\downarrow_z\rangle_B|\downarrow_z\rangle_C). \tag{12.1}$$

The GHZ state is a maximally entangled state of a spin at every site with spins in the two other sites. Therefore, the measurement of the spin at each site and in any direction can be performed, in principle, using measurements at other sites. The assumption that there is no action at a distance in nature tells us that the measurements of Alice and Bob cannot change Charley's spin. After Alice's and Bob's measurements, Charley's spin becomes known. The spin value could not have been changed by distant measurements; therefore it existed previously. This is a consequence of the celebrated EPR criterion for physical reality. According to the EPR argument, the values of the spins of Alice, Bob and Charley in all directions are elements of reality. Quantum mechanics does not provide these values. Furthermore, the uncertainty relations prevent the simultaneous existence of some of these spin values. Thus, EPR concluded that quantum theory is incomplete.

At the end of their paper, EPR expressed hope that one day quantum theory will be completed to make these elements of reality certain. It took almost thirty years before Bell showed that this cannot be done.

We need not consider many elements of reality to show the inconsistency. In the GHZ setup it is enough to consider the spin values just in two directions, x and y. Let us rewrite the GHZ state in the x basis:

$$\frac{1}{2}(|\uparrow_x\rangle_A|\uparrow_x\rangle_B|\downarrow_x\rangle_C + |\uparrow_x\rangle_A|\downarrow_x\rangle_B|\uparrow_x\rangle_C + |\downarrow_x\rangle_A|\uparrow_x\rangle_B|\uparrow_x\rangle_C + |\downarrow_x\rangle_A|\downarrow_x\rangle_B|\downarrow_x\rangle_C).$$

We see that the product of the spins measured in the x direction is -1 with certainty. Similarly, if we use the x basis for Alice and the y bases for Bob and Charley, we learn that the product of the spins measured in the x direction for Alice, and in the y direction for Bob and Charley, is 1 with certainty. Because of symmetry of the GHZ state, the product is 1 with certainty also if it is Bob or Charley, instead of Alice, who is the only one to have measured the spin in the x direction. Therefore, the following equations should be fulfilled for the results of the spin measurements:

$$\{\sigma_{Ax}\}\{\sigma_{Bx}\}\{\sigma_{Cx}\} = -1, \tag{12.2}$$

$$\{\sigma_{Ax}\}\{\sigma_{By}\}\{\sigma_{Cy}\} = 1, \tag{12.3}$$

$$\{\sigma_{Ay}\}\{\sigma_{Bx}\}\{\sigma_{Cy}\} = 1, \tag{12.4}$$

$$\{\sigma_{Ay}\}\{\sigma_{By}\}\{\sigma_{Cx}\} = 1, \tag{12.5}$$

where $\{\sigma_{Ax}\}$ signifies the outcome of the measurement of σ_x by Alice, etc. All values of spin components in the above equations are EPR elements of reality. The outcomes should exist prior to the measurement and independent of what is done to other particles. So, according to EPR, the value of Alice's spin $\{\sigma_{Ax}\}$ appearing in (12.2) should be the same as in equation (12.3), and similarly for other values of spin variables. But this contradicts the fact that equations (12.2)–(12.5) cannot be jointly satisfied: the product of the left-hand sides is a product of squares, so it is positive, while the product of the right-hand sides is -1.

12.3 From the Bell Inequality to Nonlocality

Apparently, the first conclusion that can be reached here is that nature is random.

The predictions of quantum theory, including the results (12.2)–(12.5), were tested and verified in all experiments performed to date. We have shown above that equations (12.2)–(12.5) are inconsistent with the assumption that there are definite predictions for all these results. Therefore, at least some of the results should not exist prior to the measurement: the outcome of the measurement is random!

However, there is a problem with this "proof" of randomness. Let us assume that Charley's outcome is the one that is random. This contradicts the fact that after Alice's and Bob's measurements, Charley's result is definite. A nonlocal action is then needed to fulfill the equation. But if nonlocality is accepted, the EPR concept of elements of reality loses its basis, so the proof of randomness fails.

This is a proof of nonlocality. There is no way to explain these quantum correlations by some underlying local definite values or local probability distributions. It seems that the conclusion must be that actions (measurements in the x or y directions) of Alice and Bob change the outcome of Charley's measurement performed immediately after. This is a proof that there is action at a distance. It got the name of "spooky action at a distance" since there is no known underlying mechanism, and furthermore, since it is not observable: one cannot send signals to Charley using measurements at Alice's and Bob's sites.

12.4 Against Nonlocality

The role of science is to explain how and why things happen. A bird falls because a bullet hits it. The hunter shot the bullet. He was able to point his gun since photons reflected by the bird reached his eyes. In all these explanations, objects are present in particular places and they interact with other objects by sending particles (photons, bullets) from one object to another. This allows the concept of location of an object: it is the place where it can be influenced directly by other objects and where it can influence other objects directly. Clearly, this is the picture in classical physics: only local actions exist.

It is true that classical physics also has global formulations. A minimal action principle provides a complete solution given initial conditions without presenting an explicit local mechanism. There is a logical option for existence of a world described by an action principle with nonlocal interactions as well as a world with local interactions but without a minimal action principle. I feel that the local action explanation is the most important part of our picture of nature and thus we should try to keep it even when we turn to the correct physical theory, which is not classical, but quantum.

In quantum mechanics, the Aharonov–Bohm effect [9] (AB) seems to be a counter example. An electron changes its motion as a function of the magnetic field in a region where the electron does not pass. There is a nonzero vector potential at the location of the electron, but this potential is not locally defined; only the line integral of the potential is physical (gauge invariant). Recently, I solved, at least for myself, this apparent conflict with locality. I found local explanations of both the scalar and the magnetic AB effects [10]. The electron moves in a free-field region, but the source of the potential, the solenoid or the capacitor, feels the field of the electron. I considered the latter as quantum objects and realized that there is an unavoidable entanglement between them and the electron during the AB interference experiment. I calculated the phase acquired by the solenoid in the AB experiment and found that it exactly equals the AB phase. In the AB experiment the phase is manifested in the interference shift of the electron, but it is explained by the local action of the electromagnetic field of the electron on the solenoid.

Thus, the only manifestation of nonlocal action that we know of in physics is the unexplained nonlocal Bell-type correlations. Since physics has no mechanism for nonlocal actions, their existence appears to be a "miracle". In other words, physics cannot explain it. As a physicist, I want to believe that we do understand nature. Bell apparently tells us that it is impossible, or that we need to make a large conceptual change in our views of nature. The best option I see in this situation is to reject the tacit assumption, necessary for Bell's proof, that there is only one world. There is one physical universe, but there are many outcomes in every quantum experiment, corresponding to many worlds as we perceive them.

12.5 From Bell Inequalities to the Many-World Interpretation

How exactly does the MWI resolve the difficulty due to the EPR–Bell–GHZ argument? The concept of the EPR element of reality is the core of the breakdown:

If, without in any way disturbing a system, we can predict with certainty (i.e., with probability equal to unity) the value of a physical quantity, then there exists an element of physical reality corresponding to this physical quantity.

[Einstein, Podolsky and Rosen, 1935]

It is true that Alice and Bob, after measuring and collecting the results of their spin x measurements, can predict with certainty the outcome of Charley's spin x measurement. However, believing in the MWI, Alice and Bob know that their prediction is not universally true. It is true only in their particular world. They *know* that there are parallel worlds in which Charley's outcome is different. There is no "counterfactual definiteness", which is frequently assumed in Bell-type arguments: no definite outcome exists prior to the measurement.

To see explicitly how the MWI removes the action at a distance of Bell-type experiments, consider a demonstration of the GHZ experiment that is not yet possible with current technology, but may become possible in the near future. The choice of which components of the spin are measured by Alice, Bob and Charley is made according to the "random" results of other quantum measurements [11]. I argued that a better strategy is to rely on macroscopic signals from galaxies in different parts of the Universe [12], but for my analysis here, a quantum device is more appropriate.

In the GHZ setup analysis, not all combinations of measurements are considered: spin x measurements are performed by just one observer or by all three observers. To ensure this, we distribute between Alice, Bob and Charley another GHZ set of spin-$\frac{1}{2}$ particles. So the state of all particles (in the z basis) is

$$\frac{1}{2}(|\uparrow\rangle_A|\uparrow\rangle_B|\uparrow\rangle_C - |\downarrow\rangle_A|\downarrow\rangle_B|\downarrow\rangle_C)(|\uparrow\rangle_A|\uparrow\rangle_B|\uparrow\rangle_C - |\downarrow\rangle_A|\downarrow\rangle_B|\downarrow\rangle_C). \quad (12.6)$$

Alice, Bob and Charley perform spin x measurement of their additional spins. If the outcome is 1, then the spin y measurement of the second particle, the one from the original GHZ set, is performed. If the outcome is -1, then the x component of the spin is measured instead.

Alice's measurements split the world into four worlds. According to Everett's "relative state formulation of quantum theory" [3], these are Alice's worlds: $(\uparrow_{xA})$, $(\downarrow_{xA})$, $(\uparrow_{yA})$, $(\downarrow_{yA})$. Nothing changes at Bob's and Charley's sites because of Alice's actions. The complete local descriptions of Bob's and Charley's GHZ spins remain the same mixtures: completely unpolarized spins.

Now let us add the measurements of Bob. He also splits his world into four Everett worlds. According to the EPR argument, after Alice's and Bob's measurements, there is an element of reality associated with Charley's spin. Indeed, the information about the results of their measurements tells us what Charley's spin is. For example, in the world $(\uparrow_{xA}, \uparrow_{xB})$, Charley's spin is $|\downarrow\rangle_x$. But in a parallel world $(\uparrow_{xA}, \downarrow_{xB})$, Charley's spin is $|\uparrow\rangle_x$. So there is no single element or reality of Charley's spin x in Nature.

According to the definition of a "world" that I prefer [4], in any world all macroscopic objects have well-localized states. All measuring devices show definite values, so in every world Alice, Bob and Charley have well-defined values of spins. Alice first splits the world into four different worlds according to her measurements. Each of the worlds is then split again into four worlds by Bob. Charley, however, does not do any additional splitting. In every one of the 16 worlds created by Alice and Bob, the outcomes of his two spin measurements are already fixed. Here are all 16 worlds:

$$(\downarrow_{xA}, \downarrow_{xB}, \downarrow_{xC}), \quad (\downarrow_{xA}, \uparrow_{xB}, \uparrow_{xC}), \quad (\uparrow_{xA}, \downarrow_{xB}, \uparrow_{xC}), \quad (\uparrow_{xA}, \uparrow_{xB}, \downarrow_{xC}),$$

$$(\uparrow_{xA}, \uparrow_{yB}, \uparrow_{yC}), \quad (\uparrow_{xA}, \downarrow_{yB}, \downarrow_{yC}), \quad (\downarrow_{xA}, \uparrow_{yB}, \downarrow_{yC}), \quad (\downarrow_{xA}, \downarrow_{yB}, \uparrow_{yC}),$$

$$(\uparrow_{yA}, \uparrow_{xB}, \uparrow_{yC}), \quad (\uparrow_{yA}, \downarrow_{xB}, \downarrow_{yC}), \quad (\downarrow_{yA}, \uparrow_{xB}, \downarrow_{yC}), \quad (\downarrow_{yA}, \downarrow_{xB}, \uparrow_{yC}), \quad (12.7)$$

$$(\uparrow_{yA}, \uparrow_{yB}, \uparrow_{xC}), \quad (\uparrow_{yA}, \downarrow_{yB}, \downarrow_{xC}), \quad (\downarrow_{yA}, \uparrow_{yB}, \downarrow_{xC}), \quad \downarrow_{yA}, \downarrow_{yB}, \uparrow_{xC}).$$

All the worlds fulfill equations (12.2)–(12.5). However, we do not get a contradiction, as in Section 12.2, because the equations do not have to be correct together: each of the equations is correct in four worlds to which it can be applied. Different equations are valid in different worlds, so that the values of the spins in the equations can be different. The contradiction arises if we assume that there is only one world.

12.6 The Many-Worlds Interpretation and Nonlocality

As shown above, the MWI removes action at a distance from quantum physics. As in classical relativistic physics, any local action on a system changes nothing whatsoever at remote locations at the moment of disturbance. This does not mean, however, that quantum mechanics provides a local picture similar to classical physics with particles and fields localized in 3-space.

In classical physics, the complete description is given by specifying the trajectories of the particles and values of the fields:

$$\text{universe} = \{\vec{r}_i(t),\ \vec{F}_j(\vec{r}, t)\}. \tag{12.8}$$

This is local because it can be alternatively presented as an infinite set of vectors for all space–time points $(\vec{r}, t)$ that provide values of projection operators $\{\mathbf{P}_i(\vec{r}, t)\}$ and values of all fields at this point $\{\vec{F}_j(\vec{r}, t)\}$.

In classical physics, outcomes of an experiment at every site are fully specified by the local description of this site. The measuring devices and the observers can be expressed in the same language, in terms of the locations $\vec{r}_i$ of the particles that they are made of. The final positions of the particles of the measuring devices are fully explained by their initial states and by the local interactions occurring in the interval between the initial and final times. Classical physics is deterministic (classical probability theory is relevant only for situations with incomplete knowledge of the full description), so the issue of correlations between outcomes of experiments at different places does not arise.

In a quantum world, if all particles are in a product state, then the description of the universe is similar to the classical one: it is a set of these wave functions,

$$\text{universe} = \{\Psi_i(\vec{r}, t)\}, \tag{12.9}$$

which also can be represented as an infinite set of vectors with values of the wave functions at all space–time points. (For the current analysis it is not necessary to go to field theory, which describes quantum fields.) However, if we introduce measuring devices and observers, the local coupling of the measurement process will destroy the above product state. The description of the universe then is the wave function in the configuration space of all particles,

$$\text{universe} = \Psi(\vec{r}_1, \vec{r}_2, \dots \vec{r}_i, \dots, t), \tag{12.10}$$

which cannot be represented as a set of vectors at space–time points.

In the MWI, this wave function in the configuration space is all that exists. It explains everything, but not in a simple way. It does not provide a transparent connection to our experiences. The way to connect the universal wave function to our experience is to decompose it into a superposition of terms, each corresponding to a different world. In each such term, all variables specifying states of macroscopic objects are essentially in a product state.

In each of the 16 worlds of our GHZ experiment, Alice, Bob and Charley have definite results of their spin measurements. What makes this situation nonlocal is that while all four different local options are present for all observers, i.e., there are four Everett worlds for Alice, and separately for Bob and for Charley, we do not have 64 worlds. Specifying Everett worlds of two observers fixes the world of the third. This connection between local worlds of the observers is the nonlocality of the MWI.

Is there any possibility of action at a distance in the framework of the MWI? Obviously, at the level of the physical universe that includes all the worlds, local action cannot change anything at remote locations. However, a local action splits the world, which is a nonlocal concept, and local actions can bring about splitting to worlds that differ at remote locations. Thus, an observer for whom only his world is relevant has an illusion of an action at a distance when he performs a measurement on a system entangled with a remote system. (He also has an illusion of randomness each time he performs a quantum measurement [13].)

Consider again our example where Alice and Bob finished their measurements, but Charley still did not make his measurements. He knows that Alice and Bob made the measurements, and he knows that there are 16 worlds:

$$
\begin{aligned}
&(\downarrow_{xA}, \downarrow_{xB}), \ (\downarrow_{xA}, \uparrow_{xB}), \ (\uparrow_{xA}, \downarrow_{xB}), \ \uparrow_{xA}, \uparrow_{xB}), \\
&(\uparrow_{xA}, \uparrow_{yB}), \ (\uparrow_{xA}, \downarrow_{yB}), \ (\downarrow_{xA}, \uparrow_{yB}), \ (\downarrow_{xA}, \downarrow_{yB}), \\
&(\uparrow_{yA}, \uparrow_{xB}), \ (\uparrow_{yA}, \downarrow_{xB}), \ (\downarrow_{yA}, \uparrow_{xB}), \ (\downarrow_{yA}, \downarrow_{xB}), \\
&(\uparrow_{yA}, \uparrow_{yB}), \ (\uparrow_{yA}, \downarrow_{yB}), \ (\downarrow_{yA}, \uparrow_{yB}), \ (\downarrow_{yA}, \downarrow_{yB}).
\end{aligned} \tag{12.11}
$$

Charley is in all these worlds. He is in a single Everett world that includes 16 worlds, according to my definition. There is no meaning in asking him now in which world out of 16 he is.

If Charley follows the instructions and performs the measurements according to the rules stated above, he will create four Everett worlds by creating macroscopic outcomes in his laboratory. Each of his new Everett worlds belongs to four worlds specified by Alice and Bob.

Charley has the choice of performing or not performing the measurements, and this will change the set of worlds he will belong to. He can also make spin measurements not in accordance to the instructions: for example, he can make spin y measurements instead of x measurements and vice versa. Now, for every outcome, he will end up belonging to eight, instead of four, worlds. He will also increase the total number of existing worlds to 32. If instead of following the instructions, all observers make both spin measurements in the z direction, there will be only 4 worlds:

$$
\begin{aligned}
&(\downarrow_{zA}, \downarrow_{zA}, \downarrow_{zB}, \downarrow_{zB}, \downarrow_{zC}, \downarrow_{zC}), \ (\downarrow_{zA}, \uparrow_{zA}, \downarrow_{zB}, \uparrow_{zB}, \downarrow_{zC}, \uparrow_{zC}), \\
&(\uparrow_{zA}, \downarrow_{zA}, \uparrow_{zB}, \downarrow_{zB}, \uparrow_{zC}, \downarrow_{zC}), \ (\uparrow_{zA}, \uparrow_{zA}, \uparrow_{zB}, \uparrow_{zB}, \uparrow_{zC}, \uparrow_{zC}).
\end{aligned} \tag{12.12}
$$

Each observer performing measurements in the z direction creates worlds in which other observers have definite values of spin in the z direction. It looks like nonlocal action at a distance: local measurement has changed some property in a remote location. But it is a subjective change for an observer in a particular world: he understands that in the physical universe that includes worlds with all outcomes of his local measurements, the remote spins also have all possible values.

12.7 Conclusions

Bell inequalities lead us to a hard choice: we either believe that there is some kind of action at a distance, or that there are multiple realities. My strong feeling is that accepting action at a distance has the higher price, and I am convinced that the MWI is the correct description of Nature.

In the MWI, the Bell proof of action at a distance fails in an obvious way, since it requires a single world to ensure that measurements have single outcomes. Although there is no action at a distance in the MWI, it still has nonlocality. The core of the nonlocality of the MWI is entanglement, which is manifested in the connection between the local Everett worlds of the observers. I feel that Bell inequalities can be manifested as a property of these connections, but I could not find a simple way to formulate this. I hope that it will be done in the future.

My first formulation of the MWI and arguments in its favor appeared in a preprint [14] that I sent to John Bell at the end of 1989. He was not convinced. He replied with a short paragraph saying that if there are multiple worlds, there should be one in which I do not believe in the MWI. He added, more seriously, that he did not know what is the right way to understand quantum mechanics, but the MWI did not sound plausible to him. He expressed this view in more detail in the Nobel Symposium [15]:

The "many world interpretation" seems to me an extravagant, and above all an extravagantly vague, hypothesis. I could almost dismiss it as silly. And yet . . . It may have something distinctive to say in connection to "Einstein Podolsky Rosen puzzle", and it would be worthwhile, I think, to formulate some precise version of it to see if it really so. And the existence of all possible worlds may make us more comfortable about existence of our own world . . . which seems to be in some ways a highly improbable one.

[John Bell, 1986]

For me Bell's result was the first reason to accept the MWI. Since then, the discovery of teleportation and of interaction-free measurements has turned my belief into a strong conviction [16]. I feel now that I developed "the precise version of the MWI" that John Bell was looking for [13]. I regret that I did not have this clear vision in 1989 when I discussed the interpretation of quantum mechanics with John Bell in Erice.

I thank Eliahu Cohen and Shmuel Nussinov for helpful discussions. This work has been supported in part by Israel Science Foundation Grant 1311/14 and German–Israeli Foundation Grant I-1275-303.14.

References

[1] J.S. Bell, On the Einstein–Podolsky–Rosen paradox, *Physics* **1**, 195 (1964).
[2] A. Aspect, J. Dalibard, and G. Roger, Experimental test of Bell's inequalities using time-varying analyzers, *Phys. Rev. Lett.* **49**, 1804 (1982).

[3] H. Everett, Relative state formulation of quantum mechanics, *Rev. Mod. Phys.* **29**, 454 (1957).

[4] L. Vaidman, Many-worlds interpretation of quantum mechanics, in the *Stanford Encyclopedia of Philosophy*, E.N. Zalta (ed.) (2002), http://plato.stanford.edu/entries/qm-manyworlds/.

[5] A. Einstein, B. Podolsky, and N. Rosen, Can quantum-mechanical description of physical reality be considered complete? *Phys. Rev.* **47**, 777 (1935).

[6] D.M. Greenberger, M.A. Horne, and A. Zeilinger, Going beyond Bell's theorem, in *Bell Theorem, Quantum Theory and Conceptions of the Universe*, M. Kafatos (ed.), Kluwer Academic, Dordrecht (1989), p. 69.

[7] N.D. Mermin, Quantum mysteries revisited, *Am. J. Phys.* **58**, 731 (1990).

[8] L. Vaidman, Variations on the theme of the Greenberger–Horne–Zeilinger proof, *Found. Phys.* **29**, 615 (1999).

[9] Y. Aharonov and D. Bohm, Significance of electromagnetic potentials in the quantum theory, *Phys. Rev.* **115**, 485 (1959).

[10] L. Vaidman, Role of potentials in the Aharonov–Bohm effect, *Phys. Rev. A* **86**, 040101(R) (2012).

[11] T. Scheidla, R. Ursin, J. Kofler, S. Ramelow, X.-S. Ma, T. Herbst, L. Ratschbacher, A. Fedrizzi, N.K. Langforda, T. Jennewein, and A. Zeilinger, Violation of local realism with freedom of choice, *Proc. Natl. Acad. Sci. U.S.A.* **107**, 19,708 (2010).

[12] L. Vaidman, Tests of Bell inequalities, *Phys. Lett. A* **286**, 241 (2001).

[13] L. Vaidman, Quantum theory and determinism, *Quantum Stud. Math. Found.* **1**, 5 (2014).

[14] L. Vaidman, On schizophrenic experiences of the neutron or why we should believe in the many-worlds interpretation of quantum theory, http://philsci-archive.pitt.edu/8564/ (1990).

[15] J.S. Bell, Six possible worlds of quantum mechanics, in J.S. Bell, *Speakable and Unspeakable in Quantum Mechanics* (Cambridge: Cambridge University Press, 1987), p. 173.

[16] L. Vaidman, On the paradoxical aspects of new quantum experiments, *Phil. Sci. Assoc.* 1994, pp. 211–217, www.jstor.org/stable/193026.

13

Quantum Solipsism and Nonlocality

TRAVIS NORSEN

Abstract

J.S. Bell's remarkable 1964 theorem showed that any theory sharing the empirical predictions of orthodox quantum mechanics would have to exhibit a surprising – and, from the point of view of relativity theory, very troubling – kind of nonlocality. Unfortunately, even still on this 50th anniversary, many commentators and textbook authors continue to misrepresent Bell's theorem. In particular, one continues to hear the claim that Bell's result leaves open the option of concluding *either* nonlocality *or* the failure of some unorthodox "hidden variable" (or "determinism" or "realism") premise. This mistaken claim is often based on a failure to appreciate the role of the earlier 1935 argument of Einstein, Podolsky, and Rosen in Bell's reasoning. After briefly reviewing this situation, I turn to two alternative versions of quantum theory – the "many worlds" theory of Everett and the quantum Bayesian interpretation of Fuchs, Schack, Caves, and Mermin – that purport to provide actual counterexamples to Bell's claim that nonlocality is required to account for the empirically verified quantum predictions. After analyzing each theory's grounds for claiming to explain the EPR–Bell correlations locally, however, one can see that (despite a number of fundamental differences) the two theories share a common for-all-practical-purposes (FAPP) solipsistic character. This dramatically undermines such theories' claims to provide a local explanation of the correlations and thus, by concretizing the ridiculous philosophical lengths to which one must go to elude Bell's own conclusion, reinforces the assertion that nonlocality really is required to coherently explain the empirical data.

13.1 Introduction

Fifty years ago, in 1964 [1], John Stewart Bell first proved the theorem that has become widely known as "Bell's Theorem" but that Bell himself instead referred to as the "locality inequality theorem" [2]. In Bell's own view, the theorem showed that the empirical predictions of *local* theories will be constrained by Bell's inequality (or as Bell himself preferred to call it, the "locality inequality"). Hence, *nonlocality* is a necessary feature of any theory that shares the empirical predictions of standard quantum mechanics. In recent decades, the relevant inequality-violating quantum mechanical predictions have been confirmed in a series of increasingly accurate and convincing experiments [3]. It is thus known with reasonable certainty that nonlocality is a real feature of the world.

This summary of the situation, however, remains curiously and frustratingly controversial, despite the five decades that physicists and philosophers have had to contemplate and understand Bell's arguments. There remain, for example, many commentators (including, undoubtedly, some in this very volume) who regard Bell's theorem not as a proof of nonlocality, but instead as a refutation of determinism and/or the so-called "hidden variables" program and/or some (usually ill-defined) notion of "realism." In general, that is, there remain many commentators (including, particularly troublingly, textbook authors) who assert or imply (often without even realizing that they are flatly contradicting Bell's own understanding) that Bell was simply wrong to claim that nonlocality was required by the (now well-confirmed) quantum mechanical predictions [4–7].

A systematic presentation of Bell's arguments, including some polemics against these (and other) persistent misunderstandings, can be found in Refs. [8, 9]. In the present paper, my goal is to focus on one particular thread of such disagreement with Bell, which has been especially influential in the last decade or so. In particular, I will explore the so-called quantum Bayesian ("QBist") and Everettian ("many worlds") approaches to quantum theory, both of which purport to provide counterexamples to the claim that nonlocality is required to account for the empirical data. In particular, I will develop the thesis that, although QBism and Everettism are usually thought of as almost polar opposites (with the latter being one of the more popular *realist* versions of quantum theory, and the former often being considered rather solipsistic), the two theories' grounds for claiming locality are in fact basically similar and, once brought out into the light, deeply unconvincing. Understanding how and why will then lead to a deeper appreciation of Bell's work.

13.2 Bell's argument

Before jumping in to a polemical discussion of QBism and Everettism, it will be helpful to briefly review Bell's arguments.

In Bell's original 1964 paper, the main analysis begins where the 1935 argument of Einstein, Podolsky and Rosen (EPR) [10] had left off. That is, Bell begins by recalling EPR's demonstration that (in Bell's words) "quantum mechanics could not be a complete theory but should be supplemented by additional variables [which would] restore to the theory causality and locality." [1] As Bell goes on to elaborate the argument,

[W]e make the hypothesis ... that if the two measurements are made at places remote from one another the orientation of one magnet does not influence the result obtained with the other. Since we can predict in advance the result of measuring any chosen component of [the polarization of particle 2], by previously measuring the same component of [the polarization of particle 1], it follows that the result of any such measurement must actually be predetermined. [1]

I will review the actual EPR argument shortly; for now I just want to stress that Bell's 1964 paper begins by recapitulating the 1935 EPR argument, which Bell takes to have established that a deterministic hidden variable theory was one's *only hope*, if one wanted to explain the predicted quantum correlations in a local way.

In the body of his 1964 paper, Bell then shows that this kind of deterministic hidden variable theory's predictions for a wider class of possible experimental measurements (in which the outcomes are correlated, but imperfectly so) are necessarily constrained by a Bell (or, as Bell called it, a locality) inequality. The mathematical details of this demonstration are well understood, so I will not bother to rehearse them here. The main point is just that, contrary to the impression of people who miss the role of the EPR argument in Bell's overall thesis, Bell's 1964 result already establishes the inevitability of nonlocality. It does *not* leave open some kind of choice between abandoning locality and abandoning "realism" or "hidden variables" or "determinism," because the overall argument does not *begin* with realistic/deterministic hidden variables. Instead it begins with locality alone, proceeds (via the EPR argument) to establish the necessity of deterministic hidden variables to explain locally just the subset of the quantum predictions considered by EPR, and then finally closes off that apparent possibility by showing that this kind of theory cannot reproduce the full slate of quantum predictions.

Looking back, and considering the widespread and persistent confusion about its role in his theorem, it is somewhat unfortunate that the two or three sentences I quoted above constitute basically the entirety of Bell's recapitulation of the EPR argument.[1] And of course that argument had been originally made in a not terribly rigorous way (made worse by the fact that the actual EPR paper, written by Podolsky and not seen by Einstein until after its publication, obscured the main argument, in Einstein's opinion [11, 12]). But still it is very easy to see that the EPR argument that Bell rehearses is entirely valid and provides the necessary foundation for Bell's subsequent demonstration. Consider, for example, the case of two spatially separated and polarization-entangled particles. Suppose in particular that the joint polarization state – for example,

$$|\psi\rangle = \frac{1}{\sqrt{2}}[|HH\rangle - |VV\rangle] \tag{13.1}$$

– is such that, if the polarization of both particles is measured along the same direction, the results are perfectly correlated: either both outcomes are "H" or both outcomes are "V."

The EPR argument can then be understood to proceed as follows. Suppose a measurement is made on the nearby particle, yielding some outcome, say "H." It is then certain that the distant particle, if measured along the same direction, will also yield the outcome "H." We could say that now, after the measurement on the nearby particle, it is clear that the distant particle somehow encodes this outcome, "H," in its internal structure.[2] There are then two possibilities. Either the distant particle possessed this internal structure all along (i.e., even before our measurement on the nearby particle), or it did not (i.e., it only

---

[1] That is, it would have been nice if, already in 1964, Bell had anticipated the need for a much sharper and more detailed presentation of the EPR part of the argument (including especially a more generalized and more precise formulation of the crucial "locality" premise, along the lines of what he would indeed give later, in 1976 and 1990).

[2] For example, in ordinary quantum mechanics, the quantum state of the distant particle will be $|H\rangle$.

acquired it after, and evidently as a result of, our nearby measurement). The latter option implies *nonlocality*: our nearby measurement caused the distant particle to change its physical state, to acquire a definite value ("H") for a property that was previously somehow indefinite (or just different). The former option, on the other hand, implies the existence of an outcome-determining structure in the distant particle, about which quantum theory is silent. The former option thus implies the *incompleteness* of the quantum mechanical description. EPR simply took locality for granted (no doubt on the basis of Einstein's relativity theory, which is usually taken to prohibit faster-than-light causal influences), so the established dilemma between nonlocality and incompleteness immediately suggested that quantum mechanics did not provide a complete description of microscopic reality. This was thus EPR's main conclusion.

For Bell's purposes, though, that is not really the crucial point. The important thing was instead that the *only* way to explain the perfect correlations locally is to attribute outcome-determining properties to the individual particles. These properties, evidently, would vary randomly from one particle pair to the next, but would be fixed (in an appropriately correlated way) once and for all at the source for a given particle pair. (For example, perhaps the first pair is of the type "particle 1 is 'V' and particle 2 is 'V'," the second pair is of the type "particle 1 is 'H' and particle 2 is 'H'," and so on.) It is easy to see that such deterministic hidden variables are the *only* way to account locally for the perfect correlations: any residual indefiniteness in either particle would either at least sometimes spoil the perfect correlations (if the indefiniteness were resolved locally during the subsequent measurement procedure) or would involve some physical process in which the state of one particle was nonlocally affected by the distant measurement process or result. So we have to choose between nonlocality and local deterministic hidden variables. Or equivalently, the only way to avoid nonlocality (in the face of the EPR correlations) is to embrace local deterministic hidden variables.

In any case, it should be easy to see – even without a formal definition of locality and a formally rigorous version of the argument – that the EPR argument is entirely valid and that, indeed, as of Bell's writing in 1964, it was already established that the only hope for a *local* explanation of the quantum correlations was a deterministic local hidden variable theory . . . and that, therefore, by showing that such a theory could not reproduce the quantum predictions in more general situations, Bell had proved that no local explanation of the quantum correlations was possible, full stop. Or as he put it already in the Introduction of his 1964 paper: "It is the requirement of locality . . . that creates the essential difficulty." [1]

As a matter of social/historical fact, however, none of this was clear. The vast majority of commentators simply missed, or misunderstood, the role of the EPR argument in Bell's reasoning and/or wrongly believed the EPR part of the argument to be invalid. Bell thus exerted a great deal of effort in the subsequent decades to clarify his reasoning and to make all of the relevant assumptions more rigorous and explicit. This is not the place for a systematic presentation of his subsequent clarificatory efforts, but I will mention Bell's classic 1981 "Bertlmann's Socks and the Nature of Reality," in which he remarks, in a footnote,

My own first paper on this subject [i.e., the 1964 paper] starts with a summary of the EPR argument *from locality to* deterministic hidden variables. But the commentators have almost universally reported that it begins with deterministic hidden variables.

[13, emphasis in original]

The same paper also includes the following admirably clear recapitulation of the EPR argument,

It is important to note that to the limited degree to which *determinism* plays a role in the EPR argument, it is not assumed but *inferred*. What is held sacred is the principle of 'local causality' – or 'no action at a distance'. Of course, mere *correlation* between distant events does not by itself imply action at a distance, but only correlation between the signals reaching the two places. These signals, in the idealized example of Bohm [involving the perfect polarization correlations], must be sufficient to *determine* whether the particles go up or down. For any residual undeterminism could only spoil the perfect correlations.... [13]

as well as this beautiful presentation of the overall argument for nonlocality:

Let us summarize once again the logic that leads to the impasse. The [EPR–Bohm] correlations are such that the result of the experiment on one side immediately foretells that on the other, whenever the analyzers happen to be parallel. If we do not accept the intervention on one side as a causal influence on the other, we seem obliged to admit that the results on both sides are determined in advance anyway, independently of the intervention on the other side, by signals from the source and by the local magnet setting. But this has implications for non-parallel settings which conflict with those of quantum mechanics. So we *cannot* dismiss intervention on one side as a causal influence on the other. [13]

In addition to thus clarifying the reasoning he had used in the 1964 paper, Bell's other main clarificatory innovation (in the decades after the theorem was first proved) was a careful and explicit and formal definition of "locality" (i.e., "local causality").[3] This allowed more formal and explicit demonstrations of both the EPR argument (from locality to deterministic hidden variables) and Bell's "locality inequality theorem." The next section briefly recalls Bell's formulation of "locality," as it will play an important role in our subsequent discussion of QBism and Everettism.

13.3 Bell's formulation of locality

Bell's first attempt at an explicit space–time formulation of the principle of locality (i.e., "local causality") occurs in his 1976 "The theory of local beables" [15]. The formulation is in terms of his neologism, "beable," which Bell introduced as a foil to the "observables"

---

[3] It is sometimes suggested that, for Bell, "locality" and "local causality" are logically and/or conceptually distinct notions, and that sensitivity to this distinction provides a rational basis for the types of views I have been here criticizing as mistakes/confusions. See for example Ref. [14]. In my opinion, though, there is overwhelming evidence that Bell used "locality" and "local causality" interchangeably and was only ever interested in the single, unitary concept that is roughly captured by the idea of "no faster-than-light causal influences." The Preface that Bell wrote for the (first edition) publication of *Speakable and Unspeakable* is particularly revealing on this point [2].

that play a central role in the mathematical formulation of ordinary quantum theory. As Bell points out, "It is not easy to identify precisely which physical processes are to be given the status of 'observations' and which are to be relegated to the limbo between one observation and another." The beables of a candidate theory are those elements that are assumed to correspond directly to physically real structures posited to exist (independent of any "observation") by the theory. Indeed: "'Observables' must be *made*, somehow, out of beables" [15].

As an example, Bell cites the electric and magnetic fields in classical electromagnetic theory:

In Maxwell's electromagnetic theory, for example, the fields $\mathbf{E}$ and $\mathbf{H}$ are "physical" (beables, we will say) but the potentials $\mathbf{A}$ and ϕ are "non-physical". Because of gauge invariance the same physical situation can be described by very different potentials. It does not matter that in Coulomb gauge the scalar potential propagates with infinite velocity. It is not really supposed to *be* there. [15]

The fields $\mathbf{E}$ and $\mathbf{H}$ are examples of *local* beables in the sense that they are localized in delimited space–time regions (points, actually). Such local beables are to be contrasted with nonlocal beables – objects that a candidate theory posits as physically real, but that are not localized in space-time. (The wave function of a many-particle quantum system – if it were considered a beable in some version of the theory – would be an example of a nonlocal beable: it is a function on the $3N$-dimensional configuration space of the N-particle system, not in any sense a field that realizes definite values at points in $(3 + 1)$-dimensional space-time.)

The above passage also illustrates one of the crucial (and neglected) points underlying Bell's formulation: since "locality" and "nonlocality" refer to physical processes posited by candidate theories (and in particular the issue of whether causal influences propagate always at or slower than the speed of light, or instead in some cases exceed that speed), it is simply hopeless to try to diagnose whether a theory is "local" or "nonlocal" until it is made crystal clear what the theory says *exists* in ordinary 3D physical space. The *ontology* (i.e., the *beables*) of the theory, that is, must be clearly and explicitly articulated before one can possibly make any judgment about its status vis-à-vis locality. As Bell put this point in 1976,

It is in terms of local beables that we can hope to formulate some notion of local causality. [15]

He might have written (and I think he certainly believed) that it is *only* in terms of local beables that the idea of locality can be clearly formulated. Or, as he put this point in the 1987 preface to the 1st edition of his collected papers,

If local causality in some theory is to be examined, then one must decide which of the many mathematical entities that appear are supposed to be real, and really here rather than there. [2]

It will be crucial to appreciate that "here" and "there" refer to locations in ordinary three-dimensional physical space. Bell's point is that positing a clearly articulated ontology of

Figure 13.1 "Full specification of what happens in 3 makes events in 2 irrelevant for predictions about 1 in a locally causal theory" [18].

local beables is a prerequisite for any discussion of a theory's status vis-à-vis *dynamical locality*.

In the 1976 paper, Bell first discusses "local determinism" – the idea that a complete specification of local beables in a space–time region that closes off the past light cone of some region will uniquely determine the beables in that region. (Maxwell's electromagnetism, for example, has this property.) Bell then specifically introduces "local causality" as a more general notion intended to capture the absence of faster-than-light causal influences for any theory, whether deterministic or not. For reasons (partially?) alluded to in a footnote of his 1977 remarks on "Free variables and local causality," Bell subsequently changed, in a subtle way, his formulation of local causality [16]. The updated formulation first appeared in a footnote of his 1986 "EPR correlations and EPW distributions" [17] and then received a much more careful and elaborate treatment in his 1990 "La nouvelle cuisine," the last paper to appear before his tragic and untimely death [18].

Here is Bell's brief 1986 formulation:

In a locally-causal theory, probabilities attached to values of local beables in one space-time region, when values are specified for *all* local beables in a second space-time region fully obstructing the backward light cone of the first, are unaltered by specification of values of local beables in a third region with spacelike separation from the first two. [17]

In the 1990 paper, Bell illustrates the notion with the figure I have reproduced here (with Bell's caption) as Figure 13.1. He then formulates the principle of local causality as follows:

A theory will be said to be locally causal if the probabilities attached to values of local beables in a space–time region 1 are unaltered by specification of values of local beables in a space-like separated region 2, when what happens in the backward light cone of 1 is already sufficiently specified, for example by a full specification of local beables in a space-time region 3 . . . [18]

It is perhaps also worth quoting Bell's subsequent clarificatory notes:

It is important that region 3 completely shields off from 1 the overlap of the backward light cones of 1 and 2. And it is important that events in 3 be specified completely. Otherwise the traces in region 2 of causes of events in 1 could well supplement whatever else was being used for calculating probabilities about 1. The hypothesis is that any such information about 2 becomes redundant when 3 is specified completely. [18]

Any reader not already familiar with this is urged to read Bell's 1976 and 1990 papers (as well as my own detailed analysis in Ref. [9]) to appreciate the systematic clarity of Bell's formulation.

The logical and pedagogical benefits of this explicit formulation of locality are enormous. It is straightforward, for example, to rehearse a fully rigorous version of "the EPR argument from locality to deterministic hidden variables" [17] and thus solidify the foundation of Bell's original 1964 theorem. Alternatively, the explicit formulation of locality also allows a rigorous derivation (which does not rely on perfect anticorrelations and does not introduce deterministic hidden variables as a middle step) of the so-called CHSH inequality. (See Refs. [8, 9, 18] for details.) All told, then, Bell's work to explicitly articulate the relativistic notion of "no superluminal action at a distance" makes an essentially airtight case for his conclusion that nonlocality is required by the quantum mechanical predictions (and, more importantly, we now know, by actual experiment).

It is somewhat surprising, therefore, that so many people still deny that Bell proved the necessity of nonlocality. As I have suggested above, this is largely a result of plain ignorance about what Bell actually did [19]. Hopefully, the above summary and the growing body of good literature on the subject will help improve the situation. But here I want to turn to focus on a rather different category of disagreement with Bell's claim to have demonstrated the empirical necessity of nonlocality – disagreement, that is, that is based not so much on a simple failure to understand the logical structure of Bell's arguments, but instead on an implicit rejection of Bell's point that a clearly articulated ontology of local beables in three-dimensional physical space is a *prerequisite* for any meaningful analysis of a theory's status vis-à-vis locality.

13.4 Solipsism and FAPP Solipsism

Let us then turn to analyzing several quantum worldviews that, in one way or another, fail to meet Bell's prerequisite. We begin by establishing a simple, if somewhat bizarre, point of principle. It is possible to evade Bell's conclusion (that nonlocality is required to explain the observed correlations) by adopting the philosophical viewpoint known as "solipsism," according to which nothing outside of one's own subjective conscious experience actually exists. Although it would be hard to name any actual person (philosopher or otherwise) who endorses solipsism fully, the idea can be regarded as a kind of fully consistent implementation of Berkeley's "esse est percipi" and Descartes' combination of radical doubt and the supposed prior certainty of consciousness. The idea is to regard sense experience – which is normally regarded as experience *of* a material world that exists independent of any conscious awareness – as being instead like a hallucination with no external object at all.

I do not want here to engage with the extensive philosophical literature on the reasonableness and/or refutability of solipsism. I think the situation was pretty well summed up by Bell's remark that

Solipsism cannot be refuted. But if such a theory were taken seriously it would hardly be possible to take anything else seriously. [20]

Instead I simply want to note that if one (for whatever reason) absolutely refused to allow that nonlocality might really be a feature of the external physical world, adopting solipsism would provide *a* logically possible basis for that stance: if there simply *is* no external physical world, then clearly Bell cannot have established that there are real faster-than-light causal influences in it.

The problem with this stance, of course, is that it commits one to denying quite a lot more than just the alleged nonlocal causal influences that, in the scenario I've just described, motivate its adoption. The solipsist also denies the existence of tables, trees, planets, other solipsists, and even his own physical body. Within the realm of physics, the solipsist not only denies the physical reality of the measurement outcomes that violate Bell's locality inequality and thus allegedly provide the evidence for Bell's conclusion of faster-than-light causal influences; the solipsist also denies, for example, the real existence of the facts we usually interpret as evidence for the existence of (light-speed) electromagnetic causal influences, the real existence of the facts we usually interpret as evidence for the existence of (sound-speed) verbal influences, etc. He denies, in short, the existence of *everything*. All of it, the whole apparently real physical world that constitutes the (usually assumed) subject matter for the science of physics, is instead a mere delusion or fantasy or hallucination for the solipsist. Undoubtedly this is what Bell had in mind in remarking that solipsism makes it impossible "to take anything else seriously."

A key point about the solipsist's position here is that although he eludes the need to admit the existence of Bell's nonlocal causal influences, he does not in any sense retain a local description of the world. He retains no description of the world at all – so the distinction between "local" and "nonlocal" accounts of the world simply does not apply. That is, although the solipsist's stance is not nonlocal, it is not, thereby, in any meaningful sense, local. It is neither local nor nonlocal. In denying the very existence of an external physical world, the solipsist removes any possible meaning from the question of whether causal influences in the external physical world propagate sometimes (the nonlocal case) or never (the local case) faster than light.

The entire reason we are having any of this discussion is of course that nonlocality is very difficult to reconcile with a fully relativistic account of physical goings-on in three-dimensional space and time. Someone who felt that the evidence in favor of relativity was very strong might thus be willing to trade something – even something quite significant – for a way to elude Bell's conclusion that nonlocality is real. But solipsism, as a response to Bell, seems completely crazy: the solipsist trades *everything*, including whatever possible basis he might have had for believing in the relativistic account of physical goings-on in space–time, i.e., whatever possible basis he might have had for wanting to avoid nonlocality in the first place [21]. Presumably that explains why, as I said, nobody really endorses solipsism, as a response to Bell's arguments or otherwise. Nevertheless, in principle if absurdly, one could respond to Bell's claims by adopting solipsism as a way to elude nonlocality.

Let us then contrast literal solipsism with a category of views that I will call "FAPP solipsism." "FAPP" here, following Bell, means "for all practical purposes" [22]. The most widely known version of FAPP solipsism is the notorious brain-in-a-vat scenario, memorably dramatized in the movie "The Matrix," in which it is supposed that you might be deluded about almost everything you believe because, instead of arising from causal contact with a real external world, your conscious sensations might instead result from electrical signals fed to your brain (which is kept biologically viable in a vat of appropriate fluid) by an evil scientist with a supercomputer. Such a view is clearly not literally solipsist, since its very formulation presupposes the existence of physically real brains (from which subjective conscious experiences somehow arise), vats, biologically auspicious fluids, electrical signals, evil scientists, and/or supercomputers.

And yet the brain-in-a-vat scenario just as clearly has much in common with literal solipsism. Just as with adopting solipsism, considering that one might really be a brain in a vat is a way to elude not just some particular undesirable conclusion (such as nonlocality) but anything one happens to want to elude. For example, if, for whatever reason, you are bothered not only by faster-than-light causation (à la Bell) but also by light-speed causation (à la Maxwell), it is easy enough to assert that that, too, is merely a delusion implanted in your unsuspecting brain by the evil scientists. Or if, for some reason, you have always been unable to accept that the Earth goes around the Sun (à la Copernicus) rather than vice versa, it is easy enough to concoct a story in which the evil scientists live (and set up their diabolical laboratory) on the surface of a planet which rests comfortably at the exact center of a series of concentric, rotating crystalline spheres.

The point here is that as soon as you allow the possibility that your ordinary perceptual experience might not really be *of* an external physical reality, but might instead be a hallucination fed into your brain by an evil scientist, literally every aspect of the conjectured real world (where the evil scientists are supposed to live and work) becomes purely arbitrary. You can literally make up whatever you want, because the usual epistemological burden of providing empirical (i.e., ultimately, perceptual) evidence for the various aspects of one's proposed world picture has been short-circuited by the assumption that any perception-based claim is in fact a hallucination (or is, at any rate, otherwise delusory). Note, for example, that the brain-in-a-vat theorist really has (by his own implied epistemological standards) no grounds whatever for believing in the real existence of brains, including his own. After all, whatever evidence you take yourself to have for believing that there exist real human beings, with internal organs including brains, out of whose complex electrophysiological structure conscious experience somehow emerges, is – if you are a brain-in-a-vat theorist – just another set of hallucinations fed to you by the evil scientists.

The brain-in-a-vat theorist is not exactly a solipsist, because he claims to believe that there is a physical reality out there independent of our conscious experiences. But *he might as well be a solipsist*. He claims to believe in a real physical world (with evil scientists and vats and brains in it), but he can, by his own standards, have no grounds whatsoever for those *particular* beliefs about the real world. That is, his beliefs about the physical world (including, I think it must be admitted, even the belief that conscious experience can only emerge, somehow, from an appropriately brainish sort of physical object) are entirely arbitrary. They could all be changed, or dropped entirely, without cognitive consequence. The brain-in-a-vat theorist may thus be described as a "FAPP solipsist."

We might reformulate the similarity (between literal and FAPP solipsism) as follows: insofar as they purport to account for our experiences (and thereby achieve a kind of empirical adequacy), literal solipsism and FAPP solipsism both engage directly with our subjective conscious experiences, rather than with the external material facts that those experiences are normally regarded as experiences *of*. They both, that is, purport to explain our subjective conscious experiences – that are "as if of" certain external facts – while denying the real existence of those particular external facts. The point is that whether they then posit some *other* external facts (FAPP solipsism) or not (literal solipsism) is, FAPP, irrelevant, since any other posited facts are necessarily completely arbitrary anyway and can hence play no genuine role in justifying the claimed account of conscious experience.

Considered as a possible response to Bell's claim to have established the real existence of nonlocal (faster-than-light) causal influences, the essential point is that FAPP solipsism does not attempt to provide any (local) physical explanation for a certain pattern of physically real measurement outcomes (instantiated, say, as the positions, at different times, of some physically real instrument pointers) that violate Bell's inequality. Instead, the FAPP solipsist only needs to provide a story according to which a subjective conscious experience – that is as-if-of such measurement outcomes – would arise. The FAPP solipsist's attempt at achieving empirical adequacy (i.e., consistency with experience) occurs, that is, inside of consciousness – at the level of subjective conscious experiences – rather than in the physical world itself. The real physical world, for the FAPP solipsist, may be nothing at all like the one we ordinarily believe in on the basis of ordinary perception: it need not include Bell's nonlocal causal influences, it need not include measuring instruments with pointers at all, it need not even be a world with three spatial dimensions. Any kind of world at all will do just fine, so long as one includes, as part of the story, the emergence of subjective conscious experiences that fool the inhabitants of the world into thinking they live in a three-dimensional world populated by cats, trees, planets, and Bell-inequality-violating pointer positions.

I have said that nobody is really a solipsist. I also do not know of anybody who has attempted to refute Bell's reasoning by claiming that we are all really just brains in vats. So the immediate conclusion here is a purely hypothetical one: if somebody were to attempt to refute Bell's arguments on these grounds, we would not (and should not) take them very seriously. Bell demonstrated that if you regard the familiar three-dimensional world of everyday perception as physically real, and if in particular you "include the settings of switches and knobs on experimental equipment, the current in coils, and the readings of instruments" as among the *local beables* of your theory – as, he thought, any serious and empirically viable theory *must* do [15]) – then your theory will have to involve a specific sort of nonlocality in order to achieve empirical adequacy. Responding to that demonstration by saying "Aha, but maybe there isn't any physical reality at all!" hardly constitutes a refutation. It does not even rise to the level of being a good joke. And it is exactly the same, I think, if one's response is instead, "There is a physical reality, but it is radically different from what you always thought; in particular, your knobs on experimental equipment and instrument readings are nowhere to be found there; instead these things only exist in your mind, as, in effect, hallucinations." In both cases, what is proposed is (at best) not in any sense a locally causal account of our empirical observations, but instead a hopelessly philosophical proposal to which Bell's notions of locality and nonlocality are simply inapplicable and irrelevant.

13.5 QBism and Quantum Solipsism

Let us then turn to comparing these strange philosophical views (which nobody endorses) to two more serious physical theories (that serious people do seriously endorse and that indeed are put forward, at least in large part, as ways of trying to elude Bell's nonlocality). We begin with the more straightforward case of "QBism," which originally stood for quantum Bayesianism but has apparently now quantum fluctuated its way into standing for some superposition of quantum Bayesianism, Bohrism, Bettabilitarianism, or (like "KFC", which formerly stood for "Kentucky Fried Chicken" until it was realized that the word "Fried" was a turn-off to health-conscious consumers) nothing at all [23]. This last is perhaps the most appropriate, since one of the key ideas of the theory is that quantum states merely summarize subjective beliefs/expectations about future subjective conscious experiences and hence stand for no objectively existing physical structures, i.e., no*thing*.

To elaborate on this a bit, QBism can be understood as the view that takes a personalist Bayesian interpretation of probability (understood quite broadly) and then applies this interpretation specifically to the probabilities that figure centrally in quantum theory. The result, as mentioned, is a thoroughly subjectivist account of the entire quantum calculus, with a particular emphasis on the need to understand quantum theory as a "single-user theory." That is, the probabilities that I use quantum theory to calculate should always and exclusively be understood as *my* credences that *I* assign to various possible future events in *my* subjective conscious experience. To quote one of the theory's main proponents, Chris Fuchs, answering some famous questions posed originally by Bell,

"*Whose information?* 'Mine!' *Information about what?* 'The consequences (for *me*) of *my* actions upon the physical system!' It's all 'I–I–me–me–mine,' as the Beatles sang." [24]

To understand QBism, it is perhaps helpful to compare with to the Copenhagen interpretation and its infamous "shifty split" between the unspeakable quantum world and the speakable (and, for Bohr, really existing) world of directly perceivable, macroscopic, "classical" phenomena [25]. QBism is, in effect, the result of moving this shifty split all the way in, so that the speakable "classical" realm includes just the subjective conscious experiences of the single user using the theory. Everything outside is then on the unspeakable quantum side of the split:

In QBism, the only phenomenon accessible to Alice that she does not model with quantum mechanics is her own direct internal awareness of her own private experience. This (and only this) plays the role of the 'classical objects' of Landau and Lifshitz for Alice (and only for Alice). Her awareness of her past experience forms the basis for the beliefs on which her state assignments rest. And her probability assignments express her expectations for her future experience. [25]

So, with QBism's thoroughly subjectivist understanding of the quantum state, and the idea that everything outside of one's own immediate conscious experience can only be discussed in terms of quantum states, one is left with no means whatever for saying anything definite about any part of the world external to one's own consciousness (if, indeed, there is such an external world at all).[4]

The proponents of QBism vehemently deny that their viewpoint amounts to literal solipsism [26, 27] and I for one am happy to take them at their word. They insist, for example, that there do really exist both outside physical systems (which our observations, measurements, and other interactions with the world observe, measure, and interact *with*) as well as other conscious agents with whom we can communicate (even if, due to the limitations of ordinary language, only imperfectly and approximately).

But at the end of the day, these external systems and other agents play no real role in the theory: to the extent that they can be described at all, they are described only in terms of quantum states. And these, it is insisted, do not provide anything like a direct, realistic description of those outside systems but instead have the exclusively instrumentalist/subjectivist role described above: they summarize our subjective beliefs and expectations about our own future subjective conscious experiences (which may be *as if of* external systems, but need not actually be *of* any such things at all):

[W]hen an agent writes down her degrees of belief for the outcomes of a quantum measurement, what she is writing down are her degrees of belief about her potential *personal* experiences arising in consequence of her actions upon the external world. [28]

Let us see how these ideas play out in the QBists' own writings about Bell-type nonlocality.

It is worth noting, to begin with, that the QBists' polemics against nonlocality are often not directed at Bell's theorem *per se*, but are instead directed at something like the following argument:

According to ordinary quantum mechanics, the wave function should be taken seriously (as a "beable", to use Bell's terminology) as representing some external physically real thing. But then the collapse postulate immediately implies non-locality: for example, if a single particle is prepared (as in the "Einstein's Boxes" [29] setup) with support in two well-separated spatial regions, looking to see if the particle is present in one of those regions immediately changes the wave function (and hence the physical state) in the other region, no matter how distant. Therefore ordinary quantum mechanics is non-local.

Now that is an important argument, to be sure, and it has some significant overlap with the EPR argument described earlier. But *at most* this argument would establish only that one

---

[4] For the record, I should perhaps note that QBism also has a more technical, less philosophical, foot in the realm of "quantum information" and has motivated, and developed in parallel with, a very intriguing project of re-formulating quantum mechanics exclusively in terms of *probabilities* rather than the more standard probability *amplitudes*. These developments, however, while interesting, are orthogonal to the issues we are pursuing here. But readers should be aware that, arguably, there is more to QBism than the philosophical parts I mainly address here.

particular candidate theory – "ordinary quantum mechanics" – is nonlocal...a conclusion radically different from the one Bell claimed to have established (and, in my judgment did establish), namely, that *any* theory that reproduces QM's empirical predictions will have to be nonlocal.

In any case, this is the argument against which the QBists' polemics against nonlocality are often directed. For example, Fuchs and Schack write that QBism's "thoroughgoing personalist account of *all* probabilities . . . frees up the quantum state from any objectivist obligations [and hence] wipes out the mystery of quantum-state-change at a distance...." [28] Fuchs, Mermin, and Schack explain that, according to QBism, "The notorious 'collapse of the wave-function' is nothing but the updating of an agent's state assignment on the basis of her experience" [25]. Thus, part of the QBist response to assertions of nonlocality is simply the rather obvious point that the specific nonlocality – posited by versions of quantum theory in which the wave function is a beable that sometimes collapses – can be avoided if we deny beable status to the wave function. That is of course correct as far as it goes. But since the above argument is not at all Bell's argument, it does not go very far.

But the QBists also attack the EPR argument (which they correctly recognize as undergirding Bell's nonlocality claim) and in particular the infamous EPR reality criterion:

The mistake in the 1935 argument of Einstein, Podolsky, and Rosen lies in their taking probability-1 assignments to indicate objective features of the world, and not just firmly held beliefs. Their argument uses the famous EPR reality criterion: "If, without in any way disturbing" a system, we can predict with certainty (i.e., with probability equal to unity) the value of a physical quantity then there exists an element of physical reality corresponding to this physical quantity." Without such "elements of physical reality,' there is no basis for their argument that if quantum mechanics gives a complete description of physical reality, then what is real in one place depends upon the process of measurement carried out somewhere else.

Bohr maintained that EPR's mistake lay in an "essential ambiguity" in their phrase "without in any way disturbing." But for the QBist their error is simpler. Their mistake was their failure to understand, as many physicists today continue not to understand, that $p = 1$ probability assignments are very firm personal judgments of the assigning agent, and nothing more.

The unwarranted assumption that probability-1 judgments are necessarily backed up by objective facts-on-the-ground – elements of physical reality – underlies EPR's conclusion that if quantum mechanics is complete then it must be (unacceptably to them) nonlocal. It also underlies Bell's original 1964 derivation of the Bell inequalities. [25]

In terms of my earlier recapitulation of the EPR argument, their point here is that "being certain the distant particle will come out, say, 'V' if its polarization is measured" does not (contrary to what I claimed above) imply that this outcome is somehow encoded in the really existing independent physical structure of that distant particle. Rather, they suggest, it simply means that one holds a "very firm personal judgment" that, if one later experiences a report of that particle's polarization having been measured, it will be a report of its having been measured to be "V."[5]

---

[5] As Fuchs, Mermin, and Schack explain: "[V]erbal or written reports to Alice by other agents that attempt to represent their private experiences are indeed part of *Alice's external world* and therefore suitable for her applications of quantum mechanics." Thus:

Now, it may appear that the QBists are here probing a notoriously soft spot in the EPRBell reasoning. As they point out, the criticisms of the EPR criterion of physical reality began immediately with Bohr and have never ceased. It is worth pointing out, then, two things. First, as was mentioned above, the actual EPR paper was written not by Einstein but by Podolsky, and Einstein thought Podolsky's text (and in particular the part involving the reality criterion) obscured the essential argument. And it should be understood that when Einstein himself recapitulated (something like) the EPR argument, his reasoning never included the reality criterion. So, soft or not, the reality criterion is something of a red herring. And then second, the truth is that the criterion is not nearly as soft as is often claimed [19].

But the more fundamental point here, in understanding the QBists' actual basis for rejecting Bell's nonlocality claim, is that the EPR reality criterion really has nothing to do with it. The relevant difference between their view and Bell's is not in how they interpret "$p = 1$ probability assignments" (and in particular whether they regard them as warranting the attribution of a certain particular structure to the state of the distant particle) but rather in whether they regard any such thing as the distant particle as actually, physically existing in the first place. And here the QBists are thankfully quite clear: the distant measurement outcomes (about which we sometimes make $p = 1$ probability assignments) are on the unspeakable quantum side of the shifty split; our only means of referring to them, in quantum theory, is by assigning an appropriate quantum state. But, according to QBism,

each quantum state [has] a home. Indeed, a home localized in space and time – namely, the physical site of the agent who assigns it! By this method, one expels once and for all the fear that quantum mechanics leads to 'spooky action at a distance' . . . It does this because it removes the very last trace of confusion over whether quantum states might still be objective, agent-independent, physical properties. [24]

That is, the real reason that (according to the QBists) Bell's EPR-based argument fails to establish nonlocality is not that there is some subtle logical invalidity in the EPR reality criterion, but rather that quantum theory, properly interpreted (according to the QBists), assigns all real events to the same one place: "the physical site of the agent." And therefore, since

No agent can move faster than light: the space-time trajectory of any agent is necessarily time-like. Her personal experience takes place along that trajectory. [25]

it evidently follows that

in QBism quantum correlations are necessarily between time-like separated events, and therefore cannot be associated with faster-than-light influences. This removes any tension between quantum mechanics and relativity. [25]

Indeed it does.

"A QBist does not treat Alice's interaction with Bob any differently from, say, her interaction with a Stern-Gerlach apparatus, or with an atom entering that apparatus. This means that *reality differs from one agent to another*. This is not as strange as it may sound. What is real for an agent rests entirely on what that agent experiences, and different agents have different experiences" ([25], emphasis added).

But, of course, the idea of reconciling quantum theory with relativistic locality in this way is completely absurd. The *point*, presumably, of wanting to maintain consistency with relativity is that one regards the evidence supporting relativity's picture of the world – e.g., as a 3+1-dimensional space-time with a certain structure and populated with certain material objects/fields – as conclusively established. But as I explained in the previous section, there is a huge difference between actually maintaining this picture, and simply avoiding having to assert nonlocality. The literal solipsist avoids having to assert nonlocality, but certainly does not maintain a relativistic picture of the physical world. The same is true for the brain-in-a-vat theorist or FAPP solipsist.

And, I of course want to suggest, it is the same as well for the QBist. The QBist picture of the world is one of a "single user" floating (with, for some reason I cannot begin to understand the rationale for, velocity strictly less than the speed of light, whatever that even means in this picture) through a void of immaterial nothingness, or, at least, an unspeakable haze. The very question, for example, of whether the structure of the posited space–time matches that of Einstein and Minkowski makes no sense. The QBist is free to claim that it does, but what connection would this have to any evidence in or out of QBist quantum theory? Such an assertion, from the QBist, is completely arbitrary and unconstrained, like the anti-Copernican brain-in-a-vat theorist's assertion that the evil scientists live on a stationary planet in the center of some crystalline spheres. Why *that*? It could just as easily be claimed that the real space-time through which the "single user" floats is a 2+1-dimensional Galilean space-time, or anything else. Or nothing at all.[6] There is, at the end of the day, no reason the QBist needs to even posit a physically existing body for the "single user" – just as the person who thinks he might be a brain in a vat soon realizes that he has no actual reason to believe in the existence of brains, including his own. QBism is in this respect equivalent to the brain-in-a-vat scenario, and hence meets my diagnostic criteria for FAPP solipsism.

Let me finally reiterate that I am not accusing QBists of being solipsists. The QBists claim that they are not solipsists and I believe them.[7] I think the actual situation is as follows: it's not that the QBists think there is no external reality. Rather, they do think that there is one, but they also think that quantum mechanics in no sense tells us what it is like (and, at present, neither does anything else). Fuchs and Schack, for example, write that

Quantum theory is conditioned by the character of the world, but yet is not a theory *directly* of it. Confusion on this very point, we believe, is what has caused most of the discomfort in quantum foundations . . . .

[28, emphasis added]

---

[6] Indeed, Mermin writes that "space-time is an abstraction that I construct to organize [my] experiences " [30].

[7] At least, I believe them most of the time. But there are also other times, for example: "Everything experienced, everything experiencable, has no less an ontological status than anything else. You tell me of your experience, and I will say it is real, even a distinguished part of reality. A child awakens in the middle of the night frightened that there is a monster under her bed, one soon to reach up and steal her arm – that *we-would-call-imaginary* experience has no less a hold on onticity than a Higgs-boson detection event would if it were to occur at the fully operational LHC. They are of equal status from this point of view – they are equal elements in the filling out and making of reality." [24] This really only makes any sense at all if "reality" is understood to mean "subjective conscious experiences" – and if there simply is nothing like what is *usually* meant by "reality."

But Bell had already responded to this kind of suggestion, decades before Fuchs and Schack made it:

You might say, Ψ, the wave function is just not a real thing – it is only a way of talking about something else, and then the fact that it has such funny possibilities of superposition and so on is not so disturbing. But if you say it's not real, I will ask: what *is* real in your theory? What are your kinematics? What are the possibilities that you contemplate and which you talk about when you write down a wave function?... If your wave function is not real, you must tell me what is. [31]

And that brings us back to square one. If the QBist claim is that quantum theory provides, at best, only some very *indirect* sense of what external physical reality is really like (such that, as noted before, we need not understand the collapse of the wave function as providing immediate proof of nonlocality) then we are simply endorsing a vague hidden-variables program kind of view: the quantum mechanical wave function doesn't provide a complete description of external physical reality (and perhaps plays no part whatsoever in such a complete description), but there *is* an external physical reality and someday we will hopefully figure out how to describe it. But the problem is then that any such future candidate realistic account of quantum phenomena (assuming it really accounts for real perceived events involving "knobs on experimental equipment, the currents in coils, and the readings of instruments" in a nonsolipsistic way) is subject to Bell's theorem. That is, if it genuinely reproduces the (experimentally well-verified) quantum predictions, it will have to be nonlocal in Bell's precisely formulated (and, from the point of view of relativity, very troubling) sense. That, simply put, is what Bell's theorem shows. So either QBism is really just a very long-winded (but temporary) *distraction* from Bell's conclusion (in which case it is in no sense a counterexample to Bell's claims), or it is offering a genuinely solipsistic account of a single user having conscious experiences that correspond to nothing physically real at all. Either way, in my opinion, it ceases to provide anything like a serious challenge to Bell's nonlocality claim.

13.6 Everett and Quantum Solipsism

I am hardly the first person to suggest that QBism is a rather solipsistic view [32]. But (aside from J. S. Bell! [20]) I don't know of anyone who has suggested that Everett's so-called "Many Worlds" version of quantum theory has a similarly solipsistic character. Indeed, Everettism is generally regarded (along with the de Broglie–Bohm pilot-wave theory and the several extant versions of spontaneous collapse theories) as one of the leading candidate *realistic* accounts of quantum phenomena.

I do not exactly think this is wrong. Certainly it is true that extant *proponents* of Everett's theory are motivated by a healthy sort of realism. And it is certainly true that Everettians posit various really existing physical structures as part of their theory. Nevertheless, I want to argue that – like the brain-in-a-vat kind of scenario – Everett's theory nevertheless has a FAPP solipsistic character and is thus (superficial appearances to the contrary notwithstanding) very similar to QBism as regards its relationship to Bell's theorem.

Let us start with an overview of the Everettian view. Readers will be familiar with the so-called "measurement problem" of ordinary QM, which can be understood as involving a tension between the unitary Schrödinger evolution of the quantum state, and the various measurement axioms that supposedly govern the interactions between the quantum system in question and a separate classical world. Everett's dual insight was (1) that we can avoid any such tension by simply dropping the idea of a separate classical world and letting the whole universe be governed by the unitary Schrödginer dynamics all the time, and (2) that this is not as crazy as it initially sounds. It initially sounds crazy, of course, because the unitary Schrödinger evolution produces states involving bizarre superpositions of macroscopically distinct configurations (such as living and dead cats). It is precisely to avoid such apparently embarrassing states – which do not seem to have any correlate in our empirical observations of the world – that the measurement axioms are introduced. But Everett suggested that there was in fact no conflict here at all: once it is remembered that we are parts of the universe (not outside, Godlike observers of it), we can ask what the world would look like, to us, from the inside, if Everett's postulates were right. And – or, at least, so it is claimed – things would more or less look right, due in no small part to decoherence, which renders the macroscopically distinct "branches" of the universal wave function (effectively) causally independent. And so, crucially, an observer who lives in one such decoherent branch will see and experience only the goings-on in his own branch. And so, it is said, the theory gets the (relevant) appearances right after all.

We are here interested in Everett's theory because Everettians often suggest that the theory constitutes a kind of counterexample to Bell's nonlocality claim. The basis of this suggestion is something like the following idea: the unitary Schrödinger evolution of the wave function was always perfectly local; the nonlocality of ordinary QM was always only in the nonunitary collapse dynamics; so by simply getting rid of the collapse postulate we get rid of the nonlocal part and what remains is purely local. Or the same basic idea can be expressed a little more formally as follows: in an appropriately relativistic version of quantum theory, the law governing the time evolution of the quantum state (that is, the appropriate relativistic generalization of the Schrödinger equation) is perfectly Lorentz invariant, perfectly compatible with the constraints implied by special relativity. And so (since the theory is that law and nothing more) there is simply no suggestion whatever of a violation of relativistic locality.

While it is true, however, that the fundamental dynamical equations that define Everettian QM are impeccably relativistic, this is not really the same thing as the theory being "local," at least not in Bell's sense.[8] Indeed, as soon as one considers Bell's careful formulation of "locality" – involving probabilities that the theory assigns to various local beables – one realizes that several of the formulation's key notions do not apply, or do not apply *straightforwardly*, to the Everettian theory. For example, the theory does not assign probabilities to events, at least not in a familiar kind of way. We might, for example, try to use Bell's formulation to assess the locality of some (more ordinary) candidate

---

[8] See, for example, Roderich Tumulka's "rGRWf" theory, which is manifestly relativistic yet nonlocal [33].

theory by examining the probabilities assigned by the theory to some event such as "Bob's polarization measurement has the outcome 'H'" when various other events are or are not conditioned on. But in Everett's theory, the probability that Bob's spin measurement has the outcome "H" is 100% – as is the probability that Bob's spin measurement has the alternative outcome "V"! (We assume here a general case in which the amplitudes for both events are nonzero.) The point is that, according to Everett, everything (that we normally think of as possible) is in fact deterministically guaranteed to occur, in one of the several downstream branches of the universal wave function. So, for example, the probabilities for (what we normally think of as) mutually exclusive events need not sum to unity; some of our basic ideas about what "probability" even means are thus completely inapplicable, and we must tread very carefully.

Here I do not want to get deeply into this particular issue. How to understand probability from an Everettian perspective is a big and controversial subject. I just want to mention here – in passing – that the subject relates not only to the question of whether and how Everett's theory can be understood as accounting for the usual probabilistic rules of quantum mechanics (which is the context in which the subject most often arises), but also to the question of whether and how Everett's theory can be described as "local" (in Bell's or some other sense). In particular: to whatever extent Everettians fail to provide a clear and compelling explanation of the meaning of "probability" in the many-worlds context, Bell's notion of locality will remain inapplicable and it will remain invalid to claim that the many-worlds theory provides a local (in Bell's sense) explanation of the Bell inequality violations.

13.6.1 Wave-Function Realism

But here I want instead to focus mainly on another concept that appears in Bell's formulation of "locality" but that is problematic from the point of view of Everett's theory: local beables. Bell's "locality" is at root a statement about what a theory says about (its) local beables. But in Everett's theory the only beable – the only thing posited as a real physical existent – is the quantum state of the universe. And (however exactly one understands it) this is not a local beable, i.e., in Bell's description, something that attributes physical properties to localized regions (for example, points) in ordinary, 3-dimensional, physical space and time. The quantum state can be understood as a (moving) point (or ray) in a very abstract and very high-dimensional Hilbert space, or perhaps instead as a complex-valued field in some (different, but still abstract and high-dimensional) configuration space. But neither of these representations provides, in any straightforward or clear sense, a description of particles, fields, strings, or any other type of physical "stuff" in (3-dimensional) physical space and time.

Within the school of Everettian thought known as "wave function realism" (according to which it is the universal wave function – the quantum state in position representation – that provides a kind of direct and complete description of the physical state of the universe), it has been suggested that local beables might be *emergent* rather than fundamental

[34]. In such a theory, the fundamental description of the world is in terms of a complex-
(or spinor-) valued *field* on a $3N$-dimensional space.[9] The fundamental description is simply
not of a world of physical "stuff" in 3-dimensional physical space; in Bell's terminology,
the theory posits no local beables at all. But, according to the emergence view, one may
nevertheless be able to *find* local beables – and in general a familiar-looking 3-dimensional
physical world – in the theory, much as one is able to find cats and trees and planets (and
even haircuts) in a classical theory whose fundamental ontological posits include only (say)
a number of mass points: the cats and trees and planets can be understood as being *made up
of* the posited mass points. Similarly, in the Everettian case, can we not perhaps understand
cats and trees and planets (and hence the kinds of things that are referenced in Bell's formulation of locality) to be *made up of* the universal wave function in an analogous manner?

David Wallace and Chris Timpson, for example, write that

Three-dimensional features will emerge [from the posited goings-on in the fundamental, high-
dimensional space] as a consequence of the dynamics (in large part due to the process of
decoherence) . . . each of the decohering components will correspond to a system of well-localised
(in 3-space) wavepackets for macroscopic degrees of freedom which will evolve according to the
approximately classical laws displaying the familiar three-dimensional symmetries, for all that they
are played out on a higher dimensional space. [35]

And Wallace similarly directly attacks the idea that three-dimensionality must be somehow
fundamental rather than merely emergent:

How do we know that space is three-dimensional? We look around us. How do we know that we
are seeing something fundamental rather than emergent? We don't; all of our observations . . . are
structural observations, and only the sort of a prioristic knowledge now fundamentally discredited in
philosophy could tell us more. [34]

It is my impression, though, that the proponents of Everettism have simply not appreciated
the profound difficulty associated with the claim that the 3-dimensional world could emerge
from (something like) a single field on a much-higher-dimensional space.

To be sure, it is possible for there to be a structural isomorphism between two such
realms. For example, the mathematical description of two beads on a straight wire (a system
with a two-dimensional configuration space) might be perfectly isomorphic to the mathematical description of a single pool ball bouncing around a square pool table. But just as
surely, such mathematical isomorphism does not imply that, any time a pool ball is bouncing around on a pool table, there must exist – in addition – two beads on a wire somewhere.
Universes composed of (or including) beads on a wire do not in any sense "emerge" from
the dynamical playing out (in our real universe) of pool-ball-on-table systems. At least,
there is not the slightest bit of evidence to suggest that this happens.

Or consider another example. A (three-dimensional!) box contains N gas molecules,
each of which (let us say) has at each moment a well-defined position and momentum. The

---

[9] Or a space of even higher, perhaps infinite, dimension if one considers relativistic field theories rather than N-particle quantum
mechanics.

physical state of the collection of N molecules can thus be represented, mathematically, as a single point in a $6N$-dimensional phase space. This works as a representation because there is a perfect isomorphism between the position of a single point in the $6N$-dimensional space, and the positions and velocities of N points in the three-dimensional space. But does this mean that, in any sense, an actual physical particle moving in a new $6N$-dimensional physical space *emerges* from the (let us assume) really existing collection of N molecules? Of course not. To think so is simply to reify what is in fact merely an abstract representation. And it would be exactly the same if it were (strangely) the one-particle-in-$6N$-dimensional-space story that were regarded as fundamental, and the three-dimensional-box-of-gas that were supposed to "emerge."

About just this kind of case, Wallace and Timpson seem to take a different view:

if the N-particle story were empirically adequate (which it isn't, of course) then so would the one-particle story be. For that story (by construction) is isomorphic to the N-particle story, and the emergence or otherwise of higher-level ontology from lower-level theories depends, to our minds, primarily on the structure of those theories and not on their underlying true nature (whatever that is). On the one-particle theory, three-dimensional space would turn out to be emergent, but it would be no less real for that.... [35]

But I simply do not understand how to make sense of this view, if the "emergence" of a three-dimensional space containing three-dimensional objects is supposed to mean that that space and those objects really come into existence. It is easy to understand how high-level macroscopic structures such as cats and haircuts (which involve specific patterns of three-dimensional stuff) can emerge from some fundamental ontology of three-dimensional stuff (particles, fields, whatever). But a cat, or a haircut, is an essentially three-dimensional pattern, and I do not understand how a physically real three-dimensional pattern of any kind can emerge from an underlying reality in which the physical stuff lives in a very different (much higher-dimensional) space. We could have a higher-dimensional pattern that is somehow isomorphic to the three-dimensional pattern, but this is not the same as having a really existing three-dimensional pattern.

But perhaps my incomprehension regarding such "trans-dimensional emergence" is misplaced and irrelevant. Perhaps, that is, what is being suggested (when it is suggested that three-dimensional things such as cats and planets and pointers and haircuts can emerge, for example, from a wave-function-realist ontology) is only that the *appearance* of cats and planets and pointers and haircuts will emerge. That is, perhaps the claim is merely that, for conscious inhabitants of the posited kind of world, it will *seem* as if they live in a three-dimensional world inhabited also by cats, planets, pointers, and haircuts.

Let me immediately concede that this view – what we might call "appearance-emergence" – is indeed rather plausible. The idea is that, although nobody would claim to understand the process in detail, and although many deep and profound ontological mysteries remain, we are all more or less comfortable with the idea that conscious experience somehow *emerges from* the complicated physical processes occurring in a living, functioning (three-dimensional) brain. And further, we are all reasonably confident that there is in this regard nothing unique about the specific "wet physical stuff" that happens to compose human brains. So if, for example, we could replicate the requisite *brain structure* in some other physical embodiment (say, an enormously complicated network of silicon-based transistors), consciousness would again emerge. And in particular, it seems quite likely that if the two networks (of wet neurons, on one hand, and silicon transistors, on the other) were perfectly structurally isomorphic, then not only would consciousness emerge in both cases, but, it seems quite plausible to suggest, the emergent consciousnesses would in some sense be exactly the same. But from there it is no great leap to the suggestion that in a universe that is radically different from ours – but nevertheless *structurally isomorphic to ours* – not only would the "beings" there be conscious in the same way that we (imagining ourselves to live in a 3-dimensional universe) take ourselves to be, but indeed the conscious experiences of those beings would be identical to ours.

It is in this sense, I think, that the Everettians claim – plausibly in my opinion – that an Everettian universe (consisting of, say, a field in a very high-dimensional space evolving according to Schrödinger's equation) would contain conscious beings whose conscious experiences perfectly matched ours – including, evidently, the experience of inhabiting a three-dimensional world populated by cats, trees, planets, etc. Thus, while I do not think the Everettian can plausibly claim that local beables in fact emerge (transdimensionally) from a wave-function-realist picture, I think he can plausibly claim that the subjective conscious experience of local beables might so emerge. In effect, the inhabitants of an Everettian universe would be deluded into thinking that they lived in a 3D world just like the one we take ourselves to inhabit. And so, according to the Everettians, for all we know, we might very well be those inhabitants of an Everettian universe.

And that, as far as it goes, is probably correct. There is no specific element of anyone's subjective conscious experience – there is no specific evidence – that one could point to that would show that we are *really* made of particles and fields moving and interacting in a 3-dimensional space, as opposed to being *really* made of field-stuff in a much higher-dimensional space. But then, there is not supposed to be any specific element of subjective conscious experience that one could point to, either, to prove that we are not *really* brains-in-vats. And the parallel there should by now be entirely clear. The "appearance-emergence" Everettian is offering a worldview in which the *true* physical world is radically different from the one we ordinarily take ourselves to be perceiving. In this view, every aspect of our ordinary perceptual experience (and hence also everything that is based on it) is fundamentally delusory, just as in the FAPP solipsist brain-in-a-vat scenario. And so I think all of the reasons we have for not taking those FAPP solipsist ideas too seriously (even though, as Bell noted, they cannot exactly be refuted) apply as well to Everettianism.

The Everettian will of course want to insist that his worldview – his wave function ontology – is totally unlike the brain-in-a-vat worldview, since wave functions (unlike the hypothesis that one is a brain in a vat being controlled by evil scientists) enjoy considerable empirical scientific support. But I think a little reflection will reveal that the positions are not actually so different. The Everettian here has precisely the same type of empirical grounds for believing in wave-functions as the brain-in-a-vat theorist has for believing in brains – namely, inference from a wealth of perceptual experiences that, according to the belief in question, were fundamentally misleading.

My point is not simply to tar Everettism by association with solipsism. That is, my point is not to say, "Everettian QM has a FAPP solipsistic character; this makes it *crazy*, so we shouldn't take it seriously." I actually do take it quite seriously. My point is entirely different: not that Everettian QM is crazy, but that the claim that it is a *local* theory – some kind of counterexample to Bell's claim that nonlocality is required to explain certain observed facts – is at best empty and misleading and obscurantist rhetoric. The truth is that Everettian QM (at least in the wave function-realist version, interpreted via "appearance-emergence") is neither local nor nonlocal in Bell's sense. What it posits as the fundamental nature of physical reality is so completely different from the kind of picture Bell assumes that his definition of "locality" fails entirely to apply in any meaningful way at all.

13.6.2 Space–Time State Realism

Recently, though, another way to understand the ontology of Everettian QM has been proposed, and it is worth examining that as well, lest our diagnosis be too broad or too hasty. I have in mind here Wallace and Timpson's "space–time state realism" (SSR), which, I think it is fair to say, they propose precisely to avoid the type of difficulty one gets into when one's theory contains no local beables

> Our claim, in essence, is that thinking about quantum mechanics in terms of a wavefunction on configuration space is rather like thinking about classical mechanics in terms of a point on phase space. In both cases, there is a far more perspicuous way to understand the theory, one which is connected to spacetime in a more direct way. [35]

And in particular SSR is supposed to help clarify the status of Everettian QM vis-à-vis nonlocality:

> Our project . . . might seem simple. Does Everettian quantum mechanics violate Local Action? And does it violate Separability? But things are not quite that easy: even saying what *is* the state of a given physical region in quantum theory requires us to have a more solid grasp of the physical reality that 'the quantum state' represents than is available from its abstract, Hilbert-space definition. [We need to first] get a firmer grip on just what the physical world is like according to Everettian quantum physics.
> *[36, pp. 294–5]*

Wallace, that is, seems to agree with Bell that one can only make meaningful claims about locality or nonlocality after clearly specifying a theory's ontology of local beables.

The main idea of SSR is to take the ontology of the theory to be explicitly three-dimensional: a set of properties of space–time regions, captured in particular by the set of *density operators* over all space–time regions. Wallace and Timpson suggest that this ontology is comparable to that of classical electromagnetism, whose fields can also be understood as describing properties of spacetime regions (here, points). Of course, the idea of *operators* as local beables is a little unusual. Operators don't take *values* the way that, say, an electric field does . . . but as Wallace argues, maybe objecting to operators-as-beables is simply based on the novelty and unfamiliarity of the idea, rather than any actual problem with it. Maybe. In any case, whereas it is completely clear how cats and planets and trees can be made of the local beables in (for example) Bohmian mechanics (namely: they are simply cat-, planet-, and tree-shaped constellations of particles), it remains at best obscure how to understand space–time state realism's local density operators as providing an image of the familiar everyday three-dimensional world.

But let us focus on a different point, more relevant to our overall project of questioning the Everettian claim to provide a *local* explanation of the EPR–Bell correlations. To be clear, the advocates of space-time state realism concede that the theory is nonlocal in a certain sense: it involves *nonseparable* state descriptions. This nonseparability plays a crucial role in the theory's (purported) ability to account for the EPR-Bell correlations. But, the advocates stress, nonseparability is very different from the *dynamical nonlocality* whose necessity Bell claimed to have established. And, in particular, the space–time state realists claim that their nonseparable but (dynamically) local theory is fully compatible with relativity in a way that a dynamically nonlocal theory never could be.

Let us examine these claims in some detail. First, let us understand exactly the nature of the theory's nonseparability. The reduced density operators that space–time state realism posits as (something like?) local beables, are formed by taking the density operator of the universe and doing a partial trace over the degrees of freedom from outside the space–time region in question. The density operator for a subregion (of a given, larger region) can thus always be formed from the density operator of the larger region. But we cannot recompose the density operator for a larger region, even from the density operators for a set of subregions that jointly cover the entire larger region.

More concretely, in the standard kind of EPR–Bell scenario in which spatially separated observers Alice and Bob make measurements (along the same axis, say) of the polarization of each photon from an appropriately polarization-entangled pair, the density operator for the large region (comprising both Alice's and Bob's labs) will be a mixture of operators corresponding respectively to "Alice and Bob both measure the polarization to be 'H'" and "Alice and Bob both measure the polarization to be 'V'." From this, we may compute the density operator describing the state of Alice's lab alone: it is a mixture of (operators somehow supposedly representing) "Alice measures the polarization to be 'H'" and "Alice measures the polarization to be 'V'." And similarly for the density operator describing Bob's lab alone. The point is then that, from the descriptions of Alice's and Bob's labs separately, we could not recover the fact that their outcomes are perfectly correlated. (In the Everettian picture, this perfect correlatedness of course means that the universe has split into two branches, one in which both Alice and Bob measure the polarization to be "H," and one in which they both measure the polarization to be "V"; there is no branch in which Alice's and Bob's outcomes are different.) This is the sense in which the theory involves nonseparability: there are certain facts pertaining to larger space–time regions (such as the perfect correlatedness of Alice's and Bob's polarization measurements) that cannot be decomposed into sets of facts pertaining to the subregions that jointly compose the larger region.

I think it is worth pointing out here that all of this seems to me to render these would-be local beables rather pointless. Basically this nonseparability means that any description of the states of several limited space–time regions will be decidedly *incomplete* unless we also include a description of the state of the larger region that comprises them. But once we do that, the states of the subregions are entirely redundant. Why bother positing them, as separately real existents, at all?[10] And of course this argument can be immediately repeated: if it is pointless to posit separate "space–time states" for Alice's and Bob's labs (but instead only for the larger region comprising both labs), then it is equally pointless to posit a "space–time state" for that region comprising both labs; we should instead just posit the space–time state of an even larger region comprising both labs and, say, some cat somewhere; and so on, until we realize that the only "space–time state" we really need to posit at all is the one for the ultimate space–time region – the universe as a whole. But the density operator for the universe as a whole is mathematically equivalent to the quantum state of the universe as a whole. Which leaves us right back where we started: Everettianism as positing the quantum state (perhaps in something like "position representation," i.e., wave function realism) and running into serious trouble because of the lack of local beables.[11]

It could perhaps be argued that this criticism of space–time state realism is unfair. A similar argument, for example, could be leveled against any theory in which the local beables are *functions* of the quantum state. For example, in the theory "GRWm" (which posits a universal wave function obeying the nonlinear GRW evolution equation [37]), the role of local beable (out of which cats and planets and trees are supposed to be made) is played by a "mass field" $m(\vec{x}, t)$, which is simply defined in terms of the universal wave function $\Psi(\vec{x}_1, \vec{x}_2, \ldots, \vec{x}_N, t)$. If Ψ is given at some time t, then the mass field m can be computed. So, one might argue, what is the *point* of positing the mass field as a separate existent? Is it not redundant? In my opinion, the answer here is: No. The mass field is not redundant, because of the crucial role it plays (in the context of this particular theory) in providing an image of the familiar three-dimensional world of everyday perception, including not only the cats and planets and trees I keep mentioning, but also the instrument pointers whose positions ultimately constitute so much empirical data. So after all it does, I think, make an important difference that space–time state realism's would-be local beables are (qua operators on an abstract mathematical state space) so difficult to understand as describing, in any straightforwardly comprehensible way, a (set of) familiar-looking three-dimensional world(s).

---

[10] Interestingly, the proponents of SSR sometimes seem conflicted about whether the local density operators constitute additional, separately posited ontology (beyond the quantum state), or should instead be regarded as merely parts of, or perspectives on, the quantum state (which would constitute the entire ontology by itself). For example, Wallace writes, "Everettian quantum mechanics reads the quantum state literally, as itself standing *directly* for a part of the ontology of the theory. To every different quantum state corresponds a different concrete way the world is, and the quantum state *completely* specifies the ontology" [36, p. 295]. So . . . which is it? Is the quantum state only a *part of* the ontology (with the local density operators presumably filling the ontology out)? Or does the quantum state (by itself) *completely* specify the ontology? This is not really clear. My point here is just that the ontic redundancy noted in the main text may explain the proponents' ambivalence.

[11] Wallace expresses the trouble this way: "Note that if . . . we were to treat the Universe just as one big system . . . then we would only have a single property bearer (the Universe as a whole) instantiating a single property (represented by the Universal density operator), and we would lack sufficient articulation to make clear physical meaning of what was presented" [36, pp. 299–300].

Despite the apparent impossibility (or perhaps, more generously, obscurity) of understanding space–time state realism's density operators as providing an image of the familiar physical world of ordinary perception, it should be admitted that – just as with wave function realism – there is a kind of structural isomorphism between (one "part" or "branch" of) the reduced density operators and the ordinary 3D physical world. So from a certain "functionalist" point of view, both versions of Everettism may perhaps be considered viable, at least in the FAPP solipsist sense I discussed above for the case of wave-function realism. But instead of repeating that kind of analysis, here I want to push a bit farther into the claim that SSR's (would-be) local beables allow a clear diagnosis of dynamical locality. So let us temporarily set aside the worries about the adequacy of the SSR local beables and grant to the Everettian space–time state realist more or less everything he wants to claim. I will then argue that it is *still* highly questionable whether the theory provides a genuinely (dynamically) local account of the empirically observed correlations.

Let us grant, in particular, that SSR's reduced density operators provide an unproblematic slate of local beables in terms of which we can find, in the world posited by the theory, events such as "Alice measuring her photon's polarization to be 'V'." Let us also grant that we can apply Bell's notion of "local causality" (i.e., dynamical locality) to the Everettian theory by replacing the probabilities (that appear in Bell's formulation) with the "branch weights" that play a somewhat analogous role in many-worlds theories. In particular, let us amend Bell's formulation as follows (see again the earlier figure):

A [many-worlds] theory will be said to be locally causal if the [branch weights] attached to local beables in a space–time region 1 are unaltered by specification of values of local beables in a space-like separated region 2, when what happens in the backward light cone of 1 is already sufficiently specified, for example by a full specification of local beables in a space-time region 3 . . . [18]

Does the Everettian SSR theory come out as "locally causal"? This very much depends on exactly what set of facts (posited by the theory) one regards as included in the local beables. For example, let us take the event in question (in region 1) to be "Alice's polarization measurement has outcome 'H'." The branch weight of this event is (let us say, assuming an obvious kind of setup) 0.5. And that will be true independent of whether one also considers the state of Bob's laboratory. Consider in particular the situation where Bob decides, at the last minute, whether to measure his photon along the same axis as Alice, or an orthogonal axis. In the former case, the density operator for the larger region (jointly describing both Alice's and Bob's labs) will be a 50/50 mixture of "HH" and "VV" (i.e., "Alice's outcome is 'H' and Bob's outcome is 'H'" and "Alice's outcome is 'V' and Bob's outcome is 'V'"). And so the total weight of branches in which "Alice's outcome is 'H'" will be 0.5. Whereas in the latter case, where Bob chooses instead to measure along an orthogonal axis, the density operator for the larger region will instead be a 50/50 mixture of "HV" and "VH". And so, again, the total weight of branches in which "Alice's outcome is 'H'" will be 0.5. In short, the branch weight of this (localized) event is unaffected by including the information (from Bell's region 2) about Bob's choice of measurement axes. And so one would apparently conclude that the theory is locally causal, in the modified Bell sense.

But consider another fact pertaining to Alice's measurement to which we can apply the modified Bell locality criterion: whether the (descendant of) Alice who ends up seeing the "H" outcome is in the same universe (or "branch") as the Bob who ends up seeing the "H" outcome. If Bob happens to choose to measure along the same axis as Alice, the descendant of Alice who observes the "H" outcome will definitely end up in the same universe as the descendant of Bob who observes "H." Whereas if Bob instead happens to choose to measure along the orthogonal axis, the descendant of Alice who observes "H" will definitely *not* end up in the same universe as the Bob who observes "H." The weight of branches in which the "H"-observing Alice is in the same universe as the "H"-observing Bob thus very much depends on Bob's choice of measurement axis: it is 1.0 for one of Bob's possible (free) choices, and zero for the other.

Does this constitute a clear violation of the modified Bell locality criterion and hence a demonstration that the SSR version of Everettian QM is (dynamically) nonlocal after all? I am honestly not sure. It depends on whether one allows something like "The Alice-descendant who observes 'H' is in the same universe as the Bob-descendant who observes 'H'" as a statement *about* Alice, i.e., as a local beable pertaining to Bell's space–time region 1. And I can see arguments cutting both ways there. On one hand, this is somehow clearly a statement about one of Alice's descendants who is present in region 1. On the other hand, you could not see whether this was true or not if you *only* looked at the reduced density operator pertaining to region 1. The fact in question is thus somehow clearly a non-local relational fact pertaining to both regions 1 and 2. And when one recognizes that, it becomes less surprising that Bob's free choice (which occurs in region 2) could influence it.

However, I do not think it would be right to simply dismiss the example as failing, after all, to show that there is some nonlocality in the theory. Of course it is true that no nonlocality is implied by the fact that Bob's free choice in region 2 can affect the joint state of regions 1 and 2. But remember: not only is the fact in question here (namely, whether the Alice-descendant who observes "H" is in the same universe as the Bob-descendant who observes "H") not present in region 1 alone – it is also not present in region 2 alone. So it is wrong to think that what Bob's free choice is causally influencing is merely the state of region 2. It is instead causally influencing the joint state of regions 1 and 2. And if a free choice in one region can causally influence the physical state of another region, not all of which is confined to the future light cone of the first region, that does start to sound suspiciously like a violation of relativistic causal structure – even if (because of the "nonseparability") it is impossible to pinpoint a particular space–time location for the effect.

Coming at this same point from another direction, Wallace wants to suggest the following kind of picture: Alice's measurement triggers a branching event that propagates outward from her lab at (roughly) the speed of light; similarly, but independently, Bob's spacelike separated measurement triggers another branching event that propagates outward from his lab at the speed of light; it is only when/where these two outward-propagating branchings come to overlap (that is, it is only in the intersection of the future light cones of the two measurement events) that the different branches get "connected up" in such a way as to address, for example, whether the Alice-descendant who observes "H" is in the same universe as the Bob-descendant who observes "H" [36, p. 307]. This has the superficial appearance of providing a perfectly locally causal mechanism for the kinds of correlations that are in question here (e.g., whenever Alice and Bob measure along the same axis, they find, when they compare notes later, that their outcomes are always identical).

The grain of truth in this story is that, indeed, these relational or correlational facts only become manifest in the strictly local beables in the overlapping future light cones of the individual measurement events. But – and this is the crucial point here – those facts are nevertheless quite real well before this time, as is clear from simply considering the state of the region composed jointly of 1 and 2.

Wallace of course recognizes that, in this EPR kind of case, the world-splitting events triggered by Alice's and Bob's measurements are not independent (i.e., he recognizes that there already exist non-locally-instantiated facts about which of Alice's and Bob's descendants are in the same worlds as each other):

Nor is this to be expected: ... in Everettian quantum mechanics interactions are local but states are nonlocal. The entanglement between the particle at A and the particle at B is a nonlocal property of [the joint region] $A \cup B$. That property propagates outwards, becoming a nonlocal property of the forward light cone of A and that of B. Only in their intersection can it have locally determinable effects – and it does, giving rise to the branch weights which, in turn, give rise to the sorts of statistical results recorded in Aspect's experiments and their successors: statistical results which violate Bell's inequality.

[36, p. 310]

I would summarize the situation differently: I think the SSR theory only looks dynamically local to the extent that one does not take its many-worlds character sufficiently seriously. In particular, SSR defines the local ontology as the sum total of what is happening in a given space–time region across all the worlds, rather than taking the distinct worlds (and in particular their distinctness from one another) seriously.[12] If one did take the distinct worlds more seriously – by giving them pride of place in the local ontology – it would become immediately clear that the theory is as nonlocal as it can make sense for a many-worlds theory to be: which universe Alice's descendants are born into (namely one in which Bob's measurement came out "H" or instead one in which it came out "V") depends, nonlocally, on Bob's spacelike separated choice of which axis to measure his photon's polarization along.[13]

---

[12] There is an interesting parallel here to non-Everettians who similarly define the local state as a reduced density operator and argue that there is no nonlocality, since distant interventions do not affect the local state. [4] In a non-many-worlds context, the density operator is really just a catalog of the expected statistics for an ensemble of identically prepared systems, and its not changing (as a result of distant interventions) is equivalent to the familiar inability to send superluminal signals. But it was never a violation of such "signal locality" that Bell claimed to have established. The point here is just that Wallace's defense of the supposed dynamically local status of SSR Everettianism seems to commit a kind of many-worlds analog of this standard fallacy of switching the meaning of "locality" from Bell's "local causality" to "signal locality."

[13] Note the subtly solipsistic turn in Wallace's discussion of this: "From the perspective of a given experimenter, of course, her experiment *does* have a unique, definite outcome, even in the Everett interpretation. But Bell's theorem requires more: it requires that from her perspective, her distant colleague's experiment also has a definite outcome. This is not the case in

This point is closely related to the point I raised above about the local beables (in SSR) being ultimately redundant. Much more is *physically real*, according to SSR Everettianism, than is manifest in the reduced density operators of localized space–time regions. Wallace's claim is essentially that if one assumes that these density operators (for the localized regions) capture everything that is actually going on in space–time, then there is no nonlocality. That may be true. But if one instead remembers the facts that are, according to the theory, physically real, but that are not captured by the local density operators (for example, the way that the different terms/branches at different locations *correlate* with one another), the suggestion that the theory is local looks very suspicious.

All of that said, though, let me be the first to confess that all of this is less than perfectly clear and compelling. Recall Bell's underappreciated point:

> If local causality in some theory is to be examined, then one must decide which of the many mathematical entities that appear are supposed to be real, and really here rather than there. [2]

Wallace and Timpson's SSR is a step in the right direction for Everettianism, in the sense that it represents a tacit acknowledgement that local beables are required in a theory that purports to give a realistic account of empirical (i.e., observed) reality. But the truth is that the SSR ontology remains far too obscure to allow any clear and unambiguous analysis of the sort I am attempting above.

To review, there are several issues. First, the would-be local beables are abstract mathematical operators that act on vectors in a Hilbert space. I appreciate and accept Wallace's point that

> [t]here need be no reason to blanch at an ontology merely because the basic properties are represented by such objects [namely, abstract operators]: we know of no rule of segregation which states that, for example, only those mathematical items to which one is introduced sufficiently early on in the schoolroom get to count as possible representatives of physical quantities!
>
> *[36, p. 299]*

But polemics against rejecting the proposed ontology do not really help. What is needed is a clear account of how the proposed ontology can be understood: what sort of three-dimensional thing do these density operators represent, and how exactly does the representation work? Without answers to such questions, it is simply not clear what the theory is supposed to be about, i.e., what, according to the theory, is supposed to be real.[14]

Everettian quantum mechanics – not, at any rate, until that distant experiment enters her past light cone. And from the third-person perspective from which Bell's theorem is normally discussed, no experiment has any unique definite outcome at all" [36, p. 310]. The point is that the locality or nonlocality of the theory depends on whether the theory says that what happens to Alice depends on what Bob did, or does not so depend. It has nothing to do with what things look like "from her perspective." Wallace here conflates the question of whether, according to the theory, Bob's actions influence Alice with the totally irrelevant question of when and where Alice might *find out* about Bob's actions.

[14] And, for the record, "what's real is something that can be accurately and completely represented by density operators" does not suffice here. One might as well just forget about the SSR ontology and claim that "what's real is something that can be accurately and completely represented by the quantum state." The very same thing that motivates Wallace and Timpson to propose SSR in the first place – namely the inadequacy of the kind of answer quoted in the previous sentence – should make it clear that SSR remains, at best, inadequate.

The second issue I mentioned above is that the SSR ontology of local beables seems somewhat like a metaphysical afterthought. Not only are the local beables mathematically redundant (in the sense that the density operator of any large region determines the density operator of all its subregions, but not vice versa), but also it is certainly not possible to reformulate the theory mathematically in terms of the local density operators. In short, it simply does not appear that Everettian SSR is a theory *about* local density operators. Instead, the theory looks more like (because, in fact, it is) a theory that is fundamentally *about* the universal quantum state, with some colorful but questionable and probably superfluous window dressing of an only vaguely local beablish character.

At the end of the day, then, I think the only currently available way to make any sense of Everettian quantum theory is that it describes a *physical* world radically different from the familiar three-dimensional world of everyday perception – for example, a world consisting of a dynamical field evolving in a very high-dimensional space (i.e., the wave-function realist picture). Nobody has proposed a comprehensible way of understanding how the familiar three-dimensional world of ordinary perception might be represented by, or might somehow emerge from, the universal wave function. What is (at least somewhat) plausible, though, is that the structural isomorphisms between certain dynamical degrees of freedom in the evolving wave function and human neurophysiology might imply that "information processing agents" in an Everettian world should have conscious experiences identical to those usually assumed to emerge from human neurophysiological processes. That is, the beings in an Everettian world might be expected to have conscious experiences just like ours; i.e., these beings might be deluded into thinking that they are humans who live in a three-dimensional world populated as well by cats, trees, and planets.

Extant Everettianism is, in that sense, an elaborate brain-in-a-vat proposal, and hence a clear case of what I have called "FAPP Solipsism." It is not, I think, the case at all that it provides a legitimate counterexample to Bell's claim that nonlocal dynamics is required to explain the (apparently) observed predictions of quantum mechanics. In fact, as has been pointed out before, it denies that those predictions, as usually understood, even actually occur [38]. But the radical extent of this denial has not been sufficiently appreciated. In fact, according to Everett's theory, nothing like the three-dimensional world containing Alain Aspect, his laboratory, and the pointer positions (with their famous Bell-inequality-violating correlations) actually exists. Instead we have, at best, a kind of delusional appearance of such things, in the minds of beings in a radically different kind of physical universe – a story that, even on its own premises, it is very difficult to take seriously.

13.7 Conclusions

Critics of Bell's claim to have established the reality of nonlocal causal influence in nature often just erroneously assume that his analysis *begins* with two premises: locality (i.e., no faster-than-light causal influences) and deterministic "hidden variables." Such critics are simply ignorant of Bell's actual reasoning, which begins not with an assumption of determinism but rather with a proof (due, roughly, to EPR) that, already in the case of the narrow subset of correlations considered by EPR, locality *requires* determinism. Bell's careful formulation of locality, reviewed in Section 13.3, provides a helpful corrective here, insofar as it allows the argument "*from locality to* deterministic hidden variables" [13] to be made in a more formal and rigorous way. (See, e.g., Ref. [9] for details.)

But Bell's formulation of locality is also quite helpful in confronting another rather different category of critics – namely, those (like the QBists and Everettians) purporting to offer fully worked out versions of quantum theory that elude the supposedly necessary kind of nonlocality. What is helpful, in particular, is Bell's insistence that the very idea of locality can only be formulated in terms of "local beables" – i.e., his insistence that a theory must include, in its ontology, appropriate building blocks for directly observable macroscopic things such as "the settings of switches and knobs on experimental equipment . . . and the readings of instruments" [15]. What is the sense of this alleged "must"? The point is that directly observed (macroscopic) physical objects, including the pointers that register the outcomes of physics experiments, consist of such ontological facts. A theory in which such facts are not *physically real* – but are instead some kind of hallucination or delusion – necessarily gives only a solipsistic account of observed phenomena. This means that it does not actually account for the physically realized outcomes that have been observed in various experiments, but instead only purports to account for the (delusional) subjective appearance, in consciousness, of such outcomes.

We discussed two concrete examples of such FAPP solipsistic accounts. The essence of the QBist account is an explicit and systematic denial that the theory posits any ontology at all. This denial is somewhat undercut by the proponents' protests against charges of solipsism, as well as their somewhat absurd suggestions that (what they confusingly call) "reality" (for, recall, the single user of the theory) propagates along a well-defined subluminal trajectory through physical space. But at the end of the day the theory's status is perfectly clear: everything it purports to be about, and everything we could ever hope to possibly assert on its basis, occurs inside the consciousness of some lonely agent. With no physical ontology, the solipsistic character is clear – and with it the complete and total inapplicability of concepts such as "local" and "nonlocal." So while it is certainly the case that QBism provides a way of trying to understand quantum mechanics that avoids any commitment to spooky, antirelativistic action at a distance, it is totally uninteresting as a supposed counterexample to Bell's claim that nonlocal causal influences are needed to explain what is observed in experiments. For QBism, qua FAPP solipsism, denies that anything was in fact observed in those experiments and, indeed, denies that any such experiments every actually, physically, took place at all.

Everettian quantum theory, on the other hand, is a little harder to pin down. In its so-called "space–time state realist" formulation, there is at least a nod in the direction of trying to provide an ontology of local beables. But this remains, at best, a work in progress; at present I simply cannot understand how the abstract mathematical operators (posited as "local beables") can be understood as describing the state of three-dimensional matter in the necessary table-, cat-, tree-, and switches-and-knobs-on-experimental-equipment-like configurations. (And even if the density operators could be understood as providing an appropriate image of three-dimensional physical reality, it remains far from clear that the theory is actually local in Bell's sense.) The "wave function realist" formulation, on the other hand, provides a perfectly comprehensible ontology: a dynamical field on a very high-dimensional ("configuration") space. But in this picture we are, at best, and absent some future explanation of transdimensional physical emergence, apparently like the proverbial brains in vats whose every subjective conscious experience is a hallucination or delusion. Thus, in the only way I can at present make any sense of Everettian quantum theory, it too has a FAPP solipsistic character and hence cannot really be taken seriously as providing any meaningful sort of counterexample to Bell's arguments.

In the case of QBism, I have no hope whatever that the theory might someday be developed to provide a coherent physical ontology. The denial of any such ontology is just too central to the whole motivation. In the case of Everettianism, on the other hand, I am not so sure. Indeed, there is perhaps already a hint of a more viable strategy for a coherent Everettian ontology in Ref. [39]. So I do not mean at all to claim that I have somehow once and for all diagnosed Everettian quantum theory as solipsistic. My claim is much more reserved: the ontology – the local beables – of Everettian quantum theory remains, to me, so obscure that I cannot yet understand the theory as providing a realistic description of directly observable macroscopic reality. And – *therefore* – any suggestion that the theory somehow provides a counterexample to Bell's nonlocality claim is radically premature.

The lesson of all this, as we celebrate the 50th anniversary of Bell's great achievement, is simply that we cannot forget Bell's insistence on the preconditions of discussing locality, and in particular his point that this concept can only be really *understood* "in terms of local beables" [15]. In Bell's view, the local beables of any serious theory have to "include the settings of switches and knobs on experimental equipment . . . and the readings of instruments" [15], in the sense of including some physically real stuff out of which such directly observable macroscopic objects can be coherently understood to be *made*. This, at the end of the day, is the only notion of "realism" that Bell assumes. But who could deny this? Surely even the most strident instrumentalist believes in the physical reality of directly observable macroscopic instruments! It seems that the only people who could possibly deny this very elementary sort of "realism" are solipsists and FAPP solipsists. I have suggested that there are several influential camps of FAPP solipsists involved in these quantum foundational discussions. But, revealed as such, these need not (and cannot) be taken too seriously.

What Bell established 50 years ago is really quite remarkable: as long as you understand the ordinary world of directly observable macroscopic objects in a standard, nonsolipsistic way, there is no way of filling out the microscopic details of the ontology and dynamics so that experimentally observed correlations are accounted for in a purely local way. Contrary to the assertions of many commentators, this leaves us really only two options: we can accept that nonlocality is a real feature of our world, or we can adopt a bizarre, solipsistic type of view that is far, far less palatable and (at best) "local" only in some totally arbitrary and empty sense.

Perhaps in another 50 years, when we celebrate the 100-year anniversary of Bell's discovery, the ridiculous lengths to which one must go to avoid nonlocality will finally be widely appreciated?

References

[1] J.S. Bell, "On the Einstein–Podolsky–Rosen paradox," *Physics* **1** (1964) 195–200; reprinted in J.S. Bell, *Speakable and Unspeakable in Quantum Mechanics*, 2nd ed., Cambridge, 2004.

[2] J.S. Bell, Preface to the first edition of *Speakable and Unspeakable in Quantum Mechanics*, 1987.

[3] A. Aspect, J. Dalibard, and G. Roger, Experimental test of Bell's inequalities using time-varying analyzers, *Phys. Rev. Lett.* **49**, 1804–7 (1982); G. Weihs, T. Jennewein, C. Simon, H. Weinfurter, and A. Zeilinger, Violation of Bell's inequality under strict Einstein locality conditions, *Phys. Rev. Lett.* **89**, 5039–43 (1998).

[4] Reinhard Werner, Comment on "What Bell did," *J. Phys. A: Math. Theor.* **47** (2014), 424011. See also What Maudlin replied to, arxiv:1411.2120, and Werner's blog post (and the comments thereon) at http://tjoresearchnotes.wordpress.com/2013/05/13/guest-post-on-bohmian-mechanics-by-reinhard-f-werner.

[5] Marek Zukowski and Caslav Brukner, Quantum non-locality – it ain't necessarily so … *J. Phys. A: Math. Theor.* **47** (2014), 424009.

[6] A. Aspect, Introduction to the Second Edition of *Speakable and Unspeakable in Quantum Mechanics*, by J. S. Bell, Cambridge, 2004.

[7] J.S. Townsend, *A Modern Approach to Quantum Mechanics*, 2nd ed., University Science Books, Mill Valley, CA, 2012.

[8] S. Goldstein, T. Norsen, D. Tausk, and N. Zanghí, Bell's theorem, www.scholarpedia.org/article/Bell's_theorem.

[9] T. Norsen, J.S. Bell's concept of local causality *Am. J. Phys.* **79** (12) December 2011, 1261–75.

[10] A. Einstein, B. Podolsky, and N. Rosen, *Phys. Rev.* **47** (1935), 777.

[11] A. Fine, *The Shaky Game*, University of Chicago Press, 1996.

[12] D. Howard, "Nicht sein kann, was nich sein darf," or the prehistory of EPR: Einstein's early worries about the quantum mechanics of composite systems in A.I. Miller, ed., *Sixty-Two Years of Uncertainty: Historical, Philosophical, and Physical Inquiries into the Foundations of Quantum Mechanics*, pp. 61–111, Plenum, New York, 1990.

[13] J.S. Bell, Bertlmann's socks and the nature of reality, *Journal de Physique*, Colloque C2, suppl. au numero 3, Tome 42 (1981), C2 41–61; reprinted in *Speakable and Unspeakable*.

[14] H. Wiseman, The Two Bell's Theorems of John Bell *J. Phys. A: Math. Theor.* **47**, 424001.

[15] J.S. Bell, The theory of local beables, *Epistemological Letters*, March 1976; reprinted in *Speakable and Unspeakable*.

[16] J.S. Bell, Free Variables and Local Causality, *Epistemological Letters* Feb. 1977; reprinted in *Speakable and Unspeakable*.

[17] J.S. Bell, EPR Correlations and EPW Distributions, *New Techniques and Ideas in Quantum Measurement Theory*, New York Academy of Sciences, 1986; reprinted in *Speakable and Unspeakable*.

[18] J.S. Bell, La nouvelle cuisine, in A. Sarlemihn and P. Kroes (eds.), *Between Science and Technology*, Elsevier Science Publishers, 1990; reprinted in *Speakable and Unspeakable*

[19] Tim Maudlin, What Bell Did *J. Phys. A: Math. Theor.* **47** (2014), 424010

[20] J.S. Bell, Quantum mechanics for cosmologists, in C. Isham, R. Penrose, and D. Sciama (eds.), *Quantum Gravity* **2**, Clarendon Press, Oxford (1981), 611–37; reprinted in *Speakable and Unspeakable in Quantum Mechanics*, Cambridge, 2nd ed., 2004.

[21] T. Maudlin, Space-time in the quantum world in J.T. Cushing, A. Fine, and S. Goldstein (eds.), *Bohmian Mechanics and Quantum Theory: An Appraisal*, Kluwer, 1996

[22] J.S. Bell, Against "Measurement," reprinted in *Speakable and Unspeakable in Quantum Mechanics*, Cambridge, 2nd ed., 2004.

[23] C. Fuchs, A Formalism and an Ontology for QBism, draft grant application, shared via private communication Aug. 25, 2014.

[24] C. Fuchs, QBism, the perimeter of quantum Bayesianism, arxiv:1003.5209

[25] C. Fuchs, N.D. Mermin, and R. Schack, An introduction to QBism with an application to the locality of quantum mechanics, *Am. J. Phys.* **82** (8) August 2014, pp. 749–54.

[26] N.D. Mermin, Putting the Scientist into the Science, talk at *Quantum [Un]speakables II: 50 Years of Bell's Theorem*, online at https://phaidra.univie.ac.at/detail_object/o:360625.

[27] C. Fuchs (with M. Schlosshauer and B. Stacey), My Struggles with the Block Universe, arxiv:1405.2390.

[28] C. Fuchs and R. Schack, Quantum-Bayesian coherence: The no-nonsense version' *Rev. Mod. Phys.* **85** (2013), 1693–715.

[29] T. Norsen, Einstein's boxes, *Am. J. Phys.* **73** (2) February 2005, pp. 164–76.

[30] N.D. Mermin, QBism puts the scientist back into science, *Nature* **507** (7493) (26 March, 2014).

[31] J.S. Bell, Toward an exact quantum mechanics, in S. Deser and R.J. Finkelstein (eds.), *Themes in Contemporary Physics II: Essays in Honor of Julian Schwinger's 70th Birthday*, World Scientific, 1989.

[32] www.worldsciencefestival.com/2014/05/measure-for-measure-quantum-physics-and-reality/.

[33] R. Tumulka, A relativistic version of the Ghirardi–Rimini–Weber model, *J. Stat. Phys* **125** (2006), 825.

[34] D. Wallace, Decoherence and ontology, in S. Saunders, J. Barrett, A. Kent, and D. Wallace (eds.), *Many Worlds? Everett, Quantum Theory, and Reality*, Oxford, 2010.

[35] D. Wallace and C. Timpson, Quantum mechanics on spacetime: I. Spacetime state realism, *Br. J. Phil. of Sci.* **61** (4) (2010), 697–727.

[36] D. Wallace, *The Emergent Multiverse*, Oxford, 2012.

[37] D. Bedingham, D. Dürr, G.C. Ghirardi, S. Goldstein, R. Tumulka, and N. Zanghí, Matter density and relativistic models of wave function collapse, *J. Stat. Phys.* **154** (2014), 623–31.

[38] T. Maudlin, What Bell proved: A reply to Blaylock, *Am. J. Phys.* **78** (1) (January 2010), 121–5.

[39] V. Allori, S. Goldstein, R. Tumulka, and N. Zanghí, Many-worlds and Schrödinger's first quantum theory, *Br. J. Phil, Sci.* **62**(1) (2011), 1–27.

14

Lessons of Bell's Theorem: Nonlocality, Yes; Action at a Distance, Not Necessarily

WAYNE C. MYRVOLD

14.1 Introduction

Fifty years after the publication of Bell's theorem, there remains some controversy regarding what the theorem is telling us about quantum mechanics, and what the experimental violations of Bell inequalities are telling us about the world. This chapter represents my best attempt to be clear about what I think the lessons are. In brief: There is some sort of nonlocality inherent in any quantum theory, and, moreover, in any theory that reproduces, even approximately, the quantum probabilities for the outcomes of experiments. But not all forms of nonlocality are the same; there is a distinction to be made between action at a distance and other forms of nonlocality, and I will argue that the nonlocality needed to violate the Bell inequalities need not involve action at a distance. Furthermore, the distinction between forms of nonlocality makes a difference when it comes to compatibility with relativistic causal structure.

The Bell locality condition is a condition which, if satisfied, renders possible a completely local account of the correlations between outcomes of spatially separated experiments. Bell's theorem tells us that the probabilities that we compute from quantum mechanics do not admit of such an account. Modulo auxiliary assumptions that, in my opinion, should be noncontroversial, this yields the conclusion that there must be something nonlocal about any theory that yields the quantum mechanical probabilities or anything close to them.

One way to violate the Bell locality condition – and this is the way that any deterministic theory must do it – is via straightforward action at a distance. However, we should also consider the possibility of a stochastic theory with probabilistic laws that involve irreducible correlations. The sort of nonlocality inherent in such a theory involves relations between events that are significantly different from causal relations as usually conceived. Perhaps surprisingly, such a theory, unlike its deterministic cousins, need not invoke a distinguished relation of distant simultaneity.

This makes a difference for the situation of the two sorts of theories vis-à-vis relativity. Though a deterministic theory can achieve the *appearance* of being relativistic, at the level of observable phenomena, it can do so only by introducing a preferred foliation and with it a causal structure at odds with relativistic causal structures. Things are otherwise with dynamical collapse theories.

Conclusions along these lines have, of course, been drawn by many others before, though not everyone is convinced. Travis Norsen [1], in particular, has expressed vigorous skepticism. I believe, with Ghirardi [2], that, though Norsen's criticisms of some earlier arguments are well taken, there is still a case to be made. My own understanding has been strongly influenced by the work of my teacher, Abner Shimony [3–5], but mention should also be made of Ghirardi [2], Ghirardi and Grassi [6], Jarrett [7], Redhead [8], and Skyrms [9] among others! However, the arguments presented here, though inspired by and indebted to all of the above-mentioned authors, differ in some details from each of them.

14.2 Does Relativity Preclude Action at a Distance?

One reason for making a distinction between action at a distance and other forms of non-locality has to do with compatibility with relativistic space–time structure. It is often said that relativity precludes action at a distance. This is right, I think. But, in order to get clear about what is precluded by what, a few words are in order about the notion of causation, and about what I mean by a relativistic space–time.

14.2.1 Causation

A paradigm case of causation is an intervention and its causal consequences. In the most straightforward cases, there is some variable X whose values are ones that some agent could, at least in principle, choose between, holding fixed the background state of other relevant variables, such that the probability distribution for some other variable Y depends on the value of X. It is typically assumed that the realization of Y occurs at a later time than the setting of X. If we let Λ be the background variables (besides X) relevant to Y, then, relative to a particular specification $\Lambda = \lambda$, X is causally relevant to Y if and only if, for some y, and some possible values x, x' of X,

$$\Pr(Y = y | X = x, \Lambda = \lambda) \neq \Pr(Y = y | X = x', \Lambda = \lambda).$$

We will say that the choice of X has a causal bearing on Y, relative to background conditions λ.

In the most straightforward cases, the variable that is the target of intervention is something that some agent, human or otherwise, could in principle set. But the notion of agency is not essential. When talking about causation, I will, however, assume that it makes sense to consider some process that fixes the value of X without changing the other parameters relevant to Y, and to consider the effect of such an intervention on the probability of Y. This is what distinguishes a causal relation between X and Y and situations in which X is merely informative about Y, without being causally relevant to it. To invoke a standard example, the reading of a barometer gauge may be informative about a coming storm (because it is informative about a drop in pressure which *is* causally relevant to the storm), but it is not causally relevant, because an intervention that changes the barometer reading (perhaps via a localized change in atmospheric pressure) without changing such things as large-scale atmosphere pressure does not affect the probability of Y but only disrupts the informational link between X and Y.

There are complications, familiar from the literature on causation, having to do with matters of overdetermination, preemption, and the like. But these are not, as far as I can see, important for the issue at hand, and may be safely disregarded for present purposes.[1]

Signals are special cases of cause–effect relations; the sender chooses between various settings of the signalling device, and the signal obtained by the receiver is informative about the sender's choice. But there may be causal relations that do not permit signalling. It might be the case that there are other factors, Z, imperfectly known to sender and receiver, such that, for any value of Z, the probability of Y depends on X, but it does so in different ways for different values of Z. Suppose, for example, that Alice can choose between pushing two buttons, labelled 0 and 1, and, subsequently, Bob will see a light bulb light up either red or green. But the relation between the button Alice pushes and the colour of the light Bob sees depends on the setting of a hidden switch; when the switch is set one way, button 0 yields a red light and button 1 a green, but when it is set another way, the relation between the button pushed and the colour of the light is reversed. The setting of the switch is hidden from Alice and Bob and fluctuates unpredictably. In such a case, Alice's choice of which button to push has a causal effect on the colour of the light that Bob sees, but *which* causal effect it has is unknown to Alice and Bob. (This is analogous to the cause–effect relations between the two sides of an EPR–Bohm experiment found in the de Broglie–Bohm theory; for any specification of the initial positions of the two particles, the experimental setting of the earlier experiment affects the result of the other experiment, but this relation cannot be used for signalling because neither Alice nor Bob has epistemic access to these initial positions.)

Now, if, at the time of the setting of X, the realization of Y has already occurred – that is, Y has already taken on some definite value – it is hard to see how to make sense of probabilities for Y taking on this or that value, and hence, difficult to make sense of the choice of X having an effect on the probability of Y taking on this or that value. In my view (admittedly not shared by all), these difficulties are insurmountable, and I adopt the usual conception of causation in which it involves an inherently temporally asymmetric relation. In some circles, retrocausality, or backwards-in-time causation, is invoked as a serious possibility, and one that, moreover, is meant to be relevant to quantum nonlocality.[2] Assessment of such proposals is beyond the scope of this paper.

The notion of cause that will be operative in this chapter is temporally asymmetric; the cause must precede the effect. On such a notion, the relation of temporal precedence

---

[1] See Woodward [10] and references therein.
[2] See Faye [11] for an overview; for retrocausality as a source of quantum nonlocality, see Evans et al. [12] and references therein.

constrains the relation of potential causal influence; an event x is a potential causal influence on y only if it is in the past of y. It is this feature of causation that precludes causal influences between spacelike separated events in a relativistic spacetime.

Relativistic Space–Times

From the foregoing discussion, we take it that an event x is a potential cause of another event y only if x is in the temporal past of y. There seems to be no harm in widening this to "if and only if," in which case the relations of temporal precedence and potential causal influence coincide. In Galilean space–time, the space–time manifold is partitioned into equivalence classes of simultaneity, and these equivalence classes are totally ordered by the relation of temporal precedence. Any event is either in the past of, in the future of, or simultaneous with any other. This is a marked – and to my mind, the most interesting – difference between Galilean space–time and Minkowski space–time. In Minkowski space–time, the relation of temporal precedence is given by the light-cone structure. Though the t-coordinate of any Lorentz frame partitions Minkowski space–time into equal-t equivalence classes, none of these captures the relation of temporal precedence, as there are always events that are ordered by a given reference frame's t-coordinates but are nonetheless spacelike separated, so that neither temporally precedes the other.

To speak more generally: suppose we have a space–time with a relation $\ll$ of temporal precedence. Take this to be transitive and antisymmetric (no temporal loops):

$$(x \ll y \ \& \ y \ll z) \Rightarrow x \ll z,$$
$$x \ll y \Rightarrow \neg(y \ll x). \tag{14.1}$$

Define the relation $\sim$ as the relation that holds between two events when neither temporally precedes the other:

$$x \sim y \equiv_{\text{df.}} \neg(x \ll y \lor y \ll x). \tag{14.2}$$

This relation is symmetric by construction and, by the antisymmetry of $\ll$, reflexive. It is, therefore, an equivalence relation if and only if it is transitive. In Galilean spacetime, the relation $\sim$ defined by the relation (14.2) is a transitive relation and therefore partitions the space–time into equivalence classes of simultaneous events. In Minkowski space–time, however, the relation of temporal inconnectability is not transitive.

By a relativistic space–time I will mean a space–time endowed with a causal structure such that for any x, y such that $x \sim y$, there exists z such that $x \sim z$ and $z \ll y$.

Now, we could have a theory that was formulated against a background of Minkowski space–time but was such that the causal structure depended on the distribution of matter or some other fields. One might even have it that there is a field that, for appropriate configurations of the field, yields a partition of the space–time into hypersurfaces of simultaneity that are made use of by the dynamics of the matter distribution, in such a way that these hypersurfaces yield the relation of causal connectibility. In such a case, I will say that the physical relation of causal connectibility, the one that counts, is the introduced relation, not the one given by the background space–time.[3] In such a situation, the causal structure is not relativistic.

14.3 Locally Explicable Correlations

In a classical setting, it is easy to produce setups that result in correlations (or anticorrelations) between distant events. Flip a fair coin, and, depending on the outcome, put a ball in one of two boxes, A and B. Send one of these boxes to Paris, the other to Tokyo. The setup yields a probability of $1/2$ that there is a ball in box A, a probability of $1/2$ that there is a ball in box B, but a probability of 0 that there are balls in both boxes.

In this setup, the outcomes of the experiments in Paris and Tokyo, which consist of opening the boxes and looking at them, are determined by initial conditions at the source. But this is not a necessary condition for there to be distant correlations. Suppose we have two coins, one with a bias of 2 to 1 in favour of Heads, the other with the opposite bias. We put one coin in each box, in such a way that each box has an equal chance of getting the Heads-biased coin. The boxes are sent to Paris and Tokyo, where they are opened and the coins are flipped.

This scenario yields a probability of $1/2$ that the coin in Tokyo will land Heads, since the coin is biased either 2 to 1 towards Heads or 2 to 1 against, with an equal chance of each. Similarly, the probability is $1/2$ that the coin in Paris will land Heads. But these two probabilities are not independent. Since the two tosses are independent tosses of two coins, one with probability $2/3$ of Heads, the other with probability $1/3$, the probability that they will both land Heads is $2/9$, which is less than $1/4$.

The correlation, or rather anticorrelation, between the results in Tokyo and Paris is completely explicable in local terms. The probabilistic dependence between the distant outcomes is due neither to action at a distance nor to any irreducible probabilistic dependence between distant events.

What the two scenarios have in common is that a choice is made at the source between probability distributions, each of which is one on which the outcomes of distant experiments are probabilistically independent. The resulting probability distribution for the outcomes of the distant experiments is a mixture of distributions on which the outcomes are independent. Call such correlations *locally explicable*.

Now, consider a setup such as the one envisaged by Bell. Two systems are prepared at some source and sent to distant locations A, B, where there is a choice of experiments to be formed. Let λ be a specification of local initial conditions at the source relevant to the outcomes. We assume that, for any choice a, b, of experiments at A and B, respectively, and any specification of relevant initial conditions λ, there is a probability distribution

---

[3] This is what is done, for instance, by Dürr et al. [13], who formulate a Bohmian theory against a background of Minkowski space–time by introducing an auxiliary field that picks out a distinguished foliation that is then used to formulate the dynamics of the theory. In this theory, there are causal relations between events that cannot be connected by a light signal; what counts, for the dynamics, is temporal precedence according to the introduced foliation. Though formulated in terms of Lorentz covariant equations, this is not a theory whose causal relations are relativistic in the sense used here. See [14] for discussion.

$P_{a,b}(x, y|\lambda)$ over outcomes of the experiments. We also assume that there is a probability distribution over the initial conditions λ given by $\rho_{a,b}(\lambda)$, such that

$$P_{a,b}(x, y) = \int d\lambda \, \rho_{a,b}(\lambda) \, P_{a,b}(x, y|\lambda). \tag{14.3}$$

Given $P_{a,b}(x, y|\lambda)$, we define marginals

$$P^A_{a,b}(x|\lambda) = \sum_y P_{a,b}(x, y|\lambda),$$

$$P^B_{a,b}(y|\lambda) = \sum_x P_{a,b}(x, y|\lambda). \tag{14.4}$$

We assume that it is possible to arrange things so that whatever device it is that switches between alternative experiments can be rendered effectively independent of the distribution $\rho_{a,b}(\lambda)$ of relevant initial conditions – an assumption implicit in Bell's original exposition and made explicit following Bell's exchange with Shimony, Horne, and Clauser [15, 16]. The preparation of the systems and the switching events will, of course, have events in their common past, but we assume that these can be effectively screened off. This assumption may be called the "free will" assumption, as long as one remembers that it is so called with tongue in cheek; metaphysical issues concerning the free will of the experimenters are not at stake, but only the more prosaic assumption that it is possible to set up things so that there is effective independence of state preparation and experiments subsequently performed, an assumption so pervasive that it is difficult to see how we could engage in experimental science without it.[4]

In accordance with this, we assume that the switching is done in such a way that the distribution of relevant initial conditions λ is effectively independent of the settings of the apparatus, and write

$$\rho_{a,b}(\lambda) = \rho(\lambda). \tag{14.5}$$

Now, because the experimental settings are the sorts of things that could, in principle, be manipulated in a systematic way, then, if the marginals at B depend on the setting at A, this is clearly a case of causal influence. The condition that the setting at A has no causal influence on the result at B, and vice versa, is the condition that Jarrett [7] called *locality* and Shimony [5] called *parameter independence*. This is the condition that, for all possible outcomes (x, y) of the two experiments, and for all settings a, a' and b, b', and all λ,

$$P^A_{a,b}(x|\lambda) = P^A_{a,b'}(x|\lambda),$$

$$P^B_{a,b}(y|\lambda) = P^B_{a',b}(y|\lambda). \tag{14.6}$$

When this holds, we will drop the redundant parameters and write

$$P^A_{a,b}(x|\lambda) = P^A_a(x|\lambda),$$

$$P^B_{a,b}(y|\lambda) = P^B_b(y|\lambda). \tag{14.7}$$

---

[4] See Bell [17, pp. 102–3] for a lucid discussion.

Suppose that we are in a relativistic space–time, and suppose that the two experiments are performed in space–time regions A, B such that $A \sim B$. Then, because the space–time is relativistic, there are regions X, Y such that $X \sim B$, $Y \sim A$, but $X \ll A$ and $Y \ll B$. Suppose that it is possible to arrange things so that the setting of the A-apparatus occurs in space–time region X, and the setting of the B-apparatus occurs in Y. In such a case the causal structure of our space–time *requires* that the conditions (14.6) be satisfied.

The condition that, for any given settings (a, b), any correlations between must outcomes be locally explicable is the condition that the probability distributions must be mixtures of distributions on which the outcomes are independent. This is the condition that Jarrett called *completeness* and Shimony *outcome independence*:

$$P_{a,b}(x, y|\lambda) = P_{a,b}^A(x|\lambda)\, P_{a,b}^B(y|\lambda). \tag{14.8}$$

In a relativistic space–time, for appropriate experimental arrangements, the condition of parameter independence (PI) is the condition that causal relations respect the temporal structure of the space–time. The condition of outcome independence (OI) is the condition that any correlations between outcomes be locally explicable. The conjunction of the two is the condition often called the *Bell locality* condition:

$$P_{a,b}(x, y|\lambda) = P_a^A(x|\lambda) P_b^B(y|\lambda). \tag{14.9}$$

Now, as is well known, the Bell locality condition (14.9), together with the free will assumption, entails the satisfaction of the CHSH inequality, which is violated by quantum mechanical probabilities. Furthermore, the experimental evidence vindicates the quantum mechanical probabilities.

A deterministic theory, one in which all probabilities are either zero or one, has to satisfy OI. Thus, if the theory is to produce Bell-inequality-violating statistics, it must violate the condition PI, which, we have argued, is required by relativistic space–time structures for appropriate setups. Deterministic theories that yield quantum predictions cannot respect relativistic causal structure.

It does not follow that quantum mechanical predictions and relativity are incompatible *tout court*. I have argued that PI is required by relativistic causal structure, but not that OI is. In the next section I will argue that, if we are willing to accept a stochastic theory with correlations that are not reducible to or explicable in terms of local factors, we can retain relativistic causality.

14.4 Correlations That Are Not Locally Explicable

14.4.1 A Toy Example

Let us consider another toy example. We have two boxes, A and B. On each box are a button, a light, and a switch, with two positions labelled 0 and 1. When the button on a box is pushed, its light glows either red or green.

Suppose that the physical laws governing the boxes are chancy; the probabilities do not supervene on any deeper underlying theory. And suppose that, when the button on either box is pushed, there are equal probabilities for the box's light to glow red or green, for any switch settings. Suppose also that, when the switch settings on the boxes agree, the lights on the two boxes always light up the same colour, and that, when the two switch settings disagree, there is a nonzero probability of getting a red light on one box and a green light on the other. By stipulation, these probabilities do not supervene on deeper facts about the boxes; we have given a complete description. And these probabilities, since they are part of chancy physical laws, are to be thought of as objective probabilities.

In this example, there is no causal influence of the switch setting on Box *A* on the outcome of pushing the button on either Box *A* or *B*; the probabilities are 1/2, independent of the switch setting; we have no parameter dependence. When the two switch settings agree, the outcomes of pushing the buttons on the two boxes are correlated; there is outcome dependence. Another way of saying this is that the conditional probability of one outcome, conditional on the other, is not the same as its conditional probability. But do we have, in this example, a *causal* influence of one on the other?

This example does not fall within the paradigm case of a possible intervention changing the probability of another event. By hypothesis, the outcomes are chancy, and there is no way to intervene on the system to *set* the outcome to either red or green – that is, no process that reliably produces a red or green result while holding other relevant parameters fixed. But, you may say, perhaps there are cases of causal relations that do not fall within the scope of paradigm cases such as this. Fair enough! But we should bear in mind that the example as described is symmetric under interchange of the two boxes, and so there is nothing in the setup to distinguish between cause and effect. One could, of course, expand the use of talk of causal relations to include symmetric relations that do not distinguish between cause and effect. In my opinion this would be an unfortunate terminological choice, but it is nothing more than that. What is important is that, if such a choice were made, we would lose any grounds for thinking that such a relation is prohibited by the temporal structure of a relativistic spacetime. It is only causal asymmetry, that is, the idea that causation requires temporal precedence, that leads to the conclusion that there can be no causal relations between spacelike separated events in a relativistic spacetime.

If we want to make the example more analogous to quantum mechanics, we can add in an analogue of local state preparation. Suppose that there is another knob on each box that sets the outcome to red or green, but also that (in keeping with the analogy), if we use the knob on Box *A* to set its outcome variable to red (or green), there is an equal chance of either outcome for Box *B* (and vice versa, with the roles of the two boxes switched). In such a case, there would be a possible intervention to set the outcome variable of either box, but such an intervention would have no causal effect on the outcome variable of the other box. A state preparation on one box is like a local intervention on a barometer; it sets the value of the barometer reading at the cost of destroying the informational link between the barometer reading and the distant storm.

Though there is nothing specifically quantum about this example, we do have an interesting form of nonseparability. Let *A*, *B* be variables that specify the switch setting of the two boxes, and let *X*, *Y* be the variables that specify the two outcomes, that is, whether the lights are red or green. Let $C = \langle A, B \rangle$ be a variable specifying the pair of switch settings, and let $Z = \langle X, Y \rangle$ specify the pair of outcomes. Neither A or B is causally relevant to X or Y. Nor is C, since the probability of red and green at each end of the experiment is $1/2$, regardless of the switch settings. Neither A or B, by itself, is causally relevant to Z. But C *is* causally relevant to Z.

There is in this example no nonlocal causation, because there are no causes with distant effects. But we do have a radical nonseparability of causes; C is causally relevant to Z, though its component parts, A and B, are not.

14.4.2 Probability and Becoming

Suppose, now, that our toy example lives in a relativistic space–time, and that the two button-pushings take place at spacelike separation. It might be convenient, when giving an account of events, to pick a foliation and use it to assign global times. The natural thing to do is to assign probabilities to events on a given hypersurface conditional on events to the past of that hypersurface. So, on such an account, prior to B's button being pushed, we assign a probability $1/2$ to red and green for B-outcomes on hypersurfaces to the past of the A-experiment, and, for hypersurfaces to the future of the A-experiment, probabilities conditional on the result of that experiment. So, for example, when the switch settings agree, the probability for a red light at B, conditional on the result of the A-experiment, will be either 1 or 0.

It should be clear that the transition from probability $1/2$ to the conditional probability, different from $1/2$, that occurs when we pass from a hypersurface that has the A-experiment in its future to one with the A-experiment in its past, is not a change in any intrinsic property of B. Any temptation to think of it as such results from a lingering assumption of separability. This is worth saying because there is a temptation to say that, when we do the A-experiment, it instantaneously changes some fact about B, namely, the probability of glowing red when its button is pushed, and we may wonder when this fact about B changes, and be tempted to conclude that our toy example cannot live comfortably in a relativistic spacetime after all.

Suppose, now, that the A experiment occurs in the past of the B experiment. Then, once the outcome of the A-experiment is in B's past light cone, it is to the past of every spacelike hypersurface intersecting B's world line, and, by the rule that stipulates that we associate, with any spacelike hypersurface, probabilities conditional on events to the past of that hypersurface, all hypersurfaces will agree on ascribing the conditional probabilities to B. Then, and only then, can we regard this conditional probability as a local beable.

14.5 Bell and Local Causality

Two lectures given in 1989, the last full year of his life, shed light on Bell's views on the locality conditions. One, "La nouvelle cuisine," [18] was presented in June, in Eindhoven, the other [19] in November, in Trieste.

In the June lecture, Bell formulated a principle that he called the *Principle of Local Causality*. We are first given an informal gloss,

The direct causes (and effects) of events are near by, and even the indirect causes (and effects) are no further away than permitted by the velocity of light.

[17, p. 239]

Bell followed this by what was intended to be a sharper and cleaner version of this intuitive idea:

A theory is said to be locally causal if the probabilities attached to values of local beables in a space-time region 1 are unaltered by specification of values of local beables in a space-like separated region 2, when what happens in the backward light cone of 1 is already sufficiently specified,for example by a full specification of local beables in a space-time region 3 [which, in the diagram supplied by Bell, is a cross-section of the backward light cone of region 1].

[pp. 239–40]

It seems to me that this condition is strictly stronger than the intuitive condition that motivates it. If Bell's second principle is satisfied, then (as Bell proceeds to argue) all correlations are locally explicable and there is no action at a distance. But the entailment does not go the other way, as is illustrated by the toy example of the previous section. That example is compatible with all *causal* relations being local ones; it is just that there are irreducible correlations, not explicable in terms of causal relations, local or nonlocal. In the toy example, the relation between outcomes of the two experiments are nonlocal relations, but they are not nonlocal causal relations, because they are not causal.

Again, it makes a difference because there are two ways to violate the Bell locality condition. Theories that violate parameter independence exhibit action at a distance and do not sit well with relativistic causal structures. But stochastic theories with irreducible correlations and no parameter dependence need not be at odds with relativistic causal structure.

This is something that, as pointed out by Ghirardi [2], Bell himself appreciated. In his paper "Are there quantum jumps?" [20], Bell showed that the GRW theory exhibits relative time-translation invariance – that is, for entangled pairs of particles that do not interact with each other, the theory does not require a distinguished notion of distant simultaneity, a property that he called "a residue, or at least an analogue, of Lorentz invariance." in the nonrelativistic theory. He ended with the remark,

I am particularly struck by the fact that the model is as Lorentz invariant as it could be in the nonrelativistic version. It takes away the ground of my fear that any exact formulation of quantum mechanics must conflict with fundamental Lorentz invariance.

[17, p. 209]

This is in contrast to the Bohm theory, which requires a distinguished relation of distant simultaneity for its formulation, and hence cannot be made Lorentz invariant at the fundamental level, though there is hope for a Bohm-like theory whose observable consequences *are* Lorentz invariant, because inaccessibility of the exact particle trajectories leaves the preferred foliation empirically undetectable.

In the Trieste lecture, Bell discussed the prospects for a genuinely relativistic version of a dynamical collapse theory and concluded that the difficulties encountered by Ghirardi, Grassi, and Pearle in producing a genuinely relativistic version of the continuous spontaneous localization theory (CSL), a theory that would be "Lorentz invariant, not just for all practical purposes but deeply, in the sense of Einstein, eliminating entirely any privileged reference system from the theory" (p. 2931), were "Second-Class Difficulties," technical difficulties, and not deep conceptual ones. This has been borne out by the work of Bedingham [21], who has constructed a relativistic version of the CSL theory; see also Tumulka [22] for a relativistic version of the GRW theory.

Bell's hope that there could be a deeply Lorentz invariant collapse theory that would not already be refuted by existing experiments indicates that he did not think that the requirement he called *local causality* was something required by relativity, since any theory that was locally causal in Bell's sense would have to satisfy the Bell inequalities.

14.6 Quantum State Evolution

14.6.1 Local and Nonlocal Operations: The Nonrelativistic Case

We have superluminal causation if an intervention on a variable that is a local beable pertaining to a region A affects the probability of an event that takes place at spacelike separation, that is, the realization of a variable that is a local beable pertaining to a region B, at spacelike separation from A. For this even to make sense, there must be some local beables.

When separability obtains – that is, if all beables supervene on local beables – then it is easy to distinguish between local actions and nonlocal actions. Given a composite system AB, with spatially separated parts A and B, if a complete specification of the state of the composite system is determined by local beables pertaining to A and B, then we can partition changes to the system into changes pertaining to A and changes pertaining to B.

When separability does not obtain, as is the case for quantum systems, things are a bit trickier, as the state of the system is not divisible into a part pertaining to A and a part pertaining to B. But this does not mean that we cannot distinguish between local and nonlocal action. Consider a pair of systems, A and B, with associated Hilbert spaces $\mathcal{H}_A$, $\mathcal{H}_B$. Let $\mathcal{A}$ and $\mathcal{B}$ be the algebras of operators on $\mathcal{H}_A$ and $\mathcal{H}_B$. When the evolution is unitary, it is easy to distinguish between evolution that is local and evolution that involves interaction between two separated systems, A and B. If the total Hamiltonian is a sum of Hamiltonians $H_A \in \mathcal{A}$ and $H_B \in \mathcal{B}$, then the unitary operator U that implements the evolution will factor,

$$U = U_A \, U_B, \tag{14.10}$$

with $U_A \in \mathcal{A}$ and $U_B \in \mathcal{B}$.

When dynamical collapse is involved, we can say something similar. It is generally taken to be the case that physically realizable state changes to the system A must be completely positive mappings of the state space into itself. Any such mapping can be represented by a set $\{K_i\}$ of operators, such that the density operator ρ representing the state undergoes the change

$$\rho \to \sum_i K_i \, \rho \, K_i^\dagger,\qquad(14.11)$$

where

$$\sum_i K_i^\dagger K_i \leq \mathbb{1}.\qquad(14.12)$$

If $\sum_i K_i^\dagger K_i = \mathbb{1}$, the operation is called a *nonselective operation*; if $\sum_i K_i^\dagger K_i < \mathbb{1}$, it is *selective*. Such a representation of a completely positive mapping of the state space into itself is called a *Kraus representation* of the mapping, and the operators $\{K_i\}$, *Kraus operators*.

Selective operations reduce the trace-norm of the density operator, but this is not an issue, as normalization is only a convention. With an unnormalized density operator, we compute the expectation value of an observable represented by an operator A via

$$\langle A \rangle_\rho = \mathrm{Tr}(\rho A)/\mathrm{Tr}(\rho).\qquad(14.13)$$

In the simplest case, the set of Kraus operators is a singleton, and the evolution is deterministic; this includes the case of unitary evolution. However, one can also consider stochastic processes. Suppose the state vector undergoes a stochastic transition of the following form: for some i,

$$|\psi\rangle \to K_i|\psi\rangle,\qquad(14.14)$$

with the probability for which transition it undergoes given by

$$p_i = \|K_i|\psi\rangle\|^2/\||\psi\rangle\|^2.\qquad(14.15)$$

Since these probabilities must sum to one for every vector $|\psi\rangle$, we must have $\sum_i K_i^\dagger K_i = \mathbb{1}$. For each i, we have a selective operation; one of these yields the actual state; the transition to the mixture of these candidate states, which corresponds to the proposition that some one of these transitions has occurred, without specification of which, is given by the nonselective operation obtained by summing these. It is easy to show (see the Appendix) that a nonselective operation leaves the probability distribution for results of a measurement corresponding to an operator B unchanged in all states if and only if B commutes with every K_i.

This is the general form of a dynamical collapse theory in a Galilean spacetime. For any time interval (t, t'), with $t' > t$, the theory specifies an operator-valued random variable, which takes on values from a set $\{K_i\}$ (possibly with continuous index), with

$$\sum_i K_i^\dagger K_i = \mathbb{1}.\qquad(14.16)$$

Figure 14.1 Two hypersurfaces σ, σ' sharing a common part α.

The evolution from a state at time t to a state at time t' is a stochastic process that consists of the choice of one K_i, with appropriate probabilities, and the application of the corresponding operation. Unitary evolution is a deterministic special case of this schema.

For a measurement involving apparatus that couples to observables in A, the corresponding operation will be represented by Kraus operators in $\mathcal{A}$, and similarly for B. This permits us to distinguish between local state evolutions and nonlocal state evolutions, along the lines of the distinction, common in quantum information theory, between local operations and nonlocal operations. A change of state local to A, due either to the internal dynamics of the system or to interaction with an external influence local to A, will be represented by Kraus operators in $\mathcal{A}$, and similarly for changes local to B. Changes that cannot be decomposed into local changes (such as would result from terms in the Hamiltonian that couple elements of $\mathcal{A}$ to elements of $\mathcal{B}$) count as nonlocal evolution. The condition of local action is the condition that all evolutions be composed of local evolutions. This is a condition that is satisfied by our usual relativistic quantum field theories and by relativistic versions of dynamical collapse theories.

14.6.2 Quantum State Evolution in a Relativistic Space–Time

This generalizes naturally to relativistic space–times. In Galilean space–time, if we want to talk about state evolution, we use the Schrödinger picture, and associate with any spacelike hypersurface a state which, in a collapse theory, yields probabilities for events to the future of the hypersurface, conditional on events to the past. In a relativistic context the analogue is what might be called the Tomonaga–Schwinger picture: with any spacelike hypersurface σ (which might or might not be a maximal hypersurface, as we could also consider spatially limited regions), we associate a quantum state ρ_σ, which yields probabilities for events in the $D^+(\sigma)$, the forward domain of dependence of σ, conditional on events in its past light cone.[5] We impose the condition that operators implementing the evolution from a hypersurface σ to another hypersurface σ' lying within σ's forward domain of dependence be implemented by operators that commute with all operators representing observables spacelike separated from the region between the two hypersurfaces.

Take two hypersurfaces σ, σ', with σ' in the forward domain of dependence of σ, that share a common part α (see Figure 14.1). Let $\rho_{\alpha|\sigma}$ and $\rho_{\alpha|\sigma'}$ be the reduced states of α obtained from these states, that is, the restriction of these states to observables in $D^+(\alpha)$. If

---

[5] This is the natural and most straightforward extension of the nonrelativistic Schrödinger evolution to a relativistic space–time. It was suggested, though not advocated, by Aharonov and Albert [23] and has been advocated by, among others, Fleming [24–26], Ghirardi [27, 28], Ghirardi et al. [29], Ghirardi and Grassi [30], Ghirardi and Pearle [31], and Myrvold [32, 33].

the evolution from σ to σ' is deterministic and unitary, then these reduced states coincide, because the unitary operator that implements the transition from ρ_σ to $\rho_{\sigma'}$ commutes with all observables in α's forward domain of dependence. Suppose, however, there is a collapse between σ and σ'. For some i,

$$\rho_{\sigma'} = K_i \, \rho_\sigma \, K_i^\dagger. \tag{14.17}$$

Let $\bar{\rho}_{\sigma'}$ be the mixed state

$$\bar{\rho}_{\sigma'} = \sum_i K_i \, \rho_\sigma \, K_i^\dagger. \tag{14.18}$$

This is the state that would be used by someone who did not know the particular way in which the evolution from σ to σ' turns out to be realized, only that it is given by *some* K_i, with the appropriate probabilities. Then $\rho_{\alpha|\sigma}$ and $\bar{\rho}_{\alpha|\sigma'}$ coincide. But, in general, the reduced state $\rho_{\alpha|\sigma'}$ obtained by restricting (14.17) to α will not coincide with $\rho_{\alpha|\sigma}$.

If we think that the reduced states $\rho_{\alpha|\sigma}$ and $\rho_{\alpha|\sigma'}$ are intended to be intrinsic states of the space–time region α, then it may seem that we have competing state ascriptions. And, if local beables are to be defined in terms of the state, then it might seem that there are differing accounts of the local beables ascribed to α, depending on whether we take them from a state on σ or a state on σ'.[6] For this reason, it is important to stress: $\rho_{\alpha|\sigma'}$ and $\rho_{\alpha|\sigma}$ are not local beables, nor are they intrinsic states of the region α; the intrinsic state of α is the state conditional on collapses in the past light cone of α.[7]

14.7 Local Beables for Relativistic Collapse Theories

Suppose that we had a dynamical collapse theory that produced, within a finite time, collapses to eigenstates of observables pertaining to bounded spacetime regions (perhaps a local particle number density from which one could define a local mass density). This is, after all, what the naive collapse postulate would lead one to expect. Then we could have a simple criterion for attributing local beables to systems. For a system located in a spacelike region α, we would consider the state conditional on all collapses in the past light cone of α, and attribute a property to the system if and only if that state was an eigenstate of the corresponding observable. That is, we would say that a system had the property $A = a$ if and only if the state on *every* spacelike hypersurface containing α was an $A = a$ eigenstate. These would be the local beables of the theory, and our goal would be to construct a collapse theory in such a way that the totality of local beables yielded a sensible world.

But collapse theories only approximate the naive collapse postulate; instead of collapsing to eigenstates of macroscopic observables, at best they yield approximations to such

---

[6] Something like this seems to be going on in Esfeld and Gisin [34], who characterize this sort of view as one in which "what there is in nature depends on the choice of a hypersurface – so that different facts exist in nature relative to the choice of a particular foliation of space–time" (p. 258).

[7] This aspect of the view – that, in addition to the various reduced states that might be attributed to a bounded region α as reduced states of larger hypersurfaces, there is also its intrinsic state, conditional on its past light cone – is missing from the discussion of Wallace and Timpson [35, Sect. 7]. This is a crucial point, because it enters into the identification of local beables for collapse theories; see the next section.

eigenstates. In the relativistic context, there is a matter of principle involved; it is a consequence of the Reeh–Schlieder theorem that no state of bounded energy is an eigenstate of any observable pertaining to a bounded space–time region.

The fact that we cannot expect a collapse theory to produce eigenstates of local observables led Ghirardi et al. [36] to propose, as a criterion for attribution of local objective properties to systems, one based on approximations to eigenstates:[8]

We think that the appropriate attitude is the following: when considering a local observable A with its associated support we say that an individual system has the objective property a (a being an eigenvalue of A), only when the mean value of P_a is extremely close to one, when evaluated on all spacelike hypersurfaces containing the support of A.

[p. 1310]

Ghirardi [27] formulates this criterion in terms of the state on the past light cone.

This still leaves it open *what* observable it is such that the reduction process will lead to approximate definiteness of that observable. In the original GRW theory, it is a sort of smeared position. This does not lend itself well to a relativistic extension. Beginning in the mid-1990s, Ghirardi and collaborators have favoured mass density as the preferred variable. One defines mass density operators $M(x)$; in state $|\psi\rangle$, the expectation value, or mean mass density, is given by

$$\mathcal{M}(x) = \langle \psi | M(x) | \psi \rangle. \tag{14.19}$$

Applying the criterion that an approximation to an eigenstate is good enough for property attribution, Ghirardi et al. [29] propose that we take the mass density to be "objective" if its variance is small; in subsequent works it is said to be "accessible" if the variance is sufficiently small [6, 38]. The idea is that a mass density with small variance is an acceptable stand-in for a classical mass density, which, of course, always has a definite value.

Combining the choice of mass density with the past-light-cone criterion for property attribution, the mass density that in a relativistic context is taken to be a local beable in a region α is the mass density defined by the state on the past light cone of α [39, 40].

To get a sense of how the past-light-cone mass density functions as a local beable, consider the example proposed by Einstein at the 1927 Solvay conference [41, pp. 440–42]. An electron passes through a small opening in a screen, on the other side of which is a hemispherical photographic film. The wave function of the electron takes the form of a spherical wave emanating from the opening, so that, when it reaches the film, the amplitude, and consequently the mass density, is approximately uniform on the hemisphere. When it reaches the film, the wave function of the electron begins to become entangled with the macroscopic degrees of freedom of the film, at which time a collapse is likely to occur, and the electron mass density becomes concentrated at some small region of the film. The past light cone mass density at another point p of the films remains the original, uniform density until the collapse event is in the past light cone of p, after which it is near zero. There is a spherical

---

[8] See also Ghirardi et al. [37, p. 362], Ghirardi and Grassi [30, p. 417], and Ghirardi and Pearle [31, p. 45].

region of near-zero mass density, centered on the position at which the collapse occurs, that spreads out from the collapse center at light speed.

On a collapse theory endowed with the past-light-cone mass density ontology, the mass density is a local beable, but we should not forget that this does not exhaust the ontology; there is also the quantum state, which is not a local beable. A quantum state corresponding to a single particle is entangled across spacelike separated regions, and this entanglement shows up in anticorrelation of results; there is zero probability of collapse at two distinct parts of the screen.

Concerning the ontological status of the mass density, there are two attitudes that one can take. One—and this seems to have been the attitude behind the proposal that it be taken as "objective" or "accessible" when its variance is small—is that we still have a quantum state monist ontology. The quantum state is all that there is, and the criterion has the status of a correspondence rule or meaning postulate that tells us how to understand the mathematical apparatus of quantum mechanics as representing a physical world. The other, which is Ghirardi's current view, is that "it represents an additional element which need to be posited in order to have a complete and consistent description of the world" [42, p. 2907]. In this view, collapse theories are more like the Bohm theory than might appear; in this view, a collapse theory with a mass density posits ontology above and beyond the quantum state, and the role of the quantum state is to provide dynamics for the "primitive" ontology (see Allori et al. [43]).[9]

14.8 A Comment on Everettian Theories

The mass density ontology, with approximation to an eigenstate taken as a good enough stand-in for a classical mass density, serves also for interpretations in the Oxford Everettian vein [44, 45]. Such interpretations suppose that there is no collapse, and all quantum state evolution is unitary. Suppose an experiment has taken place, and the experimental apparatus has interacted with an environment; photons have been reflected from the pointers and flown out the window, or something of the sort. Then the reduced state of the laboratory takes the form of a mixture whose components are such that each component yields a mass density corresponding to approximately well-localized macroscopic objects.

In an EPR experiment, if we consider the postdecoherence state of the joint system consisting of the laboratories at A and B, we find that this takes the form of a similar mixture, with correlations between the macroscopic experimental records in the two laboratories. These correlations emerge because of the nonseparability of the quantum state from which we have derived the mass density, and as such are not locally explicable, in the sense of being mixtures over deeper descriptions on which there are no correlations.

Thus, on Everettian theories also, distant quantum correlations represent a departure from the classical expectation that all correlations are locally explicable. But, as in collapse

---

[9] My own view is that abandoning quantum state monism is premature, but that is a discussion that will have to be reserved for a later date.

theories, there is nothing that corresponds to action at a distance, no sense in which a choice of an instruments setting or any other variable that could be the target of an intervention has an effect at a distance. We have some sort of nonlocality, but not action at a distance.

14.9 Conclusion

As many have pointed out, there is more than one way for a nonlocal theory to be nonlocal. I have defended the view that the difference makes a difference; for appropriate setups, parameter dependence involves a departure from relativistic causal structure, whereas theories, such as dynamical collapse theories, that satisfy parameter independence and exhibit only outcome dependence can satisfy the requirement that Bell hoped for, namely compatibility with relativistic causal structure at a truly fundamental level.

14.10 Acknowledgments

Work on this chapter was supported by a grant from the Foundational Questions Institute (FQXi). It was completed while the author was a Visiting Fellow at the Pittsburgh Center for the Philosophy of Science.

14.11 Appendix

14.11.1 Nonselective Operations and Commutation

It is easy to show that, *if* an operator B commutes with all elements $\{K_i\}$ of a nonselective operation, then the operation leaves the statistics of B-experiments unchanged. The converse is a little (but only a little) trickier, but, since I have found that the most straightforward proof, due to Arias et al. [46], is not as widely known as it should be, I reproduce it here.

Theorem 1 Consider a nonselective state transition

$$\rho \to \rho' = \sum_i K_i \, \rho K_i^\dagger, \tag{14.20}$$

where

$$\sum_i K_i^\dagger K_i = \mathbb{1}. \tag{14.21}$$

If, for all initial states ρ, the state transition (14.20) leaves the probabilities associated with outcomes of experiments of a observable represented by an operator B unchanged, then B commutes with each K_i.

Proof Suppose that, for all ρ, the state transition (14.20) leaves the probabilities of experiments of an observable represented by an operator B unchanged. Then it must leave the expectation value of B unchanged, as well as the variance $(\Delta B)^2$, and so we must have

$$\mathrm{Tr}[\rho \, B] = \mathrm{Tr}[\rho' \, B],$$
$$\mathrm{Tr}[\rho \, B^2] = \mathrm{Tr}[\rho' \, B^2]. \tag{14.22}$$

This means that

$$\text{Tr}[\rho \, B] = \text{Tr}[\sum_i K_i \rho K_i^\dagger \, B] = \text{Tr}[\rho \sum_i K_i^\dagger B K_i]. \tag{14.23}$$

Since this is to hold for all density operators, ρ, we must have

$$\sum_i K_i^\dagger B K_i = B. \tag{14.24}$$

Similarly, we must have

$$\sum_i K_i^\dagger B^2 K_i = B^2. \tag{14.25}$$

Now consider

$$\sum_i [K_i, B]^\dagger [K_i, B] = \sum_i \left(B K_i^\dagger K_i B - B K_i^\dagger B K_i - K_i^\dagger B K_i B + K_i^\dagger B^2 K_i \right). \tag{14.26}$$

Using (14.21), (14.24), and (14.25), we get

$$\sum_i [K_i, B]^\dagger [K_i, B] = 0. \tag{14.27}$$

Since each term in (14.27) is a positive operator, in order to sum to zero each term must be zero, and so we must have, for each i,

$$[K_i, B]^\dagger [K_i, B] = 0, \tag{14.28}$$

from which it follows that

$$[K_i, B] = 0. \tag{14.29}$$

$\square$

14.11.2 Compatibility of State Assignments

In this section, we will use the notion $\rho(A)$ for the expectation value assigned to operator A by the quantum state ρ. We define a relation, $\preccurlyeq$, between states, as follows:

Definition 1 $\rho \preccurlyeq \rho'$ if and only if, for all effects E, if $\rho(E) = 1$ then $\rho'(E) = 1$.

Equivalently,

Definition 2 $\rho \preccurlyeq \rho'$ if and only if, for all effects E, if $\rho'(E) > 0$ then $\rho(E) > 0$.

Intuitively, if $\rho \preccurlyeq \rho'$, anything that can happen, according to ρ', is possible according to ρ. This relation obtains, for example, between a mixture and its components: if ρ is a mixture of ρ' and some other states, then $\rho \preccurlyeq \rho'$.

We will say that two states are *compatible* if and only if there is no outcome of any possible experiment that is assigned probability 1 by one state and 0 by another.

Definition 3 $\rho \sim_c \rho'$ if and only if there is no effect E such that $\rho(E) = 1$ and $\rho'(E) = 0$.

We also define the relation between spacelike hypersurfaces

Definition 4 $\sigma \leqslant \sigma'$ if and only if σ' is nowhere to the past of σ.

The basic assumption about the dynamics will be the *local evolution condition*:

Let σ, σ' be two spacelike Cauchy surfaces, with $\sigma \leqslant \sigma'$. Let Γ be the region between σ and σ'. Then there exists a family of operators $\{K_x\}^{10}$ such that

$$\int dx \, K_x^\dagger K_x = \mathbb{1},$$

with the following properties:

(1) For some x, the state on σ' is

$$\rho_x(A) = \rho(K_x^\dagger A K_x)/\rho(K_x^\dagger K_x).$$

(2) The probability distribution over possible states on σ' is given by

$$\Pr(x \in \Delta) = \rho\left(\int_\Delta dx \, K_x^\dagger K_x\right).$$

We will make use of the following lemma.

Lemma 1 For any state ρ and any effect E, if $\rho(E) = 1$ then, for any operator B, $\rho(EB) = \rho(B)$.

Proof We use the Cauchy–Schwartz inequality,

$$|\rho(A^\dagger B)|^2 \leq \rho(A^\dagger A)\,\rho(B^\dagger B). \tag{14.30}$$

For any effect F, if $\rho(F) = 0$, then, because $F^2 \leq F$, $\rho(F^2) = 0$, and

$$|\rho(FB)|^2 \leq \rho(F^2)\,\rho(B^\dagger B), \tag{14.31}$$

and so $\rho(FB) = 0$. Now suppose that $\rho(E) = 1$. Then $\rho(\mathbb{1} - E) = 0$, and so, for any B,

$$\rho((\mathbb{1} - E)B) = 0, \tag{14.32}$$

from which it follows that

$$\rho(EB) = \rho(B). \tag{14.33}$$

$\square$

Theorem 2 Let α be an open subset of two spacelike hypersurfaces σ, σ', with $\sigma \leqslant \sigma'$. Let $D^+(\alpha)$ be the forward domain of dependence of α, and let $\mathcal{A}(\alpha)$ be the algebra associated

---

[10] This is written as if indexed by a real variable x. But the family could be points in a higher-dimensional space; take this as a stand-in for the condition that we have a measure space $\langle \Omega, \mathcal{A}, \mu \rangle$, and an operator-valued random variable K, satisfying the condition that

$$\int_\Omega d\mu(x) \, K_x^\dagger K_x = \mathbb{1}.$$

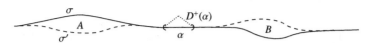

Figure 14.2 Two hypersurfaces σ, σ' sharing a common part α that has a forward domain of dependence $D^+(\alpha)$.

with $D^+(\alpha)$, whose self-adjoint elements represent observables in $D^+(\alpha)$. Let $\rho_{\alpha|\sigma}$, $\rho_{\alpha|\sigma'}$ be the restrictions of ρ_σ and $\rho_{\sigma'}$ to $\mathcal{A}(\alpha)$. Then $\rho_{\alpha|\sigma} \preccurlyeq \rho_{\alpha|\sigma'}$.

Proof If σ and σ' are two spacelike Cauchy surfaces that share α, the region between σ and σ' is spacelike separated from $D^+(\alpha)$, as shown in Figure 14.1. The state evolution from σ to σ' will be implemented by some operator K_x,

$$\rho_{\sigma'}(A) = \rho_\sigma(K_x^\dagger A K_x)/\rho(K_x^\dagger K_x), \tag{14.34}$$

where K_x commutes with all operators representing observables in $D^+(\alpha)$. Thus, for any effect $E \in \mathcal{A}(\alpha)$,

$$\rho_{\sigma'}(E) = \rho_\sigma(E K_x^\dagger K_x)/\rho(K_x^\dagger K_x). \tag{14.35}$$

If $\rho_\sigma(E) = 1$, by Lemma 1, $\rho_\sigma(E K_x^\dagger K_x) = \rho_\sigma(K_x^\dagger K_x)$, and $\rho_{\sigma'}(E) = 1$. □

The next theorem shows that, if σ and σ' are *any* two hypersurfaces with α in common (as shown, for example, in Figure 14.2), ρ_σ and $\rho_{\sigma'}$ will be compatible, in the sense of Definition 3.

Theorem 3 Let α be an open subset of two spacelike hypersurfaces σ, σ'. Let $\rho_{\alpha|\sigma}$, $\rho_{\alpha|\sigma'}$ be the restrictions of ρ_σ and $\rho_{\sigma'}$ to $\mathcal{A}(\alpha)$. Then $\rho_{\alpha|\sigma} \sim_c \rho_{\alpha|\sigma'}$.

Proof Let σ^+ be a spacelike Cauchy surface containing α, such that $\sigma \leqslant \sigma^+$ and $\sigma' \leqslant \sigma^+$. This could be constructed, for example, by the least upper bound (in the $\leqslant$ ordering) of σ and σ'. That is, it consists of the points shared by σ and σ', plus the parts of σ that are to the future of σ', and the parts of σ' that are to the future of σ:

$$\sigma^+ = \{x \in \sigma \mid \sigma' \leqslant x\} \cup \{x \in \sigma' \mid \sigma \leqslant x\}. \tag{14.36}$$

Then the evolution from σ to σ^+ is through regions spacelike separated from α, and hence is implemented by operators that commute with all observables in $\mathcal{A}(\alpha)$, and so, by Theorem 2, $\rho_{\alpha|\sigma} \preccurlyeq \rho_{\alpha|\sigma^+}$. By the same token, $\rho_{\alpha|\sigma'} \preccurlyeq \rho_{\alpha|\sigma^+}$. Therefore, if there were an effect E such that $\rho_{\alpha|\sigma}(E) = 1$ but $\rho_{\alpha|\sigma'}(E) = 0$, this would place contradictory demands on ρ_{σ^+}. □

References

[1] Norsen, Travis 2009, Local causality and completeness: Bell vs. Jarrett, *Foundations of Physics* **39**, 273–94.
[2] Ghirardi, GianCarlo 2012, Does quantum nonlocality irremediably conflict with special relativity? *Foundations of Physics* **40**, 1379–95.
[3] Shimony, Abner 1978, Metaphysical problems in the foundations of quantum mechanics, *International Philosophical Quarterly* **18**, 3–17.

[4] Shimony, Abner 1984, Controllable and uncontrollable non-locality, in S. Kamefuchi et al. (eds.), *Foundations of Quantum Mechanics in Light of New Technology*, Tokyo: The Physical Society of Japan, pp. 225–30; reprinted in Shimony [47, pp. 130–38].

[5] Shimony, Abner 1986, Events and processes in the quantum world, in R. Penrose and C.J. Isham (eds.), *Quantum Concepts in Space and Time*, Oxford: Oxford University Press, pp. 182–203; reprinted in Shimony [47, 140–62].

[6] Ghirardi, GianCarlo, and Grassi, Renata 1996, Bohm's theory versus dynamical reduction, in James T. Cushing, Arthur Fine, and Sheldon Goldstein (eds.), *Bohmian Mechanics and Quantum Theory: An Appraisal*, Berlin: Springer.

[7] Jarrett, Jon 1984, On the physical significance of the locality conditions in the Bell arguments, *Noûs* **18**, 569–89.

[8] Redhead, Michael 1987, *Incompleteness, Nonlocality, and Realism: A Prolegomenon to the Philosophy of Quantum Mechanics*, Oxford: Oxford University Press.

[9] Skyrms, Brian 1984, EPR: Lessons for metaphysics, *Midwest Studies in Philosophy* **9**, 245–55.

[10] Woodward, James 2013, Causation and manipulability, in Edward N. Zalta (ed.), *The Stanford Encyclopedia of Philosophy* (Winter 2013 ed.), http://plato.stanford.edu/archives/win2013/entries/causation-mani/.

[11] Faye, Jan 2010, Backward causation, in Edward N. Zalta (ed.), *The Stanford Encyclopedia of Philosophy* (spring 2010 ed.): http://plato.stanford.edu/archives/spr2010/entries/causation-backwards/.

[12] Evans, Peter W., Price, Huw, and Wharton, Ken B. 2013, A new slant on the EPR–Bell experiment, *British Journal for the Philosophy of Science* **634**, 297–324.

[13] Dürr, Detlef, Goldstein, Sheldon, Münch-Berndl, Karin, and Zanghì, Nino 1999, Hypersurface Bohm–Dirac models, *Physical Review A* **60**, 2729–36.

[14] Dürr, Detlef, Goldstein, Sheldon, Norsen, Travis, Struyve, Ward, and Zanghì, Nino 2014, Can Bohmian mechanics be made relativistic? *Proceedings of the Royal Society A* **470**, 20130699.

[15] Bell, John S. 1977, Free variables and local causality, *Epistemological Letters* **15**, 79–84; reprinted in *Dialectica* **39** (1985), 103–16, and in Bell [17, pp. 100–104].

[16] Shimony, Abner, Horne, Michael A., and Clauser, John F. 1976, Comment on "The theory of local beables," *Epistemological Letters* **13**, 1–8; reprinted in *Dialectica* **39** (1985), 97–102 and in Shimony [47, pp. 163–7].

[17] Bell, John S. 2004, *Speakable and Unspeakable in Quantum Mechanics* (2nd ed.), Cambridge: Cambridge University Press.

[18] Bell, John S. 1990, La nouvelle cuisine, in A. Sarlemijn and P. Kroes (eds.), *Between Science and Technology*, North-Holland; reprinted in Bell [17, pp. 232–48].

[19] Bell, John S. 2007, The Trieste Lecture of John Stewart Bell, *Journal of Physics A: Mathematical and Theoretical* **40**, 2919–33.

[20] Bell, John S. 1987, Are there quantum jumps? In C.W. Kilmister (ed.), *Schrödinger: Centenary Celebration of a Polymath*, Cambridge: Cambridge University Press, pp. 41–52; reprinted in Bell [17, pp. 201–12].

[21] Bedingham, Daniel 2011, Relativistic state reduction dynamics, *Foundation of Physics* **41**, 686–704.

[22] Tumulka, Roderich 2006, A relativistic version of the Ghirardi–Rimini–Weber model, *Journal of Statistical Physics* **125**, 825–44.

[23] Aharonov, Yakir, and Albert, David 1984, Is the usual notion of time evolution adequate for quantum-mechanical systems? II. Relativistic considerations, *Physical Review D* **29**, 228–34.

[24] Fleming, Gordon N. 1986, On a Lorentz invariant quantum theory of measurement? In Daniel M. Greenberger (ed.), *New Techniques and Ideas in Quantum Measurement Theory, Annals of the New York Academy of Sciences* **480**, 574–5.

[25] Fleming, Gordon N. 1989, Lorentz invariant state reduction, and localization, in Arthur Fine and Jarrett Leplin (eds.), *PSA 1988: Proceedings of the 1988 Biennial Meeting of the Philosophy of Science Association, Vol. Two: Symposia and Invited Papers*, East Lansing, MI: Philosophy of Science Association, pp. 112–26.

[26] Fleming, Gordon N. 1996, Just how radical is hyperplane dependence? In Rob Clifton (ed.), *Perspectives on Quantum Reality: Non-relativistic, Relativistic, and Field-Theoretic*, Dordrecht: Kluwer Academic.

[27] Ghirardi, GianCarlo 2000, Local measurements of nonlocal observables and the relativistic reduction process, *Foundations of Physics* **30**, 1337–85.

[28] Ghirardi, GianCarlo 1999, Some lessons from relativistic reduction models, in Heinz-Peter Breuer and Francesco Petruccione (eds.), *Open Systems and Measurement in Relativistic Quantum Theory: Proceedings of the Workshop Held at the Istituto Italiano per gli Studi Filosofici, Naples, April 3, 1998*, Berlin: Springer, pp. 117–52.

[29] Ghirardi, G.C., Grassi, R., and Benatti, F. 1995, Describing the macroscopic world: Closing the circle within the dynamical reduction program, *Foundations of Physics* **25**, 5–38.

[30] Ghirardi, GianCarlo, and Grassi, Renata 1994, Outcome predictions and property attribution: The EPR argument reconsidered, *Studies in History and Philosophy of Science* **25**, 397–423.

[31] Ghirardi, GianCarlo, and Pearle, Philip 1991, Elements of physical reality, nonlocality and stochasticity in relativistic dynamical reduction models, in Arthur Fine, Micky Forbes, and Linda Wessels (eds.), *PSA 1990: Proceedings of the 1990 Biennial Meeting of the Philosophy of Science Association, Volume Two: Symposia and Invited Papers*, East Lansing, MI: Philosophy of Science Association, pp. 35–47.

[32] Myrvold, Wayne C. 2002, On peaceful coexistence: Is the collapse postulate incompatible with relativity? *Studies in History and Philosophy of Modern Physics* **33**, 435–66.

[33] Myrvold, Wayne C. 2003, Relativistic quantum becoming, *British Journal for the Philosophy of Science* **53**, 475–500.

[34] Esfeld, Michael, and Gisin, Nicholas 2014, The GRW flash theory: A relativistic quantum ontology of matter in space–time? *Philosophy of Science* **81**, 248–64.

[35] Wallace, David, and Timpson, Christopher G. 2010, Quantum mechanics of spacetime: I. Spacetime state realism, *British Journal for the Philosophy of Science* **61**, 697–727.

[36] Ghirardi, GianCarlo, Grassi, Renata, and Pearle, Philip 1990, Relativistic dynamical reduction models: General framework and examples, *Foundations of Physics* **20**, 1271–316.

[37] Ghirardi, GianCarlo, Grassi, Renata, Butterfield, Jeremy, and Fleming, Gordon N. 1993, Parameter dependence and outcome dependence in dynamical models for state vector reduction, *Foundations of Physics* **23**, 341–64.

[38] Ghirardi, GianCarlo 1997, Quantum dynamical reduction and reality: Replacing probability densities with densities in real space, *Erkenntnis* **45**, 349–65.

[39] Bedingham, Daniel, Dürr, Detlef, Ghirardi, GianCarlo, Goldstein, Sheldon, Tumulka, Roderich, and Zanghì, Nino 2014, Matter density and relativistic models of wave function collapse, *Journal of Statistical Physics* **154**, 623–31.

[40] Tumulka, Roderich 2007, The "unromantic pictures" of quantum theory, in *The Quantum Universe, Journal of Physics A: Mathematical and Theoretical* **40**, 3245–73.

[41] Bacciagaluppi, Guido, and Valentini, Antony 2009, *Quantum Theory at the Crossroads: Reconsidering the 1927 Solvay Conference*, Cambridge: Cambridge University Press.

[42] Ghirardi, GianCarlo 2007, Some reflections inspired by my research activity in quantum mechanics, *Journal of Physics A: Mathematical and Theoretical* **40**, 2891–917.

[43] Allori, Valia, Goldstein, Sheldon, Tumulka, Roderich, and Zanghì, Nino 2008, On the common structure of Bohmian mechanics and the Ghirardi–Rimini–Weber theory, *British Journal for the Philosophy of Science* **59**, 353–89.

[44] Saunders, Simon, Barrett, Jonathan, Kent, Adrian, and Wallace, David (eds.) 2010, *Many Worlds? Everett, Quantum Theory, and Reality*, Oxford: Oxford University Press.

[45] Wallace, David 2012, *The Emergent Multiverse*, Oxford: Oxford University Press.

[46] Arias, A., Gheondea, A., and Gudder, S. 2002, Fixed points of quantum operations, *Journal of Mathematical Physics* **43** (12), 5872–81.

[47] Shimony, Abner 1993, *Search for a Naturalistic Worldview, Vol. II: Natural Science and Metaphysics*, Cambridge: Cambridge University Press.

15

Bell Nonlocality, Hardy's Paradox and Hyperplane Dependence

GORDON N. FLEMING

Abstract

I begin with some reminiscences of my delayed appreciation of the significance of John Bell's work. Preparatory remarks on my general perspective concerning the interpretation of quantum mechanics are presented. I argue against the conflation of hyperplane dependence with frame dependence, which occurs occasionally; and that the 'elements of reality' of Hardy's famous gedankenexperiment can retain their Lorentz invariance, i.e., their frame independence, if one recognizes the hyperplane dependence of their localization, which follows. Finally, I criticize a view of the nature of Lorentz transformations presented by Asher Peres and co-workers which conflicts with the view employed here.

15.1 Initial Reactions to Bell's Work

Like many others in the physics community, I was late to come to an appreciation of the significance of John Bell's papers on the foundations of quantum mechanics (QM). But unlike many of those, my indifference to Bell was not due to a dismissive attitude to foundational studies per se or to an intransigent commitment to some version of Copenhagenism. In 1971, for example, I was very much involved in the international conference on QM foundations at my home institution,[1] where Bell presented the paper "On the hypothesis that the Schrödinger equation is exact," published later as [1], but I paid minimal attention to his presentation. The cause of this poor judgment (as I eventually came to realize it was) was that I was already convinced that the Schrödinger equation was not exact in the sense Bell meant, but must be augmented with primordial state reductions. I was among those who, in the words of Bob Wald, as reported by Roger Penrose [2], were inclined to "take it seriously," the QM state that is, and, consequently, could not "really believe in it," i.e., really believe that purely unitary QM is a complete theory. I was delighted when the experimental tests of Bell's inequalities upheld the QM predictions. I was, furthermore, uninterested in hidden variable *reconstructions* of QM, such as Bohmian mechanics [3] and Many Worlds

---

[1] "International Colloquium on Issues in Contemporary Physics and Philosophy of Science, and their Relevance for our Society," Penn State University, September 6–18, 1971. A colloquium memorable for passionate presentations at which I first met Abner Shimony, received the withering declaration from Imre Lakatos that "Elementary particle physics is not a science!" and had the pleasure of participating in a dramatic reading of Joseph Jauch's then as yet unpublished ms. "Are Quanta Real?". I read Sagredo to C.F. von Weisacker's Salviati and, if memory serves, a young Cliff Hooker's Simplicio.

interpretations [4, 5], that were, in principle, not susceptible to empirical tests of their novel details. These attitudes, which I still largely subscribe to, were not, of course, due to any deep insights on my part into the workings of nature. They were due, rather, I suspect, to my personal allotment of psychological quirks and philosophical preferences (not to say prejudices) and they predetermined my favorable attitude to the later GRW [6] and other schemes for augmenting QM with *primordial spontaneous state reductions* (PSR).

My later recognition of the importance of Bell's work emerged only with the slow awareness that the violation of Bell's inequalities entailed a version of nonlocality different from, but not wholly unrelated to, the hyperplane dependence (HD) of many dynamical variables that I had been arguing for [7–9] on the grounds of minimal compatibility with Lorentz covariance. Continued ruminations along these lines eventually led to my arguing for HD state reductions [10] as well. Only once, however, did I engage in a direct examination of Bell-related inequalities [11], those of the GHZ type [12].

Quite apart from Bell's role in leading us to the recognition of the nonlocality of *this* world, the incisive style of his writing and his arguments are invaluable! However much the choices we make among the many competing interpretations and approaches to QM may depend on the above-mentioned personal allotments, Bell's work helps to make clear just where we stand [13].

15.2 The Ontology of Quantum Phenomena and the Quantum State

Regarding the status of the quantum state, I belong in the ψ-ontic camp [14]. But this camp is widely dispersed, with many conflicting subdivisions. In this section I try to pin down and clarify my place among the subdivisions. First, I do not hanker after determinism [15]. I regard the encounter with quantum phenomena, particularly value acquisition (or property actualization) implemented via state reduction, as the discovery of genuinely *uncaused events*! They are "governed," in their occurrence, by the basic PSRs among them, by the unitarily evolving quantum probabilities, and, in the special case of measurements, by our experimental preparations, but by nothing else. Their individual occurrences are *instances of the suspension of the principle of sufficient reason* [16]! And in turn, they cause abrupt changes in those quantum probabilities.

Quite generally, quantum *systems* can be characterized by the various possible complete sets of compatible properties (comsets), the members of which are, in principle, capable of being jointly actualized, and the PSRs may be just such uncaused actualizations. The quantum state of a system, mathematically represented (relative to an inertial frame) by a state vector in the comparatively rare case of a pure state or by a density operator in the much more common case of an "improper" mixed state, is our current best effort at conceptually grasping the ontological ground of objective physical reality. The evolving probabilities are determined by the relation of the dynamical variable operators and their eigenvectors, evolving via the Heisenberg equations of motion,[2] to the state *operators* from which the

---

[2] My preoccupation with Lorentz-covariant dynamics dictates the use of the Heisenberg picture. The Schrödinger picture is awkward, at best, in the presence of frame-dependent time coordinates and becomes even more so when time dependence is generalized to HD. Finally, in the presence of PSR, the *generalized* Heisenberg picture admits, for closed systems, a clean separation between unitary and stochastic evolution.

density *matrices*, of various representations, are formed. A generalized Heisenberg picture [17], which I am clearly employing here, restricts the dynamical variable operators to unitary evolution and the state operator, for otherwise closed systems, to stochastic evolution via PSRs and their exploitation for measurements. For open subsystems the evolution of the subsystem state operator receives a unitary contribution as well as a stochastic contribution [18].

As yet we do not have a satisfactory theory of PSR, notwithstanding heroic and encouraging efforts. I have no unequivocal favorites but am impressed with the current relativistic mass density versions [19, 20] and the efforts of [21–25, and others] to ground state reduction in quantum gravity and their search for empirical support in biological processes [26]. Regarding the contribution of environmental decoherence [27], it seems to me that while it can greatly enhance the effectiveness of PSR in avoiding or suppressing superpositions of macro-distinct states, it cannot itself supplant PSR.

I do not share the widespread inclination [28] to attribute fundamental status to the *wave function*, i.e., the position representation of a pure quantum state of a system of a definite number of quantons[3] or the local field-theoretic functional representation with diagonalized fields. In the formalism of QM, all possible comset representations of a quantum state, whether pure or mixed, are equivalent in the sense that from any one such representation on a given hyperplane, any other one on the same hyperplane is, in principle, kinematically determined. I take this as indicative of a symmetry in nature that extends very deep. Nothing remotely like it holds in classical physics. The classical mathematical equivalence of all possible choices of canonical variables or all possible choices of generalized coordinates and velocities in Hamiltonian or Lagrangian formulations, respectively, is a poor analogue.[4]

It is the quantum system and its state, mathematically represented, in relation to an inertial frame, by the abstract state vector or the abstract trace class operator, in which I tend to see ontological status or real existence. But not exclusively so; the possible potential properties of the system, mathematically represented by the dynamically evolving Heisenberg picture operators, really exist as well, or so I think. It is in the uncaused events of PSR that these two modalities of existence, the system states and their potential properties, come together in the collapse to eigenvectors of an actualized property. But which property? My hesitancy to grant primordial status to any particular representation or, equivalently, to any particular comset would seem to require a random distribution of PSR to any and all possible discrete spectrum eigenvectors. But perhaps not. In the case of systems of definite numbers of quantons, collapse to coherent states, which do not precisely actualize any commonly recognized dynamical variable, may be the way. Of course in the context of quantum field theory the number of quantons of any given type is almost always indefinite, and then what does PSR lead to? Could coherent states again, but now of the field theory variety, be candidates? Alternatively, might the PSR of quantized field states aim directly at rendering quanton numbers momentarily definite? Finally, what becomes of the whole

---

[3] Quanton = boson, fermion or anyon.

[4] For a single classical particle, neither momentary position nor momentary momentum determines the other. The whole history of momentum fails to determine position at any time. For a single, spinless quanton the momentary position or momentum state representation determines the other and all possible others!

PSR scheme in the context of a future quantum gravity theory where the "system" subject to PSR is space–time itself.

But where do the abstract system states live? it is often asked. In the generalized Heisenberg picture I have in mind they live on spacelike hyperplanes in space–time "between" the state reductions. Nonrelativistically, the state represents the history of the system literally between consecutive (collections of simultaneous) state reductions. The state lives in those time intervals. The unitarily evolving dynamical variables live, more particularly, at the individual times and, for mass-conserving fields, at points of space at those times as well.

Lorentz-covariantly, the states of closed systems live on the *collections* of spacelike hyperplanes that each divide all the state reductions into those lying in the past and those lying in the future of the collection in question. In Section 4 below, I will present yet another[5] account of the need for this HD of quantum states, this time by examining the gedankenexperiment of [29], which leads to the so-called Hardy paradox of elements of reality that must violate Lorentz invariance. If they partake of the nonlocality of HD, as I claim they must, then they regain Lorentz invariant status. This approach neither supports, necessarily, nor conflicts with other analyses such as [30] but rather indicates that the appearance of paradox in the first place is due to neglect of HD in quanton localization and state reduction.

The dynamical variables, if local fields, live at points of space–time; but if global quantities, such as total 4-momenta or angular momenta or various charges, they live on *individual* space-like hyperplanes. And then there may be nonlocal fields that live at *points on hyperplanes* [31, 32], examples of which appear to be gaining use for certain purposes [33–36]. Somewhat more conjecturally, in curved space–times, the nonlocal dynamical variables will live on spacelike hypersurfaces of zero extrinsic curvature while the states live on collections of such which separate the instances of PSR and so on.

I want to stress that I do not regard this association of states and dynamical variables with (sets of) spacelike hyperplanes as simply a possible approach to achieving manifest covariance of description under the IHLG. Rather, I claim that the association is *forced* upon us by the presumed equivalence of all inertial observers and the success such observers have in describing dynamical evolution in terms of physical states of affairs (PSA) at instants of time. The argument proceeds as follows:

(1) Any PSA that one inertial observer can describe and/or measure can, as a matter of principle, also be described and/or measured by any other inertial observer.

(2) Instants of time and the PSAs that hold at them for one inertial observer are, for any other inertial observer, spacelike hyperplanes with those PSAs holding on them.

(3) Any spacelike hyperplane and the PSA that holds on it is an instant of time, with the PSA holding at it, in some inertial frame.

(4) Consequently, all inertial observers can describe and/or measure the PSAs on any and all spacelike hyperplanes.

---

[5] For earlier arguments see [10, 37–40].

While some would extend and have extended [41, 42] these allowed descriptions to PSAs on arbitrary curvilinear spacelike hypersurfaces, such an extension in Minkowski space–time cannot be grounded in the tradition of description at instants of time and is not needed for the manifest covariance of descriptions. The extension is, therefore, artificial and I avoid it. I also note that we cannot extend this argument, *without modification*, to noninertial observers in Minkowski space–time or even freely falling observers in generally curved space–times because of the occurrence of horizons.

I might mention, in passing, that while, as claimed above, I do not favor special onto-logical status for position representation state functions for N quantons, I also do not sym-pathize with the argument that they cannot be real because they live in $3N$-dimensional configuration space. They need live there only if one insists on making the state function a *single-point field*, as is traditional. They would live just as easily in physical 3-space as an N-point field, i.e., a field assigned to each (appropriately identified) *set* of N 3-space points. Be that as it may, it is the abstract state vector or state operator that I see as the mathemati-cal representation of the objective quantum reality rather than any particular, basis-defined, number-valued representation thereof.

Finally, the results of delayed choice experiments on entangled states by Aspect's [43] and Zeilinger's [44] groups corroborating Bell-inequality-violating probabilities reveal that the pairs of measurements cannot be objectively separated into triggering and testing mea-surements of a correlation. Being spacelike separated, neither measurement is invariantly earliest or latest and the correlation is simply a single property of the state actualized by the *pair* of measurements. To paraphrase Gisin [45], "a quantum [spacelike] correlation is not a correlation between 2 events, but a single event that manifests itself at 2 locations."

15.3 Hyperplane Dependence vs. Frame Dependence

I find that some writers, including some in the minority whom I regard as sympathetic to my orientation toward HD, conflate HD with frame dependence (FD). This seeming conflation is further confused by the variety of concepts of just what an inertial frame of reference is. An older concept identifies inertial reference frames with Minkowski coordinate systems. I will call these M-frames. A more contemporary concept identifies inertial reference frames with hyperplane foliations of space-time. I will call these F-frames. The F-frames are asso-ciated with equivalence classes of M-frames, each class consisting of M-frames that are at rest relative to one another. For either kind of frame, HD is not the same as FD. I will spell this out here (in possibly unnecessary detail) with the elementary example provided by the total 4-momentum of a timelike persistent and spacelike extended system that may be open or closed. Once established here, the presence of HD and the independence of HD and FD will be exploited in Section 15.4 to defuse the Hardy paradox.

Beyond these considerations, it turns out that some workers object to the very concept of the nature of a Lorentz transformation that I will employ here. I did not appreciate this fact until rather recently. The concept I use is effectively defined (at least for 4-vectors) by equations (15.1), (15.2), and (15.12) following and I will not defend it just yet, as I believe most will not object to it. But in Section 15.5 I will defend it against the opposing view, which has rather drastic consequences.

For a quantum state, ρ, and a closed system (having conserved 4-momentum) the expectation values of the 4-momentum components will be denoted by $< P^\mu >_\rho$, and will satisfy the transformation rule

$$< P^\mu >'_\rho = \Lambda^\mu_\nu < P^\nu >_\rho \tag{15.1}$$

for two M-frames, M and M', with coordinates related by the IHLG transformation,

$$x^{\mu'} = \Lambda^\mu_\nu x^\nu + a^\mu. \tag{15.2}$$

Such expectation values are calculated using a quantum state operator, $\hat{\rho}$, and a total 4-momentum operator, $\hat{P}^\mu$. Thus,

$$< P^\mu >_\rho = Tr(\hat{\rho} \, \hat{P}^\mu). \tag{15.3}$$

While the quantum state, ρ, is objective and M-frame-independent, the quantum state *operator*, $\hat{\rho}$, represents the *relationship* of the state to the frame and thus is, itself, frame-dependent. In particular we have

$$\hat{\rho}' = \hat{U}(\Lambda, a)\hat{\rho} \, \hat{U}(\Lambda, a)^\dagger, \tag{15.4}$$

where $\hat{U}(\Lambda, a)$ is the unitary representation of the IHLG transformation. On the other hand, the operators, $\hat{P}^\mu$, are not frame-dependent, in the sense that the same set of operators are used to evaluate the expectation values in any frame. Thus,

$$< P^\mu >'_\rho = Tr(\hat{\rho}' \, \hat{P}^\mu). \tag{15.5}$$

The preceding transformation rule, (15.1), for the expectation values emerges from the property of the frame-independent momentum operators,

$$\hat{U}(\Lambda, a)^\dagger \hat{P}^\mu \, \hat{U}(\Lambda, a) = \Lambda^\mu_\nu \hat{P}^\nu, \tag{15.6}$$

but it would be double counting to identify the left-hand side as the 4-momentum operator for the frame, M'.

So far no hint of hyperplane dependence (HD) has appeared, notwithstanding the fact that the definition of $\hat{P}^\mu$ in quantum-field-theoretic terms can employ any hyperplane whatsoever. Thus, if $\hat{\theta}^{\mu\nu}(x)$ is the symmetric stress–energy–momentum (SEM) field where

$$< \hat{\theta}^{\mu\nu}(x') >' = \Lambda^\mu_\lambda \Lambda^\nu_\sigma < \hat{\theta}^{\lambda\sigma}(x) >, \tag{15.7a}$$

then

$$\hat{P}^\mu = \int d^4x \, \delta(\eta x - \tau) \hat{\theta}^{\mu\nu}(x) \eta_\nu, \tag{15.7b}$$

for any τ and any future-pointing timelike unit vector, η^μ. The integral remains hyperplane-independent due to the SEM field integrand being (for a closed system) locally conserved, i.e.,

$$\partial_\mu \hat{\theta}^{\mu\nu}(x) = 0. \tag{15.8}$$

We now consider an open subsystem of our closed system for which the local conservation of the subsystem SEM field, $\hat{\theta}_S^{\mu\nu}(x)$, does not hold. The field-theoretic definition of the open subsystem total 4-momentum operator is now HD,

$$\hat{P}_S^\mu(\eta, \tau) = \int d^4x \delta(\eta x - \tau) \hat{\theta}_S^{\mu\nu}(x) \eta_\nu, \tag{15.9}$$

and the behavior of $\hat{P}_S^\mu(\eta, \tau)$ under unitary IHLG transformations is now

$$\hat{U}(\Lambda, a)^\dagger \hat{P}_S^\mu(\eta', \tau') \hat{U}(\Lambda, a) = \Lambda_\nu^\mu \hat{P}_S^\nu(\eta, \tau), \tag{15.10a}$$

where

$$\eta' = \Lambda\eta \quad and \quad \tau' = \tau + a\eta'. \tag{15.10b}$$

Again, the left-hand side of (15.10a) is not the 4-momentum operator for the frame, M', since the frame dependence is, as before, carried by the quantum state operator. The expectation values now (using the open subsystem analogues of (15.3)–(15.6)) are both FD and HD according to

$$< P_S^\mu(\eta', \tau') >'_\rho = Tr(\hat{\rho}' \hat{P}_S^\mu(\eta', \tau')) = Tr(\hat{\rho} \Lambda_\nu^\mu \hat{P}_S^\nu(\eta, \tau)) = \Lambda_\nu^\mu < P_S^\nu(\eta, \tau) >_\rho . \tag{15.11}$$

We have seen FD quantities that were not HD quantities, (15.1), and now quantities that are both FD and HD, (15.11). Finally we consider the scalar quantities

$$< P_S^\mu(\eta, \tau) >_\rho < P_{S,\mu}(\eta, \tau) >_\rho \quad and \quad < P_S(\eta, \tau)^2 >_\rho := Tr(\hat{\rho} \hat{P}_S^\mu(\eta, \tau) \hat{P}_{S,\mu}(\eta, \tau)). \tag{15.12}$$

Being scalars, these are not FD, but they are HD for open systems and not HD for closed systems. Thus all possible combinations of FD and not FD, HD and not HD have been displayed. HD and FD are not the same.

Finally, in the presence of PSR the state operator, $\hat{\rho}$, for an otherwise closed system will acquire a stochastic HD which, strictly speaking, would eliminate the possibility of any *exactly* non-HD expectation values. This does not undermine the sharp distinction between HD and FD!

15.4 The Lorentz Covariance of Elements of Reality

I have previously presented the following kind of argument for a system of two entangled quantons subjected to spacelike separated state reductions [37] and for a system of three entangled quantons similarly subjected [38]. In each case the purpose of the argument is to demonstrate the manner in which considerations of HD can defuse (eliminate) apparent paradoxes that emerge from entanglement and EPR correlations that violate Bell inequalities in a relativistic context. This time I will again discuss the unitary evolution of a

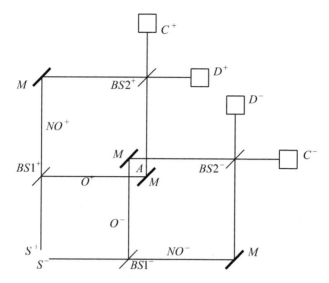

Figure 15.1 Hardy's arrangement of Mach–Zender interferometers for an electron–positron pair with sources $S^{\pm}$, overlapping arms $O^{\pm}$, nonoverlapping arms $NO^{\pm}$, beam splitters $BS1^{\pm}$ and $BS2^{\pm}$, mirrors M, annihilation region A, and detectors $C^{\pm}$ and $D^{\pm}$.

two-quanton system preceding spacelike separated state reduction. The quantons will be an electron–positron pair (e^{-}, e^{+}) and the experimental arrangement will be that of the well known gedankenexperiment due to Lucien Hardy involving two Mach–Zender interferometers (Fig. 15.1). In that case each quanton interacts unitarily with macroscopic apparatus in spacelike separated regions and *then* (i.e., timelike later), potentially, with the other quanton and *then* again with a second set of macroscopic apparatus in spacelike separated regions *before* the final state reductions.

The macroscopic apparatus that the system interacts with unitarily is four beam splitters, $BS1^{\pm}$ and $BS2^{\pm}$, and mirrors, M. The interactions of e^{-} with $BS1^{-}$ and e^{+} with $BS1^{+}$ occur in spacelike separated regions, as do the interactions of e^{-} with $BS2^{-}$ and e^{+} with $BS2^{+}$. But timelike later than the first set of interactions and timelike earlier than the second set of interactions is an annihilation region, A, where the electron–positron pair may annihilate into two photons, 2γ. The beam splitters and initial states of the quantons are arranged so that if either of the quantons traversed the arrangement alone, the quantum state evolution would follow the sequence

$$|S^{\pm} > \underset{BS1^{\pm}}{\rightarrow} (|O^{\pm} > +|NO^{\pm} >)/\sqrt{2} \underset{BS2^{\pm}}{\rightarrow} |C^{\pm} > \qquad (15.13)$$

where $|S^{\pm} >$ are the initial states of the quantons, $|O^{\pm} >$ *and* $|NO^{\pm} >$ are the partial states for the separate arms produced by interaction with $BS1^{\pm}$, and

$$|O^{\pm} > \underset{BS2^{\pm}}{\rightarrow} (|C^{\pm} > +|D^{\pm} >)/\sqrt{2}, \qquad (15.14)$$

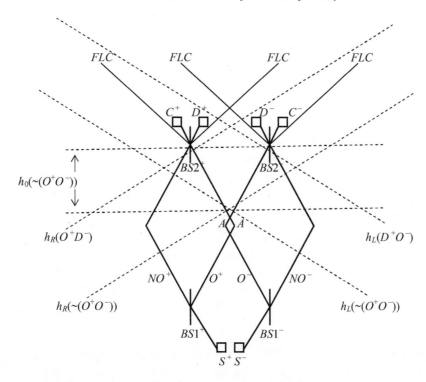

Figure 15.2 Space–time relations among various stages of Hardy's gedankenexperiment. *FLC* denotes future light cone lines. $h_{0,L,R}$ ($\sim (O^+O^-)$), $h_L(D^+O^-)$, and $h_R(O^+D^-)$ denote spacelike hyperplane lines that are instantaneous in the $F_{0,L,R}$ frames, respectively.

and

$$|NO^\pm > \xrightarrow[BS2^\pm]{} (|C^\pm > - |D^\pm >)/\sqrt{2},$$
(15.15)

where $|C^\pm >$ and $\pm |D^\pm >$, are the partial states (and relative phases) produced for the separate detectors by interaction of $|O^\pm >$ or $|NO^\pm >$ with $BS2^\pm$. Finally, the O arms can be arranged to have an overlap region, A, where the electron and positron could encounter one another and annihilate into two photons,

$$|O^+O^- > \xrightarrow[A]{} |2\gamma >$$

and in each interferometer, the final beam splitter encounter is spacelike separated from the final detection event of the other interferometer – as indicated by the hyperplane lines, $h_R(O^+D^-)$ and $h_L(D^+O^-)$, in the somewhat busy Fig. 15.2.

Let us now assume an inertial reference frame, F_0, in which the interactions of the quantons with their respective beam splitters, $BS1$, are simultaneous, as are the interactions with the beam splitters $BS2$ and the final detectors CD. If the annihilation region were *not* in place, then the two-quanton system would evolve according to the scheme

$$|S^+S^-> \;\to\; (|O^+O^-> +|O^+NO^-> +|NO^+O^-> +|NO^+NO^->)/2 \;\to\; |C^+C^->\;.$$
$$\qquad BS1^\pm \qquad\qquad\qquad\qquad\qquad\qquad\qquad\qquad\qquad BS2^\pm \quad (15.16)$$

If, instead, the overlap region, A, is in place (and we follow Hardy and consider only those instances in which the electron–positron pair is finally detected in D^+D^-), then the time evolution scheme in F_0 is

$$|S^+S^-> \;\to\; (|O^+O^-> +|O^+NO^-> +|NO^+O^-> +|NO^+NO^->)/2 \;\to$$
$$\quad BS1^\pm \qquad\qquad\qquad\qquad\qquad\qquad\qquad\qquad\qquad\qquad A$$
$$(|2\gamma> +|O^+,NO^-> +|NO^+,O^-> +|NO^+NO^->)/2 \;\to$$
$$\qquad\qquad\qquad\qquad\qquad\qquad\qquad\qquad\qquad\qquad\qquad BS2^\pm$$
$$(1/4)(2|2\gamma> +3|C^+C^-> +|C^+D^-> +|D^+C^-> -|D^+D^->) \;\to\; |D^+D^->\;.$$
$$\qquad\qquad\qquad\qquad\qquad\qquad\qquad\qquad\qquad\qquad\qquad D^+D^- \quad (15.17)$$

Note that in F_0, after the possibility of annihilation and before the BS2 interactions, the probability, *at any F_0 time*, of finding the electron in the O^- state *and* the positron in the O^+ state is zero. That is the meaning of the symbol $\sim(O^+O^-)$. Those instants of time are *invariantly* described as the hyperplanes of type $h_0(\sim(O^+O^-))$, indicated in Fig. 15.2.

In a reference frame, F_L, moving to the left relative to F_0, in which the $BS2^+$ interaction and the D^+ state reduction are earlier than the $BS2^-$ interaction, the F_L time sequence between the A event region and the $BS2^-$ interaction is

$$(|2\gamma> +|O^+NO^-> +|NO^+O^-> +|NO^+NO^->)/2 \;\to\;,$$
$$\qquad\qquad\qquad\qquad\qquad\qquad\qquad\qquad BS2^+$$
$$(\sqrt{2}|2\gamma> +2|C^+NO^-> +|C^+O^-> -|D^+O^->)/2\sqrt{2} \;\to\; -(1/2\sqrt{2})|D^+O^->$$
$$\qquad\qquad\qquad\qquad\qquad\qquad\qquad\qquad\qquad D^+ \qquad\qquad (15.18)$$

In such an F_L frame with D^+ detection there are instants of time, after A and before $BS2^-$, when, just as in F_0, there is no possibility of finding O^+O^-. See $h_L(\sim(O^+O^-))$. The probability of finding O^- at these times is $\frac{1}{4}$. This probability remains $\frac{1}{4}$ until the $BS2^+$ interaction, when the O^- probability changes to $\frac{5}{8}$, followed by the D^+ collapse, when the O^- probability jumps to unity and remains there until the $BS2^-$ interaction. Those latter instants of F_L time are *invariantly* described as hyperplanes of the type $h_L(D^+O^-)$ in Fig. 15.2.

Similarly, in a frame F_R moving to the right relative to F_0, in which the $BS2^-$ interaction and the D^- state reduction are earlier than the $BS2^+$ interaction, the F_R time sequence between the A event region and the $BS2^+$ interaction is

$$(|2\gamma > +|O^+NO^- > +|NO^+O^- > +|NO^+NO^- >)/2 \to$$
$$BS2^-$$

$$(\sqrt{2}|2\gamma > +2|NO^+C^- > +|O^+C^- > -|O^+D^- >)/2\sqrt{2} \to -(1/2\sqrt{2})|O^+D^- >$$
$$D^- \qquad \qquad (15.19)$$

In such an F_R frame with D^- detection there are instants of time, after A and before $BS2^+$, when, just as in F_0, there is no possibility of finding O^+O^-. See $h_R(\sim(O^+O^-))$. The probability of finding O^+ at these times is $\frac{1}{4}$. This probability remains $\frac{1}{4}$ until the $BS2^-$ interaction, when the O^+ probability changes to $\frac{5}{8}$, followed by the D^- collapse, when the O^+ probability jumps to unity and remains there until the $BS2^+$ interaction. Those latter instants of F_R time are *invariantly* described as hyperplanes of the type $h_R(O^+D^-)$ in Fig. 15.2.

The existence (in our subensemble of D^+D^- instances) of F_L times after A at which the electron is certain to be in the O^- arm and F_R times after A at which the positron is certain to be in the O^+ arm, while there are *no* F_0 times after A in which both localizations in O^- and O^+ are possible, constitutes the apparent paradox, or, more precisely, the frame dependence of the elements of reality, as Hardy called them.

But the two sets of hyperplanes of types $h_L(D^+O^-)$ and $h_R(O^+D^-)$ are mutually disjoint and each set is disjoint with the three sets, $h_{0,L,R}$ $(\sim(O^+O^-))$. Consequently, there are no hyperplanes, "later" than A, on which the electron–positron pair are in the $|O^+O^- >$ state. The paradox, in which $\sim(O^+O^-)$ appeared to conflict with D^+O^- *and* O^+D^-, dissolves by virtue of the absence of a common space–time substrate for the three assertions or any two of them. The *elements of reality*, as Hardy referred to them, retain their Lorentz invariance because all inertial observers agree as to the physical state of affairs on any *given* hyperplane; i.e., $F_{0,L,R}$ *all* agree that D^+O^- occurs on $h_L(D^+O^-)$, that O^+D^- occurs on $h_R(O^+D^-)$, and that O^+O^- does *not* occur on any of $h_{0,L,R}$ $(\sim(O^+O^-))$. Lorentz invariance is preserved via the nonlocality of HD, and the nonlocal HD analysis is forced upon us by the four-step argument at the end of Section 15.2. Referring back to the discussion in Section 15.3 against the conflation of HD with FD, we might emphasize here that while localization at instants of time is FD, localization, per se, is not FD. Rather, localization, per se, is HD and only which hyperplanes are instants of time is FD.

But to feel comfortable with this dissolution, one has to take seriously the HD of quanton localization, which many do not. So let us test the HD probabilities by placing a detector in the O^+ arm, say, just before $BS2^+$.

First we notice that if we do *not* find the positron in the O^+ arm (state), then the modified evolution on all three sets of hyperplanes, h_0, h_L, and h_R, yield zero probability of detecting the electron in D^-. So restricting to the final outcome of D^- guarantees finding the positron in O^+, as predicted on h_R. But then the state evolution on the h_L hyperplanes is modified *early* to ensure that the electron is in NO^-. Similar considerations apply if we place the detector in the O^- arm, or a pair of detectors in two arms. One can never find the O^+O^-

combination on any hyperplane "later" than A. Nevertheless, this does not invalidate the predictions of certainty for finding O^+ on h_R "after" D^- and "before" $BS2^+$ *or* O^- on h_L "after" D^+ and "before" $BS2^-$, respectively. And, of course, "after" *both* D^+ and D^-, no such predictions hold.

15.5 Kinematics, Dynamics and the Nature of Lorentz Transformations

The view of QM presented in broad strokes in Section 15.2 is very different from the view embraced and forcefully articulated over many years by the late Asher Peres [46] – a view that I would characterize as instrumentalist, epistemic and information-theoretic. But instead of that broad and deep disagreement, what I will discuss here is the more narrowly focused disagreement over the meaning and nature of Lorentz transformations. Throughout this section we will not be considering any kind of state reductions, neither PSR nor measurement-induced. Thus state vectors here are always FD but not HD, while the Heisenberg picture operators of interest may be HD, but are not FD. (Their expectation values may be FD. See the discussion containing (15.4–15.6).)

It is, perhaps, surprising that physicists can still disagree about the nature of something as basic, in the sense of being elementary, as Lorentz transformations! But it is so. The simplest applications of such transformations are not at issue. Everyone agrees that if M and M′ are two Minkowski coordinate systems for inertial frames, then the assignments of coordinate values in M and M′ to any given space–time point are related by equations of the form

$$x'^{\mu} = \Lambda^{\mu}_{\nu} x^{\nu} + a^{\mu}, \tag{15.20}$$

where the homogeneous coefficients Λ^{μ}_{ν} satisfy

$$\Lambda^{\mu}_{\nu} \eta^{\nu\lambda} \Lambda^{\rho}_{\lambda} = \eta^{\mu\rho}, \tag{15.21}$$

with a diagonal metric tensor that can be chosen as $\eta^{\mu\nu} = (1, -1, -1, -1)$. Agreement continues concerning the transformation rules for so-called local distribution-operator-valued fields. Denoting the general tensor–spinor components of any such field assigned to the space–time point with M coordinates, x, by $\hat{\phi}^{\alpha}(x)$, the general matrix elements must transform as

$$< \Phi'|\hat{\phi}^{\alpha}(x')|\Psi' > = S^{\alpha}_{\beta}(\Lambda) < \Phi|\hat{\phi}^{\beta}(x)|\Psi >, \tag{15.22a}$$

where x' is related to x by (15.20) and $S(\Lambda)$ is a matrix representation of the Lorentz group up to a sign, i.e.,

$$S^{\alpha}_{\beta}(\Lambda_2) S^{\beta}_{\gamma}(\Lambda_1) = \pm S^{\alpha}_{\gamma}(\Lambda_2 \Lambda_1). \tag{15.22b}$$

The state vectors ($|X >$ for M and $|X' >$ for M′), each representing the *relationship* between an objective state of affairs and the Minkowski frame in question, are themselves related by operator-valued unitary representations, up to a sign, of the Poincaré group,

$$|X' >= \hat{U}(\Lambda, a)|X >, \tag{15.23a}$$

where

$$\hat{U}(\Lambda_2, a_2)\hat{U}(\Lambda_1, a_1) = \pm\hat{U}(\Lambda_2\Lambda_1, a_2 + \Lambda_2 a_1). \tag{15.23b}$$

On all this agreement reigns! What unites the preceding examples is the assignment of quantities, coordinates or fields, to individual space–time points.

Disagreement emerges when one considers the assignment of quantities to individual *times*, but without the quantities in question also referring to specified individual space-time points. The quantities may or may not actually *vary* with time, but their assignment is to a definite time without accompanying assignment to a definite position in space. A simple example of such a quantity would be the total 4-momentum of an open subsystem S, $\hat{P}_S^\mu(t)$. The question then arises of the status of the transform of the quantity and what *part* of space–time it is assigned to?

Examples of such quantities are (1) the 4-momenta of open subsystems, which are likely to be time-dependent but cannot be associated with any definite space–time point at a given time by virtue of spacelike extension of the subsystem in question or by virtue of the Heisenberg uncertainty principle, (2) position operators of a system that "locate," at any time, some localizable property of an otherwise spacelike extended system but that can not be associated with any a priori specified space–time point, and (3) marginal probability distributions at specified times where the quantities for which the probabilities are given are marginal in the sense of not being adequate to determine their own dynamical evolution with time.

The view I wish to criticize, while not new, has recently been very well represented in an important paper [47]. While the authors do not explicitly define the view, I will characterize it here as holding (1) that some quantities do not have Lorentz transforms at all; (2) that when it exists, the Lorentz transform of any quantity assigned to instants of time in the original inertial frame, M, will itself be assigned to instants of time in the transformed inertial frame, M'; and (3) that to achieve (2) considerations of dynamical evolution as well as kinematical relations may be admitted to determine the Lorentz-transformed quantity. Peres and Terno adhere to this view in employing a time-dependent canonical formalism rather than a manifestly covariant formalism for discussing Lorentz transformations.

In contrast, the opposing view that I argue for asserts (1) that *any* quantity or any physical state of affairs *whatsoever*, described from the perspective of any Minkowski inertial frame, has Lorentz transforms to the descriptions of that quantity or that same physical state of affairs from the perspectives of *all* such frames; (2) that assignment of a quantity to an instant of time in a Minkowski inertial frame is just a special case, of no special status in principle, of the assignment of quantities to arbitrary spacelike hyperplanes, which hyperplanes do not change under Lorentz transformations but have their parameterization change, since the parameterization of a fixed hyperplane is frame-dependent; and (3) that, as a consequence of (2), the latter, generalized mode of assignment, being invariant in kind under all Lorentz transformations, will never require access to considerations of dynamical evolution for the evaluation of the transformed quantities; Lorentz transformations are purely kinematical.

15.5.1 Existence of Lorentz Transforms

For our first example of the clash between these views, consider the claim of PT that, already within classical theory, a partial or marginal Liouville phase space density function in M cannot determine a Lorentz transform for M' because the integrated variables, which in general will interact with the remaining variables, are required to determine a transformed density function in M'. This conclusion follows from defining the transformed function as associated with a definite time in M' if the original function was so in M. But a purely kinematical transform exists in M' requiring no access to the integrated variables nor any details about their interactions. The kinematical transform is associated with the same spacelike hyperplane that was an instant of time in M but is not so in M'. If the remaining variables are the positions and momenta of N point particles (the q'_n and p'_n are 4-vectors orthogonal to the timelike unit 4-vector, η', while the $\mathbf{q}_n$ and $\mathbf{p}_n$ are spatial 3-vectors), and using the abbreviations $\mathbf{q} := (\mathbf{q}_1, --, \mathbf{q}_N)$, $\mathbf{p} := (\mathbf{p}_1, --, \mathbf{p}_N)$, $q' := (q'_1, --, q'_N)$, $p' := (p'_1, --, p'_N)$, the original partial Liouville density, $L(\mathbf{q}, \mathbf{p}\, t)$, in M has the kinematical Lorentz transform $L'(q', p'; \eta', \tau')$ in M', which satisfies

$$L'(q', p'; \eta', \tau')\Pi\delta(\eta'q')d^4q'\,\Pi\delta(\eta'p')d^4p' = L(\mathbf{q}, \mathbf{p}; t)\Pi d^3\mathbf{q}\Pi d^3\mathbf{p}, \qquad (15.24a)$$

where (in, hopefully, self-explanatory notation), $\eta' = \Lambda[1, \mathbf{0}]$, $\tau' = ct + \eta'a$, $q' = \Lambda[0, \mathbf{q}] + a - \eta'(\eta'a)$ and $p' = \Lambda[0, \mathbf{p}]$. The hyperplane parameterized by (η', τ') in M' is the *same* hyperplane parameterized by $([1, \mathbf{0}], ct)$ in M; i.e., it is the instant of time t in M.

   More generally, if we begin with the partial Liouville density on the hyperplane, (η, τ), in M, $L(q, p; \eta, \tau)$, the kinematical transform in M' (on the same hyperplane) is $L'(q', p'; \eta', \tau')$, where now $\eta' = \Lambda\eta$, $\tau' = \tau + \eta'a$, $q' = \Lambda q + a - \eta'(\eta'a)$, $p' = \Lambda p$ and

$$L'(q', p'; \eta', \tau')\Pi\delta(\eta'q')d^4q'\,\Pi\delta(\eta'p')d^4p' = L(q, p; \eta, \tau)\Pi\delta(\eta q)d^4q\Pi\delta(\eta p)d^4p.$$
$$(15.24b)$$

The change, in M, from any $L(q, p; \eta_1, \tau_1)$ to any other $L(q, p; \eta_2, \tau_2)$ is one of dynamical evolution and, in general, requires access to all the interacting variables for its calculation. Similarly for the change, in M', from any $L'(q', p'; \eta'_1, \tau'_1)$ to any other $L'(q', p'; \eta'_2, \tau'_2)$. But the change from $L(q, p; \eta, \tau)$ in M to the corresponding $L'(q', p'; \eta', \tau')$ in M', given by (15.24b), is purely kinematical and is the proper content of the IHLG transform of L from M to M'.

15.5.2 Spin and Spin Entropy under Kinematical Lorentz Transformations

We turn now from consideration of a denial of the existence of a Lorentz transform (and a second such denial will be encountered here as well) to consideration of a claim of very different behavior under Lorentz transformations for two quantities which have the same tensorial rank, i.e., two 4-vectors. Our system of interest will be closed (originally, for PT, a free single quanton, but all the points to be made here apply to arbitrary closed systems) with self-adjoint generators of the IHLG, $\hat{M}^{\mu\nu}$ and $\hat{P}^{\mu}$. The latter is the total 4-momentum of the system and the former contains the total angular momentum and the total *internal* angular momentum (spin) of the system, which will be our main concern.

By definition, a global 4-vector observable will be represented by a self-adjoint operator, $\hat{V}^{\mu}(\eta, \tau)$, that is assigned to entire individual spacelike hyperplanes (parameterized by (η, τ)) rather than to limited or bounded portions of such hyperplanes. Under arbitrary transformations of the IHLG, the expectation values of such an observable in arbitrary pure states will transform as classical global 4-vectors do, i.e.,

$$< \Psi' | \hat{V}^{\mu}(\eta', \tau') | \Psi > = \Lambda_{\nu}^{\mu} < \Psi | \hat{V}^{\nu}(\eta, \tau) | \Psi >, \qquad (15.25a)$$

where, as above, $\eta' = \Lambda\eta$, $\tau' = \tau + \eta'a$, and $|\Psi' > = \hat{U}(\Lambda, a)|\Psi >$, where

$$\hat{U}(\Lambda, a) = \exp\left((i/\hbar)\hat{P}a\right)\exp\left(-(i/2\hbar)\hat{M}\omega\right) \qquad (15.25b)$$

and, $\Lambda = e^{\omega}$. The transformation, (Λ, a), between the Minkowski frames, M and M' say, relates the expectation values evaluated in the two frames but referring to the *same hyperplane* differently, parameterized by (η, τ) and (η', τ'), respectively. Because of the generality of the state vector, $|\Psi >$, the expectation value equation implies the operator-valued equation,

$$\hat{U}^{\dagger}(\Lambda, a)\hat{V}^{\mu}(\eta', \tau')\hat{U}(\Lambda, a) = \Lambda_{\nu}^{\mu}\hat{V}^{\nu}(\eta, \tau). \qquad (15.25c)$$

Our interest here is the applicability of this transformation equation to the system 4-momentum, $\hat{P}^{\mu}$, the system Pauli–Lubyanski (PL) operator,

$$\hat{W}^{\mu} = -(1/2)\varepsilon^{\mu\alpha\beta\gamma}\hat{M}_{\alpha\beta}\hat{P}_{\gamma} \qquad (15.26)$$

(where $\varepsilon^{0123} = -\varepsilon_{0123} = +1$) and the system spin operator, $\hat{S}^{\mu}(\eta)$, which will be identified shortly. All three of these operators satisfy (15.25a, 15.25c) when substituted for $\hat{V}^{\mu}(\eta, \tau)$. They all transform under the IHLG in exactly the same way. To maximize the distinction between the spin operator on one hand and the 4-momentum and PL operator on the other, we will restrict the discussion to the case of a closed system. Extension of the results to the open system case in which $\hat{P}^{\mu}$ and $\hat{W}^{\mu}$ are HD is straightforward.

But PT identify $\hat{P}^{\mu}$ as a "primary" operator and the system spin as a "secondary" operator which is "entangled" with the 4-momentum and transforms differently (PT do not mention $\hat{W}^{\mu}$ and I am hard put to assess how they would classify it). Besides the profound difference between $\hat{P}^{\mu}$ and $\hat{S}^{\mu}(\eta)$ that one is a linear momentum and the other an internal angular momentum, they also differ in that, for a closed system, $\hat{P}^{\mu}$ is completely independent of hyperplane assignment while $\hat{S}^{\mu}(\eta)$ retains, as we will see, a dependence on the orientation of the hyperplane. But, remembering again the discussion in Section 15.3, this is a difference in how they change (or not) from hyperplane to hyperplane *from the perspective of a given frame*. It does not undermine the identity of the way they transform from frame to frame *on a given hyperplane*! The difference in HD is significant and may well justify the "primary/secondary" classification, but it has nothing to do with Lorentz transformation.

Now just what is the spin operator, $\hat{S}^{\mu}(\eta)$? It is given, in terms of $\hat{P}^{\mu}$ and $\hat{W}^{\mu}$, by

$$\hat{S}^{\mu}(\eta) := \frac{\hat{W}^{\mu}}{\hat{M}} - \frac{\hat{P}^{\mu} + \eta^{\mu}\hat{M}}{\eta\hat{P} + \hat{M}} \frac{\eta\hat{W}}{\hat{M}}, \tag{15.27}$$

where $\hat{M} = |\sqrt{\hat{P}^2}|$. The justification of this identification is provided by the following considerations (in order of increasing importance):

(1) The constraint, $\eta\hat{S}(\eta) = 0$, which follows from (15.27), allows only three independent spacelike components, which, for an instantaneous hyperplane, $\eta^{\mu} = [1, \mathbf{0}]$,[6] yields the familiar form $\hat{S}^{\mu}([1, \mathbf{0}]) = [0, \hat{\mathbf{S}}]$ for a spin operator.

(2) From the commutation relations for the PL operator, $[\hat{W}^{\mu}, \hat{P}^{\nu}] = 0$ and

$$[\hat{W}^{\mu}, \hat{W}^{\nu}] = i\hbar\,\varepsilon^{\mu\nu\alpha\beta}\hat{W}_{\alpha}\hat{P}_{\beta}, \tag{15.28a}$$

we obtain $[\hat{S}^{\mu}(\eta), \hat{P}^{\nu}] = 0$ and

$$[\hat{S}^{\mu}(\eta), \hat{S}^{\nu}(\eta)] = i\hbar\,\varepsilon^{\mu\nu\alpha\beta}\hat{S}_{\alpha}(\eta)\eta_{\beta}. \tag{15.28b}$$

If we now choose ξ_1^{μ}, ξ_2^{μ}, and $\xi_3^{\mu} = \varepsilon^{\mu\alpha\beta\gamma}\xi_{1,\alpha}\xi_{2,\beta}\eta_{\gamma}$ so that $\xi_1^2 = \xi_2^2 = -1$ and $\xi_1\xi_2 = \xi_1\eta = \xi_2\eta = 0$, then we find $\xi_3^2 = -1$, $\xi_1\xi_3 = \xi_3\xi_2 = \xi_3\eta = 0$, and

$$[\xi_1\hat{S}(\eta), \xi_2\hat{S}(\eta)] = i\hbar\xi_3\hat{S}(\eta), \tag{15.29}$$

along with the cyclic permutations thereof. For instantaneous hyperplanes these are just the familiar spin commutation relations for spatial Cartesian components.

(3) The square of the spin operator is equal to the ratio of the square of the PL operator and the square of the 4-momentum. For closed systems this is the ratio of the Casimir invariants for the IHLG and is not HD; i.e.,

$$\hat{S}(\eta)^2 = \hat{S}^2 = \hat{W}^2/\hat{M}^2 \to -\hbar^2 s(s+1), \tag{15.30}$$

the last arrow indicating the value for an irreducible representation of the IHLG.

If a Lorentz transformation were to relate the spin on an instantaneous hyperplane in Bob's frame to the spin on an instantaneous hyperplane in Alice's frame (to use the popular Alice–Bob language employed by PT), then the combined momentum dependence and HD of the spin operator would require access to the momentum structure of the quantum state to calculate the spin transform expectation value. For example, consider Bob's reduced spin density operator,

$$\hat{\rho}_B := Tr_{P/S}(|\Psi_B> < \Psi_B|), \tag{15.31}$$

---

[6] For *any* 4-vector, V^{μ}, $V^{\mu} = [V^0, V]$.

where the subscript, P/S, denotes that the trace is taken over all those dynamical variables, including the total 4-momentum, required to define a system comset, but excluding only the spin, S, and with the spin evaluated at $\eta_B^\mu = [1, \mathbf{0}]$. This quantity would have no Lorentz transform, as PT conceive it, that could be evaluated in Alice's frame, based solely on its value in Bob's frame and the relation between the frames. In particular, PT point out, the spin entropy, $Tr_S(\hat\rho_B \ln \hat\rho_B)$, could have value zero for Bob, due to $|\Psi_B >$ being a spin eigenstate for Bob, while $|\Psi_A >= \hat U(\Lambda_{AB}, a_{AB})|\Psi_B >$ would not be a spin eigenstate at $\eta_A^\mu = [1, \mathbf{0}]$ for Alice and $Tr_S(\hat\rho_A \ln \hat\rho_A)$ could have various values depending on details of the state vectors washed out by the partial tracing.

But the state vector does not lose the spin eigenstate character due to passing from Bob's frame to Alice's. It loses the spin eigenstate character due to dynamical evolution from hyperplanes that are instantaneous in Bob's frame to those that are instantaneous in Alice's frame. This dynamical evolution occurs within Bob's frame alone in going from $\eta_B^\mu = [1, \mathbf{0}]$ to $\eta_B^\mu = (\Lambda_{AB}^{-1})[1, \mathbf{0}]$ and in Alice's frame alone in going from $\eta_A^\mu = \Lambda_{AB}[1, \mathbf{0}]$ to $\eta_A^\mu = [1, \mathbf{0}]$. In any hyperplane for which the state is a spin eigenstate for Bob, it will also be a spin eigenstate for Alice with the same eigenvalue and for the "same" component of spin. To see this suppose that

$$\xi_B \hat S(\eta_B)|\Psi_B >= |\Psi_B > \hbar\mu, \tag{15.32a}$$

where ξ_B is a spacelike unit vector orthogonal to η_B. Then with $\eta_A = \Lambda_{AB}\eta_B$ and $\xi_A = \Lambda_{AB}\xi_B$, we have

$$\begin{aligned}
\xi_A \hat S(\eta_A)|\Psi_A > &= (\Lambda_{AB}\xi_B)\hat S(\Lambda_{AB}\eta_B)\hat U(\Lambda_{AB}, a_{AB})|\Psi_B > \\
&= \hat U(\Lambda_{AB}, a_{AB})(\Lambda_{AB}\xi_B)\hat U^\dagger(\Lambda_{AB}, a_{AB})\hat S(\Lambda_{AB}\eta_B)\hat U(\Lambda_{AB}, a_{AB})|\Psi_B > \\
&= \hat U(\Lambda_{AB}, a_{AB})(\Lambda_{AB}\xi_B)(\Lambda_{AB}\hat S(\eta_B))|\Psi_B >= U(\Lambda_{AB}, a_{AB})\xi_B \hat S(\eta_B)|\Psi_B > \\
&= \hat U(\Lambda_{AB}, a_{AB})|\Psi_B > \hbar\mu = |\Psi_A > \hbar\mu, \tag{15.32b}
\end{aligned}$$

as claimed.

Dynamical evolution from one hyperplane to another changes the spin entropy in both frames. Lorentz transformation between the frames on a given hyperplane does not. Entropy is a Lorentz scalar. Failure to employ the HD language can obscure these matters!

Acknowledgement

I wish to thank Chet Sapalio for valuable discussions concerning aspects of this work.

References

[1] Bell, J.S. (1978), On the hypothesis that the Schrödinger equation is exact, *Epistemological Letters* July, 1–28; Revised version: Quantum mechanics for cosmologists, in (Bell 1987), 117–38.

[2]   Penrose, R., A. Shimony, N. Cartwright and S. Hawking (1997), *The Large, the Small and the Human Mind*, Cambridge University Press, pp. 72–3.

[3]   Bohm, D. (1952), A suggested interpretation of the quantum theory in terms of "hidden" variables, I and II, *Physical Review* **85**, 166–93.

[4]   Everett, H., III (1957), "Relative state" formulation of quantum mechanics, *Reviews of Modern Physics* **29**, 463–5.

[5]   DeWitt, B.S. and N. Graham (eds.) (1973), *The Many Worlds Interpretation of Quantum Mechanics*, Princeton University Press.

[6]   Ghirardi, G.C., A. Rimini and T. Weber (1986), Unified dynamics for microscopic and macroscopic systems, *Physical Review D* **34**, 470–91.

[7]   Fleming, G.N. (1965), Covariant position operators, spin and locality, *Physical Review B* **137**, 188–97.

[8]   Fleming, G.N. (1965), Nonlocal properties of stable particles, *Physical Review B* **139**, 963–8.

[9]   Fleming, G.N. (1966), A manifestly covariant description of arbitrary dynamical variables in relativistic quantum mechanics, *Journal of Mathematical Physics* **7**, 1959–81.

[10] Fleming, G.N. (1985), Towards a Lorentz invariant quantum theory of measurement, in A. Rueda (ed.), *Proceedings of the First Workshop on Fundamental Physics at the University of Puerto Rico* (Universidad of Puerto Rico at Humacao), pp. 8–114; e-print at https://scholarsphere.psu.edu/files/pg15bd999.

[11] Fleming, G.N. (1995), A GHZ argument for a single spinless particle, in D.M. Greenberger and A. Zeilinger (eds.), Fundamental Problems in Quantum Theory, *Annals of the New York Academy of Sciences* **755**, 646–53.

[12] Greenberger, D.M., M. Horne and A. Zeilinger (1989), Going beyond Bell's theorem, in M. Kafatos (ed.), *Bell's Theorem, Quantum Theory and Conceptions of the Universe*, Kluwer Academic Publishers. E-print at arXiv:0712.0921.

[13] Maudlin, T. (2014), What Bell did, *Journal of Physics: Mathematical and Theoretical* **47**, 424010.

[14] Leifer, M.S. (2014), Is the quantum state real? A review of ψ-ontology theorems, arXiv:1409.1570.

[15] Vaidman, L. (2014), Quantum theory and determinism, *Quantum Studies: Mathematics and Foundations* **1**, 5–38, E-print at arXiv:1405.4222v1.

[16] Melamed, Y. and M. Lin (2010), Principle of sufficient reason, in E. N. Zalta (ed.), *The Stanford Encyclopedia of Philosophy* (Summer 2014 ed.), available at http://plato.stanford.edu/archives/sum2014/entries/sufficient-reason/.

[17] Weinberg, S. (2012), Collapse of the state vector, *Physical Review A* **85**, 062116.

[18] Fleming, G.N. (1996). Just how radical is hyperplane dependence?, in R. Clifton (ed.), *Perspectives on Quantum Reality*, Kluwer Academic, pp. 11–28.

[19] Bassi, A., K. Lochan, S. Satin, T.P. Singh and H. Ulbricht (2013), Models of wavefunction collapse, underlying theories and experimental tests, *Reviews of Modern Physics* **85**, 471–528.

[20] Bedingham, D., D. Dürr, G.C. Ghirardi, S. Goldstein, R. Tumulka and N. Zanghi (2014), Matter density and relativistic models of wave function collapse, *Journal of Statistical Physics* **154**, 623–31.

[21] Diósi, L. (1987), A universal master equation for the gravitational violation of quantum mechanics, *Physics Letters A* **120**, 377–81.

[22] Diósi, L. (1989), Models for universal reduction of macroscopic quantum fluctuations, *Physical Review A* **40**, 1165–74.

[23] Diósi, L. (2007), Notes on certain Newton gravity mechanisms of wavefunction localization and decoherence, *Journal of Physics A: Mathematical Theory* **40**, 2989–95.

[24] Penrose, R. (1996), On gravity's role in quantum state reduction, *General Relativity and Gravitation* **28**, 581–600.

[25] Penrose, R. (2009), Black holes, quantum theory and cosmology, *Journal of Physics: Conference Series* **174**, 012001–16.

[26] Hameroff, S. and R. Penrose (2013), Consciousness in the universe: A review of the 'Orch OR' theory, *Physics of Life Review*, http://dx.doi.org/10.1016/j.plrev.2013.08.002.

[27] Schlosshauer, M. (2008), *Decoherence and the Quantum-to-Classical Transition*, Berlin: Springer-Verlag.

[28] Albert, D. and A. Ney (eds.) (2014), *The Wave Function: Essays on the Metaphysics of Quantum Mechanics*, Oxford University Press.

[29] Hardy, L. (1992), Quantum mechanics, local realistic theories, and Lorentz-invariant realistic theories, *Physical Review Letters* **68**, 2981–4.

[30] Aharonov, Y., A. Botero, S. Popescu, B. Reznik and J. Tollaksen (2001), Revisiting Hardy's paradox: Counterfactual statements, real measurements, entanglement and weak values, *Physics Letters A* **301**, 130–38.

[31] Boyer, C. and G.N. Fleming (1974), Quantum field theory on a seven-dimensional homogeneous space of the Poincare group, *Journal of Mathematical Physics* **15**, 1007–24.

[32] Ardalan, F. and G.N. Fleming (1975), A spinor field theory on a seven-dimensional homogeneous space of the Poincaré group, *Journal of Mathematical Physics* **16**, 478–84.

[33] Cacciatori, S., F. Costa and F. Piazza (2009), Renormalized thermal entropy in field theory, *Physical Review D* **79**, 025006.

[34] Piazza, F. and F. Costa (2007), Volumes of space as subsystems, in *Proceedings of Science: From Quantum to Emergent Gravity: Theory and Phenomenology, June 11–15, 2007, Trieste, Italy*, E-print at arXiv:0711.3048v1.

[35] Schuster, P. and Natalia Toro (2013), On the theory of continuous-spin particles: Wavefunctions and soft-factor scattering amplitudes, at arXiv:1302.1198.

[36] Schuster, P. and Natalia Toro (2013), On the theory of continuous-spin particles: Helicity correspondence and forces, at arXiv:1302.1577.

[37] Fleming, G.N. (1989), Lorentz invariant state reduction and localization, in A. Fine and J. Leplin (eds.), *PSA 1988*, Vol. 2, East Lansing: Philosophy of Science Association, pp. 112–26.

[38] Fleming, G.N. (1992), The objectivity and invariance of quantum predictions, in D. Hull, M. Forbes and K. Okruhlik (eds.), *PSA 1992*, pp. 104–13.

[39] Fleming, G.N. (2003), Observations on hyperplanes: I. State reduction and unitary evolution, e-print at http://philsci-archive.pitt.edu/1533/.

[40] Fleming, G.N. and J. Butterfield (1992), Is there superluminal causation in quantum theory?, in A. van der Merwe, F. Selleri and G. Tarozzi (eds.), *Bell's Theorem and the Foundations of Modern Physics*, World Scientific, pp. 203–7.

[41] Schwinger, J. (1948), Quantum electrodynamics, I, *Physical Review* **74**, 1439–61.

[42] Tomonaga, S. (1946), On a relativistically invariant formulation of the quantum theory of wave fields, *Progress of Theoretical Physics* **1**, 27–40.

[43] Jacques, V., E. Wu, F. Grosshans, F. Treussart, P. Grangier, A. Aspect and J.-F. Roch (2007), Experimental realization of Wheeler's delayed-choice gedanken experiment, *Science* **315**, 966–8.

[44] Ma, S., S. Zotter, J. Kofler, R. Ursin, T. Jennewein, C. Bruckner and A. Zeilinger (2012), Experimental delayed-choice entanglement swapping, *Nature Physics* **8**, 479–84.

[45] Gisin, N. (2005), Can relativity be considered complete? From Newtonian nonlocality to quantum nonlocality and beyond, arXiv:quant-ph/0512168.

[46] Peres, A. (1995), *Quantum Theory: Concepts and Methods*, Kluwer Academic Publishers, pp. 249–56.

[47] Peres, A. and D. Terno (2004), Quantum information and relativity theory, *Reviews of Modern Physics* **76**, 93–123.

16

Some Thoughts on Quantum Nonlocality and Its Apparent Incompatibility with Relativity

It may well be that a relativistic version of the theory, while Lorentz invariant and local at the observational level, may be necessarily non-local and with a preferred frame (or aether) at the fundamental level.

– John Bell [1]

16.1 Introduction

Bell's theorem establishes a contradiction between locality (and a few other assumptions) and certain predictions of quantum mechanics [2]. There have been two main issues surrounding the deep implications of this celebrated theorem. The first issue is basic, and it concerns whether the theorem really establishes that nonlocality is a necessary feature of any empirically viable theory and hence a feature of nature itself. One aspect of the issue concerns exactly what the underlying assumptions of Bell's theorem are. For example, some argue that these assumptions include the assumption of counterfactual definiteness [3–5], while others disagree [6–8]. The other aspect of the issue concerns which assumption should be dropped due to the resulting contradiction. Some believe that standard quantum mechanics may avoid nonlocality by denying counterfactual definiteness [3–5].

The second issue is much deeper, and it concerns how to make sense of the strange nonlocality when assuming that Bell's theorem establishes that our world is nonlocal. In particular, it has yet to be determined whether such quantum nonlocality is compatible with the theory of relativity, e.g., whether the nonlocality requires the existence of a preferred Lorentz frame [7, 9, 10]. It seems that "there is a preferred frame of reference, and in this preferred frame of reference things do go faster than light . . . Behind the apparent Lorentz invariance of the phenomena, there is a deeper level which is not Lorentz invariant" [11]. But it also seems possible that the existence of a preferred Lorentz frame is an illusion, since it cannot be detected according to standard quantum mechanics. A more subtle issue is whether quantum nonlocality permits superluminal signaling. It is widely thought that the answers to these questions will lead to a deeper understanding of Bell's theorem and quantum nonlocality.

In this paper, we will present a new analysis of quantum nonlocality and its apparent incompatibility with relativity. First of all, we will give a new, simpler proof of nonlocality in standard quantum mechanics. The proof avoids the controversial assumption of counterfactual definiteness. Next, we will argue that the new proof may imply the existence of a preferred Lorentz frame. After giving a few arguments for the detectability of the frame, we further show that the frame can be detected in a recently suggested model of energy-conserved wave function collapse. Third, we will analyze the possible implications of quantum nonlocality for simultaneity of events. We give two pieces of evidence supporting absolute simultaneity. One is to show that absolute simultaneity is possible within the framework of special relativity. The other is to demonstrate that it is not relative simultaneity but absolute simultaneity that is more suitable for describing nonlocal processes such as the dynamical collapse of the wave function. Last, we will introduce and discuss a possible mechanism for nonlinear quantum evolution and superluminal signaling. Conclusions are given in the last section.

16.2 A New Proof of Nonlocality in Standard Quantum Mechanics

In this section, we will propose a new strategy to prove the existence of nonlocality in standard quantum mechanics. The proof is different from existing proofs of Bell's theorem, and in particular, it may avoid certain controversial assumptions in these proofs, such as the assumption of counterfactual definiteness [7, 10].

Consider an EPRB experiment in which there is an ensemble of two spatially separated spin-1/2 particles (labelled 1 and 2) prepared in a spin singlet state. The spin of particle 1 is measured along two possible directions, a and a', and the spin of particle 2 is always measured along the same direction b. Moreover, the two spin measurements on particles 1 and 2 are spacelike separated. Then, when the spin of particle 1 is measured along direction a, the outcomes of the spin measurements on particle 2 have a certain statistical distribution, such as 1/2 (↑) and 1/2 (↓). Similarly, when the spin of particle 1 is measured along direction a', the outcomes of the spin measurements on particle 2 also have a certain statistical distribution. If these two distributions are different, then there will be superluminal signaling, by which information is transmitted nonlocally and faster than the speed of light and in which there is definitely nonlocality. However, according to the predictions of quantum mechanics, these two distributions for particles 2 are precisely the same. Therefore, it appears that in this case there is no nonlocality, or even though there is nonlocality as required by Bell's theorem, the nonlocality is not as explicit as superluminal signaling.

Our new proof of nonlocality is as follows. According to standard quantum mechanics, after each spin measurement on particle 1 (along direction a or a'), the wave function of particle 2 collapses from the initial entangled state to a definite spin state. The collapsed wave functions are different for different measuring settings a and a'. Then by introducing a possible nonlinear evolution for the wave function of particle 2 such as that suggested in Refs. [12, 13], the two statistical distributions of the measurement outcomes of particle 2 corresponding to the settings a and a' may be different. This leads to superluminal signaling, in which there is definitely nonlocality. Since the introduced nonlinear evolution for particle 2 is *local*, there must exist nonlocality in the original EPRB experiment. Moreover, the nonlocality must assume the same form as the superluminal signaling resulting from the nonlinear evolution.

It should be noted that this proof of nonlocality does not depend on the meaning of the wave function, which is a merit of the proof. No matter if the wave function is epistemic or ontic or something else, it may be changed in a linear way by a usual external potential according to the Schrödinger equation, and it is also possible that it may be changed in a nonlinear way by a special hypothetical external potential.[1] The above proof relies only on the possible existence of the nonlinear evolution of the wave function (besides the collapse postulate).[2] As we will argue later, such nonlinear quantum evolution might also exist in reality.

Certainly, this new proof of the existence of nonlocality in standard quantum mechanics does not establish that our world is nonlocal. But the above result may still be a surprise for some people. For the proof shows that no matter how the wave function is interpreted, nonlocality always exists in standard quantum mechanics. Moreover, one cannot resort to a possible lack of counterfactual definiteness or noncontextuality to avoid the nonlocality.

It is obvious that the above proof also applies to dynamical collapse theories (though it is unnecessary, since in the theories the wave function is ontic and the collapse of the wave function is already a nonlocal physical process). But it is not clear whether the strategy can be applied to other no-collapse quantum theories such as Everett's theory. The difficulty is as follows. Since the entangled wave function of the two particles in the EPRB experiment does not collapse, according to these theories, it seems that in order to change the statistical distribution of the measurement outcomes of particles 2, the hypothetical nonlinear interaction also needs to obtain the information about particle 1. Then such interaction will already be nonlocal, and thus it seems that the strategy cannot apply to these no-collapse quantum theories.

16.3 Existence and Detectability of Preferred Lorentz Frame

If quantum nonlocality indeed exists, then a natural question is whether it is compatible with the theory of relativity, e.g., whether the nonlocality requires the existence of a preferred Lorentz frame. This question seems difficult to answer, and it is still a controversial issue [7, 9, 11]. In this section, we will argue that the above strategy to prove the existence of quantum nonlocality may also help answer this longstanding question.

Let us first consider the above hypothetical case of superluminal signaling. It is noncontroversial that the existence of superluminal signaling is incompatible with the theory of relativity, and it will lead to the existence of a preferred Lorentz frame. If the invariance of the one-way speed of light or standard synchrony is assumed as usual, then superluminal signaling will single out a preferred Lorentz frame, in which the signaling is transferred

---

[1] If the proof depends on the meaning of the wave function, then the possible evolution of the wave function will depend on its interpretation, which seems to be implausible.

[2] It is not necessary that nonlinear quantum evolution must exist in reality. But the proof certainly requires that the evolution is possible in principle, that is, it is not logically and mathematically prohibited.

instantaneously in space. Similarly, if superluminal signaling is transferred instantaneously in every inertial frame, then the one-way speed of light will be not isotropic in all but one inertial frame, and the noninvariance of the one-way speed of light will single out a preferred Lorentz frame, in which the one-way speed of light is isotropic. In the final analysis, the existence of a preferred Lorentz frame is the inevitable result of the combination of the constancy of two-way speed of light and the existence of superluminal signaling. Thus, no matter which assumption or convention is adopted, the preferred Lorentz frame can always be defined as the inertial frame in which the one-way speed of light is isotropic and superluminal signaling is transferred instantaneously in space.

By its definition, the above superluminal signaling is composed of two processes: a nonlocal process and a local subluminal process. Obviously, the local subluminal process does not lead to the existence of a preferred Lorentz frame. Thus the process leading to the existence of the preferred Lorentz frame must be the nonlocal process. This then means that quantum nonlocality also leads to the existence of a preferred Lorentz frame.[3] Here we use again the previous strategy to prove nonlocality.

If a preferred Lorentz frame indeed exists, the next question will be whether the frame can be detected. It is usually thought that the answer to this question is negative. For example, although the de Broglie–Bohm theory and certain dynamical collapse theories assume the existence of a preferred Lorentz frame, the frame is undetectable in these theories. In the following, we will first give a few arguments for the detectability of the preferred Lorentz frame and then show that the frame can be detected in a recently suggested model of energy-conserved wave function collapse [14].

First of all, it should be noted that the no-signaling theorem does not prohibit the detectability of the preferred Lorentz frame. If the preferred Lorentz frame can only be detected by using superluminal signaling, then the no-signaling theorem will indeed prohibit the detectability of the frame. However, there are other possible methods for detecting the preferred Lorentz frame, one of which will be given below.

Next, the detectability of the preferred Lorentz frame is supported by one of our basic scientific beliefs, the so-called minimum ontology. It says that if a certain thing cannot be detected in principle, then it does not exist, whereas if a certain thing does exist, then it can be detected. According to this view, the preferred Lorentz frame should be detectable in principle if it exists. Imagine that there exists some kind of fundamental particles around us, but they cannot be detected in principle. How unbelievable this is!

However, it seems that there are two common objections to this view. First, one might refute this view by resorting to the fact that the objects beyond the event horizon of an observer cannot be detected by the observer. Our answer is that although these objects cannot be detected by the observer, they can be detected locally by other observers. Second, one may refute this view based on the fact that an unknown quantum state cannot be measured. This objection seems to have a certain force, and it probably makes some people believe in

---

[3] For a different point of view, see Chapters 5 and 14 in this volume. In our view, in collapse theories, even if the underlying process such as the flash involves no action at a distance, the resulting collapse of the *nonlocal* wave function does involve action at a distance, which is the origin of quantum nonlocality and preferred Lorentz frame.

the undetectability of the preferred Lorentz frame. But this objection is arguably invalid too. To begin with, a preferred Lorentz frame is a classical system, not a quantum system, and its state of motion is described by a definite velocity, not by a superposed quantum state, while the unknown velocity of a classical system can be measured. Next, although the unknown state of a quantum system cannot be measured, the other definite properties of the system, such as its mass and charge, can still be detected by gravitational and electromagnetic interactions. Moreover, an unknown quantum state being not measurable does not mean that a known quantum state is not measurable either. A known quantum state of a single system can be measured by a series of protective measurements, and even an unknown nondegenerate energy eigenstate can also be measured by protective measurements [15–17].

In the following, we will demonstrate that the preferred Lorentz frame can be detected by measuring the (average) collapse time of the wave function in a model of energy-conserved wave function collapse [14]. According to the model, the (average) collapse time formula for an energy superposition state is

$$\tau_c \approx \frac{\hbar^2}{t_P (\Delta E)^2}, \tag{16.1}$$

where t_P is the Planck time and ΔE is the rms energy uncertainty of the state.[4] We assume that this collapse time formula is still valid in an inertial frame in the relativistic domain. This assumption seems reasonable, as the collapse time formula already contains the speed of light c via the Planck time t_P.[5] Since the formula is not relativistically invariant, its relativistically invariant form must contain a term relating to the velocity of the experimental frame relative to a preferred Lorentz frame. In other words, there must exist a preferred Lorentz frame according to the collapse model. We define the preferred Lorentz frame, denoted by S_0, as the inertial frame in which the above formula is valid. Then in another inertial frame the collapse time will depend on the velocity of the frame relative to S_0. According to the Lorentz transformation, in an inertial frame S' with velocity v relative to the frame S_0 we have

$$\tau_c' = \frac{1}{\sqrt{1 - v^2/c^2}} \cdot \tau_c, \tag{16.2}$$

$$t_P' = \frac{1}{\sqrt{1 - v^2/c^2}} \cdot t_P, \tag{16.3}$$

$$\Delta E' \approx \frac{1 - v/c}{\sqrt{1 - v^2/c^2}} \cdot \Delta E. \tag{16.4}$$

Here we only consider the situation where the particle has very high energy, namely $E \approx pc$, and thus Eq. (16.4) holds. Further, we assume the Planck time t_P is the minimum time in the preferred Lorentz frame, and in another frame the minimum time (i.e., the duration of a

---

[4] As argued in Ref. [14], when energy conservation and the existence of discrete Planck time are assumed, this collapse time formula seems to be an inevitable result.

[5] In contrast, the dynamical collapse theories in which the collapse time formula does not contain c are not directly applicable in the relativistic domain.

discrete instant) is connected with the Planck time t_P by the time dilation formula required by special relativity. Then by inputting these equations into Eq. (16.1), we can obtain the relativistic collapse time formula for an arbitrary experimental frame with velocity v relative to the frame S_0:

$$\tau_c \approx (1 + v/c)^{-2} \frac{\hbar^2}{t_P (\Delta E)^2}. \tag{16.5}$$

This formula contains a factor relating to the velocity of the experimental frame relative to the preferred Lorentz frame. It can be argued that this velocity-dependent factor originates from the relativistic equation of collapse dynamics, which contains a velocity term in order to be relativistically invariant (see [14] for more discussions).

Therefore, according to the energy-conserved collapse model, the collapse time of a given wave function will differ in different inertial frames. For example, considering the maximum difference of the speed of revolution of the Earth with respect to the Sun is $\Delta v \approx 60$ km/s, the maximum difference of collapse times measured at different times (e.g., spring and fall) on the Earth will be $\Delta \tau_c \approx 4 \times 10^{-4} \tau_c$. As a result, the collapse dynamics will single out a preferred Lorentz frame in which the collapse time of a given wave function is longest, and the frame can also be determined by comparing the collapse times of a given wave function in different frames. It may be expected that this preferred Lorentz frame is the CMB-frame in which the cosmic background radiation is isotropic, and the one-way speed of light is also isotropic in this frame.

16.4 Simultaneity: Relative or Absolute?

We have argued that quantum nonlocality requires the existence of a preferred Lorentz frame. This is an implication of nonlocality for relativity of motion. In this section, we will further analyze the possible implications of quantum nonlocality for simultaneity of events.

The relativity of simultaneity has been often regarded as one of the essential concepts of special relativity. However, it is not necessitated by experimental facts but is a result of the choice of standard synchrony.[6] As Einstein already pointed out in his first paper on special relativity, whether or not two spatially separated events are simultaneous depends on the adoption of a convention in the framework of special relativity. In particular, the choice of standard synchrony, which is based on the constancy of one-way speed of light and results in the relativity of simultaneity, is only a convenient convention.

Standard synchrony can be described in terms of the following thought experiment. There are two spatial locations A and B in an inertial frame. Let a light ray, traveling in vacuum, leave A at time t_1 (as measured by a clock at rest there) and arrive at B coincident with the event E at B. Let the ray be instantaneously reflected back to A, arriving at time t_2 (as measured by the same clock at rest there). Then the standard synchrony is defined by saying that E is simultaneous with the event at A that occurred at time $(t_1 + t_2)/2$. This

---

[6] For more discussions about this issue, see Ref. [18] and references therein.

definition is equivalent to the requirement that the one-way speeds of light are the same on the two segments of its round-trip journey between A and B.

According to special relativity, no causal influence can travel faster than the speed of light in vacuum; thus it seems reasonable that any event at A whose time of occurrence is in the open interval between t_1 and t_2 could be defined to be simultaneous with E. However, if there exists quantum nonlocality, which permits the possibility of arbitrarily fast causal influences (at least in the preferred Lorentz frame), then it is arguable that the influences will be able to single out a unique event at A that would be simultaneous with E. Moreover, if there is a causal influence connecting two distinct events, then it seems that these events not being simultaneous will also have a nonconventional basis [18].

In the following, we will discuss two further pieces of evidence supporting absolute simultaneity. One is to show that absolute simultaneity is possible within the framework of special relativity. The other is to demonstrate that it is not relative simultaneity but absolute simultaneity that is more suitable for describing nonlocal processes such as the dynamical collapse of the wave function.

To show the possibility of absolute simultaneity, we need to analyze the general space–time transformation required by the constancy of the two-way speed of light. The transformation is not the familiar Lorentz transformation but the Edwards–Winnie transformation [19, 20]:

$$x' = \eta(x - vt), \tag{16.6}$$

$$t' = \eta[1 + \beta(k + k')]t + \eta[\beta(k^2 - 1) + k - k']x/c, \tag{16.7}$$

where x, t and x', t' are the coordinates of inertial frames S and S', respectively, v is the velocity of S' relative to S, c is the invariant two-way speed of light, $\beta = v/c$, and $\eta = 1/\sqrt{(1 + \beta k)^2 - \beta^2}$. k and k' represent the directionality of the one-way speeds of light in S and S', respectively, and they satisfy $-1 \leqslant k, k' \leqslant 1$. Concretely speaking, the one-way speeds of light along the x and $-x$ directions in S are $c_x = \frac{c}{1-k}$ and $c_{-x} = \frac{c}{1+k}$, respectively, and the one-way speeds of light along the x' and $-x'$ directions in S' are $c_{x'} = \frac{c}{1-k'}$ and $c_{-x'} = \frac{c}{1+k'}$, respectively.

If we adopt the standard synchrony convention, namely assuming that the one-way speed of light is isotropic and constant in every inertial frame, then $k, k' = 0$ and the Edwards–Winnie transformation will reduce to the Lorentz transformation, which leads to the relativity of simultaneity. Alternatively, one can adopt the nonstandard synchrony convention that makes simultaneity absolute. In order to do this, one may first synchronize the clocks at different locations in an arbitrary inertial frame by Einstein's standard synchrony, that is, one may assume that the one-way speed of light is isotropic in this frame, and then let the clocks in other frames be directly regulated by the clocks in this frame when they coincide in space. The corresponding space–time transformation can be derived as follows. Let S be the preferred Lorentz frame in which the one-way speed of light is isotropic; namely, let $k = 0$. Then we get

$$k' = \beta(k^2 - 1) + k = -\beta. \tag{16.8}$$

Moreover, since the synchrony convention leads to the absoluteness of simultaneity, we also have in the Edwards–Winnie transformation

$$\beta(k^2 - 1) + k - k' = 0. \tag{16.9}$$

Thus the space–time transformation that restores absolute simultaneity is

$$x' = \frac{1}{\sqrt{1 - v^2/c^2}} \cdot (x - vt), \tag{16.10}$$

$$t' = \sqrt{1 - v^2/c^2} \cdot t, \tag{16.11}$$

where x, t are the coordinates of the preferred Lorentz frame, x', t' are the coordinates of another inertial frame, and v is the velocity of this frame relative to the preferred frame. In this frame, the one-way speed of light along the x' and $-x'$ directions is $c_{x'} = \frac{c^2}{c-v}$ and $c_{-x'} = \frac{c^2}{c+v}$, respectively.

The above analysis demonstrates the possibility of keeping simultaneity absolute within the framework of special relativity. One can adopt the standard synchrony that leads to the relativity of simultaneity, and one can also adopt the nonstandard synchrony that restores the absoluteness of simultaneity. This is permitted because there is no causal connection between two spacelike separated events in special relativity. Here one may argue that the definition of standard synchrony makes use only of the relation of equality (of the one-way speeds of light in different directions), so that simplicity dictates its choice. However, even in the framework of special relativity, since the equality of the one-way speed of light is a convention, this choice does not simplify the postulational basis of the theory but only gives a symbolically simpler representation [21]. On the other hand, as we will demonstrate below, when going beyond the framework of special relativity and considering quantum nonlocality, standard synchrony is not simple but complex and will lead to serious distortions in describing nonlocal processes such as the dynamical collapse of the wave function.

Consider a particle that is in a superposition of two Gaussian wavepackets $\frac{1}{\sqrt{2}}(\psi_1 + \psi_2)$ in an inertial frame S. The centers of the two wavepackets are located in x_1 and x_2 ($x_1 < x_2$), respectively, and the width of each wavepacket is much smaller than the distance between them. A projective measurement may randomly collapse this superposition state to ψ_1 or ψ_2 with the same probability 1/2. Suppose the collapse happens at different locations at the same time in S. This means that when the superposition state collapses to the branch ψ_1 near position x_1, the other branch ψ_2 near position x_2 will disappear simultaneously.

Now let us analyze the above collapse process in another inertial frame S' with velocity v relative to S. Suppose the superposition state $\frac{1}{\sqrt{2}}(\psi_1 + \psi_2)$ collapses to the branch ψ_1 near position x_1 at instant t in S. This process contains two events happening simultaneously in two spatially separated regions. One event is the disappearance of the branch $\frac{1}{\sqrt{2}}\psi_2$ near position x_2 at instant t, and the other is the change from $\frac{1}{\sqrt{2}}\psi_1$ to ψ_1 happening near position x_1 at instant t.[7] According to the Lorentz transformation for standard simultaneity, the times

---

[7] Strictly speaking, since the collapse time is not zero, these events happen not at a precise instant but during a very short time, which may be much shorter than the time of light propagating between x_1 and x_2.

of occurrence of these two events in S' are

$$t_1' = \frac{t - x_1 v/c^2}{\sqrt{1 - v^2/c^2}}, \tag{16.12}$$

$$t_2' = \frac{t - x_2 v/c^2}{\sqrt{1 - v^2/c^2}}. \tag{16.13}$$

It can be seen that $x_1 < x_2$ leads to $t_1' > t_2'$. Then during the period between t_1' and t_2', the branch $\frac{1}{\sqrt{2}}\psi_2'$ near position x_2' has already disappeared, but the branch $\frac{1}{\sqrt{2}}\psi_1'$ near position x_1' has not changed to ψ_1'. This means that at any instant between t_1' and t_2', there is only a non-normalized state $\frac{1}{\sqrt{2}}\psi_1'$. Similarly, if the superposition state $\frac{1}{\sqrt{2}}(\psi_1 + \psi_2)$ collapses to the branch ψ_2 near position x_2 at instant t in frame S, then in S', during the period between t_1' and t_2', the branch $\frac{1}{\sqrt{2}}\psi_2'$ near position x_2' has already turned to ψ_2', while the branch $\frac{1}{\sqrt{2}}\psi_1'$ near position x_1' has not disappeared and is still there. Therefore, there is only a non-normalized state $\frac{1}{\sqrt{2}}\psi_1' + \psi_2'$ at any instant between t_1' and t_2'.

However, although the state of the particle in S' is not normalized, the total probability of *detecting* the particle in the whole space is still 1, not 1/2 or 3/2, in the frame. In other words, although the collapse process is seriously distorted in S', the distortion cannot be measured. The reason is that in S' the collapse process happens at different instants in different locations,[8] and the superposition of the branches in these locations and at these instants is always normalized. In the following, we will give a more detailed explanation.

As noted above, in frame S' the collapse first happens at t_2' for the branch $\frac{1}{\sqrt{2}}\psi_2'$ near position x_2', and then happens at t_1' for the branch $\frac{1}{\sqrt{2}}\psi_1'$ near position x_1', after a delay. If we measure the branch $\frac{1}{\sqrt{2}}\psi_2'$, then the resulting collapse will influence the other branch $\frac{1}{\sqrt{2}}\psi_1'$ only after a delay of $\Delta t' = \frac{|x_1 - x_2|v/c^2}{\sqrt{1 - v^2/c^2}}$, while if we measure the branch $\frac{1}{\sqrt{2}}\psi_1'$, then the resulting collapse will influence the other branch $\frac{1}{\sqrt{2}}\psi_2'$ in advance by the same time interval $\Delta t'$, and the influence is backward in time. Now suppose we make a measurement on the branch $\frac{1}{\sqrt{2}}\psi_2'$ near position x_2' and detect the particle there (i.e., the collapse state is ψ_2'). Then before the other branch $\frac{1}{\sqrt{2}}\psi_1'$ disappears, which happens after a delay of $\Delta t'$, we can make a second measurement on this branch near position x_1'. It seems that the probability of detecting the particle there is not zero but 1/2, and thus the total probability of finding the particle in the whole space is greater than one and it is possible that we can detect two particles. However, this is not the case. Although the second measurement on the branch $\frac{1}{\sqrt{2}}\psi_1'$ near position x_1' is made later than the first measurement, it is the second measurement that collapses the superposition state $\frac{1}{\sqrt{2}}(\psi_1' + \psi_2')$ to ψ_2' near position x_2'; the local branch $\frac{1}{\sqrt{2}}\psi_1'$ near position x_1' disappears immediately after the measurement, while the influence of the resulting collapse on the other branch $\frac{1}{\sqrt{2}}\psi_2'$ near position x_2' is backward in time and happens before the first measurement on this branch. Therefore, the second measurement near position x_1' must obtain a null result, and the reason the first measurement detects the

[8] Concretely speaking, the time order of the collapses happening at different locations in S' is connected with that in S by the Lorentz transformation.

particle near position x_2' is that the superposition state already collapses to ψ_2' near position x_2' before the measurement due to the second measurement.

To sum up, we have demonstrated that the picture of the dynamical collapse of the wave function is seriously distorted by the Lorentz transformation (in all but the preferred frame), though the distortion is unobservable in principle. The crux of the matter lies in the relativity of simultaneity. Only when simultaneity is absolute can the picture of wave function collapse be kept perfect in every inertial frame. This provides another support for absolute simultaneity.

Here one might further argue that quantum nonlocality requires absolute simultaneity, i.e., that when there is an arbitrarily fast causal influence connecting two spacelike separated events, then simultaneity must be absolute and nonconventional. In our opinion, this view seems to be problematic. We agree that if there is an arbitrarily fast causal influence connecting two spacelike separated events, then these two events will be simultaneous and simultaneity will be nonconventional. But we think this argument may hold true only in the preferred Lorentz frame. For it is still possible that in other frames these two events are not simultaneous and the nonsimultaneity is merely an illusion resulting from standard synchrony. Moreover, we think the time order of two events, causally related or otherwise, need not be invariant, and it may be physically meaningless.

16.5 A Possible Mechanism for Superluminal Signaling

Our new strategy to prove quantum nonlocality depends on the possible existence of superluminal signaling. It has been an intriguing and deep question whether superluminal signaling exists in reality. It is widely thought that the no-signaling theorem in quantum mechanics strongly suggests or even implies a negative answer to this question. Admittedly, however, this is still an unsolved issue. For one, although superluminal signaling has not been discovered, no firm evidence indicates its nonexistence either (and it seems that one can never verify its nonexistence by evidence). John Bell, in his last paper, 'La nouvelle cuisine' [22], also expresses his worries about the fundamentality of the no-signaling theorem in standard quantum mechanics. He thinks the theorem suffers from a lack of conceptual sharpness, since the involved notions of measurement and preparation contain an anthropocentric element. In this section, we will argue that a further analysis of the anthropocentric element may indeed make the theorem invalid, since it might lead us to a possible mechanism for nonlinear quantum evolution and superluminal signaling.

Imagine that a microscopic particle, like a conscious observer, also has conscious awareness and can be aware of the change of its state. Then, in the above EPRB experiment, particle 2 will be able to know the collapse of its state and identify the nonlocal influence introduced by the measuring device near particle 1. In other words, particle 2 will be able to distinguish nonorthogonal states with certainty and decode the superluminally transferred information. Moreover, if particle 2 can manifest its awareness by an action detectable by a measuring device, then superluminal signaling can also be achieved (by us). Certainly, it is a commonsense assumption that microscopic particles have no conscious awareness. However, when particle 2 is replaced with a conscious observer, it seems that a similar argument for superluminal signaling still exists [23, 24]. In the following, we will introduce and discuss this result.

Let ψ_1 and ψ_2 be the physical states of two definite perceptions of a conscious being, and $\psi_1 + \psi_2$ the quantum superposition of these two states. For example, ψ_1 and ψ_2 are triggered respectively by small numbers of photons with a certain frequency entering into the eyes of the conscious being from two directions, and $\psi_1 + \psi_2$ is triggered by the superposition of these two input states (see also [25]). Suppose the conscious being satisfies the slow collapse condition that the collapse time of the superposition state $\psi_1 + \psi_2$, denoted by t_c, is longer than the conscious time t_p of the conscious being for forming the perceptions ψ_1 or ψ_2, and the time difference is large enough for him to identify. This condition ensures that consciousness can take part in the process of wave function collapse; otherwise consciousness can only appear after the collapse and will surely have no effects during the collapse process.

It can be argued that under a reasonable assumption the conscious being can distinguish with certainty the definite perception state ψ_1 or ψ_2 from the superposition state $\psi_1 + \psi_2$ and can thus achieve superluminal signaling. The assumption is that a definite conscious perception appears only after the superposition state $\psi_1 + \psi_2$ collapses into ψ_1 or ψ_2. Under this assumption, only after the collapse time t_c can the conscious being have a definite perception for the superposition state $\psi_1 + \psi_2$, while the conscious being can have a definite perception after the conscious time t_p for the states ψ_1 and ψ_2. Since the conscious being satisfies the slow collapse condition and can distinguish the times t_p and t_c, he can distinguish with certainty the definite perception state ψ_1 or ψ_2 from the superposition state $\psi_1 + \psi_2$. A similar argument was first given by Squires [26].

We can also give a reduction-to-absurdity argument that does not depend on the above assumption [23, 24]. It is well known that if the mixture of ψ_1 and ψ_2 with probabilities $1/2$ and $1/2$ and the mixture of $\psi_1 + \psi_2$ and $\psi_1 - \psi_2$ with probabilities $1/2$ and $1/2$ can be distinguished with nonzero probability, then superluminal signaling can be achieved by a EPRB setting. Now assume that a conscious being, who satisfies the slow collapse condition, cannot distinguish these two mixtures with nonzero probability. This requires that for the superposition state $\psi_1 + \psi_2$ or $\psi_1 - \psi_2$ the conscious being must have a perception corresponding to ψ_1 or ψ_2 immediately after the conscious time t_p, and moreover, the perception must be exactly the same as his perception corresponding to the result of the collapse of the superposition state. Since the conscious being satisfies the slow collapse condition and the conscious time t_p is shorter than the collapse time t_c, this requirement means that the conscious being knows the collapse result beforehand. This is impossible due to the essential randomness of the collapse process.

The above result relies on a very stringent condition, the slow collapse condition, which says that for a conscious being the collapse time of a superposition of his definite conscious perceptions is longer than his normal conscious time. Whether this condition is readily available for human brains depends on concrete models of consciousness and wave function collapse. However, it should be pointed out that the collapse time of a single superposition state is an essentially stochastic variable, which value can range between zero and infinity. As a result, the slow collapse condition can always be satisfied with nonzero probability in some collapse events, in which the collapse time of the single superposition state is much longer than the (average) collapse time and the normal conscious time. This means that the above mechanism for superluminal signaling is available in principle.

Last, we discuss a potential problem of the above mechanism for superluminal signaling. It can be seen that when able to distinguish between two nonorthogonal states, consciousness will introduce one kind of nonlinear quantum evolution. The nonlinearity is definite, not stochastic (cf. [27]). This is not unexpected, since it has been shown that a small nonlinearity in quantum dynamics permits superluminal communication [13]. However, nonlinear quantum mechanics has a general characteristic, namely that the description of composite systems depends on a particular basis in a Hilbert space. This is a potential difficulty for the theory because it may make such theories inconsistent. Although it does not seem impossible to solve the problem by some other method, consciousness may help avoid this difficulty more directly. The reason is that the consciousness of an observer will naturally select a privileged basis in its state space; for physically definite perception states there is a one-to-one correspondence between the physical state and its conscious content, while for superpositions of physically definite perception states there is none. In other words, a conscious observer can be aware of the content of physically definite perceptions, but can be aware of none of the content of these perceptions when in a superposition of them.

16.6 Conclusions

The concept of superluminal signaling has been useful in analyzing the fundamental problems of quantum mechanics by its negation, the no-signaling theorem. In this paper, we have argued that an analysis of the concept itself might also lead to a deeper understanding of quantum nonlocality and its apparent incompatibility with special relativity. First of all, we give a new, simpler proof of nonlocality in standard quantum mechanics based on the possibility of superluminal signaling. The proof avoids certain controversial assumptions in existing proofs, such as counterfactual definiteness. Next, we argue that the new proof may imply the existence of a preferred Lorentz frame. After arguing for the detectability of the frame, we further show that the frame can be detected by measuring the collapse time of the wave function according to a recently suggested model of energy-conserved wave function collapse. In addition, we analyze the possible implications of quantum nonlocality for simultaneity of events by giving two arguments supporting absolute simultaneity. Last, we also introduce and discuss a possible mechanism for superluminal signaling. Whether these speculative analyses are valid needs to be further examined.

Acknowledgements

I wish to thank Sheldon Goldstein and Richard Healey for helpful discussions. This work was partly supported by the Top Priorities Program of the Institute for the History of Natural Sciences, Chinese Academy of Sciences, under Grant Y45001209G.

References

[1] J.S. Bell, Quantum mechanics for cosmologists, in C. Isham, R. Penrose, and D. Sciama (eds.), *Quantum Gravity* 2 Clarendon Press, Oxford (1981), pp. 611–37.

[2] J.S. Bell, On the Einstein–Podolsky–Rosen paradox, *Physics* **1** (1964), 195–200.

[3] A. Peres, Unperformed experiments have no results, *Am. J. Phys.* **46**(7) (1978), 745.

[4] W.M. de Muynck, W. De Baere and H. Martens, Interpretations of quantum mechanics, joint measurement of incompatible observables, and counterfactual definiteness, *Found. Phys.* **24** (1994), 1589–664.

[5] M. Zukowski and C. Brukner, Quantum non-locality – It ain't necessarily so ..., *J. Phys. A: Math. Theor.* **47** (2014), 424009.

[6] T. Norsen, Bell locality and the nonlocal character of nature, *Found. Phys. Lett.*, **19**(7) (2006), 633–55.

[7] S. Goldstein, T. Norsen, D. Tausk and N. Zanghí, Bell's Theorem, *Scholarpedia* **6**(10), 8378

[8] T. Maudlin, What Bell did, *J. Phys. A: Math. Theor.* **47** (2014), 424010

[9] T. Maudlin, *Quantum Non-locality and Relativity*, Blackwell, Cambridge, 1st ed., 1994, 3rd ed., 2011.

[10] A. Shimony, Bell's theorem, in Edward N. Zalta (ed.), *The Stanford Encyclopedia of Philosophy* (Winter 2013 Edition), http://plato.stanford.edu/archives/win2013/entries/bell-theorem.

[11] J.S. Bell, in: P. Davies and J. Brown (eds.), *The Ghost in the Atom*, Cambridge University Press, Cambridge, 1986.

[12] S. Weinberg, Testing quantum mechanics, *Ann. Phys.* **194** (1989), 336.

[13] N. Gisin, Weinberg's non-linear quantum mechanics and superluminal communication, *Phys. Lett. A* **143** (1990), 1–2.

[14] S. Gao, A discrete model of energy-conserved wave function collapse, *Proc. R. Soc. A* **469** (2013), 20120526.

[15] Y. Aharonov and L. Vaidman, Measurement of the Schrödinger wave of a single particle, *Phys. Lett. A* **178** (1993), 38.

[16] Y. Aharonov, J. Anandan and L. Vaidman, Meaning of the wave function, *Phys. Rev. A* **47** (1993), 4616.

[17] S. Gao (ed.), *Protective Measurement and Quantum Reality: Towards a New Understanding of Quantum Mechanics*, Cambridge University Press, Cambridge, 2014.

[18] A. Janis, Conventionality of Simultaneity, in Edward N. Zalta (ed.), *The Stanford Encyclopedia of Philosophy* (Fall 2014 Edition), http://plato.stanford.edu/archives/fall2014/entries/spacetime-convensimul/.

[19] W.F. Edwards, Special relativity in anisotropic space, *Am. J. Phys.* **31** (1963), 482–9.

[20] J. Winnie, Special relativity without one-way velocity assumptions: I and II, *Philosophy of Science* **37** (1970), 81–99, 223–38.

[21] A. Grünbaum, *Philosophical Problems of Space and Time* (Boston Studies in the Philosophy of Science, Vol. 12), 2nd enlarged ed. Reidel, Dordrecht/Boston, 1973.

[22] J.S. Bell, La nouvelle cuisine, in A. Sarlemihn and P. Kroes (eds.), *Between Science and Technology*, Elsevier Science, North-Holland, 1990, pp. 97–115.

[23] S. Gao, Quantum collapse, consciousness and superluminal communication, *Found. Phys. Lett.* **17** (2004), 167–82.

[24] S. Gao, On the possibility of nonlinear quantum evolution and superluminal communication, *International Journal of Modern Physics: Conference Series* **22** (2014), 1–6.

[25] G.C. Ghirardi, Quantum superpositions and definite perceptions: Envisaging new feasible experimental tests, *Phys. Lett. A* **262** (1999), 1–14.

[26] E.J. Squires, Explicit collapse and superluminal signaling, *Phys. Lett. A* **163** (1992), 356–8.

[27] E.P. Wigner, *Symmetries and Reflections*, Indiana University Press, Bloomington and London, 1967.

17

A Reasonable Thing That Just Might Work

DANIEL ROHRLICH

Abstract

In 1964, John Bell proved that quantum mechanics is "unreasonable" (to use Einstein's term): there are nonlocal bipartite quantum correlations. But they are not the most nonlocal bipartite correlations consistent with relativistic causality ("no superluminal signalling"); maximally nonlocal "superquantum" (or "PR-box") correlations are also consistent with relativistic causality. I show that – unlike quantum correlations – these correlations do not have a classical limit consistent with relativistic causality. The generalization of this result to all stronger-than-quantum nonlocal correlations is a derivation of Tsirelson's bound – a theorem of quantum mechanics – from the three axioms of relativistic causality, nonlocality, and the existence of a classical limit. But is it reasonable to derive (a part of) quantum mechanics from the unreasonable axiom of nonlocality?! I consider replacing the nonlocality axiom with an equivalent axiom that even Bell and Einstein might have considered reasonable: an axiom of local retrocausality.

In 1964, John Bell [1] proved that quantum mechanics is "unreasonable," as defined by Einstein, Podolsky and Rosen [2] in 1935: "No reasonable definition of reality could be expected to permit this." "This" (i.e., violation of "Einstein separability," to use a technical term, or "spooky action at a distance," as Einstein put it) turns out to be endemic to quantum mechanics. For example, pairs of photons measured at spacelike separations may yield nonlocal quantum correlations, i.e., correlations that cannot be traced to any data or "programs" the photons carry with them. As Bell [3] put it 20 years later, "For me, it is so reasonable to assume that the photons in those experiments carry with them programs, which have been correlated in advance, telling them how to behave. This is so rational that I think that when Einstein saw that, and the others refused to see it, *he* was the rational man. The other people, although history has justified them, were burying their heads in the sand. I feel that Einstein's intellectual superiority over Bohr, in this instance, was enormous; a vast gulf between the man who saw clearly what was needed, and the obscurantist. So for me, it is a pity that Einstein's idea doesn't work. The reasonable thing just doesn't work." True, history has confirmed nonlocal quantum correlations; but has history passed judgment on Einstein and "the others"? Consider what Newton [4] wrote about his own theory of gravity:

"That gravity should be innate, inherent and essential to matter so that one body may act upon another at a distance through a vacuum without the mediation of anything else, by and through which their action or force may be conveyed from one to another, is to me so great an absurdity that I believe no man who has in philosophical matters any competent faculty of thinking can ever fall into it." More than two centuries years later, Einstein confirmed Newton's misgivings: gravitational interactions are indeed local. Einstein's theory of gravity is free of the "absurdity" of action at a distance. Well, if it took centuries for history to justify Newton's rejection of action at a distance, couldn't history yet justify Einstein's rejection of action at a distance?

This paper offers, not *the* reasonable thing, but *a* reasonable thing that just might work. Section 17.1 reviews the search for simple physical axioms from which to derive quantum mechanics. Ideally, such a search could help us understand the theory and make it seem more reasonable. But, while we can derive part of quantum mechanics from three simple physical axioms, one of the three axioms is the unreasonable axiom of nonlocality! Apparently, we are no better off than before. However, Section 17.2 considers replacing the axiom of nonlocality with an axiom that even Bell and Einstein might have considered reasonable: an axiom of local retrocausality (microscopic time-reversal symmetry). Then Section 17.3 rewrites the derivation in Section 17.1 using the three axioms of causality, local retrocausality, and the existence of a classical limit.

17.1 Nonlocal Correlations in the Classical Limit

It is convenient to discuss Bell's inequality in the form derived by Clauser, Horne, Shimony and Holt [5] (CHSH) for spacelike separated measurements by "Alice" and "Bob" on a bipartite system:

$$|C(a, b) + C(a, b') + C(a', b) - C(a', b')| \leq 2, \tag{17.1}$$

where a, a', b and b' are observables with eigenvalues in the range $[-1, 1]$; Alice measures a or a' in her laboratory, Bob measures b or b' in his laboratory, and the correlations $C(a, b)$, etc., emerge from their measurements. Correlations that violate Eq. (17.1) by *any* amount are nonlocal. But it is a curious fact, discovered by Tsirelson [6], that the violation of Eq. (17.1) by quantum correlations $C_Q(a, b)$, etc., is bounded by $2\sqrt{2}$:

$$\left|C_Q(a, b) + C_Q(a, b') + C_Q(a', b) - C_Q(a', b')\right| \leq 2\sqrt{2}, \tag{17.2}$$

even though it is straightforward to define "superquantum" correlations

$$C_{SQ}(a, b) = C_{SQ}(a, b') = C_{SQ}(a', b) = 1 = -C_{SQ}(a', b') \tag{17.3}$$

that violate Eq. (17.1) maximally. A good guess is that superquantum (or "PR-box" [7]) correlations are too strong to be consistent with relativistic causality, but this guess is easily disproved [7, 8]. Just assume that when Alice measures a or a' she gets ± 1 with equal probability, and likewise when Bob measures b or b'; this assumption is consistent with

Eq. (17.3), and it implies that Alice and Bob cannot signal to each other, since in any case Alice and Bob obtain ± 1 with equal probability from their measurements.

Nonlocal quantum correlations are unreasonable, and not even maximally unreasonable!

But perhaps we should not be so surprised that PR-box correlations are consistent with relativistic causality. After all, we have set up quite an artificial comparison. We have not compared two theories. We have compared nonlocal quantum correlations belonging to a complete theory – quantum mechanics – with ad hoc super-duper nonlocal correlations that do not belong to any theory we know. We are not even comparing apples and oranges. We are comparing a serial Nobel prize winner and a lottery winner. Quantum mechanics, as a complete theory, is subject to constraints. In particular, quantum mechanics has a classical limit. In this limit there are no complementary observables; there are only macroscopic observables, all of which are jointly measurable. This classical limit – our direct experience – is an inherent constraint, a kind of boundary condition, on quantum mechanics and on any generalization of quantum mechanics. Thus stronger-than-quantum correlations, too, must have – as a minimal requirement – a classical limit.

And now the fun begins [9]. Consider the PR box and note that if Alice measures a and obtains 1, she can predict with certainty that Bob will obtain 1 whether he measures b or b'; if she obtains -1, she can predict with certainty that he will obtain -1 whether he measures b or b'. (In contrast, quantum correlations would allow Alice to predict with certainty only the result of measuring b or the result of measuring b' but not both [2].) If Alice measures a', she can predict with certainty that Bob will obtain her result if he measures b and the opposite result if he measures b'. Thus, all that protects relativistic causality is the (tacitly assumed) complementarity between b and b': Bob cannot measure both, although – from Alice's point of view – no uncertainty principle governs b and b'.

Next, suppose that Alice measures a or a' consistently on N pairs. Let us define macroscopic observables B and B':

$$B = \frac{b_1 + b_2 + \cdots + b_N}{N}, \quad B' = \frac{b'_1 + b'_2 + \cdots + b'_N}{N}, \tag{17.4}$$

where b_m and b'_m represent b and b', respectively, on the mth pair. Alice already knows the values of B and B', and there must be *some* measurements that Bob can make to obtain partial information about *both* B and B'; for, in the classical limit, there can be no complementarity between B and B'. Now it is true that $a = 1$ and $a = -1$ are equally likely, and so the average values of B and B' vanish, whether Alice measures a or a'. But if she measures a on each pair, then typical values of B and B' will be $\pm 1/\sqrt{N}$ (but possibly as large as ± 1) and correlated. If she measures a' on each pair, then typical values of B and B' will be $\pm 1/\sqrt{N}$ (but possibly as large as ± 1) and *anti*correlated. Thus Alice can signal a single bit to Bob by consistently choosing whether to measure a or a'. This claim is delicate because the large-N limit in which B and B' commute is also the limit that suppresses the fluctuations of B and B'. We cannot make any assumption about the *approach* to the classical limit; all that we assume is that it exists, e.g., that the uncertainty product $\Delta B \Delta B'$ can be made as small as desired, for large enough N. On the other hand, the axiom of relativistic causality cannot grant Bob even the slightest indication about both B and B'. Hence all we need is that when Bob detects a correlation, it is more likely that Alice measured a than when he detects an anticorrelation. If it were not more likely, it would mean that Bob's measurements yield zero information about B or about B', contradicting the fact that there is a classical limit in which B and B' are jointly measurable.

To ensure that Bob has a good chance of measuring B and B' accurately enough to determine whether they are correlated or anticorrelated, N may have to be large and therefore the fluctuations in B and B' will be small. However, Alice and Bob can repeat this experiment (on N pairs at a time) as many times as it takes to give Bob a good chance of catching and measuring large enough fluctuations. Alice and Bob's expenses and exertions are not our concern. Relativistic causality does not forbid superluminal signalling only when it is cheap and reliable. Relativistic causality forbids superluminal signalling altogether.

For example, let us suppose Bob considers only those sets of N pairs in which $B = \pm 1$ and $B' = \pm 1$. The probability of $B = 1$ is 2^{-N}. But if Alice is measuring a consistently, the probability of $B = 1$ *and* $B' = 1$ is also 2^{-N}, and not 2^{-2N}, while the probability of $B = 1$ and $B' = -1$ vanishes. If Alice is measuring a' consistently, the probabilities are reversed. (These probabilities must be folded with the scatter in Bob's measurements, but the scatter is independent of what Alice measures.) Thus, with unlimited resources, Alice can send a (superluminal) signal to Bob. Superquantum (PR-box) correlations are *not* consistent with relativistic causality in the classical limit.

We have ruled out superquantum correlations [9]. To derive quantum correlations, however, we have to rule out all stronger-than-quantum correlations; i.e., we have to derive Tsirelson's bound from the three axioms of nonlocality, relativistic causality, and the existence of a classical limit. The proof appears elsewhere [10].

The existence of a classical limit is not the only axiom we can consider adding to the axioms of nonlocality and relativistic causality. Alternative axioms [11] (or a stronger axiom of relativistic causality called "information causality" [12]) have been shown to rule out PR-box correlations, and come close to ruling out all stronger-than-quantum correlations. However, the physical significance of these axioms requires clarification. Navascués and Wunderlich [13] consider an axiom for a classical limit, but define the classical limit via the "wiring" [14] of entangled systems, and not via complementary measurements that become jointly measurable as the number N of systems grows without bound.

17.2 Local Retrocausality as an Axiom

Bell's theorem rules out any locally causal account of quantum mechanics. But a number of authors, most notably Price [15], have suggested a locally causal–retrocausal account of quantum mechanics. Here "causality" means "relativistic causality" as before (i.e., no superluminal signalling); what is "retro" is that the effect precedes the cause. If the retrocausality is *local* – no action at a distance – then the order of cause and effect is independent of the reference frame. Retrocausality is an expression of a fundamental time-reversal symmetry in physics. While time-reversal symmetry is not manifest at the macroscopic

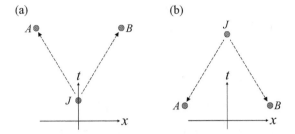

Figure 17.1 Configurations in which Jim can (a) causally and (b) retrocausally put pairs of particles shared by Alice and Bob in product or entangled states, as he chooses. The dashed arrows connect cause with effect.

level – for example, a star emits more light than it absorbs – we explain the asymmetry by saying that the universe has not reached a state of maximum entropy. At the same time, almost all fundamental physical processes at the microscopic level exhibit time-reversal symmetry. Aharonov, Bergmann and Lebowitz (ABL) [16] derived an explicitly time-symmetric formula for intermediate quantum probabilities, conditioned on initial and final states; they suggested that quantum mechanics has no arrow of time of its own and that time asymmetry (e.g., in measurements) originates in macroscopic physics. While the ABL formula is not manifestly local, it opens the way to a local account of quantum mechanics via local retrocausality. Such an account would replace nonlocality with something not only local, but even palatable: a fundamental time-reversal symmetry of microscopic causality and retrocausality. Moreover, if the account includes the quantum correlations that violate Bell's inequality, we can replace the axiom of nonlocality assumed in Section 17.1 with an axiom of local retrocausality, and try to derive quantum mechanics from the three axioms of (relativistic) causality, local retrocausality, and the existence of a classical limit. Section 17.3 begins such a derivation.

Remarkably, retrocausality is intrinsic to quantum mechanics, as we see if we consider three observers, Alice, Bob and Jim, who share an ensemble of triplets of spin-1/2 particles in the Greenberger, Horne and Zeilinger (GHZ) [17] state $|GHZ\rangle = (|\uparrow_A\uparrow_B\uparrow_J\rangle - |\downarrow_A\downarrow_B\downarrow_J\rangle)/\sqrt{2}$. (See Fig. 17.1; Alice, Bob and Jim each get one particle in each triplet.) Let Alice and Bob, at space–time points A and B, measure spin components $\sigma^{(A)} \cdot \hat{\mathbf{n}}_A$ and $\sigma^{(B)} \cdot \hat{\mathbf{n}}_B$, respectively, on their particles. For simplicity, let the unit vectors $\hat{\mathbf{n}}_A$ and $\hat{\mathbf{n}}_B$ (which may change from particle to particle) lie in the xy-plane. Let Jim have the special role of the "jammer" [18]; he chooses whether to put the particles held by Alice and Bob in a product state or an entangled state. To put them in a product state, Jim (at spacetime point J) measures $\sigma_z^{(J)}$, the z-component of the spin of his particles. To put them in an entangled state, he measures, say, $\sigma_x^{(J)}$. It does not matter when Jim makes his measurements. In Fig. 17.1(a), Jim's measurements precede Alice's and Bob's by a timelike separation; but in Fig. 17.1(b), Alice's and Bob's measurements precede Jim's by a timelike separation. Either way, Jim cannot send a superluminal signal to Alice and Bob, because his measurements leave the pairs held by Alice and Bob in a mixed state – either a mixture of the product

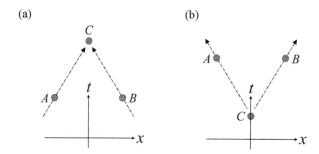

Figure 17.2 (a) Alice and Bob independently measure (prepare) spin states of a pair of converging spin-1/2 particles; when the particles meet, Claire's measurement leaves them in a product state or a Bell state. (b) In the reverse sequence, Claire's measurement prepares a product state or a Bell state of the two (diverging) particles, for Alice and Bob to measure independently.

states $|\uparrow_A\uparrow_B\rangle$ and $|\downarrow_A\downarrow_B\rangle$ or a mixture of the entangled states $(|\uparrow_A\uparrow_B\rangle - |\downarrow_A\downarrow_B\rangle)/\sqrt{2}$ and $(|\uparrow_A\uparrow_B\rangle + |\downarrow_A\downarrow_B\rangle)/\sqrt{2}$. Without access to the *results* of Jim's measurements, Alice and Bob cannot distinguish between these two mixtures. But *with* access to the results, they can bin their data accordingly and verify that their results either do or do not violate Bell's inequality in accordance with whether Jim chooses to entangle their pairs or not.

Thus, Fig. 17.1(b) nicely illustrates the fact that quantum mechanics is retrocausal – even if we take quantum nonlocality at face value without considering retrocausality. On the one hand, there is no reason to doubt that Alice, Bob and Jim have free will. Indeed, the results of Alice and Bob's measurements are consistent with whatever Jim chooses, right up to the moment when he decides to measure $\sigma_z^{(J)}$ or $\sigma_x^{(J)}$ on each of his particles and record the results. On the other hand, there is no doubt about the effect (in Jim's past light cone) of Jim's choice. After Alice and Bob obtain the results of Jim's measurements (within his forward light cone) they can reconstruct from their data whether their particles were entangled or not at the time they measured them. Thus quantum mechanics is retrocausal (though not necessarily *locally* retrocausal).

Now, to illustrate how local retrocausality could obviate nonlocality, consider Fig. 17.2. Figure 17.2(a) represents what might be called a "reverse EPR experiment." Alice and Bob, at space–time points A and B, measure spin components $\sigma^{(A)} \cdot \hat{\mathbf{n}}_A$ and $\sigma^{(B)} \cdot \hat{\mathbf{n}}_B$, respectively, on spin-1/2 particles approaching in pairs, in opposite directions, along the x axis. For each pair, Alice and Bob are completely free to choose the unit vectors $\hat{\mathbf{n}}_A$ and $\hat{\mathbf{n}}_B$ independently. The spin states of the particles before they reach Alice and Bob are irrelevant, and the pairs leave in products of eigenstates of $\sigma^{(A)} \cdot \hat{\mathbf{n}}_A$ and $\sigma^{(B)} \cdot \hat{\mathbf{n}}_B$ with eigenvalues ± 1. Let $|\xi_A \xi_B'\rangle$ denote these product states. At the point where the particles meet, Claire is free to make one of two measurements. She either measures a nondegenerate operator P with the product eigenstates $|\uparrow_A\uparrow_B\rangle$, $|\downarrow_A\uparrow_B\rangle$, $|\uparrow_A\downarrow_B\rangle$, and $|\downarrow_A\downarrow_B\rangle$ (i.e., products of the eigenstates of $\sigma_z^{(A)}$ and $\sigma_z^{(B)}$), or she measures a nondegenerate operator B with the Bell states $(|\uparrow_A\uparrow_B\rangle) \pm |\downarrow_A\downarrow_B\rangle)/\sqrt{2}$ and $(|\uparrow_A\downarrow_B\rangle \pm |\uparrow_A\downarrow_B\rangle)/\sqrt{2}$ as eigenstates. The Bell states are entangled. Let Alice and Bob send Claire the results of their measurements. Suppose Claire chooses to measure the Bell operator each time; she can bin the data for each pair that arrives in her laboratory according to the Bell state that she finds for it. Over time, she will be able to measure the quantum correlations between Alice's and Bob's measurements from the binned data, for each Bell state. These quantum correlations are precisely the non-local quantum correlations that violate Bell's inequality, since for quantum probabilities, and hence for correlations, time order does not matter: $|\langle \xi_A \xi'_B | B_i \rangle|^2 = |\langle B_i | \xi_A \xi'_B \rangle|^2$, where $|B_i\rangle$ is any one of the four Bell states. Yet nothing even slightly nonlocal is going on here. The results of Alice's and Bob's measurements propagate locally and causally to Claire, who "clarifies" the overall state of each pair of particles that arrives in her laboratory with her measurement. The reason that Fig. 17.2(a) is locally causal is that local causality brings the results of Alice's, Bob's and Claire's measurements all together at the spacetime point C. In Fig. 17.2(b), local causality *cannot* bring the results of Alice's, Bob's and Claire's measurements all together at any point, because the particles in each pair diverge to space–time points A and B. Thus the conditions for Bell's theorem hold and the quantum correlations of Fig. 17.2(b), which are also the quantum correlations of Fig. 17.2(a), are nonlocal.

Nevertheless, time-reversal symmetry suggests that Figs. 17.2(a) and 17.2(b) are analogous. Perhaps local retrocausality could play the role in Fig. 17.2(b) that local causality cannot play: local retrocausality could propagate the results of Alice's and Bob's measurements at A and B, respectively, *backwards* in time to bring the results of Alice's, Bob's and Claire's measurements all together at the space–time point C. Using the ABL formula, we can express the conditional probability that Claire's measurement at space–time point C yields the Bell state $|B_j\rangle$ as

$$\text{prob}(|B_j\rangle) = \frac{|\langle \xi_A \xi'_B | U(t_{AB} - t_C)|B_j\rangle\langle B_j|U(t_C - t_0)|0\rangle|^2}{\sum_{i=1}^{4} |\langle \xi_A \xi'_B | U(t_{AB} - t_C)|B_i\rangle\langle B_i|U(t_C - t_0)|0\rangle|^2}, \tag{17.5}$$

where $|0\rangle$ is the state of the two spin-1/2 particles at time t_0, before Claire's measurement, t_C is the time of her measurement, and t_{AB} is the time of Alice's and Bob's measurements, which we can take (for simplicity and without lost of generality) to be simultaneous. The unitary operator $U(t_{AB} - t_C)$ can be rewritten as $U^\dagger(t_C - t_{AB})$ to remove any arrow of time from the ABL formula:

$$\text{prob}(|B_j\rangle) = \frac{|\langle \xi_A \xi'_B | U^\dagger(t_C - t_{AB})|B_j\rangle\langle B_j|U(t_C - t_0)|0\rangle|^2}{\sum_{i=1}^{4} |\langle \xi_A \xi'_B | U^\dagger(t_C - t_{AB}))|B_i\rangle\langle B_i|U(t_C - t_0)|0\rangle|^2}; \tag{17.6}$$

here the initial state $|0\rangle$ and final state $|\xi_A \xi'_B\rangle$ both evolve *locally* towards the intermediate time of Claire's measurement.

The ABL formula realizes the time-reversal symmetry between Figs. 17.2(a) and 17.2(b). But we have already noted that time-reversal symmetry holds only for microscopic physics, and not for macroscopic physics. In particular, the Born rule belongs to the realm of macroscopic physics. In Fig. 17.2(b), we can use $|\langle \xi_A \xi'_B | B_i \rangle|^2 = |\langle B_i | \xi_A \xi'_B \rangle|^2$ to predict the probability of the state $|\xi_A \xi'_B\rangle$ given the state $|B_i\rangle$; we *cannot* use it to retrodict the probability of the state $|B_i\rangle$ given the state $|\xi_A \xi'_B\rangle$. More concretely, Claire in Fig. 17.2(b) could certainly entangle two spin-1/2 particles by measuring on them an operator B with the Bell states as eigenstates; but in another experiment to test Bell's inequality, the particles might be photons in a singlet state, produced by the decay of an excited state of an atom. If so, the time-reversed experiment – Fig. 17.2(a), in which the photons converge so precisely as to excite an atom – is much less likely. However, the ABL approach [16, 19] is still valid: quantum mechanics (microscopic physics) contains no arrow of time, and the macroscopic arrow of time derives from thermodynamics and boundary conditions on the universe. If so, perhaps we can overlook the imperfect analogy between Figs. 17.2(a) and 17.2(b) and let retrocausality evolve the states at A and B back to C, where quantum probabilities determine the actual sequence of results. This retrocausal description fits naturally with the "two-state-vector" formulation of quantum mechanics [16, 20].

It is also consistent with free will, in the following sense. There would be a problem regarding free will if, say, Alice could obtain any information about what she measured *before* the measurement. Any physical theory that allowed such a causal loop would be inconsistent. But suppose Alice could not obtain any such information before the measurement, but someone else could. No causal loop could arise, but would we still say that Alice has free will? The question does not apply to Fig. 17.2(b) because no one has access to information about Alice's measurement before the event A: a normal ("strong") measurement between C and A would eliminate the causal/retrocausal connection between the two events, and a "weak" measurement [21] could yield a result only *after* Alice's measurement.

17.3 PR-Box Correlations from Local Retrocausality

We can now define a toy model for the PR box as a retrocausal box rather than a nonlocal box (as Argaman [22] defined a toy model for bipartite singlet correlations). Returning to Fig. 17.2(b), we let Alice's and Bob's choices of what to measure (a or a' and b or b', respectively) propagate retrocausally to C, where (for the PR box) choices (a, b), (a, b') and (a', b) yield values $(1, 1)$ or $(-1, -1)$ with equal probability, while choice (a', b') yields values $(1, -1)$ or $(-1, 1)$ with equal probability. (By analogy with the previous section, we could let Claire clarify if the box is a PR box or a different but equivalent box.) Then Alice and Bob's measured correlations are PR-box correlations, i.e., the retrocausal box is equivalent to the nonlocal box. And now the conclusions of Section 17.1 apply to the retrocausal box just as they apply to the nonlocal box: Alice and Bob can violate the axiom of no superluminal signalling in the classical limit (the limit of arbitrarily many boxes). In other words, the PR-box is *not* causal in the classical limit. Just as in Section 17.1, we can eliminate PR-box correlations as not satisfying the three axioms of causality, local retrocausality and the existence of a classical limit. Likewise, from these three axioms alone we can expect to derive Tsirelson's bound – a theorem of quantum mechanics.

To conclude, local retrocausality offers us an alternative to "spooky action at a distance." Would Einstein have accepted it? Is local retrocausality a deep principle worthy of being an axiom? It is appropriate to let Bell have the last word [23]:

I think Einstein thought that Bohm's model was too glib – too simple. I think he was looking for a much more profound rediscovery of quantum phenomena. The idea that you could just add a few variables and the whole thing [quantum mechanics] would remain unchanged apart from the interpretation, which was a kind of trivial addition to ordinary quantum mechanics, must have been a disappointment to him. I can understand that – to see that that is all you need to do to make a hidden-variable theory. I am sure that Einstein, and most other people, would have liked to have seen some big principle emerging, like the principle of relativity, or the principle of the conservation of energy. In Bohm's model one did not see anything like that.

Acknowledgments

I thank Huw Price and Ken Wharton for stimulating correspondence and Yakir Aharonov, Sandu Popescu and Nathan Argaman for critical comments. I acknowledge support from the John Templeton Foundation (Project ID 43297) and from the Israel Science Foundation (Grant 1190/13). The opinions expressed in this publication are mine and do not necessarily reflect the views of either of these supporting foundations.

References

[1] J.S. Bell, On the Einstein–Podolsky–Rosen paradox, *Physics* **1**, 195 (1964).

[2] A. Einstein, B. Podolsky and N. Rosen, Can quantum-mechanical description of reality be considered complete? *Phys. Rev.* **47**, 777 (1935); see also N. Bohr, Can quantum-mechanical description of reality be considered complete? *Phys. Rev.* **48**, 696 (1935).

[3] J. Bernstein, *Quantum Profiles* (Princeton, NJ: Princeton Univ. Press), 1991, p. 84.

[4] I. Newton, letter to R. Bentley, 25 February 1693, in *The Correspondence of Isaac Newton*, Vol. **III**, ed. H.W. Turnbull (Cambridge: Cambridge Univ. Press), 1961, pp. 253–6, cf. p. 254; punctuation and spelling edited.

[5] J.F. Clauser, M.A. Horne, A. Shimony and R.A. Holt, Proposed experiment to test local hidden-variable theories, *Phys. Rev. Lett.* **23**, 880 (1969).

[6] B.S. Tsirelson (Cirel'son), Quantum generalizations of Bell's inequality, *Lett. Math. Phys.* **4**, 93 (1980).

[7] S. Popescu and D. Rohrlich, Quantum nonlocality as an axiom, *Found. Phys.* **24**, 379 (1994).

[8] L. Khalfin and B. Tsirelson, Quantum and quasi-classical analogs of Bell inequalities, in P. Lahti et al. (eds.), *Symposium on the Foundations of Modern Physics '85* (Singapore: World Scientific), 1985, p. 441; P. Rastall, Locality, Bell's theorem, and quantum mechanics, *Found. Phys.* **15**, 963 (1985); G. Krenn and K. Svozil, Stronger-than-quantum correlations, *Found. Phys.* **28**, 971 (1998).

[9] D. Rohrlich, PR-box correlations have no classical limit, in D.C. Struppa and J.M. Tollaksen (eds.), *Quantum Theory: A Two-Time Success Story* [Yakir Aharonov Festschrift], (Milan: Springer), 2013, pp. 205–11.

[10] D. Rohrlich, Stronger-than-quantum bipartite correlations violate relativistic causality in the classical limit, arXiv:1408.3125. See also N. Gisin, Quantum measurement of spins and magnets, and the classical limit of PR-boxes, arXiv:1407.8122.

[11] W. van Dam, *Nonlocality and Communication Complexity* (Ph.D. thesis), Oxford University (2000); Implausible consequences of superstrong nonlocality, quant-ph/0501159 (2005); D. Dieks, Inequalities that test locality in quantum mechanics, *Phys. Rev. A* **66**, 062104 (2002); H. Buhrman and S. Massar, Causality and Tsirelson's bounds, *Phys. Rev. A* **72**, 052103 (2005); J. Barrett and S. Pironio, Popescu–Rohrlich

correlations as a unit of nonlocality, *Phys. Rev. Lett.* **95**, 140401 (2005); G. Brassard, H. Buhrman, N. Linden, A.A. Méthot, A. Tapp and F. Unger, Limit on nonlocality in any world in which communication complexity is not trivial, *Phys. Rev. Lett.* **96**, 250401 (2006); J. Barrett, Information processing in generalized probabilistic theories, *Phys. Rev. A* **75**, 032304 (2007); D. Gross, M. Müller, R. Colbeck and O.O. Dahlsten, All reversible dynamics in maximally nonlocal theories are trivial, *Phys. Rev. Lett.* **104**, 080402 (2010).

[12] M. Pawłowski et al., Information causality as a physical principle, *Nature* **461**, 1101 (2009).

[13] M. Navascués and H. Wunderlich, A glance beyond the quantum model, *Proc. R. Soc. A* **466**, 881 (2010).

[14] N. Brunner and P. Skrzypczyk, Nonlocality distillation and postquantum theories with trivial communication complexity, *Phys. Rev. Lett.* **102**, 160403 (2009).

[15] H. Price, *Time's Arrow and Archimedes' Point: New Directions for the Physics of Time* (New York: Oxford Univ. Press), 1996; P.W. Evans, H. Price and K.B. Wharton, New slant on the EPR–Bell experiment, *Brit. J. Phil. Sci.* **64**, 297 (2013); H. Price and K. Wharton, Dispelling the quantum spooks – A clue that Einstein missed?, arXiv:1307.7744v1.

[16] Y. Aharonov, P.G. Bergmann and J.L. Lebowitz, Time symmetry in the quantum process of measurement, *Phys. Rev.* **134** (1964), B1410. See also Y. Aharonov and D. Rohrlich, op. cit., Chap. 10.

[17] D.M. Greenberger, M. Horne and A. Zeilinger, Going beyond Bell's theorem, in M. Kafatos (ed.), *Bell's Theorem, Quantum Theory, and Conceptions of the Universe* [Proceedings of the Fall Workshop, Fairfax, Virginia, October 1988], (Dordrecht: Kluwer Academic), 1989, pp. 69–72.

[18] J. Grunhaus, S. Popescu and D. Rohrlich, Jamming nonlocal quantum correlations, *Phys. Rev. A* **53** (1996) 3781; D. Rohrlich, Three attempts at two axioms for quantum mechanics, in Y. Ben-Menahem and M. Hemmo (eds.), (*The Frontiers Collection*) *Probability in Physics* (Berlin: Springer), 2012, pp. 187–200. The latter paper notes that jamming arises in quantum mechanics, contrary to what the former paper assumes.

[19] Y. Aharonov and D. Rohrlich, op. cit., Sect. 18.2.

[20] Y. Aharonov and L. Vaidman, The two-state vector formalism: An updated review, in J.G. Muga, R.S. Mayato and Í. Egusquiza (eds.), *Time in Quantum Mechanics*, Vol. **1** [Lecture Notes in Physics **734**], 2nd ed. (Berlin: Springer), 2008, pp. 399–447. See also Y. Aharonov and D. Rohrlich, op. cit., Chap. 18.

[21] Y. Aharonov, D.Z. Albert and L. Vaidman, How the result of a measurement of a component of the spin of a spin-$\frac{1}{2}$ particle can turn out to be 100, *Phys. Rev. Lett.* **60**, 1351 (1988); see also Y. Aharonov and D. Rohrlich, op. cit., Chaps. 16–17.

[22] N. Argaman, Bell's theorem and the causal arrow of time, *Am. J. Phys.* **78**, 1007 (2010).

[23] J. Bernstein, op. cit., pp. 66–67.

18

Weak Values and Quantum Nonlocality

YAKIR AHARONOV AND ELIAHU COHEN

Entanglement and nonlocality are studied in the framework of pre-/postselected ensembles with the aid of weak measurements and the two-state-vector formalism. In addition to the EPR–Bohm experiment, we revisit the Hardy and Cheshire Cat experiments, whose *entangled* pre- or postselected states give rise to curious phenomena. We then turn to even more peculiar phenomenon suggesting "emerging correlations" between *independent* pre- and postselected ensembles of particles. This can be viewed as a quantum violation of the classical "pigeonhole principle."

18.1 Introduction

It is seldom acknowledged that seven years before the celebrated Bell paper [1], Bohm and Aharonov [2] published an analysis of the EPR paradox [3]. They suggested an experimental setup, based on Compton scattering, for testing nonlocal correlations between the polarizations of two annihilation photons. In 1964, Bell proposed his general inequality, thereby excluding local realism. During the same time, Aharonov et al. constructed the foundations of a time-symmetric formalism of quantum mechanics [4]. While Bell's proof utilizes entanglement to demonstrate nonlocal correlations, we will describe in what follows the emergence of nonlocal correlations between product states. For this purpose, however, we shall invoke weak measurements of pre- and postselected ensembles.

In classical mechanics, initial conditions of position and velocity for every particle fully determine the time evolution of the system. Therefore, trying to impose an additional final condition would lead either to redundancy or to inconsistency with the initial conditions. This is radically different in quantum mechanics. Because of the uncertainty principle, an initial state vector does not fully determine, in general, the outcome of a future measurement. However, adding a final (backward-evolving) state vector results in a more complete description of the quantum system in between these two boundary conditions, which has a bearing on the determination of measurement outcomes.

The basis for this time-symmetric formulation of quantum mechanics was laid down by Aharonov, Bergman, and Lebowitz (ABL), who derived a symmetric probability rule concerning measurements performed on systems, while taking into account the final state of the system, in addition to the usual initial state [4]. Such a final state may arise due to postselection, that is, performing an additional measurement on the system and considering only the cases with the desired outcome. Since then, the time-symmetric formalism has been further generalized (see for instance [5, 6]) and has been shown to be very helpful for understanding conceptual ideas in quantum mechanics, such as the past of the quantum particle [7] and the measurement problem [8].

To verify the two-state description without intervening with the final (postselected) boundary condition, a subtle kind of quantum measurement was suggested – *weak measurement* [9]. Weak measurements are based on the von Neumann scheme for performing quantum measurements, albeit with a very small coupling compared with the measurement's uncertainty. The weak coupling created between the measured system and the measuring (quantum) pointer does not change the measured state significantly, yet provides robust information when an ensemble of states is discussed [9, 10]. Given an operator A that we wish to measure on a system $|\psi\rangle$, the coupling to the measuring pointer is achieved through the Hamiltonian

$$H_{int} = \epsilon g(t)AP_d, \tag{18.1}$$

where $\epsilon \ll 1$ is a small parameter, $\int_0^T g(t)dt = 1$ for a measurement of duration T, and P_d is the pointer's momentum. The result of this coupling to a pre- and postselected ensemble $\langle\phi|\ |\psi\rangle$ (i.e., the reading of the pointer) is known as a *weak value* [9]:

$$\langle A\rangle_w = \frac{\langle\phi|A|\psi\rangle}{\langle\phi|\psi\rangle}. \tag{18.2}$$

Weak measurements were shown to be very useful in analyzing a variety of problems [11–14]. We will focus henceforth on entanglement and nonlocality.

The outline of the paper is as follows: Section 18.2, describes three experiments with entangled pre- or postselected states: Hardy's paradox, the Cheshire Cat, and finally an EPR–Bohm experiment. Section 18.3 presents the analysis of "emerging correlations" within nonentangled pre- and postselected systems.

18.2 Entangled Pre- and Postselected Systems

We shall revisit three gedankenexperiments that highlight the unique features of weak values between entangled pre- and postselected states.

18.2.1 Hardy's Experiment

An interesting demonstration of weak values between an entangled preselected state and a product postselected state, as well as a conceptual success of the TSVF, is given by the Hardy experiment [15, 16]. Two Mach–Zehnder interferometers overlap at one corner (see Fig. 18.1). Their length is tuned so that the electron entering the first will always arrive at detector C_-, while a positron entering the second will always arrive at detector C_+. Hence, when an electron and a positron simultaneously traverse the setup, they might annihilate

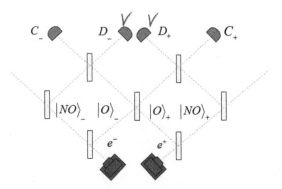

Figure 18.1 Hardy's experiment.

or make their partner reach the "forbidden" detector D_-/D_+. In case no annihilation was recorded, we know that the state of the particles is

$$|\psi_i\rangle = \frac{1}{\sqrt{3}}[|O\rangle_+|NO\rangle_- + |NO\rangle_+|O\rangle_- + |NO\rangle_+|NO\rangle_-]; \qquad (18.3)$$

i.e., at least one of the particles took the nonoverlapping (NO) state, thereby excluding the case where they both took the overlapping path (O). The interferometers were tuned so that C_- clicked for the $\frac{1}{\sqrt{2}}(|O\rangle_- + |NO\rangle_-)$ state, D_- clicked for the $\frac{1}{\sqrt{2}}(|O\rangle_- - |NO\rangle_-)$ state, and similarly for C_+ and D_+. Therefore, choosing the case of clicks at D_- and D_+ amounts to postselection of the state

$$|\psi_f\rangle = \frac{1}{2}(|O\rangle_+ - |NO\rangle_+)(|O\rangle_- - |NO\rangle_-). \qquad (18.4)$$

This postselection is possible, because ψ_i is not orthogonal to ψ_f, but it is peculiar nevertheless: Detection of the electron at D_- naively tells us that the positron took its overlapping path, while detection of the position at D_+ naively tells us that the electron took its overlapping path. This scenario, however, is impossible, because we know annihilation did not take place. The paradox is resolved using the TSVF. When we calculate the weak values of the various projection operators we find that

$$\langle \Pi_O^- \Pi_O^+ \rangle_w = 0 \qquad (18.5)$$

and

$$\langle \Pi_{NO}^- \Pi_O^+ \rangle_w = \langle \Pi_O^- \Pi_{NO}^+ \rangle_w = +1, \qquad (18.6)$$

while

$$\langle \Pi_{NO}^- \Pi_{NO}^+ \rangle_w = -1. \qquad (18.7)$$

This leads us to conclude that although the number of pairs is 1, we have two "positive" pairs and one "negative" pair – a pair of particles with opposite properties. The pair in the

"NO–NO" path creates a negative "weak potential" [17]; that is, when it interacts weakly with any other particle in the intermediate time, its effect will have a negative sign.

18.2.2 The Cheshire Cat

The second demonstration is the "Cheshire Cat" [18]. Let a particle (the "Cat") have two degrees of freedom: spatial $|L\rangle,|R\rangle$ (the cat is in the left/right box) and spinorial $|\uparrow\rangle,|\downarrow\rangle$ (the Cat is smiling/frowning). The Cat is preselected at $t = 0$ in the entangled state

$$|\psi_i\rangle = \frac{1}{2}(|\uparrow\rangle + |\downarrow\rangle)|L\rangle + \frac{1}{2}(|\uparrow\rangle - |\downarrow\rangle)|R\rangle \qquad (18.8)$$

and postselected at $t = T$ in the product state

$$|\psi_f\rangle = \frac{1}{2}(|\uparrow\rangle - |\downarrow\rangle)(|L\rangle + |R\rangle). \qquad (18.9)$$

At $0 < t < T$ the Cat is in the right box, since

$$\langle\Pi_L\rangle_w = 0 , \ \langle\Pi_R\rangle_w = 1, \qquad (18.10)$$

and it is smiling, since

$$\langle\sigma_z\rangle_w = 1, \qquad (18.11)$$

but its smile is in the left box (!), since

$$\langle\sigma_z\Pi_L\rangle_w = 1 , \ \langle\sigma_z\Pi_R\rangle_w = 0. \qquad (18.12)$$

This can be understood as the failure of the product rule for weak values between pre- and postselected states. Weak values reveal a perplexing phenomenon: the spin of a quantum particle can be separated from its mass.

18.2.3 An EPR–Bohm Experiment

The third demonstration is an EPR–Bohm experiment [19] (for a GHZ-like demonstration, where a set of N particles is faced with the four GHZ mutually exclusive requirements, see [20]). Alice and Bob share an ensemble of spin-1/2 entangled particles prepared in

$$|\psi_i\rangle = \frac{1}{\sqrt{2}}(|\uparrow\rangle_A|\downarrow\rangle_B - |\downarrow\rangle_A|\uparrow\rangle_B). \qquad (18.13)$$

Alice and Bob measure their particles along axes that they choose at random from a finite set. Suppose Alice measures her spin along the x-axis and Bob measures his spin along the y-axis, and the outcomes are

$$|\sigma_x\rangle_A = 1, \ , |\sigma_y\rangle_B = 1, \qquad (18.14)$$

i.e., the postselected state is

$$|\psi_f\rangle = \frac{1}{2}(|\uparrow\rangle_A + |\downarrow\rangle_A)(|\uparrow\rangle_B + i|\downarrow\rangle_B). \tag{18.15}$$

According to the EPR paradox, the results of Alice's measurement cannot depend on Bob's choice of axes and vice versa. Therefore, σ_x^A, σ_y^A, σ_x^B, σ_y^B are all elements of reality (in the EPR sense). We now note that $|\psi_i\rangle$ is an eigenvalue of the three operators $\sigma_x^A \sigma_x^B$, $\sigma_y^A \sigma_y^B$ (with eigenvalue -1) and $\sigma_x^A \sigma_y^B + \sigma_y^A \sigma_x^B$ (with eigenvalue 0). Therefore, the postselection accords, on one hand, with $\sigma_x^B = \sigma_y^A = -1$, but on the other hand, with $\sigma_y^A \sigma_x^B = -1$. An apparent contradiction! To resolve this paradox we turn again to the weak values of the various projection operators,

$$\langle \Pi_{\uparrow y}^A \Pi_{\uparrow x}^B \rangle_w = -1/2,$$
$$\langle \Pi_{\uparrow y}^A \Pi_{\downarrow x}^B \rangle_w = 1/2,$$
$$\langle \Pi_{\downarrow y}^A \Pi_{\uparrow x}^B \rangle_w = 1/2,$$
$$\langle \Pi_{\downarrow y}^A \Pi_{\downarrow x}^B \rangle_w = 1/2. \tag{18.16}$$

Hence, for Alice's system,

$$\langle \Pi_{\uparrow y}^A \rangle_w = \langle \Pi_{\uparrow y}^A \Pi_{\uparrow x}^B \rangle_w + \langle \Pi_{\uparrow y}^A \Pi_{\downarrow x}^B \rangle_w = 0,$$
$$\langle \Pi_{\downarrow y}^A \rangle_w = \langle \Pi_{\downarrow y}^A \Pi_{\uparrow x}^B \rangle_w + \langle \Pi_{\downarrow y}^A \Pi_{\downarrow x}^B \rangle_w = 1, \tag{18.17}$$

consistent with the requirement $\sigma_y^A = -1$. Similarly, for Bob,

$$\langle \Pi_{\uparrow x}^B \rangle_w = \langle \Pi_{\uparrow y}^A \Pi_{\uparrow x}^B \rangle_w + \langle \Pi_{\downarrow y}^A \Pi_{\uparrow x}^B \rangle_w = 0,$$
$$\langle \Pi_{\downarrow x}^B \rangle_w = \langle \Pi_{\uparrow y}^A \Pi_{\downarrow x}^B \rangle_w + \langle \Pi_{\downarrow y}^A \Pi_{\downarrow x}^B \rangle_w = 1, \tag{18.18}$$

consistent with $\sigma_x^B = -1$. In addition,

$$\langle \sigma_y^A \sigma_x^B \rangle_w = \langle \Pi_{\uparrow y}^A \Pi_{\uparrow x}^B \rangle_w - \langle \Pi_{\uparrow y}^A \Pi_{\downarrow x}^B \rangle_w - \langle \Pi_{\downarrow y}^A \Pi_{\uparrow x}^B \rangle_w + \langle \Pi_{\downarrow y}^A \Pi_{\downarrow x}^B \rangle_w = -1, \tag{18.19}$$

consistent with $\sigma_y^A \sigma_x^B = -1$.

Hence we see an alternative way of understanding quantum nonlocality. The classical limitation on correlations can be violated by quantum weak values that are negative. As in the case of Hardy's experiment, these weak values should be understood as reversing the interaction sign, rather than as negative probabilities. In fact, each weak value (not necessarily a peculiar one) defines a "weak potential" within a pre-/postselected ensemble [17].

For a more complex setup of an EPR–Bohm experiment with weak measurements, see [21].

18.3 Nonentangled Pre- and Postselected Ensembles

Let two spins be independently prepared at $t = 0$ in the state $|\sigma_x = +1\rangle$, to create the product state

$$|\sigma_x = +1\rangle_1 |\sigma_x = +1\rangle_2. \tag{18.20}$$

Suppose that later, at time $t = T$, they are independently measured along the y-axis and found at

$$|\sigma_y = +1\rangle_1 |\sigma_y = +1\rangle_2. \tag{18.21}$$

Could there be correlations between these two independent spins at times $0 < t < T$? The correlation between operators A and B is defined according to

$$\mathrm{Corr}(A, B) \equiv \langle AB \rangle - \langle A \rangle \langle B \rangle. \tag{18.22}$$

If measured strongly, σ_z between $t = 0$ and $t = T$ would be clearly found to have a zero expectation value, $\langle \sigma_z \rangle = 0$, for both particles, since they were prepared in an eigenstate of σ_x and postselected in an eigenstate of σ_y. But what is the product of their spins along the z-axis? Eq. (18.2) tells us that the weak value of σ_z is

$$\langle \sigma_z \rangle_w = i. \tag{18.23}$$

We saw in Sect. 18.2 the breakdown of the product rule for entangled states, but now the pre- and postselected states are not entangled and hence

$$\left\langle \sigma_z^{(1)} \sigma_z^{(2)} \right\rangle_w = i \cdot i = -1. \tag{18.24}$$

In addition, for dichotomic operators, we know that if the weak value equals one of the eigenvalues, then it also equals the strong value. Hence,

$$\left\langle \sigma_z^{(1)} \sigma_z^{(2)} \right\rangle = -1, \tag{18.25}$$

and

$$\mathrm{Corr}\left(\sigma_z^{(1)}, \sigma_z^{(2)}\right) = -1. \tag{18.26}$$

We thus see that the two spins were anticorrelated along the z-axis, but they were also correlated along the x-axis and along the y-axis, so they must have been maximally entangled. But alas, they were pre- and postselected in a product state! To better understand why these particles seem to be maximally entangled, we can represent their initial and final states in the z basis:

$$|\sigma_x = +1\rangle_1 |\sigma_x = +1\rangle_2 = \frac{1}{2}(|\uparrow\uparrow\rangle + |\uparrow\downarrow\rangle + |\downarrow\uparrow\rangle + |\downarrow\downarrow\rangle) \tag{18.27}$$

and

$$|\sigma_y = +1\rangle_1 |\sigma_y = +1\rangle_2 = \frac{1}{2}(|\uparrow\uparrow\rangle + i|\uparrow\downarrow\rangle + i|\downarrow\uparrow\rangle - |\downarrow\downarrow\rangle). \tag{18.28}$$

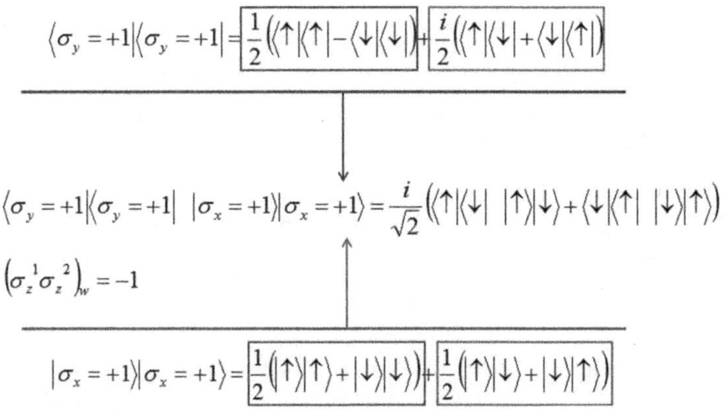

$$\langle \sigma_y =+1 | \langle \sigma_y =+1 | = \boxed{\frac{1}{2}\left(\langle\uparrow|\langle\uparrow| - \langle\downarrow|\langle\downarrow| \right)} + \boxed{\frac{i}{2}\left(\langle\uparrow|\langle\downarrow| + \langle\downarrow|\langle\uparrow| \right)}$$

$$\langle \sigma_y =+1 | \langle \sigma_y =+1 | \; |\sigma_x =+1\rangle |\sigma_x =+1\rangle = \frac{i}{\sqrt{2}}\left(\langle\uparrow|\langle\downarrow| \; |\uparrow\rangle|\downarrow\rangle + \langle\downarrow|\langle\uparrow| \; |\downarrow\rangle|\uparrow\rangle \right)$$

$$\left(\sigma_z^1 \sigma_z^2 \right)_w = -1$$

$$|\sigma_x =+1\rangle |\sigma_x =+1\rangle = \boxed{\frac{1}{2}\left(|\uparrow\rangle|\uparrow\rangle + |\downarrow\rangle|\downarrow\rangle \right)} + \boxed{\frac{1}{2}\left(|\uparrow\rangle|\downarrow\rangle + |\downarrow\rangle|\uparrow\rangle \right)}$$

Figure 18.2 Emerging correlations between independent ensembles.

Hence, the correlated part in the pre- and postselected states cancels due to orthogonality, and only the anticorrelated part remains (see Fig. 18.2). These correlations can be verified, for example, by performing nonlocal measurements [22, 23]. This in fact is a very general phenomenon occurring each time the pre- and postselected states do not coincide. In these cases the states will have orthogonal parts whose cancelation would yield correlations. Repeating this procedure for an ensemble of *N* particles, we find each pair to be maximally entangled in an apparent violation of "entanglement monogamy." However, this entanglement is of a subtle kind, since both the pre- and postselected states were not entangled in the first place. Moreover, it cannot be verified on each particle alone (only on pairs) and cannot be used for teleportation.

It turns out that the weak values contain, in some sense, even more information. In the above scenario the two experimenters need not know what are the pre- and postselected outcomes; they only need to know which particles had the same outcomes, and then, performing weak measurements along the *x*, *y*, and *z* axes, they would know which direction was chosen and which outcome was measured. Indeed, it can be shown that

$$(\sigma_x)_w^2 + (\sigma_y)_w^2 + (\sigma_z)_w^2 = 1 \tag{18.29}$$

for any pre- and postselected ensemble. We thus understand that quantum correlations underlie almost any experiment, but are only visible upon postselection and grouping of similar results.

This gedankenexperiment can be viewed as demonstrating the breakdown of the classical "Pigeonhole principle." It was previously shown [24] that special pre- and postselection of a quantum system lead to unusual correlations between its parts. Here we witness once more the appearance of emerging correlations in a pre-/postselected quantum ensemble. Thinking about σ_z as denoting the position of a particle in one of two boxes, we can see that within a group of three particles with the above pre- and postselection, every pair is anticorrelated, that is, no pair resides in the same box. This clearly stands in contrast with the classical principle, according to which at least one pair of pigeons within a group of three pigeons must share the same hole. The result can be trivially generalized to N particles.

18.4 Discussion

Bell's proof inclined us to think of entangled states as concealing nonlocal correlations. Within pre- and postselected ensembles, these correlations are responsible for intriguing effects. Yet the truly curious result we have just seen is the emergence of nonlocal correlations in practically every pre- and postselected ensemble of product states.

Weak measurements were demonstrated once more to provide us with a richer description of the quantum reality. Negative weak values were shown to be essential for understanding the Hardy, Cheshire Cat, and EPR–Bohm experiments, while imaginary weak values indicated emerging correlations in a product state. These results accord well with a previous work of Marcovitch, Reznik, and Vaidman [25], where correlations within pre- and postselected ensembles were shown to exceed Tsirelson's bound [26] and reach the Popescu–Rohrlich bound [27].

We feel that the current research is not over yet. Weak values between entangled states might have even a more crucial role in understanding fundamental questions such as the information paradox in black holes [28] and time [29].

Acknowledgements

This work has been supported in part by Israel Science Foundation Grant 1311/14. We would like to thank Aharon Brodutch and Tomer Landsberger for helpful comments and discussions.

References

[1] J.S. Bell, On the Einstein–Podolsky–Rosen paradox, *Physics* **1**, 195–200 (1964).
[2] D. Bohm, Y. Aharonov, Discussion of experimental proof for the paradox of Einstein, Rosen, and Podolsky, *Phys. Rev.* **108**, 1070–76 (1957).
[3] A. Einstein, B. Podolsky, N. Rosen, Can quantum-mechanical description of physical reality be considered complete, *Phys. Rev.* **47**, 777–80 (1935).
[4] Y. Aharonov, P.G. Bergmann, J.L. Lebowitz, Time symmetry in the quantum process of measurement, *Phys. Rev. B* **134**, 1410–16 (1964).
[5] Y. Aharonov and L. Vaidman, The two-state vector formalism of quantum mechanics, in J.M. Muga et al. (eds.), *Time in Quantum Mechanics*, Springer, 369–412 (2002).
[6] B. Reznik and Y. Aharonov, Time-symmetric formulation of quantum mechanics, *Phys. Rev. A* **52**, 2538–50 (1995).
[7] L. Vaidman, Past of a quantum particle, *Phys. Rev. A* **87**, 052104 (2013).
[8] Y. Aharonov, E. Cohen, E. Gruss, T. Landsberger, Measurement and collapse within the two-state-vector formalism, *Quantum Stud.: Math. Found.* **1** , 133–46 (2014).
[9] Y. Aharonov, D.Z. Albert, L. Vaidman, How the result of a measurement of a component of the spin of a spin-1/2 particle can turn out to be 100, *Phys. Rev. Lett.* **60**, 1351–4 (1988).

[10] Y. Aharonov, E. Cohen, A.C. Elitzur, Foundations and applications of weak quantum measurements, *Phys. Rev. A* **89**, 052105 (2014).

[11] O. Hosten, P. Kwiat Observation of the spin Hall effect of light via weak measurements, *Science* **319**, 787–90 (2008).

[12] D.J. Starling, P.B. Dixon, A.N. Jordan, J.C. Howell, Optimizing the signal-to-noise ratio of a beam-deflection measurement with interferometric weak values, *Phys. Rev. A* **80**, 041803 (2009).

[13] X.Y. Xu, Y. Kedem, K. Sun, L. Vaidman, C.F. Li, G.C. Guo, Phase estimation with weak measurement using a white light source, *Phys. Rev. Lett.* **111**, 033604 (2013).

[14] A.N. Jordan, J. Tollaksen, J.E. Troupe, J. Dressel, Y. Aharonov, Heisenberg scaling with weak measurement: A quantum state discrimination point of view, arXiv:1409.3488 (2014).

[15] L. Hardy, Quantum mechanics, local realistic theories, and Lorentz-invariant realistic theories, *Phys. Rev. Lett.* **68**, 2981 (1992).

[16] Y. Aharonov, A. Botero, S. Popescu, B. Reznik, J. Tollaksen, Revisiting Hardy's paradox: Counterfactual statements, real measurements, entanglement and weak values, *Phys. Lett. A* **301**, 130–38 (2002).

[17] Y. Aharonov, E. Cohen, S. Ben-Moshe, Unusual interactions of pre- and past-selected particles, *EPJ Web Conf.* **70**, 00053 (2014).

[18] Y. Aharonov, S. Popescu, D. Rohrlich, P. Skrzypczyk, Quantum Cheshire cats, *New J. Phys.* **15**, 113015 (2013).

[19] Y. Aharonov, D. Rohrlich, *Quantum Paradoxes: Quantum Theory for the Perplexed*, Weinheim: Wiley-VCH, Ch. 17 (2005).

[20] Y. Aharonov, S. Nussinov, S. Popescu, and L. Vaidman, Peculiar features of entangled states with postselection, *Phys. Rev. A* **87**, 014105 (2012).

[21] Y. Aharonov, E. Cohen, D. Grossman, A.C. Elitzur, Can weak measurement lend empirical support to quantum retrocausality? *EPJ Web Conf.* **58** (2013).

[22] Y. Aharonov, D.Z. Albert, L. Vaidman, Measurement process in relativistic quantum theory, *Phys. Rev. D* **34**, 1805–13 (1986).

[23] A. Brodutch, E. Cohen, Nonlocal measurements via quantum erasure, *Phys. Rev. Lett.* **116**, 070404 (2016).

[24] Y. Aharonov, F. Colombo, S. Popescu, I. Sabadini, D.C. Struppa, J. Tollaksen, Quantum violation of the pigeonhole principle and the nature of quantum correlations, *Proc. Natl. Acad. Sci. USA* **113**, 532–5 (2016).

[25] S. Marcovitch, B. Reznik, L. Vaidman, Quantum-mechanical realization of a Popescu–Rohrlich box, *Phys. Rev. A* **75**, 022102 (2007).

[26] B.S. Cirel'son, Quantum generalizations of Bell's inequality, *Lett. Math. Phys.* **4**, 93–100 (1980).

[27] S. Popescu, D. Rohrlich, Quantum nonlocality as an axiom, *Found. Phys.* **24**, 379–85 (1994).

[28] G. Horowitz, J. Maldacena, The black hole final state, *J. High Energy Phys.* **2004.02**, 008 (2004).

[29] Y. Aharonov, S. Popescu, J. Tollaksen, Each instant of time a new universe, in *Quantum Theory: A Two-Time Success Story*, Springer, Milan, pp. 21–36 (2014).

Part IV
Nonlocal Realistic Theories

19

Local Beables and the Foundations of Physics

TIM MAUDLIN

19.1 Introduction: The Theory of Local Beables

John Bell's most celebrated contribution to the foundations of physics is his famous theorem. The theorem demonstrates that any physical theory capable of generating the predictions of the standard quantum-mechanical algorithm, in particular the prediction of violations of Bell's inequality for experiments done at spacelike separation, cannot be local. The sense of "locality" used here is the same sense that Einstein had in mind when he pointed out that the standard interpretation of the quantum algorithm was committed to "spooky action at a distance" To this day, the import of Bell's theorem is not universally appreciated. I have written about this elsewhere [1], and others in this volume will take up that task. It is properly the main focus during this 50th anniversary of that great achievement.

But it is also important to recall and celebrate Bell's other achievements. In many of his later writings, including "The Theory of Local Beables," "Quantum Mechanics for Cosmologists," "On the Impossible Pilot Wave," "Beables for Quantum Field Theory," "Six Possible Worlds of Quantum Mechanics," "Are There Quantum Jumps?" and "Against 'Measurement,'" [1] Bell turned his attention to the more general problem of physically construing the mathematical formalism used to derive these predictions. This activity is often denominated "interpreting quantum theory" as if there were some precise physical theory that might somehow be supplemented with an "interpretation" Once the issue is framed this way it is easy to ask: But if I already have a theory in hand, what can be gained by supplementing it with an "interpretation"? Many physicists, at this juncture, are happy to conclude that "interpretations" are not a matter of physics at all – maybe they are only of interest to philosophers – and that therefore the whole enterprise of "interpreting quantum theory" is not within the purview of physics per se.

What then is in the purview of physics proper? One answer to this question goes under the banner "instrumentalism": all physics as such, is concerned about is *predicting the outcomes of experiments*. In the service of making these predictions physicists may invent various mathematical formalisms, together with rules for their use as prediction-generating instruments. It is neither necessary, nor perhaps even desirable, to accompany these prediction-generating algorithms with any "picture" or "account" or "story" of what exists *beside* the

---

[1] All of these are reproduced in Bell [2], from which the page citations will be taken in this paper.

instruments. Indeed, a common myth about quantum theory is that it is actually *impossible* to provide any such accompanying story, and that the progress of physics requires the positive renunciation of the desire for one. If this were correct then the desire for anything more than such a prediction-generating set of rules must arise from concerns outside of physics proper.

Bell rejected this account of physics root and branch. As usual, he expressed his dissatisfaction so clearly and elegantly that there is nothing to do but quote him:

> In the beginning, natural philosophers tried to understand the world around them. Trying to do that they hit upon the great idea of contriving artificially simple situations in which the number of factors involved is reduced to a minimum. Divide and conquer. Experimental science was born. But experiment is a tool. The aim remains: to understand the world. To restrict quantum mechanics to be exclusively about piddling laboratory operations is to betray the great enterprise. A serious formulation will not exclude the big world outside the laboratory.
>
> *[2, pp. 216–17]*

Physics itself aims at more than just predicting the outcomes of experiments. What more is easily stated: physics aims at a complete and accurate account of the physical structure of the universe. Of course! And the different "interpretations of quantum theory" are really different physical theories, which happen to make exactly, or nearly, the same predictions as the standard quantum-mechanical algorithm. But what general features should such a physical theory have?

One of Bell's signal contributions to this problem is what he called the *theory of local beables*. There is a certain irony here. For while his most famous achievement was to show that the *non*locality that Einstein long ago identified in the standard "interpretation" of the quantum formalism (the Copenhagen interpretation) could not be eliminated, his attention to local beables highlighted just the opposite problem: the standard story *fails* to be clear about what exists locally. So the standard account, if one tries to take it seriously, both *contains* a nonlocality that was not acknowledged and *lacks* a different kind of locality that it requires. It is this second sort of locality I want to discuss here.

Any clearly formulated and articulated physical theory should contain an *ontology*, which is just a statement of what the theory postulates to exist. The word "ontology" can perhaps look a little intimidating, or overly "philosophical" so Bell invented his own terminology for this: the "beables" of the theory. Stating what the beables of the theory are is nothing more nor less than stating what the theory postulates as being physically real. Once what the ontology *is* has been made clear, then (and only then) can one go on to ask what the ontology *does*, how it behaves. This question is answered by a *dynamics*: a mathematically precise characterization of how the beables change through time. The dynamics might be deterministic or might be stochastic. But according to the professional standards of mathematical physics it ought to be precise. It should be specified in sharp equations relating the beables, rather than by using vague words (such as "measurement").

Within the ontology (the beables) of the theory yet another distinction can be made. Some of the beables (but not necessarily all) are *local* beables. The local beables are those which (unlike for example the total energy) can be assigned to some bounded space-time region. For example, in Maxwell's theory the beables local to a given region are just the fields **E** and **H** in that region, and all the functionals thereof. It is in terms of local beables that we can hope to formulate some notion of local causality. Of course, we may be obliged to develop theories in which there *are* no strictly local beables. That possibility will not be considered here.[2]

The obvious *non*local beable that arises in the context of quantum theory is, of course, the "wave function" or "quantum state" of a system. My own preference is to use "wave function" for the mathematical representative of this piece of physical ontology and "quantum state" for the beable itself. Any theory that can properly be called a quantum theory must have some such part of its ontology. But it is not a *local* beable. As Bell wrote "[I]t makes no sense to ask for the amplitude or the phase or whatever of the wavefunction at a point in ordinary space. It has neither amplitude nor phase nor anything else until a multitude of points in ordinary space are specified."[3] So the wave function – or the quantum state it represents – is not a local beable. Any theory that only commits itself to the existence of the quantum state faces the challenge of providing an account of the physical universe without any local beables at all.

At the end of the quote above Bell mentions such a possibility, only to dismiss it from present consideration. And he never returned to discuss the possibility in any of his other writings. This is already prima facie evidence that he considered the local beables to play a central role in any physical theory. Our main challenge now is to clearly articulate what that role is, and what sorts of local beables might play it.

19.2 Two Examples Proposed by Bell

Bell clarifies the methodological role he takes the local beables of a theory to play when discussing some possible choices for them. The most explicit discussion – which is characteristically quite compact – occurs in "Beables for Quantum Field Theory." Since the topic is quantum field theory, one might well expect that the local beable would be some sort of *field*, i.e., something continuously distributed in space–time. But it is not. He first reminds us what beables are:

In particular we will exclude the notion of "observable" in favor of that of "*be*able." The beables of a theory are those elements that might correspond to elements of reality, to things which exist. Their existence does not depend on "observation." Indeed, observation and observers must be made out of beables.

[2, p. 174]

Of course, specifying how an observer such as a human being is "made out of beables" would be a monumental task, requiring detailed physiology and biology. But Bell makes clear that he does not (yet) demand this. What then is the criterion for an adequate choice of beable?

---

[2] From "The Theory of Local Beables" [2, p. 53].   [3] From "Are There Quantum Jumps?" [2, pp. 204–5].

The following passage provides the key:

> Not all "observables" can be given beable status, for they don't all have simultaneous eigenvalues, i.e. do not all commute. It is important to realize therefore that most of these "observables" are entirely redundant. What is essential is to be able to define the positions of things, including the positions of pointers or (the modern equivalent) of ink of computer output.
>
> *[2, p. 175]*

Bell first considers a quantity that might well be fieldlike – the energy density – but rejects it on technical grounds.

> We fall back then on a second choice – fermion number density. The distribution of fermion number in the world certainly includes the positions of instruments, instrument pointers, ink on paper . . . and much, much more.
>
> *[2, p. 175]*

This last sentence holds the key to the method of local beables.

We began by insisting that a properly articulated physical theory must have an ontology, a set of entities that (according to the theory) exist. Further, some of these entities (but not all) must be local in the sense that they exist in delimited regions of space–time. Why should it be essential for a theory to postulate such things? Because, as the sentence above indicates, it is here that *the language of the theory makes contact with reports of the empirical data.*

What, after all, does a report from a lab, or indeed any observation report, look like? It may describe the lab setup using macroscopic language. It probably describes the outcome of the experiment in terms of macroscopic situations of things (pointers). Or if not pointers, the experimental apparatus certainly outputs its result as a certain disposition of ink on paper. So if a theory can accurately predict the behavior of the pointers and the distribution of ink on paper, then it correctly predicts the reported data.

And, of course, much, much more. A theory's specification of the fermion density in every region of the universe entails the distribution of matter at macroscopic scale. And if what it predicts at macroscopic scale matches everything we think we know about the world at macroscopic scale (including where the pointers ended up pointing, where the ink is on the paper, the shape of the earth, the dimensions of the Empire State Building, etc., etc., etc.) then the theory is *empirically adequate* in any reasonable sense. There may be objections to such a theory, but they cannot rightfully be called empirical objections.

Why the emphasis here on macroscopic scale? Because although we take ourselves to know a lot about the microscopic structure of many things, the evidence for that knowledge always appeals to macroscopically observable facts. A theory that gets the macroscopics right accounts for all our evidence about microscopic structure *even if it also contradicts what we believe about microscopic structure*

This is not just an idle possibility. We have seen that Bell made the somewhat surprising choice of fermion number density (rather than a continuous field quantity) for a local beable in quantum field theory. Even more striking was his choice for a local beable in his version of the GRW collapse theory. At a purely mathematical level (but not an *ontological* level) the theory is specified by just a wave function and its dynamics. Since the wave function is a continuous fieldlike mathematical object, one might again anticipate that any local beable postulated by the theory would also be fieldlike. But Bell's choice was not. Instead he proposed the completely novel "flash" ontology. Continuing the passage about the wave function cited above, Bell writes,

> However, the GRW jumps (which are part of the wavefunction, not something else) are well localized in ordinary space. Indeed, each is centered on a particular spacetime point $(\mathbf{x}, t)$. So we can propose these events as the basis of the "local beables" of the theory. These are the mathematical counterparts in the theory to real events at definite places and times in the real world (as distinct from the many purely mathematical constructions that occur in the working out of physical theories, as distinct from things that may be real but not localized, and as distinct from the "observables" of other formulations of quantum mechanics, for which we have no use here). A piece of matter is then a galaxy of such events.
>
> *[2, p. 205]*

These localized point-events have come to be called "flashes."

To what extent does the distribution of flashes in space–time constitute an adequate representation of a piece of matter? A calculation of the density of these events in space–time, according to the GRW dynamics, reveals that they reflect very, very, very little of the detailed spatial structure that we attribute to things at microscopic scale. To take a simple example, an average DNA molecule contains about 2×10^{12} atoms. A generous rough estimate is therefore 10^{14} quarks. But each quark, according to the standard GRW dynamics, suffers a collapse only once every 10^{15} seconds. So each strand of DNA would manifest itself in space–time with less than a single point-event a second. No double-helix structure could be inferred from full information about these flashes without centuries of data (and even then one could not easily sort out which flashes belonged to which molecules, how the molecules might have "moved" in the interim, etc.). The spatial structure of a cell would, according to this theory, be inexpressibly less detailed than we think it is.

Nonetheless, the motion of a whole human body at macroscale would come out just right. The whole body would be associated with over 10^{13} flashes per second: much more information than required to convey everything we think we can observe with the naked eye. And when we used, say, a scanning electron microscope the theory would make accurate predictions about the macroscopically observable images produced by the machinery. This is how the theory of local beables allows us to determine what the macroscopic predictions of a theory are, and hence whether the theory should be counted as consistent with all available data.

Note that we have, *en passant*, completely solved the "measurement problem" or the problem of Schrödinger's cat. The measurement problem is best formulated not as a problem about "measurement" per se, or about laboratory operations, but about the *universal scope* of the quantum theory. As a proposal for the fundamental physical structure of everything that exists, quantum theory ought to be applicable to everything, including stars, planets, laboratory equipment, and cats. The macroscopic structure and behavior of all these things ought to be just the cumulative or aggregate structure and behavior of their microscopic parts. If the beables of a cat, for instance, are provided by the fermion density of particles in the cat, then it is easy to tell from that whether the cat is alive and kicking or has expired. Similarly for the distribution of flashes in the flash ontology. The postulation of local beables is exactly what allows this result without appeal to any "shifty split" between a "classical" and a "quantum" domain. The universe, according to these theories, is quantum-mechanical through and through. And the local beables, existing at microscopic scale, automatically determine also the macroscopic local spatio-temporal properties of things.

19.3 Pilot Wave Theory and Bell's Many Worlds

Beside inventing his own novel proposals for the local beables of quantum field theory and of the GRW collapse theory, Bell also expressed great admiration for another theory with a clear commitment to local beables: pilot wave theory. The local beables here are more familiar than flashes. They are plain old vanilla particles. It is simplest to consider them as point particles – giving them a finite volume would not contribute to the role they play in the theory – which always have definite positions and always move on continuous trajectories. Indeed, the concept of a point particle is so familiar that Bell expends no words at all on further elaboration. In the pilot wave picture matter is made of particles that move in precisely specified ways. There is, of course, also a real, physical nonlocal beable, represented by the wave function. This nonlocal object determines the motions of the particles via the *guidance equation*,

$$\frac{dq_k}{dt} = \frac{\hbar}{m_k} Im \left(\frac{\psi^*(x.t)\nabla_k \psi(x, t)}{\psi^*(x.t)\psi(x.t)} \right).$$

One of the virtues of Bell's presentation of the theory was to avoid the talk of a "quantum potential" and the appeal to Newtonian dynamics that sometimes obscures the fundamental architecture of the theory. Given an initial wave function, an initial configuration of particles, a dynamics for the wave function (e.g., Schrödinger's equation) and the guidance equation, one has a completed physics. With these four pieces of information, it becomes a matter of pure mathematical analysis to determine how the particles will move, and therefore what the macroscopic behavior of material objects will be. That can then be compared with the reported results of observation.

It cannot be overstated both how simple this physical theory is and how hard it is to get it across to the average physicist. Somehow, the very notion of a point particle moving around in some precisely specified way is taken to be incoherent, or impossible. Even if the basic ontology is understood, strange objections are raised such as this: according to the theory, an electron in the ground state of an atom is *static*; it does not move. True. But why anyone takes this to be an *objection* to theory is completely obscure. If those atoms and their electrons (which are of course not directly visible) demonstrably move *en masse* in such a way as to produce the macroscopic motion of matter that we think we see, then the theory has no empirical problems. All of this was so transparently obvious to Bell that he makes no comment about it.

Bell expressed his admiration for the pilot wave theory in many places. In "Six Possible Worlds of Quantum Mechanics," we find this:

The last unromantic picture I will present is the "pilot wave" picture. It is due to de Broglie (1925) and Bohm (1952). While the founding fathers agonized over the question

"particle" *or* "wave"

de Broglie in 1925 proposed the obvious answer

"particle" *and* "wave."

Is it not clear from the smallness of the scintillation on the screen that we have to do with a particle? And is it not clear, from the diffraction and interference patterns, that the motion of the particle is directed by a wave? De Broglie showed in detail how the motion of a particle, passing through just one of two holes in the screen, could be influenced by waves propagating through both holes. And so influenced that the particle does not go where the waves cancel out, but is attracted to where they cooperate. This idea seems to me so natural and simple, to resolve the wave-particle dilemma in such a clear and ordinary way, that it is a great mystery that it was so generally ignored.

[2, p. 191]

Even more telling, perhaps, is his account of the evolution of his own interest in foundational matters. As a young man, Bell had been taught that the observable quantum-mechanical phenomena (such as the double-slit experiment) simply *could not* be accounted for by any theory that postulates beables existing independent of observation and measurement, and certainly not by any theory in which those beables are governed by deterministic equations. As he relates in "On the Impossible Pilot Wave,"

But in 1952, I saw the impossible done. It was done in papers by David Bohm. Bohm showed explicitly how parameters could indeed be introduced, into nonrelativistic quantum mechanics, with the help of which the indeterministic description could be transformed into a deterministic one. More importantly, in my opinion, the subjectivity of the orthodox version, the necessary reference to "the observer," could be eliminated.

Moreover, the essential idea was one that had been advanced already by de Broglie in 1927, in his "pilot wave" picture.

But why then had Born not told me of this "pilot wave"? If only to point out what was wrong with it? Why did von Neumann not consider it? More extraordinarily, why did people go on producing "impossibility" proofs, after 1952, and as recently as 1978? When even Pauli, Rosenfeld, and Heisenberg, could produce no more devastating criticism of Bohm's version than to brand it as "metaphysical" and "ideological"? Why is the pilot wave picture ignored in text books? Should it not be taught, not as the only way, but as an antidote to the prevailing complacency? To show that vagueness, subjectivity, and indeterminism are not forced on us by experimental facts, but by deliberate theoretical choice?

[2, p. 160]

Bell's concern about fundamental interpretive problems of quantum theory is brilliantly expressed in this passage. It is clear that his concern was inspired, from the beginning, by reflection on what the pilot wave theory had accomplished. And it is equally clear that the main accomplishment for Bell was not the recovery of determinism. It was rather the recovery of *objectivity* and *clarity*. Reference to "the observer," and hence to "observation," "measurement," and so on does not occur in the articulation of the theory. There are rather a clear set of beables and a mathematically precise dynamics for them. *Extracting the empirical predictions from the theory is a matter of appreciating what it implies about how matter will move, nothing else.* And it is exactly the commitment to a precise set of local beables – the point particles – that enables all of the conceptual progress.

Indeed, Bell regarded the role of the particles in the pilot wave theory as so central to understanding its account of the physical world that he even inserted the same set of local beables into theories whose originators had no such thing in mind. This occurs in "Quantum theory for cosmologists." Bell there provides his account of the many worlds interpretation, which he denominates "Everett(?)." The question mark is well deserved. Bell presents the many worlds idea in a completely idiosyncratic way, essentially as a version of the pilot wave theory (whose wave function, like that of the many worlds approach, never collapses) but with a radically stochastic evolution law for the local beables. While the pilot wave particles move continuously through space as a consequence of the guidance equation, Bell's many worlds particles have their configuration chosen randomly at each moment, with the probability supplied by the wave function. The configuration of particles at any given time is completely uninfluenced by the configuration at any earlier time, and provides no information (beyond what can be extracted from the uncollapsed universal wave function alone) about what the future configuration might be. Here are two salient passages from that paper:

Then it could be said that classical variables x do not appear in Everett and De Witt. However, it is taken for granted there that meaningful reference can be made to experiments having yielded one result rather than another. So instrument readings, or the numbers on computer output, and things like that, are the classical variables of the theory. We have argued already against the appearance of such vague quantities at a fundamental level. There is always some ambiguity about an instrument reading; the pointer has some thickness and is subject to Brownian motion. The ink can smudge in computer output and it is a matter of practical human judgment that one figure has been printed rather than another. These distinctions are unimportant in practice, but surely the theory should be more precise. It was for that reason that the hypothesis was made of fundamental variables x, from which instrument readings and so on can be constructed, so that only at the stage of this construction, of identifying what is of interest to gross creatures, does an inevitable and unimportant vagueness intrude. I suspect that Everett and De Witt wrote as if instrument readings were fundamental only in order to be intelligible to specialists in quantum measurement theory.

[2, p. 134]

On Bell's reading, the x variable that appears in the wave function really is a variable representing the possible spatial locations of point particles, which are the fundamental local beables. Thus the Everett(?) theory solves the measurement problem with the very same resources as the pilot wave theory

The difference between the two theories, in Bell's telling, is solely in the dynamics for the *x*'s. In Everett(?) they jump around in the most astonishing fashion, from one configuration to another. Each configuration is drawn from a possible pilot wave theory history, but there is no consistency over time about which of these many possible histories is being sampled. There is, for example, almost no chance according to this theory that one's present apparent memories (which are encoded in the present configuration of particles) provide accurate information about anything that actually occurred in the past. The Everett(?) theory is a sort of physical blueprint for a Cartesian demon, but one where the subject is deceived about the past (and even her own past). As Bell puts it,

> Thus in our interpretation of the Everett theory there is no association of the particular present with any particular past. And the essential claim is that this does not matter at all. For we have no access to the past. We have only our "memories" and "records." But these records and memories are in fact *present* phenomena. The instantaneous configuration of the *x*s can include clusters which are markings in notebooks, or in computer memories, or in human memories. These memories can be of the initial conditions in experiments, among other things, and of the results of those experiments. The theory should account for the present correlations between these present phenomena. And in this respect we have seen it to agree with ordinary quantum mechanics, insofar as the latter is unambiguous . . .
>
> Everett's replacement of the past by memories is a radical solipsism – extending to the temporal dimension the replacement of everything outside my head by my impressions, of ordinary solipsism or positivism. Solipsism cannot be refuted. But if such a theory were taken seriously it would hardly be possible to take anything else seriously. So much for the social implications. It is always interesting to find that solipsists and positivists, when they have children, have life insurance.
>
> *[2, pp. 134–5]*

Bell's Everett(?) theory is certainly not a theory any contemporary Everettians would recognize as their own. Indeed, contemporary Everettians rather claim that they can see no important function for the particles in the pilot wave picture! David Deutsch has famously characterized pilot wave theories as "parallel universes theories in a state of chronic denial" [3, p. 225]. To borrow yet another phrase from Bell, misunderstanding could hardly be more complete. For Bell, the particles in the pilot-wave theory, the local beables, are exactly what provide the physical foundation of all claims about the macroscopic behavior of material objects in space. Eliminate the particles and you eliminate the heart of the theory. Deutsch and his fellow Everettians, in contrast, are convinced that the empirical consequences flow already just from the behavior of the quantum state, and the particles serve merely as a "pointer" to one or another branch of the wave function [4, p. 527]. But the wave function, as we have repeatedly mentioned, does not represent any sort of local beable at all. So a theory with only a wave function (or more precisely, with only a quantum state represented by a wave function) has no local beables. It is therefore unclear how such a theory could entail any claims about anything happening in ordinary space or space–time.

The standard contemporary Everettian response to this worry is to gesture at the idea of "functional definitions" A functional definition, if framed at a sufficiently high level of abstraction, is supposed to be the sort of thing that a quantum state (or parts of it), all on its own, can satisfy. But the exact details of how this works, beyond slogans such as "A tiger, instead, is to be understood as a pattern in the physical state" [5, p. 92], are not easy to pin down. Specifying, even in the most broad-brush terms, the sort of "pattern" in *a nonspatial* object that would make a tiger is a conundrum. Contrast this with the corresponding question: what sort of behavior of microscopic beables in space–time (particles, say, or flashes) would correspond, at macroscale, to what we take ourselves to know about tigers? It would not be difficult to make possible answers to such a question, and to judge whether a proposed distribution of local beables could possibly be a tiger, without engaging in a speck of functional analysis of anything. Tigers have characteristic shapes and sizes and move in characteristic ways. All of this could be read off of the behavior of the local beables without further ado.

This methodological role of local beables as the part of the ontology of a theory whose behavior determines the theory's observable consequences underpins Bell's concern. It is this role that explains why Bell never even considers a theory bereft of local beables. He simply would not know what to make of it as an account of the world we take ourselves to inhabit. But with the local beables in place the empirical consequences of a theory are easy, in principle, to identify.

The requirement that a theory posit some sort of local beables in "ordinary space" is not very stringent. As we have already seen by example, many different sorts of local beable are available: particles, fermion number densities and flashes can work, as could local fields (as in electromagnetic theory) or microscopic strings. There is a similar wealth of options for the exact specification of "ordinary space." Classical space–time works, of course, as do the space–times of special and general relativity. Extra compactified spatial dimensions, such as those postulated by string theory, would pose no difficulty. Nor would a space–time structure that became discrete at microscopic scale.

The latitude in the choice of both local beables and the space–time they inhabit arises from the same source: the data against which the theory is tested are ultimately reported as the behavior of things at macroscopic scale. So the detailed, precise microscopic geometry and distribution of local beables according to the theory merely need to *coarse-grain* the right way to provide an adequate account of the data. It is obvious, and has been from antiquity, that both a fundamentally continuous and a fundamentally discrete space–time structure are consistent with the world as we experience it. Not for nothing is detail below a certain scale called "microscopic." Since it is too small to see or otherwise access directly, our evidence about it is always mediated by macroscale events and objects. Many different microscopic structures could coarse-grain into the four-dimensional space–time occupied by macroscopic objects to which we have direct perceptual access.

19.4 Bell and Bohr

The insistence that for a physical theory to be empirically acceptable it needs to account for the macroscopic structure and motions of material bodies not only is central to Bell's account of local beables, but also was central to the Copenhagen interpretation. Bell's commentary about Bohr's position here is notable. He strongly approves of Bohr's insistence that there *must* be a "classical" part of the ontology of a physical theory if the theory is to be comprehensible. "Classical" here does not mean "obeying the laws of classical physics." Of course, if any physical theory is to be empirically adequate it must somehow entail the existence of macroscopic bodies that very nearly obey the laws of classical physics because the laws of classical physics, in many, many circumstances, describe the behavior of bodies to very, very high precision. But Bell's use of "classical" is essentially identical to his use of "local beable": the classical part of the ontology is the local stuff that is *just there* in space–time, independent of whether it is being "observed" or "measured." Observation and measurement themselves, as Bell insists, must be built out of the things that are just there on their own.

Bell's complaint about Bohr and Copenhagen, then, is not that the account fails to have any local beables at all. It is rather that the observation–independent, locally existing objects postulated by Bohr are all *fundamentally macroscopic*. That is, Bohr is committed to macroscopic local beables *while simultaneously denying the existence of any microscopic local beables*. Bell finds such a position conceptually unacceptable. One part of its unacceptability arises from the fact that "macroscopic" is a vague term: it has no precise meaning. But that does not really get to the heart of the matter. Even if one were to propose some precise criterion of "macroscopic," it is just not at all clear how macroscopic items can have any spatiotemporal characteristics apart from those they inherit in the obvious way from their microscopic local parts. If a tiger, as a whole, has no small parts that are situated in parts of space–time, it is extremely obscure how *it* can be so situated. But if a tiger has microscopic parts situated in space–time, then its spatiotemporal structure is nothing more than the collective spatio-temporal structure of those parts.

Regarded in this way the abstract structure of the Copenhagen theory, as presented by Bell, is identical to the abstract structure of the pilot wave theory, and of the Everett(?) theory, and of GRW flash theory. Each of these theories has a bipartite division of its ontology. One part, the nonlocal beable(s) of the theory, is the quantum state or quantum states. The other part is some sort of local beable. The nonlocal part must, of course, have some influence on the behavior of the local items, since the observable consequences of the theory all flow from the behavior of the local items. In the pilot wave and Everett(?) theories the local ontology is a collection of particles, in the GRW flash theory it is a set of flashes, in the Copenhagen theory it is a collection of macroscopic "classical" objects including laboratory equipment. These theories may disagree about the dynamics of the nonlocal part (pilot wave and Everett(?) vs. GRW) and may disagree about how the behavior of the local items is constrained by the nonlocal part (pilot wave vs. GRW vs. Everett(?)). But these are disagreements of detail between theories with the same general architecture. And the articulation of these theories (pilot wave, Everett(?) and GRW flash theory) is uniformly clear and mathematically precise.

Bell understands Copenhagen to have exactly the same overall architecture, but implemented in a vague and imprecise way. Here is the description from his masterpiece of foundational discussion, "Against 'Measurement'":

Then came the Born interpretation. The wavefunction gives not the density of *stuff*, but rather (on squaring its modulus) the density of probability. Probability of *what* exactly? Not of the electron *being* there, but of the electron being *found* there, if its position is "measured."

Why this aversion to "being" and insistence on "finding"? The founding fathers were unable to form a clear picture of things on the remote atomic scale. They became very aware of the intervening apparatus, and for the need for a "classical" base from which to intervene on the quantum system. And so the shifty split.

The kinematics of this world, on the orthodox picture, is given by a wavefunction (maybe more than one?) for the quantum part, and classical variables – variables which *have* values – for the classical part: $(\Psi(t, q, \ldots), X(t), \ldots)$. The Xs are somehow macroscopic. This is not spelled out very explicitly. The dynamics is not very precisely formulated either. It includes a Schrödinger equation for the quantum part, and some sort of classical mechanics for the classical part, and "collapse" recipes for their interaction.

It seems to me that the only hope of precision with the dual (Ψ, x) kinematics is to omit completely the shifty split, and let both Ψ and x refer to the world as a whole. Then the xs must not be confined to some vague macroscopic scale, but must extend to all scales. In the picture of de Broglie and Bohm, every particle is attributed a position $x(t)$. Then instrument pointers – assemblies of particles – *have* positions, and experiments *have* results.

[2, p. 228]

The foundational significance of Bell's insistence on the postulation of local beables in any precise theory cannot be given clearer or more concise expression.

Even "instrumentalists" who want to abjure all speculation about the microscopic are committed to the real existence of their laboratory apparatus. Any physical theory that purports to describe the physical world must be able to encompass the apparatus within its scope. The postulation of macroscopic localized bodies thereby justifies the search for the microscopic local beables of which they are composed.

Equally important in this discussion are the consequences for the status of the quantum state. For just as elimination of the "shifty split" drives the demand for local beables down to the microscopic scale, so too it drives the scope of the quantum state up to the universal or cosmological scale. The physical universe, as a whole, is a quantum system. And the entanglement of the wave function implies that the quantum state of the whole cannot be reduced to a collection of quantum states of the parts. This invites a seldom asked question: if, fundamentally, there is only one quantum state, the quantum state of the entire universe, how do we come to be able to ascribe separate quantum states to subsystems and use them so effectively for making predictions? Whereas the relation of the fundamental microscopic local beables to the local characteristics of macroscopic objects is trivial (the macroscopic object is just where its microscopic parts are), the relation of the wave function ascribed to small parts of the universe to the wave function of the universe as a whole is much more opaque. But that is a topic for another paper.

19.5 Conclusion

Bell's analysis of how the local beables postulated by a theory function in deriving its empirical consequences stands on its own as a contribution to the foundations of physics. From this perspective we can understand both what the "measurement problem" of quantum theory is and how it can be solved in a principled way. We can also see how theories that appear on the surface to be quite different – pilot wave theory, GRW, Everett(?), and even Copenhagen – all appeal to the same basic architectonic.[4] From this perspective, it is obvious why, for example, the decoherence of the quantum state, all on its own, could not solve the measurement problem. For it cannot, all on its own, bring any local beables into existence. It also highlights a challenge for the orthodox Everettian position. Orthodox Everettians would certainly renounce Everett(?) as their theory: they will have no truck with real particle configurations at all. But what, then, are the local beables that their theory is committed to? No answer is readily forthcoming. This is a completely different problem from the problems about probability and unique outcomes of experiments that are usually discussed. But it is a problem to be solved if the theory's relation to observational data is to be comprehensible.

Bell's theory of local beables is a signal accomplishment in the foundations of physics completely independent of his famous theorem. But also, at the end of the day, it is not irrelevant to understanding the theorem. Violations of Bell's inequality in experiments done at spacelike separation indicate some sort of nonlocality in the world, exactly the sort of nonlocality that Einstein abhorred. There is a lively debate about how exactly to define the nonlocality at issue. (It has nothing to do with signaling, for example, as is clear from Einstein's complaint about the standard theory.) But no clear and comprehensible account of any sort of locality or nonlocality can even proceed without the postulation of some sort of local beables in space–time. If *nothing* definite ever happens *anywhere* in space–time, how could any question of locality even be posed?

Bell was aware of this connection as well. We have seen that in "The Theory of Local Beables," he remarks,

It is in terms of local beables that we can hope to formulate some notion of local causality.

[2, p. 53]

Skeptics about the significance of Bell's theorem (of which there are still distressingly many) might try to seize on this remark. "So," they might argue, "if I *refuse* to recognize any local beables at all, no question of the holding (or not holding) of local causality in my theory can even arise." Is this then an escape route from the claim that the violations of Bell's inequality by experiments done at spacelike separation show that that actual physics is nonlocal?

Only in the Pickwickian sense that such a blanket refusal to admit any local beables implies that *no such experiments were ever done at spacelike separation at all*. For the results of the experiments – the disposition of the laboratory apparatus at the end of the day – are taken to be local physical facts about what happened in the lab. Throw out all local beables and you throw out the labs and their results altogether. Bohr would not countenance such a thing, as his remarks about the necessity of a classical description of the laboratory operations shows. Metaphors about babies and bathwater spring to mind.

---

[4] For elaboration on this theme, see Alloriet et al. [6].

So not only does the theory of local beables stand on its own as a contribution to the foundations of physics, but also it underpins all analyses of the significance of violations of Bell's inequality. Looking back at that epochal proof from the distance of half a century, the view is sharpened by the other great contributions John Bell made since then

References

[1] Maudlin, T. (2014), What Bell did, *Journal of Physics A: Mathematical and Theoretical* **47**, 424010

[2] Bell, J.S. (2004), *Speakable and Unspeakable in Quantum Mechanics*, 2nd ed., Cambridge: Cambridge University Press.

[3] Deutsch, D. (1996), Comment on Lockwood, *British Journal for the Philosophy of Science* **47**(2), 222–8.

[4] Brown, H.R. and D. Wallace (2005), Solving the measurement problem: De Broglie–Bohm loses out to Everett, *Foundations of Physics* **35**, 517–40.

[5] Wallace, D. (2003), Everett and structure. *Studies in the History and Philosophy of Modern Physics* **34**, 86–105.

[6] Allori, V.S., Goldstein, R. Tumulka, et al. (2008), On the common structure of Bohmian mechanics and the Ghirardi–Rimini–Weber theory, *British Journal for the Philosophy of Science* **59**(3), 353–89.

20

John Bell's Varying Interpretations of Quantum Mechanics: Memories and Comments

H. DIETER ZEH

Abstract

Various interpretations of quantum mechanics, favored (or neglected) by John Bell in the context of his nonlocality theorem, are compared and discussed.

20.1 Varenna 1970

I met John Bell for the first time at the Varenna conference of 1970 [1]. I had been invited on the suggestion of Eugene P. Wigner, who had already helped me to publish my first paper on the concept that was later called decoherence – to appear in the first issue of *Foundations of Physics* a few months after the conference [2]. This concept arose from my conviction, based on many applications of quantum mechanics to composite systems under various conditions, that Schrödinger's wave function in configuration space, or more generally the superposition principle, is universally valid and applicable. In particular, stable narrow wavepackets can represent classical configurations, while their superpositions define *novel individual* properties – such as "momentum," defined as a plane wave superposition of different positions. Superpositions of macroscopically different properties, on the other hand, are regularly irreversibly "dislocalized" (distributed over many degrees of freedom) by means of interactions described by the Schrödinger equation. The corresponding disappearance of certain *local* superpositions ("decoherence") seems to explain the phenomenon of a classical world as well as the apparent occurrence of quantum jumps or stochastic "events" – see Sect. 20.4 or [3] for a historical overview of the conceptual development of quantum theory. So I had never felt any motivation to think of "hidden variables" or any other physics *behind* the successful wave function.

Therefore, I was very surprised on my arrival in Varenna to hear everybody discuss Bell's inequality. It had been published a few years before the conference, but I had either not noticed it or not regarded it as particularly remarkable until then. As this inequality demonstrates that the predictions of quantum theory would require any *conceivable* reality possibly underlying the nonlocal wave function to be nonlocal itself, I simply found my conviction that the latter describes individual reality confirmed. For example, I had often discussed the conservation of total spin or angular momentum in an individual decay process, which requires nonlocal entanglement between the fragments at any distance in a form that was later called a "Bell state". Therefore, this entanglement cannot represent "just information"; information must be physical – anything else would be homeopathy.

Although the first results from crucial Bell experiments (presented at Varenna by Clauser, Horne, Shimony and others) were still preliminary, they assured me that everybody would share my conviction as soon as Bell's argument had become generally known and understood. I certainly did not expect that fifty years later many physicists would still search for loopholes in the experiments or for justifications of nonlocality beyond the wave function, or even deny any microscopic reality in order to avoid contradictions or absurd consequences that result from the prejudice of a reality that has to be local (such as in terms of particles or fields).

So I was particularly glad to hear about John's announcement of a talk "On the Assumption That the Schrödinger Equation is Exact" one or two years after Varenna at a meeting Bernard d'Espagnat had organized in Paris. This title seemed to represent my own ideas, but I will have to come back in Sect. 20.2 to what he really meant.

Before he published his inequality in 1964, John Bell had shown von Neumann's refutation of hidden variables to be insufficient. (The publication of this paper had been delayed until 1966 by some accidents.) Von Neumann's claim had often been cited to defend the Copenhagen interpretation with its irrational "complementarity" concept against such proposals. In Varenna, Bell began his talk by arguing that all physical systems are described by means of two different concepts: classical properties Λ and a wave function ψ. The latter he suspected to be merely "subjective" (the traditional argument for searching for hidden variables). Today we would then call it an epistemic concept, representing incomplete information, but "information" would only make sense for him with respect to the essential questions "about what?" and "by whom?" This remains true although an objective "state" of incomplete information (an unspecified ensemble of unknown elementary states) may then be *operationally* defined by a certain (incomplete) preparation procedure.

It was this kind of clarity in pointing out misconceptions that always impressed me in discussions with John, or in his talks and publications [4]. He never shared the "pragmatic logic" of many physicists who regard arguments as correct just because they somehow lead to the expected or empirically known result. Another example was his objection to some operational arguments used at the conference by axiomatic quantum theorists who suggested the application of certain "superselection rules" in order to replace superpositions by ensembles whenever the required observables were not realizable for some general reason. He insisted that not being able in practice to distinguish between a superposition and an ensemble consisting of its components does not prove them to be the same. This conceptual confusion may also occur in connection with decoherence when one uncritically interprets the reduced density matrix of a subsystem as representing an ensemble rather than entanglement (see Sect. 20.3). Bernard d'Espagnat had already coined the terms "proper" and "improper mixtures," respectively, to distinguish these two cases. A related third example that comes to mind is his very politely formulated criticism of Hepp's attempt to justify ensembles of measurement outcomes by means of the purely formal limit of an infinite number of subsystems or degrees of freedom [4, Ch. 6].

The assumption of two different realms of physics (quantum and classical: ψ and Λ) represented the consensus among most physicists at that time – even though one knew from the early Bohr–Einstein debate that macroscopic variables, too, had to obey the uncertainty principle in order to avoid inconsistencies. However, in contrast to the majority of physicists, most participants at the conference agreed that the absence of a well-defined borderline between these two realms represented a severe defect that seemed to call for new physics. Decoherence was not yet known as a possible *effective* borderline, while mesoscopic quantum physics had hardly been seriously studied yet. In fact, when I began presenting decoherence arguments to my colleagues in those years, the usual objection was that "quantum mechanics does not apply to the environment."

John then continued his talk by explaining his arguments against von Neumann's exclusion of hidden variables, gave an outline of David Bohm's theory (which had motivated these arguments), and finally derived his inequality, whose violation, predicted by quantum theory, would exclude *local* hidden variables if confirmed by experiment. This conclusion seemed to be a great surprise and to appear almost unacceptable to many participants. Some young and also some not-so-young physicists there were strongly motivated by dialectical materialism (this conference took place in Italy, two years after 1968!). They could not accept any "idealistic" interpretation of physical phenomena, and sometimes tried to propose very naïve classical models that somewhere had to be in conflict with quantum theory. However, Bohr had correctly concluded already in 1924 (when his attempt with Kramers and Slater had failed) that "there can be no simple solution" to the problems presented by the quantum phenomena. Nonetheless, in Bell's (and my) opinion this was no reason to abandon the whole concept of reality, which in a theory must be reflected by a consistent, universally valid and successful description of Nature. For him, the renunciation of reality would be the end of physics (as I understood him). Very probably, this conviction was the major motivation in all his endeavors regarding the foundation of quantum mechanics, but his theorem revealed that quantum reality is in strong contrast to traditional concepts.

At Varenna, I was particularly interested in Bryce DeWitt's talk on the many worlds interpretation, because I had mentioned Everett's ideas myself as the only remaining (but possible) solution if the Schrödinger equation was assumed to be exact, universal and complete. But I felt a bit confused when I saw DeWitt translate Everett into the Heisenberg picture. For me, Everett's main point was an evolving wave function of the universe. He had attended lectures given in Princeton by von Neumann, who had described the measurement process in purely wave mechanical terms, assuming the pointer position to be represented by a moving narrow wavepacket rather than a classical variable. This picture of quantum mechanics (which Wigner always meant when he spoke of its "orthodox interpretation") seems to have also influenced Richard Feynman [5]. Only much later did I understand that for DeWitt and David Deutsch, "many worlds" meant many trajectories in configuration space (or Feynman paths), while for Everett and me, this concept meant many, in excellent approximation dynamically "autonomous," wavepackets, which may possibly even form an overcomplete set (see Sect. 20.4). For example, while Deutsch regards a quantum computer as an example of many worlds in action, in Everett's sense they must all remain part of one "world" in order to lead to one quasi-classical result that can be observed and used by humans. Only if there were *macroscopically* different intermediate states of the computer could their superposition give rise to different "worlds" by their decoherence – and thus ruin the quantum computer. These different formal representations of "reality" (by classical configurations or by wavepackets) are also relevant to Bell's different interpretations of quantum mechanics, which I will now discuss.

20.2 Bell on Bohm's Theory

Although Bohm and Hiley were present at Varenna (as well as Andrade e Silva, who represented Louis de Broglie), I first understood Bohm's theory [6] when studying Bell's Varenna contribution [4, Ch. 4]. He presented this theory as a "simple example" for hidden variables, even though it was in contrast to his introductory remarks: it neither *replaced* the wave function ψ nor explained it in terms of an ensemble of hidden variables. In more recent language, this theory is "ψ-ontic," but in addition it assumes the existence of hidden variables λ that are here identified with the prequantum variables (such as particle positions and field amplitudes): it is not "ψ-complete." So these variables are isomorphic to the arguments of his wave function, while the appearance of particles (such as photons) for quantum *fields* remains an open problem. Nonetheless, this theory allowed Bohm to assume the Schrödinger equation to be exact (the same as later in Everett's theory!), and a classical configuration of the world to be dynamically guided by the arising wave function instead of obeying Hamilton's equations. Bell meant essentially Bohm's theory by the title of his talk that I first heard in Paris, where the Schrödinger equation is assumed not only to be exact, but also to be universal. There are no strictly classical variables Λ any more (they are simply functions of the λ's), but Bell regarded it as an important advantage that Bohm's theory does not need the "notoriously vague concept of a reduction of the wave packet."

However, he also remarked that "what happens to the hidden variables during and after a measurement is a delicate matter." In my opinion this is a serious weak point of the theory, since the λ's have to be *postulated* to form a statistical distribution with probabilities given by $|\psi(\lambda)|^2$, while only one of the trajectories is assumed to describe reality. Although this distribution is dynamically consistent with Bohm's dynamics, (1) no plausible motivation for this statistical assumption (in contrast to the individually treated wave function) is given, and (2) the probabilities would have to change under a change of information by measurements, although no physical carrier of this information is taken into account ("information by whom?"). This comes close to the crucial assumption of an external (human?) observer in the Copenhagen interpretation.

Supporters of Bohm's theory often present special applications in order to illustrate it, although it is evident already from its general construction that *all* its observable predictions must be in accord with traditional quantum mechanics. Therefore, they cannot serve to confirm Bohm's theory. However, these trajectories appear plausible only in simple cases, such as single-particle scatterings. In general, they may have very surprising properties, and little to do with what one would expect or what we *seem* to observe [7]. In my opinion, this fact eliminates the major motivation for this theory, since its "traditionalistic" trajectories can neither be observed nor remembered: they are observationally meaningless.

On the other hand, Bohm was perhaps the first physicist to seriously consider entangled wave functions for macroscopic systems. Shelly Goldstein even claimed that Bohm anticipated the decoherence concept when discussing measurements in his theory. This is a bit of an overstatement and a misunderstanding. In order to describe *successions* of measurements, Bohm had to discuss how the probability distribution of his classical configurations λ is restricted by all previous measurements to the carrier of some small "effective" component of the wave function (essentially identical with "our" Everett branch), and this means first of all that these branches have to remain dynamically autonomous for some time (the way we are using wave functions in practice). This is similar to Mott's early treatment of α-particle tracks in the Wilson chamber, which did *not* yet take into account subsequent decoherence of the droplet positions by their entanglement with an unbounded environment. Only this *real* (irreversible) decoherence explains why different "quasi-classical" wavepackets forming one superposition never meet again in configuration space in order to interfere, that is, why the required autonomy holds "forever" in practice. Within these autonomous branches, wave functions for macroscopic variables are restricted to narrow wavepackets that resemble classical points. Bohm might then have noticed that his presumed fundamental variables λ would become obsolete if these branches were "selected" in some sense. In mesoscopic cases, decoherence theory requires detailed calculations for realistic environments, which were performed during the eighties by Wojciech Zurek, Erich Joos, and many others.

During the decade following Varenna, John Bell presented various versions of his talk about the "assumption that the Schrödinger equation is exact." Like many other fundamental papers at that time, they were often first published in the informal *Epistemological Letters*, since established journals were still reluctant to accept papers on interpretational issues of quantum theory. Only after his inequality had become known to allow crucial experiments to be performed in laboratories did this situation slowly change – one of John's historically most important achievements.

A slightly modified and extended version of these talks (for a special purpose) was published in 1981 under the new title "Quantum Mechanics for Cosmologists" [4, Ch. 15]. It contains a number of important statements. Talking about Bohm, he says that "nobody can understand this theory until he is willing to think of ψ as a real objective field rather than a probability amplitude." This is in explicit contrast to his introductory remarks about ψ as a "subjective" concept. As only one set of λ's is assumed to be real (representing one point somewhere in the myriads of branches of the universal wave function), he compares ψ with the Maxwell fields, which are similarly assumed to exist even where no charged "test particles" are present. But he adds that "it is in terms of the λ" (which he now calls x) "that we would define a psycho-physical parallelism – if we were pressed to go so far." Therefore, he now called Bohm's "hidden" variables "exposed," although their exposure (together with their very existence) remains a model-specific hypothesis. The λ's may appear "more real" than ψ to the traditional mind because they are defined as *local* "beables." This is also why observable quantum nonlocality is often understood as requiring spooky action at a distance rather than the consequence of a nonlocal beable: the "real" wave function. (In a classical context, we similarly prefer to believe that we see objects rather than – more realistically – the light reflected by them, or even the nerve cells excited by the light in the retina and in the brain. In this classical picture, however, all these physical elements and their interactions are local and can be regarded as empirically well established.)

When mentioning Everett's interpretation as another possibility for the Schrödinger equation to be exact, John usually disregarded it as "extravagant" – not for being wrong [4, Ch. 20]. This position appears natural from a traditional point of view. Similarly, Stephen Weinberg declared in an interview about his recent book on quantum mechanics (for *Physics Today Online* of July 2013) that "this effort [of not conceptually distinguishing the apparatus or the physicist from the rest of the world] may lead to something like a 'many worlds' interpretation, which I find repellent." But he had to add, "I work on the interpretation of quantum mechanics from time to time, but have gotten nowhere." In fact, there are strong emotions but hardly any convincing arguments against Everett. This has even led to some "mobbing" by traditionalists of all kinds against Everettians or even against a fundamental role of decoherence. In [4, Ch. 11], Bell raised the objection that Everett's branches are insufficiently defined or arbitrary, but precisely this ambiguity has been removed by decoherence (see Sects. 20.3 and 20.4).

After John had given a version of his talk at Heidelberg in about 1980, we had a brief correspondence, where I tried to point out to him that Bohm's theory is just as extravagant as Everett's, in the sense that Bohm's wave function contains the same components that are regarded as "many worlds" by Everettians. The only difference is that all but one of them are called "empty" by Bohmians. Nonetheless, all the empty parts of the wave function are assumed to *exist*, too, in order to avoid the collapse! We also debated the relation between the concept of reality and that of "heuristic fictions" in physics on this occasion, but the correspondence led to no obvious result. However, it may have had some consequences a few years later (see Sect. 20.3).

When rereading Bell's "Quantum Theory for Cosmologists" for the preparation of this paper, I discovered another astonishing remark about Everett. Bell initially points out that he is not quite sure whether he understands Everett correctly, but then claims a "previously unknown close relationship between Everett and Bohm." He says that "all instantaneous classical configurations λ are supposed to exist" in Everett's theory (the assumption that he regarded as extravagant). This interpretation would come close to Deutsch's identification of (many) "worlds" with a continuum of trajectories in configuration space (cf. Sect. 20.1). Deutsch has indeed repeatedly called Bohm's theory a "many-worlds theory under permanent denial." Bell's remark indicates that he, too, would prefer *beables* to be local – probably a major reason for his favoring Bohm's theory. So Bohm and Deutsch *presume* classical concepts (points in configuration space); this explains why they both do not need decoherence to justify them. If we instead defined "worlds" as consisting of trajectories for macroscopic objects plus wave functions for microscopic ones, we would be back searching for Bohr's borderline between two different realms of physics, using much improved but nonetheless as yet unsuccessful experimental techniques. In contrast, Everett interpreted the world completely and solely in terms of wave functions (he was von Neumann's student). A relation to classical concepts may then be provided only in terms of wave packets in configuration space. This means that Everett is conceptually *not* closely related to Bohmian mechanics with its classical variables λ, but rather to Bell's favorite to come: collapse theories.

20.3 Collapse Theories

In 1987, John Bell surprised his admirers by a drastic change of mind. Inspired by Ghirardi et al. [8], he now advocated collapse theories [4, Ch. 22]. That is, he dropped the assumption that the Schrödinger equation is exact (as it is in Bohm's mechanics) and instead supported what he had previously called the "notoriously vague collapse" – although in a newly specified, hypothetical form. If correct, this proposal would avoid all those myriads of "other" branches of the wave function which he found extravagant in Everett's interpretation, and which had to be regarded as "empty" in Bohm's. This does not necessarily mean that he abandoned Bohm completely. He may simply have started an independent attempt to search for a solution of the quantum problems in terms of a realistic theory, but his radical change of concepts may also indicate that he was not quite happy any more with his previous favorite.

When John von Neumann first formulated his collapse or reduction of the wave function in Chapters V and VI of his book [9], he felt motivated by the need not only to explain definite pointer positions, but also to facilitate a psychophysical parallelism that is applicable to local observers in spite of the nonlocality of the generic wave function. These two different, though related intentions reflect Bohr's and Heisenberg's slightly different understandings of quantum measurements. While the former insisted that indeterministic measurement outcomes have to be described in terms of classical pointer states (which could thereafter be observed in a traditional way by interaction with classical media and observers), the latter had originally regarded the measured properties (including particle positions) as being *created by their observation by humans*. This difference left many traces in the history of quantum measurement theory, but both aspects seem to be relevant in some way even for an ontic interpretation of the wave function (see Sect. 20.4).

The GRW collapse was clearly meant to describe an objective physical process (for a phenomenon that Bohr had regarded as *not* dynamically analyzable). Therefore, these authors concentrated on a process of "spontaneous localization" for the wave functions of macroscopic variables. For this purpose, they postulated a nonunitary and irreversible "master equation" instead of the von Neumann equation for the density matrices of all isolated physical systems. The precise form and strength of its nonunitarity had to be adjusted in order to describe Born's probabilities as part of this new fundamental dynamical law. Von Neumann's equation is equivalent to the linear Schrödinger equation, which they now assumed to apply only approximately in a microscopic limit. They also assumed tacitly that this density matrix describes an (ever growing) ensemble of *potential* wave functions, but the problem is that such an ensemble is not uniquely determined by the density matrix, nor can the latter distinguish between ensembles and an entanglement of the considered system with others.

Indeed, immediately after their paper had appeared, Erich Joos was able to demonstrate [10] that their master equation can be well understood, and even be made precise, within unitary quantum mechanics as a consequence of the unavoidable interaction of macroscopic systems with their realistic environment (later called decoherence). This mechanism could be quantitatively confirmed in several mesoscopic cases, whereas it obviously applies to all macroscopic ones. However, it describes growing entanglement rather than a transition from pure states into ensembles (such as those of different measurement outcomes). Therefore, two questions arise: (1) how would GRW's master equation have to be interpreted in order to describe measurements, and (2) what does the undeniable environmental decoherence, which can hardly *accidentally* lead precisely to the required density matrix (*as though* it were an ensemble), mean for the measurement process?

In order to answer the first question, John Bell proposed a nonlinear stochastic ("quantum Langevin") equation for the dynamics of individual wave functions. This new equation would thus have to replace the Schrödinger equation. The ensemble of potential future wave functions thereby arising can be represented by a density matrix that would then obey GRW's master equation. His specific model postulated that jumps of single-particle wave functions into slightly more localized partial waves occur spontaneously with Born-type probabilities. He assumed the time scale for these jumps to be of order 10^{15} s, but for the center of mass of a multiparticle object this time scale would have to be divided by the number of particles, and so become sufficiently short for such collective variables. However, he also noticed and listed a number of problems, such as the entanglement between particles or the generalization of his proposal to QFT (others have later been added), while he expressed hope that they can be overcome. I doubt that this has ever been achieved for this kind of model, but there exist a wealth of similar and also quite different possibilities for a collapse mechanism, which can be falsified only one after another and when defined exactly. In that case, they would share this property of being falsifiable with Everett's interpretation, which would be ruled out by the discovery of a corresponding violation of global unitarity [11]. Therefore, the *possibility* of a dynamical collapse still exists in principle, and in spite of the fact that none of its proposed versions has ever been verified by experiments. While the confirmation of such hypothetical nonunitarity *would* close this fundamental debate forever, environmental decoherence must remain important, as it seems to describe all as yet observed *apparent* (local) nonunitarities.

The major reason for this undecided situation is that any conceivable collapse dynamics would have to be carefully shielded against all competing decoherence effects in order to be confirmed – an almost impossible requirement in the macroscopic realm. As the reduced density matrix arising from decoherence cannot be locally distinguished from that of an ensemble, it is sufficient FAPP (for all practical purposes – an often misused term that Bell used heavily in his last paper, "Against Measurement" [12]).

However, interaction with the environment can never describe the transition of a global pure state into an ensemble of possible outcomes – it merely describes the dislocalization of all macroscopic superpositions by means of their spatially spreading entanglement. Even if the environment or the apparatus were described by an *ensemble of different initial states* (incomplete information about microscopic initial conditions), as often suggested in the hope that this ensemble might then lead to the expected ensemble of different outcomes, the conclusion of a resulting *superposition* of different outcomes would remain valid for each of its individual members. This very general argument has often been emphasized by Eugene Wigner, who pointed out that the density matrix characterizing the initial ensemble merely *hides* the lasting entanglement. The latter must ultimately give rise to a global superposition of "many worlds" for each *individual* state. For similar reasons, no kind of classical "noise" (represented by an uncertain Hamiltonian, for example) would explain the required ensemble, whereas the often mentioned interaction with quantized gravity is quantum mechanically just a special (though not very relevant) form of environmental decoherence. A genuine collapse has to be explicitly postulated and empirically confirmed as a fundamental deviation from unitarity. The omnipresent formation and spreading of initially absent entanglement, on the other hand, seems to form the general "master arrow of time" characterizing our universe. In this way, it also forms the physical basis for the general concept of time-asymmetric "causality" [13].

In contrast to Bohr's above-mentioned understanding, collapse theories assume the wave function (though not the Schrödinger equation) to apply universally, and therefore to form a complete ontic concept. This wave function must then in principle also describe the brain with its expected specific role in a psychophysical parallelism. In the absence of Bohm's λ's, and under his new assumption of spontaneous jumps, Bell now suggested that consciousness be related to such (again model-specific) "events," while von Neumann had assumed consciousness to be related in the sense of a psychophysical parallelism to the observers' quantum *states*, which he assumed to arise by means of his vaguely defined collapse. Some kind of a psychophysical connection is certainly required in order to understand how our subjective observations are related to the hypothetical real world (that is, how subjective observations come about in objective terms). Einstein spoke of the "whole long chain of interactions from the object to the observer" that we must understand in order to know *what* we have observed.

20.4 Consequences of Decoherence

Already in my first paper on decoherence [2], I had explicitly pointed out that entanglement with the environment (in that paper called "quantum correlations") does *not* lead to the apparently required ensemble of possible outcomes (a proper mixture). It merely explains the absence of certain *local* superpositions, which until then was attributed to fundamental "superselection rules" or a global stochastic collapse. Since no hints of a collapse had ever been observed in controllable quantum systems, I suggested an interpretation similar to Everett's – see also [3, Sect. 4] for details. (Some Oxford philosophers have recently "rediscovered" this successful combination of decoherence and Everett.) This very possibility is sufficient to demonstrate that Bell's claim that "the wave function must either be incomplete or not always right" cannot be upheld. In several subsequent papers during the seventies I even tried to learn more about the physical localization of consciousness in quantum mechanical terms by using the single-sum Schmidt canonical representation for entangled states (assuming a fundamental local observer system in the brain, for example), but this attempt did not turn out to be particularly helpful. Objectively, we can argue only in terms of "robust" properties, such as memory (physically realized in data storage devices or by decohered variables in the brain – even though these are only intermediary concepts in the long chain between object and subject). Such robust states describe quasi-classical properties, which may form discrete sets for digital devices or neuronal frameworks. In the general case, they were later called "pointer states" by Zurek [14], whose readers, however, may have misunderstood them as defining proper mixtures after a decoherence process [15]. Precisely this "naïve" misunderstanding of decoherence seems to have considerably contributed to its popularity, while the introduction of new names, such as "einselection," "quantum Darwinism" or "existential interpretation," did not add any new content to the theory of decoherence.

Most workers in the field of decoherence have indeed restricted their interest to the effects of the environment on the density matrices of local systems. This is sufficient for *most* practical purposes, and it allows one to understand the pseudoconcept of "complementarity" simply as a consequence of different couplings to the environment by means of different measurement devices. Because of this decoherence phenomenon, existing proposals for a fundamental collapse are mostly attempts to mimic what can be explained as environmental decoherence. They are thus based on a prejudice that arose before decoherence was understood as a unitary process. However, there would be far more *other* possibilities for a collapse along Einstein's "whole long chain" in order to explain individual states of awareness (all we know with certainty) in an objective dynamical way. As decoherence describes *apparent* transitions into ensembles and *apparent* quantum jumps, which can even be experimentally confirmed as forming smooth processes, it has indeed often been misinterpreted as a derivation of the probabilistic collapse from the Schrödinger equation. (I remember authors claiming that decoherence saves us from the conclusion of many worlds, although precisely the opposite is true!) This misuse of the density matrix has a long tradition in measurement theory, and John was certainly right to object to it also in connection with decoherence. But his critical position is itself often misused as an argument to entirely dismiss the essential role of decoherence in measurements. In order to understand what decoherence "really" means, one has to analyze the consequences of interactions with the environment on the *individual* global wave function (similar to Bell's formulation of the stochastic collapse mechanism for individual states described in Sect. 20.3).

The essential insight that originally led to the concept of decoherence was that entanglement must be far more common for dynamical reasons than had ever been envisioned or taken into account. It had been overlooked as long as unitarity was believed to apply only to microscopic systems. Ironically, though, global unitarity is able to explain local nonunitarity. However, it is *not* the formal diagonalization of the thereby arising reduced density matrices of all systems that explains "effective ensembles," but rather the fact that an ever-increasing number of components of the wave function become dynamically independent of one another ("autonomous"). In particular, this progressive entanglement means that nonlocal superpositions cannot, in practice, be relocalized any longer in order to describe some "recoherence." This situation is analogous to the justification of Boltzmann's H-theorem in deterministic particle mechanics, where the μ-space distribution is very unlikely to be later affected in a relevant way by nonlocal many-particle correlations that were previously created in chaotic collisions. These correlations must nonetheless persist in order to preserve Gibbs's ensemble entropy for distributions on Γ-space. The permanent "branching" of the wave function into autonomous components that can never interfere any longer is similarly a consequence of Schrödinger's determinism and the high dimensionality of configuration space (using plausible cosmic initial conditions). As long as there are no empirical indications for a collapse mechanism, the existence of other branches of the wave function ("worlds") in addition to the observed one is thus not a matter of "philosophy," but of dynamical consistency.

One may then easily recognize that different pointer positions, dead and alive cats, *and* different states of awareness of an observer (but *none* of their superpositions) can only exist separately within such autonomous branches of the wave function. Therefore, the various autonomous "versions" of each observer that arise from decoherence can only define *separate* "subjective identities" ("many minds") – even though they may have shared their *early* histories. Because of their emerging dynamical autonomy, they can only remember the histories of their own branches, including consequences of superpositions that arose *before* the latter's decoherence. Statistical weights according to the squared norms of the branch wave functions have to be postulated for the sole purpose of correctly "selecting" *our subjectively observed* version by chance; they have no meaning from the objective "birds perspective" that is described by unitary dynamics. These empirically observed weights, which explain the frequencies of results in series of measurements *observed by us* ("Born's rule"), cannot be derived from the objective part of the theory, but they are the only dynamically consistent ones, in the sense that they are conserved under the Schrödinger dynamics. In contrast to other conceivable weights, they are thus not changed by later branchings any longer – and so give rise to "consistent histories" in terms of effective wave functions.

So we have to conclude that the indeterminism we observe in quantum phenomena does not reflect an objective dynamical law. Rather, it is a consequence of the branching histories of all observers in a quantum world. While, in the classical description, the subjective decision about who is "you" among all conscious observers existing in the deterministically evolving universe occurs only once for your lifetime, it is repeated many times every second with respect to your permanently arising new "versions." At least, this would be the consequence of global unitarity; it is not mere fiction. It is also the reason that a wave function often appears to us as representing "just information." You are free to regard a theory as incomplete if it does not determine the individual subjective observer, but this "defect" would then also apply to classical (Laplacean) theories.

Slightly generalizing John Bell's terminology, one may say that the autonomy of individual branches arising from decoherence allows us to replace the formal "plus" of their superposition with an effective phase-independent "and," while an "or" is meaningful only with respect to the individual versions of potential observers. In this Everettian sense, Heisenberg's subjective interpretation of measurement outcomes as being *created* by their observation may thus be justified at last – although not in terms of fundamental particle or other classical concepts. Without taking into account this "subjective individualization" of measurement results (an effective projection in Hilbert space), we could not even prepare pure states for microscopic systems in the laboratory, where we perform series of measurements for this purpose in order to continue the experiment only with the appropriate outcomes.

Unfortunately, John Bell never seriously considered this version or variant of Everett (as far as I know). He may still have regarded it as "extravagant" because of the myriad versions of each observer that have to arise according to the Schrödinger equation. Collapse models, in contrast, may *postulate* that similarly defined branches of the wave function all but one disappear from reality, but an observer, who can only exist in a branch, does not have to bother whether many other versions of himself do exist in other autonomous branches, or rather have disappeared from reality – unless he is one of those rare nonpragmatic physicists such as John Bell.

A pragmatic physicist will in any case *use* the collapse FAPP as soon as decoherence (understood as the dislocalization of initially microscopic superpositions) has become "real" (irreversible in practice) rather than virtual after a measurementlike process, in which a microscopic difference led to different macroscopic consequences. This onset of environmental decoherence defines a natural position for the Heisenberg split. Decoherence also allows us approximately to describe all robust properties of macroscopic systems in classical terms, while most others can locally be treated by conventional statistical methods, such as retarded master equations [13].

The popular objection "but this apparent collapse applies only FAPP" means no more than that this solution of the problem is not what John Bell (and many other physicists) had expected and hoped for. It is nonetheless sufficient, without being based on any novel and speculative postulates. We have learned from Everett that we do not have to expect the collapse to represent an objective physical process that may some day be further specified and confirmed by experiments (although the possibility of such a genuine nonunitarity can never be excluded with certainty). The "effective collapse FAPP" is just a convenient picture, representing the situation of the subjective observer in his branch. So it cannot (and need not) be defined *exactly*, and it may even be assumed to "act" superluminally. Events may consistently be assumed to "occur" even in cases where they will never be observed: the phase relations between different "world" components, where they must still exist according to the Schrödinger equation, become irrelevant for all *potential* observers in their branches as a consequence of decoherence. However, this reasonable convention (in a dynamically consistent quantum world that may be assumed to describe reality) seems to be the source of many misconceptions, such as counterfactuals, new logics, complementarity, an "uncertain" reality, "it from bit," and similar terminology that John Bell never found very helpful or meaningful.

Acknowledgements

I wish to thank Shelly Goldstein, Basil Hiley, Erich Joos and Claus Kiefer for their criticism, comments, and discussions regarding earlier versions of the manuscript.

References

[1] d'Espagnat, B. (1971), *Proceedings of the International School of Physics "Enrico Fermi,"* Course IL, New York: Academic Press.

[2] Zeh, H.D. (1970). On the interpretation of measurement in quantum theory. *Found. Phys.* **1**, 69.

[3] Zeh, H.D. (2013), The strange (hi)story of particles and waves, arXiv:1304.1003.

[4] Bell, J.S. (1987), *Speakable and Unspeakable in Quantum Mechanics*, Cambridge: Cambridge University Press.

[5] Zeh, H.D. (2011), Feynman's interpretation of quantum theory, *Eur. Phys. J.* **H36**, 147.

[6] Bohm, D. (1952), A suggested interpretation of the quantum theory in terms of "hidden" variables, *Phys. Rev.* **85**, 166.

[7] Zeh, H.D. (1999), Why Bohm's quantum theory? *Found. Phys. Lett.* **12**, 197.

[8] Ghirardi, G.C., Rimini, A., and Weber, T. (1986), Unified dynamics for microscopic and macroscopic systems, *Phys. Rev. D* **34**, 470.

[9] von Neumann, J. (1932), *Mathematische Grundlagen der Quantenmechanik,* Berlin: Springer.

[10] Joos, E. (1987). Comment on "Unified dynamics for microscopic and macroscopic systems," *Phys. Rev. D* **36**, 3285.

[11] Arndt, M., and Hornberger, K. (2014), Testing the limits of quantum mechanical superpositions, *Nature Physics* **10**, 271.

[12] Bell, J.S. (1990), Against measurement, *Physics World*, August, 33.

[13] Zeh, H.D. (2007), *The Physical Basis of the Direction of Time*, Heidelberg: Springer.

[14] Zurek, W.H. (1981), Pointer basis of quantum apparatus: Into what mixture does the wave packet collapse? *Phys. Rev. D* **24**, 1516.

[15] Camilleri, K. (2009), A history of entanglement: Decoherence and the interpretation problem, *Studies in the History and Philosophy of Modern Physics* **40**, 290.

21

Some Personal Reflections on Quantum Nonlocality and the Contributions of John Bell

BASIL J. HILEY

Abstract

I present the background of the Bohm approach that led John Bell to a study of quantum nonlocality, from which his famous inequalities emerged. I recall the early experiments done at Birkbeck with the aim of exploring the possibility of 'spontaneous collapse', a way suggested by Schrödinger to avoid the conclusion that quantum mechanics is grossly nonlocal. I also review some of the work that John did which directly impinged on my own investigations into the foundations of quantum mechanics and report some new investigations towards a more fundamental theory.

21.1 Introduction

My first encounter with quantum nonlocality was in discussions with David Bohm when I joined him as an assistant lecturer at Birkbeck in the sixties. Although my PhD thesis had been in condensed matter physics under the supervision of Cyril Domb, I had always been fascinated and puzzled by quantum phenomena, and when the opportunity to study the subject with Bohm came up, I took it. I would listen to the discussions between Bohm and Roger Penrose, who was at Birkbeck at the time, and it soon became clear that the prescribed interpretation of quantum mechanics that I was taught as an undergraduate left too many questions unanswered.

At that time there was a very strange atmosphere in the physics community. I was often informed that there were no problems with the interpretation of the quantum formalism. If I did not follow the well-prescribed rules of the quantum algorithm, I would be 'wasting my time'. There was no alternative, just do it! But Louis de Broglie and David Bohm had shown there was another way. However, their views were strongly opposed by many, and as far as I could judge their reasons did not seem to be based on mathematics or on logic, but on a preconceived notion that the formalism was, in principle, the best we could do. The experimental data were bizarre when looked upon from the standpoint of classical physics, but the uncertainty principle somehow prevented us from examining the actual process in detail. Reality was somehow veiled (see [1]). Were we condemned to use a mere algorithm or was there an underlying ontology?

In the late sixties I learnt that there was another physicist working at CERN, a mecca of orthodoxy no less, who shared our worries about quantum mechanics, and that was John Bell. I used to meet him at various conferences in Europe, where we would exchange ideas. As a reflection of the attitudes of the time, one day John remarked that he felt he had to keep a low profile at these meetings because his CERN colleagues might disapprove!

Our discussions would be centred on the work of David Bohm [2]. His proposals had met not merely with scepticism but also, surprisingly, with open hostility. Yet all he had shown was that if we focus on the real part of the Schrödinger equation, we find an equation that has the same form as the classical Hamilton–Jacobi equation, but with the addition of an extra term. This term was called the 'quantum potential', and with it one could account for quantum behaviour without giving up the notion of a particle following a trajectory. More importantly, this could also be done without giving up the uncertainty principle.

The story of the Bohm approach is by now well known through either our book, "The Undivided Universe" [3], or the excellent book by Peter Holland [4], or alternatively, through an offshoot of what is now called "Bohmian mechanics" [5]. "The Undivided Universe" contains Bohm's last words on his own approach, as he died just as we finished it. By now the approach is well known, so I do not need to go into the details here. I want to concentrate on one aspect of the approach that inspired John Bell to reconsider the foundation of quantum theory, namely, the appearance of nonlocality. As John remarked about Bohm's work, "he had seen the impossible done", but it was the appearance of the nonlocality in the original paper [2] that caught his attention, and he went on to reflect deeply on this phenomenon before he finally produced his famous inequality [6].

21.2 Nonlocality and Bohm

To see how nonlocality arises in the Bohm approach, we need to look at the two-body entangled state, with wave function $\psi(x_1, x_2, t)$, where x_1 and x_2 are the coordinates of the two particles at a given time, t. Under the polar decomposition $\psi(x_1, x_2, t) = R(x_1, x_2, t) \exp[iS(x_1, x_2, t)/\hbar]$, R and S being real, the real part of the Schrödinger equation becomes what we call the quantum Hamilton–Jacobi equation,

$$\partial_t S + \frac{(\nabla_1 S)^2}{2m_1} + \frac{(\nabla_2 S)^2}{2m_2} + Q + V = 0. \tag{21.1}$$

The similarity to the two-particle classical Hamilton–Jacobi equation is clear, except for an additional term,

$$Q_\psi(x_1, x_2, t) = -\frac{\hbar^2}{2m_1} \frac{\nabla_1^2 R(x_1, x_2, t)}{R(x_1, x_2, t)} - \frac{\hbar^2}{2m_2} \frac{\nabla_2^2 R(x_1, x_2, t)}{R(x_1, x_2, t)}. \tag{21.2}$$

This is the quantum potential, and it is clearly nonlocal, since it depends on the positions of both particles, x_1 and x_2, at the *same* time. Notice that if the wave function is a product $\psi_1(x_1, t)\psi_2(x_2, t)$, the quantum potential becomes local, as can easily be seen by substitution into the Schrödinger equation.

Although this result was clearly stated in Bohm's original 1952 paper, the main discussion focussed on whether the ideas proposed by Bohm violated the von Neumann 'no hidden variables' theorem (see von Neumann [7]). This claimed to show that it is not possible to reproduce the results of standard quantum mechanics by adding so-called 'hidden variables'. But Bohm had not added any new variables; he had simply *interpreted* the coordinate $x(t)$ that appears in the wave function as being the actual position of the particle, and *interpreted* ∇S as the momentum of the particle, so no new variables were added to the formalism.

Clearly this interpretation assumes that a particle has, simultaneously, a well-defined position *and* a well-defined momentum, so surely this must violate the uncertainty principle? Actually no, because the uncertainty principle only claims that it is not possible to *measure* simultaneously the position and momentum. It does not claim that it is impossible for the particle to *possess simultaneously* a position and a momentum. Note that if simultaneous measurement is ruled out, then it is impossible to experimentally determine whether the particle actually has or does not have simultaneous values of x and p. The standard approach merely *assumes* that a particle cannot possess well-defined values of these variables simultaneously, whereas Bohm *assumes* the particle can have simultaneous x and p, but we do not know what the values are.

So how does the uncertainty principle enter in? Here we follow Wheeler [8] and assume that measurement does not merely reveal those values present; the measuring instrument becomes active in the measuring process, producing the appropriate eigenvalue of the operator that is being measured. In this participatory act, the complementary variables are changed in an uncontrollable way. Thus, the statistical element of a quantum process remains.

There is much evidence that the Bohm approach, either in the form discussed by Bohm and Hiley [3] and Holland [4] or in the form of its offshoot, Bohmian mechanics, as discussed by Dürr, Goldstein and Zanghì [9] and Dürr and Teufel [5], does not lead to any internal inconsistencies and, in fact, contains all the same experimental predictions as does the standard approach, so there are no experimental differences between the two approaches. However because particles are assumed to have well-defined positions at all times, the approach is able to bring out very clearly the nonlocality that appears with entangled states.

It is this feature that John Bell noticed and asked, "Are all theories, which attribute properties to local entities while reproducing the results of quantum mechanics, nonlocal?" Before answering that question, let us consider another question, "Were there any clues of nonlocality in the usual formulation of quantum theory?"

21.3 Nonlocality in the Standard Quantum Formalism?

In this section I would like to highlight some key arguments that suggested in fact that nonlocality was at the heart of quantum phenomena. It is not merely a feature that arises only when the real part of the Schrödinger equation is considered. To bring this out, I begin with a quotation from a review paper by Dirac [10]:

For an assembly of the particles we can set up field quantities which do change in a local way, but when we interpret them in terms of probabilities of particles, we get something which is non-local . . .

He goes on,

I think one ought to say that the problem of reconciling quantum theory and relativity is not solved. The concepts which physicists are using at the present time are not adequate.

We can go back much earlier to find unease at the appearance of some form of nonlocality. History treats the famous Einstein, Podolsky, Rosen paper [11] as the beginning of the debate on nonlocality, but as the title emphasises, their paper concentrated on the *completeness problem*, that is, whether we need additional parameters (hidden variables) to complete the quantum formalism and rid it of its statistical features. The question of nonlocality was not directly confronted. In reaching their conclusion, EPR added a caveat:

Indeed, one would not arrive at our conclusion if one insisted that two or more physical quantities can be regarded as simultaneous elements of reality *only when they can be simultaneously measured or predicted*. On this point of view, since either one or the other, but not both simultaneously, of the quantities *P* and *Q* can be predicted, they are not simultaneously real. This makes the reality of *P* and *Q* depend upon the process of measurement carried out on the first system, which does not disturb the second system in any way. No reasonable definition of reality could be expected to permit this.

Thus they were ruling out the possibility that measurement could be nonlocal and participatory.

One should note that Schrödinger [12] had pointed out in a much earlier paper that there was a problem with the quantum description of a two-body system. He noticed that, in the presence of a small coupling between two systems, the state of system # 1 becomes entangled with the state of system # 2, no matter how far apart they are spatially. Schrödinger [13] emphasised this again, but more clearly, in a later paper when he wrote,

If for a system which consists of two entirely separated systems the representative (or wave function) is known, then the current interpretation of quantum mechanics obliges us to admit not only that by suitable measurements, taken on one of the two parts only, the state (or representative or wave function) of the other part can be determined without interfering with it, but also that, in spite of this non-interference, the state arrived at depends quite decidedly on what measurements one chooses to take – not only on the results they yield.

This control of the state of a distant system by an experimenter was to Schrödinger

rather discomforting that the theory should allow a system to be steered or piloted into one or the other type of state at the experimenter's mercy in spite of his having no access to it.

Schrödinger's papers did not seem to attract much attention at the time. Rather it was Bohr who dominated the debate with Einstein. The key element of Bohr's response to what Einstein called "spooky action at a distance" appears in at least two places with identical wording, wording that I believe John Bell found, to say the least, unclear. The exact wording used by Bohr [14, p. 700] is,

From our point of view we now see that the wording of the above mentioned criterion of physical reality proposed by Einstein, Podolsky and Rosen contains an ambiguity as regards the meaning of the expression *without in any way disturbing a system*. Of course there is no question of a mechanical disturbance of the system under investigation during the last critical stage of the measuring procedure. But even at this stage there is essentially the question of an *influence* [my emphasis] on the very conditions which define the possible types of predictions regarding the future behaviour of the system.

The puzzle is clear: 'What does the word "influence" actually mean?' Notice it is not a mechanical disturbance, but an 'influence'. That is an extraordinary word to use in twentieth century science. Could it be that Bohr had already anticipated the appearance of a term like the quantum potential in the real part of the Schrödinger equation?

It is the quantum potential in equation (21.1) that clearly brings out the nonlocality. Thus there is a new quality of energy emerging only at the quantum level, and it is this energy that is responsible for the nonlocality that arises in entangled states.

Schrödinger's equation is nonrelativistic, so could it be that this nonlocal effect is an artifact of a nonrelativistic theory? The problem with that claim is that we know entanglement is a necessary and indispensable feature of the standard quantum formalism. For example, we know that in helium, the ground state wave function of the two electrons is an entangled state, and this is necessary to produce the correct energy levels. Any relativistic corrections will merely produce small modifications, fine structure effects, and not overrule the general result. Perhaps, as Schrödinger argued, when the particles are separated by a sufficient distance, the nonlocality *spontaneously localises*. Schrödinger writes,

It seems worth noticing that the paradox could be avoided by a very simple assumption, namely … the *phase relations* between the complex coupling coefficients become entirely lost as a consequence of the process of separation. This would mean that not only the parts, but the whole system, would be in the situation of a mixture, not of a pure state. It would not preclude the possibility of determining the state of the first system by *suitable* measurements in the second one or vice versa. But it would utterly eliminate the experimenter's influence on the state of that system which he does not touch.

This 'spontaneous localisation' process would remove the unwelcome feature that 'steers or pilots the distant system'. Clearly the theory has these unpleasant features, but are such effects found in Nature? Surely we must now turn to experiment.

21.4 Spontaneous Localisation–New Physics?

21.4.1 Existence of Entangled State for Separated Systems

It was in the mid-sixties – as, unknown to us, John Bell was developing his ideas on nonlocality – that David Bohm, experimentalist David Butt and I discussed the possibility of looking for evidence of spontaneous localisation experimentally. Rather than using position and momentum, we decided to use the polarisation properties of γ-rays, and confront directly the spin version of the EPR argument originally introduced by Bohm [15].

At this stage we were not aware that John was already in the process of publishing his inequalities paper in a new journal which, ironically, ceased publication after the first one or two volumes [6]. This all happened in the days before the Internet, when ideas filtered down very slowly.

Be that as it may, our interest was in the experimental verification of the existence of entangled states over macroscopic distances. Specifically, would the entanglement remain as the particle detectors were moved further apart or would there be some form of as yet unknown phase randomisation, as suggested by Schrödinger himself?

At that time, in the mid-sixties, the best experimental evidence for entanglement existing over larger distances [~0.5 m] was supplied by Langhoff [16]. This experiment involved measurements of the correlation between the planes of polarization of the correlated γ-rays produced by the annihilation of the s-state positronium produced by some radioactive source. Langhoff used two different sources, Na^{22} and Cu^{64}, in his experiment. The normalized ket for a pair of γ-photons immediately after annihilation is

$$|\psi\rangle = \frac{1}{\sqrt{2}} [|x_1\rangle|y_2\rangle - |y_1\rangle|x_2\rangle], \tag{21.3}$$

where $|x\rangle$ and $|y\rangle$ are the kets of the individual γ-photons plane-polarized in the x and y directions, respectively (see [17]). Clearly equation (21.3) is an entangled state. The photons propagate in opposite directions along the z-axis.

After travelling about 0.5 m, the emerging photons undergo Compton scattering in silicon crystals. The scattered photons are then detected by a pair of scintillation counters. Coincidences between the two counters are recorded when the azimuthal angles of the two counters are identical ($\phi = 0$) and when they differ by 90° ($\phi = \pi/2$). For the wave function (21.3), this ratio of coincidences scattered through the same scattering angle θ is given by

$$\rho = \frac{N_{\phi=\pi/2}}{N_{\phi=0}} = 1 + \frac{2\sin^4\theta}{\gamma^2 - 2\gamma\sin^2\theta}, \tag{21.4}$$

where

$$\gamma = 2 - \cos\theta + \frac{1}{2 - \cos\theta}.$$

This gives a theoretical value $\rho = 2.60$ at $\theta = 90°$, with a maximum of 2.85 at $\theta = 82°$. These are the values before corrections are made for the angular resolution of the apparatus and before corrections for the geometrical size of the equipment are taken into account. For the Langhoff experiment, these corrections produced a theoretical maximum value for $\rho = 2.48 \pm 0.2$. The two sets of experimental results gave $\rho_{exp} = 2.50 \pm 0.03$ for positrons produced by Na^{22} and $\rho_{exp} = 2.47 \pm 0.07$ for positrons produced by Cu^{64}. Thus there is a good agreement between theory and experiment on a 0.5 m separation.

21.4.2 Is the Distance Between the Scattering Centres Large Enough?

Clearly nonlocality exists over such a separation, but is this distance large enough to make sure randomisation of the phase relation between the two photons has had time to take place? Again, is this distance far enough to notice any reduction in the 'influence' that Bohr talks about? To answer these questions, we need to speculate about what factors could maintain these correlations over greater distances.

The first thing that comes to mind is the notion of a coherence length. It is well known that interference effects depend on this feature, so it could be that we require that the photons be separated by distances greater than their coherence length before we see any evidence of a change in the pair wave function. Thus we need to ensure that the scattering centres of the two annihilation γ-rays are separated by a distance that is greater than their combined coherence lengths.

The decay of s-state electron–positron pairs in Cu^{64} has been shown to occur with only a one-component lifetime and is found to be (191 ± 3) ps [18], giving a coherence length of 0.057 m. From this we can assume that the width of the wavepacket associated with each photon should be less than about 0.12 m. However, this width is a somewhat ambiguous notion, and (191 ± 3) ps therefore it is necessary to explore situations in which the source–scatterer distances are an order of magnitude greater than this to ensure that the overlap of the wavepackets is indeed negligible. This suggests that a distance of over 1 m is required to make absolutely sure that the γ-rays are sufficiently separated so that their individual coherent lengths do not overlap. As we have already indicated, in the Langhoff experiment, the separation between the scatters was only about 0.5 m, so there is room for doubt, even though this doubt is small.

There is also another feature that we must consider, namely to ensure the scattering events are *spacelike separated*. If this condition is satisfied and the pure state predictions are confirmed, then we can rule out the possibility of any local hidden-variable theory that might involve some form of signalling at luminal or subluminal velocities as one of the quanta is being measured.

Ruling out this possibility will obviously depend upon the resolving time of the detection counters. In the Langhoff experiment [16], the resolving time of the scintillation counters was 5 ns. This gives a photon path length of 1.5 m, implying that the separation of the detectors must be at least this distance to ensure that the detection events are spacelike separated. Unfortunately, as the separation in this case is only 0.5 m, the detection events are not spacelike separated. This again means that we cannot rule out some kind of coupling mechanism, no matter how unlikely.

Although in the Langhoff experiment the events were not spacelike separated, some later results obtained by Faraci et al. [19] reported that a decrease in the anisotropy was found for greater separation distances (5.6 m). Such a distance would satisfy both criteria discussed above, suggesting that there was indeed some form of spontaneous localisation taking place, so Wilson, Lowe and Butt [WLB] [20] decided to repeat these experiments, but this time, systematically increasing the source–scatterer distances.

Unfortunately, as the separation increases, the detection efficiency of the scatterers decreases dramatically, varying approximately as the square of the scatterer dimensions and the square of the solid angle subtended by the second detector at the scatterer for a fixed source–polarimeter separation. Thus there is a need to start with physically larger scatterers. Unfortunately, with large scatterers, the geometric corrections become more difficult, which makes it hard to calculate absolute values of the anisotropy, so that the final results will be less meaningful. Thus the WLB experiment was designed not to measure absolute values of the anisotropy factors, but to look for changes in the anisotropy as the separation distances were systematically increased to spacelike separations and beyond.

To estimate the order of magnitude of change expected, WLB used a suggestion of Furry [21] who argued that the final state will not be given by (21.3) but by the mixture

$$\text{either} \quad |x_1\rangle|y_2\rangle \quad \text{or} \quad |y_1\rangle|x_2\rangle. \tag{21.5}$$

A more general final state was suggested by [22], namely,

$$\rho' = \int \int P(\theta, \phi)|1, +\theta, \phi\rangle|2, -\theta, \phi\rangle\langle 1, +\theta, \phi|\langle 2, -\theta, \phi|d\theta d\phi. \tag{21.6}$$

In both cases the anisotropy should decrease to about 1.0. With geometric corrections in all these cases, the maximum anisotropy factor is considerably less than 2. Hence, if there is any form of spontaneous localisation process taking place, we would expect to find a significant decrease in the anisotropy as the source–scatterer distances were increased.

The experimental results show that there is no detectable change in anisotropy as the distances between the scatterers are systematically increased to 4.9 m. Two asymmetric arrangements of the detectors were also investigated. In the first, the left-hand scatterer was placed 0.6 m from the centre, while the right-hand one was placed at 1.6 m from the centre. In the second measurement, the respective separations were chosen to be 1.5 and 2.45 m. In neither case was any significant change in the anisotropy detected.

The distances chosen by WLB ensured that the detector separation was well beyond the coherence length criteria and the detection events were spacelike separated. The resolving time of the detectors used by WLB was 1 ns, giving a light length of 0.3 m as against a final separation of 4.9 m. In these sets of experiments, they were unable to confirm the results of Faraci et al. [19]. The details of the WLB experiment are found in their paper and will not be discussed further here. A later experiment was performed by Paramananda and Butt [23], with the separation distance between the two scatterers extended to 23 m. Again, no change in the anisotropy was seen. Today experiments show that these correlations hold for distances of up to 144 km [24].

21.5 Quantum Nonlocality

These experimental results clearly show that the possibility that entangled states spontaneously localise to mixed product states as suggested by Schrödinger [25] and Furry [21] must be ruled out. This leaves us with either Bohr's original explanation using his notion of 'influence', or with the Bohm interpretation where the appearance of nonlocal polarisation correlations was maintained by the quantum potential energy. Of course, there may be other explanations, but whichever way we choose to go, quantum mechanics contains an element of nonlocality totally foreign to classical physics. Naturally the appearance of nonlocality in physical theories has been strongly resisted, as it appears to deny the possibility of doing science at all if, in the extreme, everything is inseparable from everything else. Fortunately, not all states are entangled and there are many product states which are local.

From the above discussion, we see that the clearest indication of nonlocality appears in the Bohm treatment of the quantum formalism. In this approach, it is the reintroduction of the notion of a localised particle to which independent properties can be attached that shows clearly the presence of nonlocality. But attaching properties to particles immediately reignites the debate about 'hidden variables' and the completeness problem raised by EPR. In addition to this, criticisms have been raised on the ground that it was a blatant attempt to return to the ideas of classical physics. However, this is a curious criticism, because it is this approach that highlights nonlocality, a feature that is totally foreign to classical thinking.

Fortunately, John Bell [26] very quickly understood the significance of what Bohm [2] had done, writing,

Bohm's 1952 papers on quantum mechanics were for me a revelation . . . I have always felt since that people who have not grasped the ideas of those papers (and unfortunately they remain the majority) are handicapped in any discussion of the meaning of quantum mechanics.

Bohm's approach, in Bell's own words, had replaced 'unprofessionally vague and ambiguous' features of the standard approach with a 'sharpness that brings into focus some awkward questions'. In fact, it was the appearance of nonlocality in equation (21.1) that triggered the question, "Will all theories that attach properties to individual particles and reproduce the results of quantum mechanics be nonlocal?"

In a famous paper in 1964, his inequalities first appeared. These inequalities became the focus of an intense theoretical debate as to their precise meaning and implications. This discussion also produced a wealth of experiments showing that the inequalities were, in fact, violated, confirming the early results – the existence of nonlocality in quantum phenomena. A discussion of the inequality and its experimental consequences will be found elsewhere in this volume, so I will not make any further comments here.

21.6 Global Properties and Quantum Nonlocality

21.6.1 Orthogonal Groups, Spin Groups and the Clifford Algebra

For me, John Bell's most important contribution to quantum physics has been his constant criticism of the standard interpretation of the formalism itself. This criticism often echoed some of my own worries about the interpretation. Contrary to classical physics, which concerns the description of what is actually going on in an evolving physical process, quantum physics focusses on the measurement and what the observer 'knows'. Why should nature care about what we, products of nature, know or do not know in order to evolve?

Traditionally physics tried to find descriptions that are entirely independent of the observer and, more importantly, of what the observer knows or does not know about the unfolding process. In quantum mechanics, we seem to have given up this tradition. The standard interpretation talks about the results of measurement *A*, followed by the results of measurement *B*, but then declares that it is not possible talk about what goes on in between measurements. In other words, one only speaks about the *results of external intervention.* Indeed, as Bohr [27] writes,

As regards the specification of the conditions for any well-defined application of formalism, it is more-over essential that the *whole experimental arrangement* be taken into account. In fact, the introduction of any further piece of apparatus like a mirror, in the way of a particle might imply new interference effects essentially influencing the predictions as regard the results to be eventually recorded.

Bell [28] came out very strongly against giving measurement such a prominent position. For him, the 'measurement problem', which still plagues quantum theory, arises because the measuring instrument is singled out to play a special role. But surely any instrument is simply a collection of atoms governed by the very same laws that we are investigating, so what makes it special? All other physical processes end up in linear superpositions of eigenstates; the exception is the measuring instrument. Why? One essentially has to resort to a tautology: 'a measuring instrument is a system that reveals an eigenvalue of some operator'. We seem to be arguing that what the measuring instrument does is to replace the logical 'and' with the logical 'or', without giving a physical reason to justify such a step.

Bell drew attention to two possible ways of avoiding this measurement problem. The first way was to use the de Broglie–Bohm theory, the theory that leads to equation (21.1). Here the theory assumes the particles have sharp position values, $X(t)$, and these trace out a trajectory which the particle is assumed to follow. Thus we have a way of characterising individual quantum processes. The ensemble is characterised by a set of trajectories that are fully determined by the quantum Hamilton–Jacobi equation, namely, the real part of the Schrödinger equation. The trajectory characterising a particular individual particle is contingent on its initial position, a value that cannot be precisely controlled by the observer.

The second way is to maintain that we do not need to worry about whether a particle follows a trajectory or not and to continue using the wave function, which has an additional feature, namely that, from time to time, it undergoes some new random collapse process as proposed by Ghirardi, Rimini and Weber [29]. In other words, this collapse process can be regarded as a spontaneous spatial localisation of the microconstituents occurring at random times, a process of which Schrödinger would approve. The mean frequency of the localisation is assumed to be extremely small, with the localisation width being large on an atomic scale. In this way, no prediction of standard quantum formalism is changed in any appreciable way.

Clearly my own preference is the theory presented in Bohm and Hiley [3], simply because the experiments of WLB [20] and their extension by Paramananda and Butt [23], together with a considerable number of later experimental results, provide no evidence for any kind of spontaneous localisation over very large distances. However, there is one caveat:

both theories are nonrelativistic. It would be very enlightening to find some relativistic generalisation so that we could examine more closely how this nonlocality sits in 'peaceful coexistence' with relativity.

21.6.2 Relativistic Considerations

The obvious approach for an investigation of a relativistic generalisation of the Bohm approach would be to separate the Klein–Gordon equation into its real and imaginary parts. However, there are some serious difficulties with this procedure, which were discussed in detail in Chapter 11 of Bohm and Hiley [3]. To avoid these difficulties, it was found necessary to go to field theory before a satisfactory treatment of bosons in general could be found [30, 31]. We will not go into this approach, as it will take us into an area not directly relevant to the rest of this paper. We will, instead, turn to consider fermions, even though these also present difficulties.

For ferimions two new features appear, spin and relativity. The preliminary attempt to apply the original method used by Bohm [2] for the Schrödinger equation to the Pauli equation [32] met with some success, although the method did not inspire full confidence. Nevertheless, some extremely illuminating illustrations of the model were published by Dewdney, Holland, Kyprianidis and Vigier [33, 34]. In spite of these successes, the extension to the relativistic Dirac equation presented considerable difficulties, although several early unsuccessful attempts were made [3, 4, 35, 36]. It was possible to replace the guidance condition,[1] $p_\psi = \nabla S$, with the Dirac current (see, for example, [3]), but no satisfactory relativistic generalisation of the quantum Hamilton–Jacobi equation (21.1) was found even for the case of a single particle.

John Bell [26] himself attempted to rectify this situation by developing a quantum-field-theoretic approach, but his approach was not very satisfactory and his ideas have not been developed any further, as far as I am aware. More recently, following on from the earlier work of Hestenes [37], I have, together with Bob Callaghan, developed an approach to the Dirac equation using the full structure of the orthogonal Clifford algebra [38, 39].

We also applied the method to the Pauli equation, removing some of the unsatisfactory features of the earlier attempts. Fortunately we found that this attempt is a special case, so the results of Dewdney, Holland and Kyprianidis [33] still stand, giving us the best illustration of nonlocality for two nonrelativistic particles with spin in an entangled state.

In this more general approach, Hilbert space representations are not used; rather we make full use of the orthogonal Clifford algebra. This means replacing the wave function with what can essentially be regarded as a density matrix, ρ, but now defined algebraically in terms of elements of suitable left and right ideals. These elements contain all the information carried by the wave function and so describe the state of the system. To anyone who had read the third edition of Dirac's classic book, *The Principles of Quantum Mechanics* [40], carefully, this will come as no surprise and explains the meaning of his *standard ket*.

---

[1] I prefer to call p_ψ the Bohm momentum, for reasons that will become clear in Section 21.6.

Furthermore, as Fröhlich [41] has already pointed out, this density matrix is not the statistical matrix introduced by von Neumann [7]. The formalism can even be used for a single particle in a pure state, in which case $\rho^2 = \rho$.

John Bell [42] himself attempted to use the density matrix in a different context, namely, to analyse the delayed-choice experiment, and concluded that the density matrix could not be used in the de Broglie–Bohm theory because the theory gives fundamental significance to the wave function. However, it turns out that this conclusion is not correct. The density matrix can be used, but this means generalising the whole approach [43]. I believe that it is the *insistence* on using the wave function that has held up progress in the development not only of a relativistic generalisation of the de Broglie–Bohm approach, but also of standard quantum mechanics. In the algebraic approach the wave function appears as a special case of a more general structure and is not basic.

In the algebraic approach, the choice of what ideals to use is determined by the physics of the system under consideration. This has been explained in detail in Hiley and Callaghan [38]. Essentially the necessity of using the Clifford algebra for the Pauli and Dirac equations arises because the wave function does not capture the full implications of the noncommutative structure, which only become significant in these algebras. To bring out these features, we must go to the algebra. In such algebras one must distinguish between left and right operations. This means that when we come to consider time development we need two equations, a left translation and a right translation–the equation and its dual. In the case of the Schrödinger equation, the dual is simply the complex conjugate equation, which is dismissed, since it seems to add nothing new. This cannot be done for the Pauli or Dirac equations, where the full noncommutative structure is necessary.

Given these two equations, if we subtract the equation from its dual, we get immediately the quantum Hamilton–Jacobi equation. On the other hand, if we add the two equations, we get the Liouville equation. Using this generalisation, we give a complete description of the one-particle dynamics of both the Pauli and Dirac equations [43]. We find in the generalised quantum Hamilton–Jacobi equation expressions for the quantum potential energy in all cases. Thus not only have we generalised the approach to a nonrelativistic particle with spin, removing the unsatisfactory features of the Bohm–Schiller–Tiomno [32] approach, but also we are able to obtain its relativistic generalisation, namely, the Dirac equation. This immediately opens the way to treat fermion fields, a question in which Bell himself was interested.

If we examine the expression for the Dirac quantum potential energy, we find that it is a Lorentz scalar, a result I found rather surprising, although it provides a clue as to why quantum nonlocality and relativity can exist in 'peaceful coexistence'. A scalar is frame-independent and does not propagate. Furthermore, in the Dirac case, we find *two* currents appearing. One is the Dirac current used in Bohm and Hiley [3]. The other is the generalisation of the Schrödinger current, which gives raise to the guidance condition $p = \nabla S$. This is a feature that has already been pointed out by Tucker [44] using a different approach to the Clifford algebra, so that it is a general feature of the relativistic particle with spin. The implications of this unexpected result have still to be fully understood.

So far our study has been confined to the one-body problem. Of much greater interest is the two-body problem, where the question of quantum nonlocality will directly confront relativistic locality. Unfortunately, the generalisation is proving difficult at this present time. This situation is very tantalising, as the two-Pauli-particle case has been beautifully analysed by Dewdney et al. [34]. However, even in this nonrelativistic case, I have not been able to extend the algebraic approach to the two-body system, but the investigation into the whole method is still on going at the time of writing.

In one sense the success achieved using the orthogonal Clifford algebra should not be too surprising, because the conventional wave functions of the Pauli and Dirac particles are simply Hilbert space representations of elements contained in the Clifford algebra itself.

One of the important features of the Clifford algebra is that it automatically contains the spin group, the double cover of the orthogonal group. Normally this is treated by the somewhat abstract method of constructing spin bundles. However, the Clifford algebra was originally created, before the advent of quantum theory, to understand the properties of space and to discuss movement in space. The Clifford group, the spin group, describes the global features of the rotation group and gives a natural account of the difference between the 2π and 4π rotations. This is a feature of the rotational properties of space and has, a priori, little to do with quantum phenomena. It is rather that physical processes exploit these global properties, with quantum nonlocality being merely a consequence of these global features. This would account for the appearance of nonlocality in spin correlations, but what about the nonlocality revealed in the $x - p$ correlations originally proposed in the EPR discussion?

21.6.3 The Symplectic Group, Its Double Cover and Another Clifford Algebra

If we are to understand the $x - p$ nonlocality as a global feature in the same spirit as the rotational nonlocality discussed in the previous section, we need to express the translation dynamics in an analogous algebraic structure. Thus while the orthogonal Clifford algebra is the geometric algebra of rotations, there ought to be a geometric algebra dealing with translations.

At first sight this seems a nonstarter because translations in space form an Abelian group, but we want a noncommutative structure if we are to find an analogue of the orthogonal Clifford algebra. One early candidate was proposed by von Neumann [45], the famous paper that provides the foundation of the Stone–von Neumann theorem, which establishes the uniqueness of the Schrödinger representation. But we are not interested here in representations; we are interested in the algebraic structure itself.

Translations in space involve movement, that is, momentum, so that not only must we consider translation in space, but also we must consider changes of momentum, that is, translations in momentum space. These sets of translations do not commute, and it is this feature that gives rise to a noncommutative symplectic space. This leads us into the rich field of symplectic geometry, a geometry that contains subtle topological structure having a very relevant significance for quantum processes. I am deeply grateful to the mathematician

Maurice de Gosson [46, 47], an expert in symplectic geometry, who has helped me understand some of the difficult aspects of this geometry.

For the purposes of this paper, I will give a brief account of this structure, because John
himself tried to use this approach to explore quantum nonlocality in more detail. Let us
start with von Neumann, who, following Weyl [48], introduces a pair of noncommuting
translation operators, $U(\alpha) = e^{i\alpha\widehat{P}}$ (translations in space) and $V(\beta) = e^{i\beta\widehat{X}}$ (translations in
momentum space). These operators satisfy the relations

$$U(\alpha)V(\beta) = e^{i\alpha\beta}V(\beta)U(\alpha),\tag{21.7}$$

together with

$$U(\alpha)U(\beta) = U(\alpha + \beta); \qquad V(\alpha)V(\beta) = V(\alpha + \beta).$$

Von Neumann then defines an operator,

$$\widehat{S}(\alpha, \beta) = e^{i(\alpha\widehat{P}+\beta\widehat{X})} = e^{-i\alpha\beta/2}U(\alpha)V(\beta) = e^{i\alpha\beta/2}V(\beta)U(\alpha),$$

and proves that the operator $\widehat{S}(\alpha, \beta)$ can be used to define uniquely any bounded operator
$\widehat{A}$ on a Hilbert space through the relation

$$\widehat{A} = \iint a(\alpha, \beta)\widehat{S}(\alpha, \beta)d\alpha d\beta.\tag{21.8}$$

Here $a(\alpha, \beta)$ is a function on a Schwartz space spanned by two variables, α and β, in $\mathbb{R}^{2N}$.
Thus we can establish a one-to-one relationship between the operator algebra of quantum
mechanics and a set of real-valued functions, $a(\alpha, \beta)$ on a noncommutative symplectic
space.

Having used an abstract mathematical structure to establish the uniqueness of the
Schrödinger representation, physicists were handed a convenient mathematical tool using
differential operators and wave functions, with which they were familiar and which enabled
them to perform calculations more easily. Naturally the representational structure became
established and accepted, independent of how it arose.

It was only later that Moyal [49] identified this symplectic space with a generalised phase
space by identifying $\alpha = x$, $\beta = p$ and interpreting the algebra[2] as providing a description
of a generalised noncommutative statistics. Unfortunately Moyal's approach was misunderstood, and it became known as a semiclassical approach, even though the mathematical
structure it uses is exactly the full quantum formalism.

One of the key results of the von Neumann–Moyal algebra is that the quantum expectation value of an operator $\widehat{A}$ is given by the relation

$$\langle\psi|A|\psi\rangle = \int\int a(x, p)F_\psi(x, p)dxdp,\tag{21.9}$$

---

[2] Notice that the Moyal algebra is isomorphic to the quantum formalism and not, as sometimes interpreted, an approximation to
the quantum formalism. It should strictly be called the von Neumann–Moyal algebra.

where $F_\psi(x, p)$ is the Fourier transform of the Wigner function. The algebra provides an explanation of the exact mathematical origins of the Wigner function. The $a(x, p)$ are functions in $\mathbb{R}^{2N}$ and are subject to the noncommutative twisted or Moyal product. These functions completely characterise the quantum state of a *single particle* and a priori are not specifically a feature of many-body statistics, as is often assumed.

The form of equation (21.9) suggests that we can interpret $F_\psi(x, p)$ as a probability density, albeit in a noncommutative space. This is indeed the way Moyal suggested it should be interpreted. However, it turns out that it is not a positive definite quantity, taking negative values in the regions where interference takes place. This adds to the belief that the Wigner function approach is some quasiclassical approximation to quantum theory. But the above results show that this conclusion in not correct. The von Neumann–Moyal algebra is central to the quantum formalism. Indeed Bohm and Hiley [50] and Hiley [51] have shown how a two-point density matrix of configuration space can be transformed into the density function $F_\psi(x, p)$, again showing that this approach is an exact alternative formulation of the operator approach. The Hilbert space formalism is merely a representation of this deeper structure.

21.6.4 Nonlocality and the Wigner Function

Realising the significance of the Wigner function, John Bell [52] suggested that the two-body Wigner function might provide another way of exploring the original EPR $x - p$ correlations directly rather than only looking at spin correlations. His analysis was based on the assumption that all quantum interference effects will produce negative values for the Wigner function. The question would then be simply to examine the Wigner function constructed from the original EPR wave function,

$$\psi_\delta = \delta\left((x_1 + a/2) - (x_2 - a/2)\right).$$

He found that the Wigner function was nowhere negative, implying under his assumption that there is no nonlocality problem in the phase space approach.

However, earlier, Bohm and Hiley [53] had shown that if, for mathematical convenience, we replace the delta function with a real function ψ that is very sharply peaked at $x_1 - x_2 = a$, the quantum potential takes the form

$$Q_\psi = -\frac{\hbar^2}{2m} \frac{\nabla^2 f(x_1 - x_2)}{f(x_1 - x_2)}.$$

This expression shows there is a nonlocal connection between the two particles, so they are coupled and not free. Clearly something is wrong with Bell's assumption, as both the Moyal and Bohm approaches give the same quantum expectation values for all operators.

My reasoning was that if two seemingly very different formalisms lead to the same expectation values, then there must be a relationship between the two approaches. I returned to a closer examination of the Moyal paper [49] and discovered that the key formulae of the Bohm interpretation already appeared in the appendix.

The Bohm momentum, $p_\psi = \nabla S$, is identical to a momentum defined by Moyal as a conditional expectation value,

$$p_\psi = \int p F_\psi(x, p) dp, \tag{21.10}$$

while the transport of this momentum produced the one-body equivalent to equation (21.1),

$$\partial_t S + \frac{(\nabla S)^2}{2m} + Q + V = 0, \tag{21.11}$$

where

$$Q(x, t) = -\frac{\hbar^2}{2m} \frac{\nabla^2 R(x, t)}{R(x, t)}. \tag{21.12}$$

These results show that, once again, there is a deep connection between the mathematics used in the Bohm approach and the approach based on the von Neumann–Moyal algebra.

In order to see this connection in a wider context, we need to be aware of some important aspects of the von Neumann–Moyal algebra that have been discussed by Crumeyrolle [54]. He has shown that the von Neumann–Moyal algebra is actually a symplectic Clifford algebra, the algebraic analogue of the more familiar orthogonal Clifford algebra. In other words, it is the geometric algebra of a symplectic space containing a Clifford group that describes the double cover of the symplectic group, namely, the metaplectic group and its nonlinear extension. Thus the metaplectic group and its generalisation play a role analogous to that of the spin groups; namely the metaplectic group is the covering group of the symplectic group. It is this covering group that describes the global features of the symplectic geometry. This once again suggests that quantum nonlocality is a global feature of the dynamics that quantum processes exploit. It is not a signalling process.

What is even more important from our point of view is that the Schrödinger equation appears as the lift onto the covering space of the classical Hamiltonian flow in the symplectic space [55, 56] and it is in the covering space that the global properties of the geometry appear. This suggests that the nonlocal properties of quantum phenomena emerge from the global properties of the covering groups in general, opening up another avenue for exploring quantum nonlocality.

21.7 Conclusion

I am forever grateful to John Bell for his tireless energy in drawing the world's attention to quantum nonlocality, this truly radical feature of Nature. I saw him forcefully defending his views against the attacks of the more conservative elements of the physics community. I admired his courage. I also gained energy for my own investigations from his fierce defence of the right of physicists to investigate further the full implications of different approaches to quantum phenomena, approaches that had been wrongly criticised by the founding fathers of quantum mechanics. Whenever I met him on conferences at various venues in Europe, he would always be encouraging, defending the right to critically explore quantum phenomena in new ways. I certainly needed that encouragement. John, you left us too soon, but thank you for your support.

References

[1] d'Espagnat, B. (2003), *Veiled Reality: An Analysis of Present-Day Quantum Mechanical Concepts*. Westview Press, Boulder, CO.

[2] Bohm, D. (1952), A suggested interpretation of the quantum theory in terms of hidden variables, I, *Phys. Rev.* **85**, 166–79; and II, **85** 180–93.

[3] Bohm, D., and Hiley, B.J. (1993), *The Undivided Universe: An Ontological Interpretation of Quantum Mechanics*, Routledge, London.

[4] Holland, Peter R. (1995), *The Quantum Theory of Motion: An Account of the de Broglie–Bohm Causal Interpretation of Quantum Mechanics*, Cambridge University Press, Cambridge.

[5] Dürr, D., and Teufel, S. (2009), *Bohmian Mechanics: The Physics and Mathematics of Quantum Theory*, Springer, Dordrecht.

[6] Bell, J.S. (1964), On the Einstein–Podolsky–Rosen paradox, *Physics* **1**, 195–200.

[7] von Neumann, J. (1955), *Mathematical Foundations of Quantum Mechanics*, Princeton University Press, Princeton, NJ.

[8] Wheeler, J.A. (1991), *At Home in the Universe*, AIP Press, New York, p. 286.

[9] Dürr, D., Goldstein, S., and Zanghì, N. (1992), Quantum equilibrium and the origin of absolute uncertainty, *J. Stat. Phys.* **67**, 843–907.

[10] Dirac, P.M. (1973), in J. Mehra (ed.), *The Physicist's Conception of Nature*, Reidel, Dordrecht-Holland, p. 10.

[11] Einstein, A., Podolsky, B., and Rosen, N. (1935), Can quantum-mechanical description of physical reality be considered complete, *Phys. Rev.* **47**, 777–80.

[12] Schrödinger, E. (1927), Energieaustausch nach der Wellenmechanik, *Annalen der Physik* (4) **388**, 956–68.

[13] Schrödinger, E. (1936), Probability relations between separated systems, *Proc. Cam. Phil. Soc.* **32**, 446–52.

[14] Bohr, N. (1935), Can quantum-mechanical description of physical reality be considered complete? *Phys. Rev.* **48**, 696–702; repeated in a letter to *Nature* **36**, 65 (July 1935).

[15] Bohm, D. (1951), *Quantum Theory*, Prentice-Hall, Englewood Cliffs, NJ.

[16] Langhoff, H. (1960), Die Linearpolarisation der Vernichtungsstrahlung von Positronen, *Zeit. Phys.* **160**, 186–93.

[17] Kasday, L. (1972), Foundations of quantum mechanics, in *Proc. Intern. School of Physics Enrico Fermi*, Academic Press, New York, pp. 195–210.

[18] Hautojärvi, P., and Jauho, P. (1971), Positron annihilation, in *International Conference on Queens University, Kingston, Canada*, Kingston.

[19] Faraci, G., Gutkowski, D., Notarrigo, S., and Pennisi, A.R. (1974), An experimental test of the EPR paradox, *Lett. Nuovo Cimento* **9** (15), 607–11.

[20] Wilson, A.R., Lowe, J., and Butt, D.K. (1976), Measurement of the relative planes of polarization of annihilation quanta as a function of separation distance, *J. Phys. G: Nucl. Phys.* **2** (9), 613.

[21] Furry, W.H. (1936), Note on the quantum-mechanical theory of measurement, *Phys. Rev.* **49**, 393–9.

[22] Bohm, D., and Aharonov, Y. (1957), Discussion of experimental proof for the paradox of Einstein, Rosen, and Podolsky, *Phys. Rev.* **108** (4), 1070.

[23] Paramananda, V., and Butt, D.K. (1987), Quantum correlations, non-locality and the EPR paradox, *J. Phys. G: Nucl. Phys.* **13**, 449–52.

[24] Ursin, R., Tiefenbacher, F., Schmitt-Manderbach, T., Weier, H., Scheidl, T., Lindenthal, M., Blauensteiner, B., Jennewein, T., Perdigues, J., Trojek, P., Ömer, B., Fürst, M., Meyenburg, M., Rarity, J., Sodnik, Z., Barbieri, C., Weinfurter, H., and Zeilinger, A. (2007), Entanglement-based quantum communication over 144 km, *Nat. Phys.* **3**, 481–6.

[25] Schrödinger, E. (1935), Discussion of probability relations between separated systems, *Proc. Cam. Phil. Soc.* **31**, 555–63.

[26] Bell, J.S. (1987), Beables for quantum field theory, in B.J. Hiley and D. Peat (eds.), *Quantum Implications: Essays in Honour of David Bohm*, Routledge & Kegan Paul, London, pp. 227–34.

[27] Bohr, N. (1961), *Atomic Physics and Human Knowledge*, Science Editions, New York, p. 50.

[28] Bell, J.S. (1990), Against measurement, in A.I. Miller (ed.), *Sixty-Two Years of Uncertainty: Historical, Philosophical, and Physical Inquiries into the Foundations of Quantum Mechanics*, NATO ASI Series **226**, pp. 17–31.

[29] Ghirardi, G.C., Rimini, A., and Weber, T. (1986), Unified dynamics for microscopic and macroscopic systems, *Phys. Rev. D* **34**, 470–91.

[30] Bohm, D.J., Hiley, B.J., and Kaloyerou, P.N. (1987), An ontological basis for the quantum theory: II. A causal interpretation of quantum fields, *Phys. Rep.* **144**, 349–75.

[31] Kaloyerou, P.N. (1994), The causal interpretation of the electromagnetic field, *Phys. Rep.* **244**, 287–385.

[32] Bohm, D., Schiller, R. and Tiomno, J. (1955), A causal interpretation of the Pauli equation (A) and (B), *Nuovo Cim. Supp.* **1**, 48–66 and 67–91.

[33] Dewdney, C., Holland, P.R., and Kyprianidis, A. (1987), A causal account of Non-local Einstein–Podolsky–Rosen spin correlations, *J. Phys. A: Math. Gen.* **20**, 4717–32.

[34] Dewdney, C., Holland, P.R., Kyprianidis, A., and Vigier, J.-P. (1988). Spin and non-locality in quantum mechanics, *Nature* **336**, 536–44.

[35] Bohm, D. (1962), Classical and nonclassical concepts in the quantum theory, *Brit. J. Phil. Sci.* **12** (48), 265–80.

[36] de Broglie, L. (1960), *Non-linear Wave Mechanics: A Causal Interpretation*, Elsevier, Amsterdam.

[37] Hestenes, D. (2003), Space–time physics with geometric algebra, *Am. J. Phys.* **7**, 691–714.

[38] Hiley, B.J., and Callaghan, R.E. (2011), The clifford algebra approach to quantum mechanics: B. The Dirac particle and its relation to the Bohm approach, arXiv: Maths-ph:1011.4033.

[39] Hiley, B.J., and Callaghan, R.E. (2012), Clifford algebras and the Dirac–Bohm quantum Hamilton–Jacobi equation, *Foundations of Physics* **42**, 192–208.

[40] Dirac, P.M. (1947), *The Principles of Quantum Mechanics*, Oxford University Press, Oxford, p. 80.

[41] Fröhlich, H. (1967), Microscopic derivation of the equations of hydrodynamics, *Physica* **37**, 215–26.

[42] Bell, J.S. (1980), De Broglie–Bohm, delayed-choice double slit experiment and the density matrix, *Int. J. Quant. Chem, Quant. Chem. Symp.* **14**, 155–9.

[43] Hiley, B.J. (2013), Bohmian non-commutative dynamics: History and new developments, arXiv 1303.6057.

[44] Tucker, R.W. (1988), Closed currents, vector fields and phenomena, *Found. Phys.* **18**, 8–20.

[45] von Neumann, J. (1931), Die Eindeutigkeit der Schrödingerschen Operatoren, *Mathematische Annalen* **104**, 570–87.

[46] de Gosson, M. (2001), *The Principles of Newtonian and Quantum Mechanics: The Need for Planck's Constant*, Imperial College Press, London.

[47] de Gosson, M. (2010), *Symplectic Methods in Harmonic Analysis and in Mathematical Physics,* Birkhäuser Verlag, Basel.

[48] Weyl, H. (1928), Quantenmechanik und Gruppentheorie, *Zeit. für Physik* **46**, 1–46.

[49] Moyal, J.E. (1949), Quantum mechanics as a statistical theory, Proceedings of the Cambridge Philosophical Society 45, 99–123.

[50] Bohm, D., and Hiley, B.J. (1981), On a quantum algebraic approach to a generalised phase space, *Found. Phys.* **11**, 179–203.

[51] Hiley, B.J. (2015), On the relationship between the Moyal algebra and the quantum operator algebra of von Neumann, *Journal of Computational Electronics* **14**, 869–78; arXiv 1211.2098.

[52] Bell, J.S. (1986), EPR correlations and EPW distributions, *Ann. N.Y. Acad. Sci.* **480**, 263–6.

[53] Bohm, D., and Hiley, B.J. (1975), On the intuitive understanding of nonlocality as implied by quantum theory, *Found. Phys.* **5**, 93–109.

[54] Crumeyrolle, A. (1990), *Orthogonal and Symplectic Clifford Algebras: Spinor Structures*, Kluwer, Dordrecht.

[55] Guillemin, V.W., and Sternberg, S. (1984), *Symplectic Techniques in Physics*, Cambridge University Press, Cambridge.

[56] de Gosson, M., and Hiley, B.J. (2011), Imprints of the quantum world in classical mechanics, *Found. Phys.* **41**, 1415–36.

22

Bell on Bohm

SHELDON GOLDSTEIN

22.1 Prologue

A memorial conference for John Bell, open to the public, was held at Rutgers University around 20 years ago. I gave a short talk there on Bell's views about David Bohm's "hidden variables" formulation of quantum mechanics, a version of quantum mechanics often called the de Broglie–Bohm theory or Bohmian mechanics. This theory was in fact discovered by Louis de Broglie in 1927. In 1952, it was rediscovered and developed by Bohm, who was the first to appreciate its connection to the predictions of standard quantum mechanics. I did not publish the talk.

I have decided that it would be appropriate to publish a very lightly edited version of it here, in this volume devoted to the work of Bell on the foundation of quantum mechanics. One reason for doing so is that the connection between Bell and Bohm continues to be somewhat underplayed, with the strength of his advocacy of Bohmian mechanics not properly appreciated. For example, about half of the papers in Bell's collected works on the foundations of quantum mechanics deal with Bohmian mechanics. But in his fine introduction to the revised edition of this great book [1], Alain Aspect mentions this theory in only a single sentence, and parenthetically at that.

For several decades after Bell proved his nonlocality theorem, based in part on Bell's inequality, it was widely claimed that Bell had shown that hidden variables – and Bohmian mechanics in particular – were impossible, that they were incompatible with the predictions of quantum mechanics. For example, the great physicist Eugene Wigner, who, unlike most of his contemporaries, was profoundly concerned with the conceptual foundations of quantum mechanics and usually wrote on the subject with great clarity and insight, has written that

This [hidden variables] is an interesting idea and even though few of us were ready to accept it, it must be admitted that the truly telling argument against it was produced as late as 1965, by J. S. Bell ... This appears to give a convincing argument against the hidden variables theory.

[2, p. 53]

More recently, we find Stephen Hawking writing that

Einstein's view was what would now be called a hidden variable theory. Hidden variable theories might seem to be the most obvious way to incorporate the Uncertainty Principle into physics. They form the basis of the mental picture of the universe, held by many scientists, and almost all philosophers of science. But these hidden variable theories are wrong. The British physicist, John Bell, who died recently, devised an experimental test that would distinguish hidden variable theories. When the experiment was carried out carefully, the results were inconsistent with hidden variables. [3]

So let us look at what Bell actually thought about the matter. Here is the talk.

22.2 Introduction

John Stewart Bell is best known for his discovery of the theorem that bears his name. This theorem establishes the impossibility of any explanation of quantum phenomena in terms of what are called local hidden variables.[1] And since one might well imagine that any account in terms of *nonlocal* hidden variables would have to be artificial – cooked up just to do the job – and generally unacceptable, Bell's theorem is widely regarded as precluding any hidden variable account worthy of our consideration. (As far as the meaning of a "hidden variable account" is concerned, for now let me just say, somewhat imprecisely, that a hidden variable formulation of quantum theory would eliminate quantum craziness while retaining the quantum predictions.) In other words, Bell's theorem is widely used to support the proposition that quantum phenomena demand radical epistemological and metaphysical innovations – precisely what hidden variables promise to avoid.

Now Bell wrote much and lectured much about his theorem and its implications. But he wrote and lectured as much, if not more, concerning the virtues of what is the most famous of all hidden variable theories, that of David Bohm. The question thus naturally arises, why would Bell spend so much time and effort expounding upon a theory of just the sort that he himself had shown to be, if not impossible, unworthy of consideration?

Indeed, some physicists have spoken of two Bells, and have suggested that Bell must have been schizophrenic [4].

I wish to argue that there was, unfortunately for us, but one Bell, and he was the sanest and most rational of men.

There is something else that I would like to do: I would like to convey a small sense of Bell's wonderful style, wit, and clarity. So to the extent possible I shall allow Bell to speak for himself. I shall read excerpts from Bell's articles on the foundations of quantum mechanics that pertain to our question. These articles are all collected in a marvelous book, *Speakable and Unspeakable in Quantum Mechanics* [1]. I would urge all of you, and, indeed, anyone with an interest in physics, to read this book, and then read it again. I shall also have occasion to read from an interview Bell gave several years ago, to the philosopher Renee Weber.

---

[1] I now regret having written that sentence. While not wrong, it is nonetheless misleading. What Bell established was the impossibility of any local account of quantum phenomena (that does not involve many worlds), not just the impossibility of an explanation in terms of local hidden variables. That this is what Bell himself thought is clear, for example, from the quotations of Section 22.8.

22.3 The Impossibility of Hidden Variables

Bell's "On the impossible pilot wave" begins thus:

When I was a student I had much difficulty with quantum mechanics. It was comforting to find that even Einstein had had such difficulties for a long time. Indeed they had led him to the heretical conclusion that something was missing in the theory: "I am, in fact, rather firmly convinced that the essentially statistical character of contemporary quantum theory is solely to be ascribed to the fact that this (theory) operates with an incomplete description of physical systems."

[5; 1, p. 59]

Einstein is expressing here the conviction that the supposedly novel quantum randomness will ultimately turn out to be of the same character as the familiar, normal, down-to-earth randomness exhibited, for example, in the behavior of a roulette wheel or a coin flip. The behavior appears random because there are too many relevant details to keep track of. If the quantum description could be completed by the incorporation of such details, the result would be called a hidden variable theory.

However, soon after the advent of quantum theory, any hidden variable account of quantum phenomena was mathematically "proven" to be impossible. Bell continues,

Einstein did not seem to know that this possibility, of peaceful coexistence between quantum statistical predictions and a more complete theoretical description, had been disposed of with great rigour by J. von Neumann. I myself did not know von Neumann's demonstration at first hand, for at that time it was available only in German, which I could not read. However I knew of it from the beautiful book by Born, *Natural Philosophy of Cause and Chance,* which was in fact one of the highlights of my physics education. Discussing how physics might develop Born wrote: "I expect . . . that we shall have to sacrifice some current ideas and to use still more abstract methods. However these are only opinions. A more concrete contribution to this question has been made by J. v. Neumann in his brilliant book, *Mathematische Grundlagen der Quantenmechanik.* . . . The result is that . . . no concealed parameters can be introduced with the help of which the indeterministic description could be transformed into a deterministic one. Hence if a future theory should be deterministic, it cannot be a modification of the present one but must be essentially different. How this could be possible without sacrificing a whole treasure of well established results I leave to the determinists to worry about."

Having read this, I relegated the question to the back of my mind and got on with more practical things.

[5; 1, p. 159]

22.4 The Impossible Accomplished

Bell continues,

But in 1952 I saw the impossible done. It was in papers by David Bohm. Bohm showed explicitly how parameters could indeed be introduced, into nonrelativistic wave mechanics, with the help of which the indeterministic description could be transformed into a deterministic one. More importantly, in my opinion, the subjectivity of the orthodox version, the necessary reference to the "observer," could be eliminated.

Moreover, the essential idea was one that had been advanced already by de Broglie in 1927, in his "pilot wave" picture.

[5; 1, p. 160]

Let me very briefly try to indicate the sort of thing Bell had in mind when objecting to the subjectivity of orthodox quantum theory, by means of a perhaps extreme example. Concerning the implications of quantum theory, in fact of Bell's theorem itself (about which more later), a very distinguished physicist once wrote that "the moon is demonstrably not there when nobody looks" [6].[2]

More Bell:

Bohm's 1952 papers on quantum mechanics were for me a revelation. The elimination of indeterminism was very striking. But more important, it seemed to me, was the elimination of any need for a vague division of the world into "system" on the one hand, and "apparatus" or "observer" on the other. I have always felt since that people who have not grasped the ideas of those papers ... and unfortunately they remain the majority ... are handicapped in any discussion of the meaning of quantum mechanics.

[8; 1, p. 173]

Interview:

In my opinion the picture which Bohm proposed then completely disposes of all the arguments that you will find among the great founding fathers of the subject – that in some way, quantum mechanics was a new departure of human thought which necessitated the introduction of the observer, which necessitated speculation about the role of consciousness and so on.

All those are simply refuted by Bohm's 1952 theory. In that theory you find a scheme of equations which completely reproduces all the experimental predictions of quantum mechanics and it simply does not need an observer ... So I think that it is somewhat scandalous that this theory is so largely ignored in textbooks and is simply ignored by most physicists. They don't know about it.

22.5 How to Do the Impossible

What does Bohm add to the standard quantum description? In a word, the particles themselves: For Bohm the so-called hidden variables are simply the positions of the particles

---

[2] I can not resist including a mildly polemical rejoinder, from the late philosopher David Stove [7, p. 99]: "If philosophy or religion prompts a person to deny or doubt that humans, or that kangaroos, are land-mammals, the only rational thing to do is to ignore him; and the same holds for science, too, whether past, present, or future.

I may be reminded that some respected physicists have said in recent years that something like Berkeleian idealism is actually a logical consequence of their best fundamental theories. (One of them wrote, for example: 'We now know that the moon is demonstrably not there when nobody looks.') It would be irrational to believe this logical claim, but if it is true then it would be irrational to believe these physicists' best theories. Fundamental physical theories never say anything about a particular macroscopic physical object, such as the moon; but if they did say something about the moon, then they would say the same thing about all macroscopic physical objects, hence about all land-mammals, and hence about the particular land-mammal, Professor N.D. Mermin, who wrote the sentence I have just quoted. Now it may perhaps be true that Professor Mermin depends for his ease of mind on being an object of attention. This would not even be especially surprising, in view of the powerful emotional root which idealism has in common with religion. But that he depends for his very existence on being an object of attention, is entirely out of the question: it is much more likely (to say the least) that one or more of his scientific theories is wrong. Mammals are very complex, of course, and depend for their existence on a great many things; but somebody's looking at them is not among those things, and everybody knows this."

of the quantum system, say the electrons of an atom. These particles move in a manner that is naturally choreographed by the wave function of the system. From the perspective of Bohm's theory, orthodox quantum mechanics leaves out the guts of the description, the very particles that combine to form everything we see around us.

Thus as applied to Bohm's theory, the terminology "hidden variables" seems rather inappropriate, suggesting as it does something exotic, artificial, and ad hoc. Bell:

Absurdly, such theories are known as "hidden variable" theories. Absurdly, for there it is not in the wavefunction that one finds an image of the visible world, and the results of experiments, but in the complementary "hidden"(!) variables. Of course the extra variables are not confined to the visible "macroscopic" scale. For no sharp definition of such a scale could be made. The "microscopic" aspect of the complementary variables is indeed hidden from us.

[9; 1, p. 201]

Here Bell refers to the fact that in Bohm's theory the detailed trajectories of the microscopic particles are not observable. While this unobservability is a *consequence* of the very structure of Bohm's theory, many physicists quickly objected. After all, physics is about prediction, about observations, not about things which cannot be observed. Bell continues,

But to admit things not visible to the gross creatures that we are is, in my opinion, to show a decent humility, and not just a lamentable addiction to metaphysics.

[9; 1, p. 202]

22.6 The Accomplished Impossible Is Ignored

The very existence of Bohm's theory, agreeing as it did in its predictions with those of orthodox quantum theory, quite naturally, under the circumstances, raised many questions for Bell:

But why then had Born not told me of this "pilot wave"? If only to point out what was wrong with it? Why did von Neumann not consider it? More extraordinarily, why did people go on producing "impossibility" proofs, after 1952, and as recently as 1978? When even Pauli, Rosenfeld, and Heisenberg, could produce no more devastating criticism of Bohm's version than to brand it as "metaphysical" and "ideological"? Why is the pilot wave picture ignored in text books? Should it not be taught, not as the only way, but as an antidote to the prevailing complacency? To show us that vagueness, subjectivity, and indeterminism, are not forced on us by experimental facts, but by deliberate theoretical choice?

[5; 1, p. 160]

22.7 What Went Wrong?

Of course, the most immediate question raised was, or should have been, What went wrong with the "proof"? Bell:

The realization that von Neumann's proof is of limited relevance has been gaining ground since the 1952 work of Bohm. However, it is far from universal. Moreover, the writer has not found in the literature any adequate analysis of what went wrong. Like all authors of noncommissioned reviews, he thinks that he can restate the position with such clarity and simplicity that all previous discussions will be eclipsed.

[10; 1, p. 2]

And Bell proceeded to do just that!

Bell analyzed von Neumann's proof as well as other proofs, found that they were based upon rather arbitrary assumptions or axioms, and focused on the manner in which Bohm's theory violates these assumptions. In so doing he noticed that

in this theory an explicit causal mechanism exists whereby the disposition of one piece of apparatus affects the results obtained with a distant piece.

Bohm of course was well aware of these features of his scheme, and has given them much attention. However, it must be stressed that, to the present writer's knowledge, there is no *proof* that *any* hidden variable account of quantum mechanics *must* have this extraordinary character. It would therefore be interesting, perhaps, to pursue some further "impossibility proofs," replacing the arbitrary axioms objected to above by some condition of locality, or of separability of distant systems.

[10; 1, p. 11]

22.8 Quantum Nonlocality

No sooner said than done! In fact, if we follow the publication dates, done before said – the EPR–Bell's theorem paper [11] in which it was done appeared almost two years before the paper [10] from which I was just quoting. Publication delay!

Bell interview:

as a professional theoretical physicist I like the Bohm theory because it is sharp mathematics. I have there a model of the world in sharp mathematical terms that has this non-local feature. So when I first realized that, I asked: "Is that inevitable or could somebody smarter than Bohm have done it differently and avoided this non-locality?" That is the problem that the theorem is addressed to. The theorem says: "No! Even if you are smarter than Bohm, you will not get rid of non-locality," that any sharp mathematical formulation of what is going on will have that non-locality.

Moreover, the nonlocality of Bohm's theory derives solely from the nonlocality built into the structure of standard quantum theory, as provided by a wave function on configuration space, an abstraction that, roughly speaking, combines – or binds – distant particles into a single irreducible reality. Bell:

That the guiding wave, in the general case, propagates not in ordinary three-space but in a multidimensional-configuration space is the origin of the notorious "nonlocality" of quantum mechanics. It is a merit of the de Broglie–Bohm version to bring this out so explicitly that it cannot be ignored.

[12; 1, p. 115]

Now the relevant experiments have been done [13], confirming the strange predictions to which Bell was led by his analysis of Bohm's theory. Where does this now leave us?

22.9 Lorentz Invariance

There is a basic problem: Bohm's theory violates Lorentz invariance, a central principle of physics. Nor can Bohm's theory be easily modified so that it becomes Lorentz invariant. The difficulty here arises from the fundamental tension, the *apparent* incompatibility, between nonlocality and Lorentz invariance. Bell interview:

Now what is wrong with this theory, with David's theory? What is wrong with this theory is that it is not Lorentz-invariant. That's a very technical thing and most philosophers don't bother with Lorentz-invariance and in elementary quantum mechanics books the paradoxes that are presented have nothing to do with Lorentz-invariance.

Those paradoxes are simply disposed of by the 1952 theory of Bohm, leaving as *the* [my emphasis] question, the question of Lorentz-invariance. So one of my missions in life is to get people to see that if they want to talk about the problems of quantum mechanics – the real problems of quantum mechanics – they must be talking about Lorentz-invariance.

And from the last sentence of (to my knowledge) Bell's last publication[3] – the LAST WORD, as it were:

Referring to Bohm's theory and to GRW theory [9, 14] – a modification of quantum theory in which he became interested in his last years – Bell said,

The big question, in my opinion, is which, if either, of these two precise pictures can be redeveloped in a Lorentz invariant way.

[15; 1, p. 230]

I believe that this really is the big question. And I urge it upon you. But I am afraid that in trying to answer this question, we shall miss Bell's help and inspiration very much indeed! And we shall miss Bell's marvelous style, his penetrating wit, and his brilliant clarity!

22.10 Epilogue

That is what I said some twenty years ago. Much the same could be said today. Bell's views on Bohm are more widely appreciated now. But they are not nearly as well appreciated as they should be. Bell was, after all, not an obscure writer, and he made his views perfectly clear. But he treated his readers seriously, and expected them to read him with the same care with which he wrote. In our age this sort of reading is all too rare.

As to the "big question," about Lorentz invariance, of which Bell spoke, significant progress has been made, of some of which, I'd like to think, Bell would have approved. Employing a "flash" ontology, Roderich Tumulka has formulated a fully Lorentz invariant generalization of the GRW theory for entangled but noninteracting particles [17]. Along rather different lines, by suitably employing past light cones, Bedingham et al. have provided a scheme for defining a Lorentz invariant redevelopment of GRW-type theories of a

---

[3] At the time I wrote this my knowledge was in fact incomplete: It seems that Bell's last publication was La nouvelle cuisine [16] and not Against measurement [15].

rather general character, including interacting quantum field theories [18]. And for Bohmian mechanics there have been several proposals for Lorentz invariant generalizations, for example, exploiting the possibility of a covariant map from wave functions to spacelike foliations of space–time [19].

Tumulka's model is based on the ideas of Bell himself, and the peculiar flash ontology of the model was Bell's rather strange proposal for an ontology of the GRW theory [9; 1, p. 205]. Concerning this theory, Bell wrote that

I am particularly struck by the fact that the model is as Lorentz invariant as it could be in the nonrelativistic version. It takes away the ground of my fear that any exact formulation of quantum mechanics must conflict with fundamental Lorentz invariance.

[9; 1, p. 209]

References

[1] Bell, J.S. (2004), *Speakable and Unspeakable in Quantum Mechanics*, Cambridge: Cambridge University Press.

[2] Wigner, E.P. (1983), Review of quantum mechanical measurement problem, In P. Meystre and M.O. Scully (eds.), *Quantum Optics, Experimental Gravity and Measurement Theory*, New York: Plenum Press, pp. 43–63.

[3] Hawking, S. (1999), Does God play dice? www.hawking.org.uk/does-god-play-dice .html.

[4] Speiser, D. (1988), Ce qu'il est possible de dire et ce qui ne peut être dit en mécanique quantique (d'après J.S. Bell), *Revue des questions scientifiques* **159**, 365–9.

[5] Bell, J.S. (1982), On the impossible pilot wave, *Foundations of Physics* **12**, 989–99. Reprinted in (Bell, 2004), pp. 159–68.

[6] Mermin, N.D. (1981), Quantum mysteries for anyone, *Journal of Philosophy* **78**, p. 397.

[7] Stove, D.C. (1991), *The Plato Cult*, Cambridge: Basil Blackwell.

[8] Bell, J.S. (1986), Beables for quantum field theory, *Physics Reports* **137**, 49–54. Reprinted in (Bell, 2004), pp. 173–80.

[9] Bell, J.S. (1989), Are there quantum jumps?. In C.W. Kilmister (ed.), *Schrödinger. Centenary Celebration of a Polymath*, Cambridge: Cambridge University Press. Reprinted in (Bell, 2004), pp. 201–12.

[10] Bell, J.S. (1966), On the problem of hidden variables in quantum mechanics, *Reviews of Modern Physics* **38**, 447–52. Reprinted in (Bell, 2004), pp. 1–13.

[11] Bell, J.S. (1964), On the Einstein–Podolsky–Rosen paradox, *Physics* **1**, 195–200. Reprinted in (Bell, 2004), pp. 14–21.

[12] Bell, J.S. (1980), De Broglie–Bohm, delayed-choice double-slit experiment, and density matrix, in Quantum Chemistry Symposium, *International Journal of Quantum Chemistry* **14**, 155–9. Reprinted in (Bell, 2004), pp. 111–16.

[13] Aspect, A., Grangier, P. and Roger, G. (1982), Experimental realization of Einstein–Podolsky–Rosen–Bohm Gedankenexperiment: A new violation of Bell's inequalities, *Physical Review Letters* **49**, 91–4.

[14] Ghirardi, G.C., Rimini, A., and Weber, T. (1986), Unified dynamics for microscopic and macroscopic systems, *Physical Review D* **34**, 470–91.

[15] Bell, J.S. (1990a), Against "measurement," in A.I. Miller (ed.), *Sixty-Two Years of Uncertainty: Historical Philosophical, and Physical Enquiries into the Foundations of Quantum Mechanics*, New York: Plenum Press. Reprinted in (Bell, 2004), pp. 213–31.

[16] Bell, J.S. (1990b), La nouvelle cuisine, in A. Sarlemijn and P. Kroes (eds.), *Between Science and Technology*, Amsterdam: Elsevier Science Publishers. Reprinted in (Bell, 2004), pp. 232–48.

[17] Tumulka, R. (2006), A relativistic version of the Ghirardi–Rimini–Weber model, *Journal of Statistical Physics* **125**, 821–40.

[18] Bedingham, D., Dürr, D., Ghirardi, G.C., Goldstein, S. and Zanghì, N. (2014), Matter density and relativistic models of wave function collapse, *Journal of Statistical Physics* **154**, 623–31.

[19] Dürr, D., Goldstein, S., Norsen, T., Struyve, W., and Zanghì, N. (2013), Can Bohmian mechanics be made relativistic? *Proceedings of the Royal Society A* **470**, 20130699.

23

Interactions and Inequality

PHILIP PEARLE

Abstract

John Bell was a staunch supporter of the dynamical wave function collapse approach to making a well-defined quantum theory. Through letters from him, I reminisce on my handful of interactions with him, all of which were memorable to me. Then I discuss nonlocality, violation of Bell's inequality and some further implications within the framework of the CSL (continuous spontaneous localization) model of dynamical collapse.

23.1 Interactions

I was an instructor at Harvard with a newly minted Ph.D. from MIT, earned with an indifferent thesis on particle theory. It took me two years to come to terms with, dare I say it, the fact that I *did not care* about the S-matrix. I was going to be out of a job anyway in 1966 so, in the fall of 1965, I put my full energy into writing my very first paper, about what I did care: that, to my eyes there was something deeply wrong with the quantum theory I had been taught; that the rules of how to use it were inadequate. The title of the paper [1] was "Elimination of the reduction postulate from quantum theory and a framework for hidden variable theories." It was a long title, because it was really two rather separate papers.

The first was based upon the perceived inadequacy that the collapse (reduction) postulate of the theory is ill-defined.

The second stated some postulates I thought a good hidden variable theory ought to obey, and gave a model for their satisfaction in a two-dimensional (i.e., spin-1/2) Hilbert space. Try as I might, I was not able to generalize my model to a higher-dimensional (i.e., spin-1) Hilbert space, but I put it out there for someone more clever than I to achieve. I sent the paper to a very few people who I thought might be interested. One was John Bell, and he replied (see Figure 23.1)!

So someone more clever than I pointed out that generalizing my model could not be achieved. I therefore excised the model and, after rewriting and retitling, published my very first paper [2]!

After a three-year stint at Case Institute of Technology, which graduated to become Case Western Reserve University while I was there, in 1969 I became ensconced at Hamilton

ORGANISATION EUROPÉENNE POUR LA RECHERCHE NUCLÉAIRE

CERN EUROPEAN ORGANIZATION FOR NUCLEAR RESEARCH

Adresser la correspondance:

CERN 1211 Genève 23

Télex: GENÈVE - 2 25 48

Téléphone: (022) 41 98 11

Télégramme: CERNLAB-GENÈVE

Votre référence:

Notre référence: mj

Meyrin-Genève, le 10 February 1966

Dr. Philip Pearle
Harvard University
Lyman Laboratory of Physics
C a m b r i d g e   38
Mass. - U S A

Dear Dr. Pearle :

Thank you for your interesting paper. I would like to comment on the question of whether your programme could be realized in state spaces of higher dimension. As shown by Gleason, the two dimensional space is quite special. I reviewed his reasoning in Section 5 of my paper on hidden variables, and I think one can see from that that your programme would have to be modified in the general case. A possible modification would be to let P depend not only on a particular state γ_k, but also on the choice of orthogonal basis for the rest of the Hilbert space.

your function

With best regards.

John Bell

John S. Bell

Figure 23.1 Letter from John Bell, February 1966.

College. In 1973, through good behavior, it was time for my first sabbatical. In the spring of 1972, I wrote to John Bell, asking him if I would be able to spend it at CERN. He wrote the note in Figure 23.2.

So my family and I ended up spending a glorious year at the University of Geneva, under the benevolent eye of Josef Jauch (who, tragically, died at age 60 a couple of months after

ORGANISATION EUROPÉENNE POUR LA RECHERCHE NUCLÉAIRE

CERN EUROPEAN ORGANIZATION FOR NUCLEAR RESEARCH

Téléphone :   (022) 41 98 11   **1211 GENÈVE 23**

Télex :   GENÈVE - 2 36 98

Télégramme :   CERNLAB-GENÈVE   Genève,   27 April 1972

Dr. Philip Pearle
Department of Physics
Hamilton College
CLINTON, N.Y. 13323

Dear Professor Pearle,

       If you were to come to CERN I think you would find
in me a kindred spirit, but you might not find any others. CERN
is quite intensely and narrowly dedicated to high energy elementary
particle physics, and our library, our seminar programs, our
personnel, are all specialized quite thoroughly in this direction.
My own occasional excursions into broader fields are tolerated
aberrations rather than normal activities. If you are attracted
to this region and to occasional contact with CERN, you could
consider the Institute of Physics of the University of Geneva,
which has some solid state physics, high energy physics, and
research into fundamental problems in the style of Jauch – to whom
you could write. The University people are of course welcome to
come out here as often as they like, for as much elementary particle
physics as they can stand.

                           Yours sincerely,

John Bell

                         John S. Bell

Figure 23.2 Letter from John Bell, April 1972.

we left Switzerland). While there, I visited with John Bell at CERN only once, and we discussed my beginning excursions into constructing a dynamical theory of wave function collapse, which emerged a few years later [3]. He was interested and supportive, but I felt diffident enough not to bother him again without something new to report.

While on sabbatical once more, this time in 1981–2 at Oxford with the group of Roger Penrose (whom gravitational considerations had led to argue for dynamical collapse [4]), it looked as if John Bell might come by, but it fell through. I attach a note he sent (Figure 23.3), which contains an interesting concluding remark.

Sabbatical time came yet again in 1987–8, which proved to be the most productive and exciting time of my professional life: July–October at South Carolina working with Yakir Aharonov and Lev Vaidman (which resulted in a joint paper on the Aharonov–Casher effect) and getting to know Shmuel Nussinov and Frank Avignone (with whom I subsequently wrote papers) and Jeeva Anandan, November–January at Cambridge with Michael Redhead, Jeremy Butterfield and Rob Clifton (who subsequently edited a book to which I contributed), February–April at Trieste with GianCarlo Ghirardi, and May–June at Pavia with Alberto Rimini.

The last two engagements, which led to my discovering CSL, I owed to John Bell. He had sent me a preprint of "Are there quantum jumps?" [5], from which I learned about Ghirardi, Rimini, and Weber's spontaneous localization (SL) theory of wave function collapse. I wrote to him on Nov. 5, 1986, expressing my appreciation for their work, and I added that I had a sabbatical coming up, ending with:

I teach at a small liberal arts college in the U.S. and thus have relatively little time for research and interaction with other physicists, so these sabbaticals are precious to me. I recall the one time I visited you at CERN, in 1973, when I was on sabbatical at the University of Geneva with Prof. Jauch, you let me know that the Powers at CERN were not too encouraging of research in Foundational questions but, nonetheless, I will ask you if you think CERN might be possible or appropriate. Or, perhaps you might be able to recommend a suitable place, for example, do you think Professors Ghirardi, Rimini or Weber might be interested in inviting me for part of that time? I do not know to which one of them it would be appropriate for me to write. I would appreciate any advice or help you could offer.

I received an immediate response (Figure 23.4).

GianCarlo and I worked hard at making a relativistic version of the SL theory, but did not succeed in finding one (subsequently one was found by Euan Squires and Chris Dove [6] and later by Rodi Tumulka [7]). But during this time, I had an inkling of what turned out to be CSL.

A week before I was to leave Trieste for Pavia, GianCarlo and I drove two hours to the University of Padua, to hear John Bell giving a talk entitled "Six Possible Realities." Afterward, we had a chance to talk with him. In trying to explain my new ideas, I realized that I didn't understand them very well.

So, arriving at Pavia, I asked Alberto if he would mind my trying to clarify my thoughts, instead of working on a project he had proposed, and he graciously acquiesced. In 17 days, I had finished my paper [8]. Alberto read it, and invited GianCarlo to visit over the weekend

ORGANISATION EUROPÉENNE POUR LA RECHERCHE NUCLÉAIRE •

CERN EUROPEAN ORGANIZATION FOR NUCLEAR RESEARCH

1211 GENÈVE 23
SUISSE / SWITZERLAND

Téléphone: (022) 41 98 11
Telex GENÈVE : 2 36 98
Télégramme: CERNLAB-GENÈVE

Votre référence
Your reference

Notre référence
Our reference

▶ A rappeler dans la réponse
Please quote in your reply

Genève. 1982 July 7

Dear Professor Pearle,

thank you for letter and paper. It is now clear that I will not get to England again while you are there, so we unfortunately miss one another on this occasion. Your nonlinear work is the most serious that I know of and I wish you great success in continuing it. It is not so much relativistic invariance that I hope for eventually (I suspect it is impossible) but a somehow theory covering nevertheless the phenomena of relativistic quantum field theory.

with best wishes
John Bell

Figure 23.3 Letter from John Bell, July 1982.

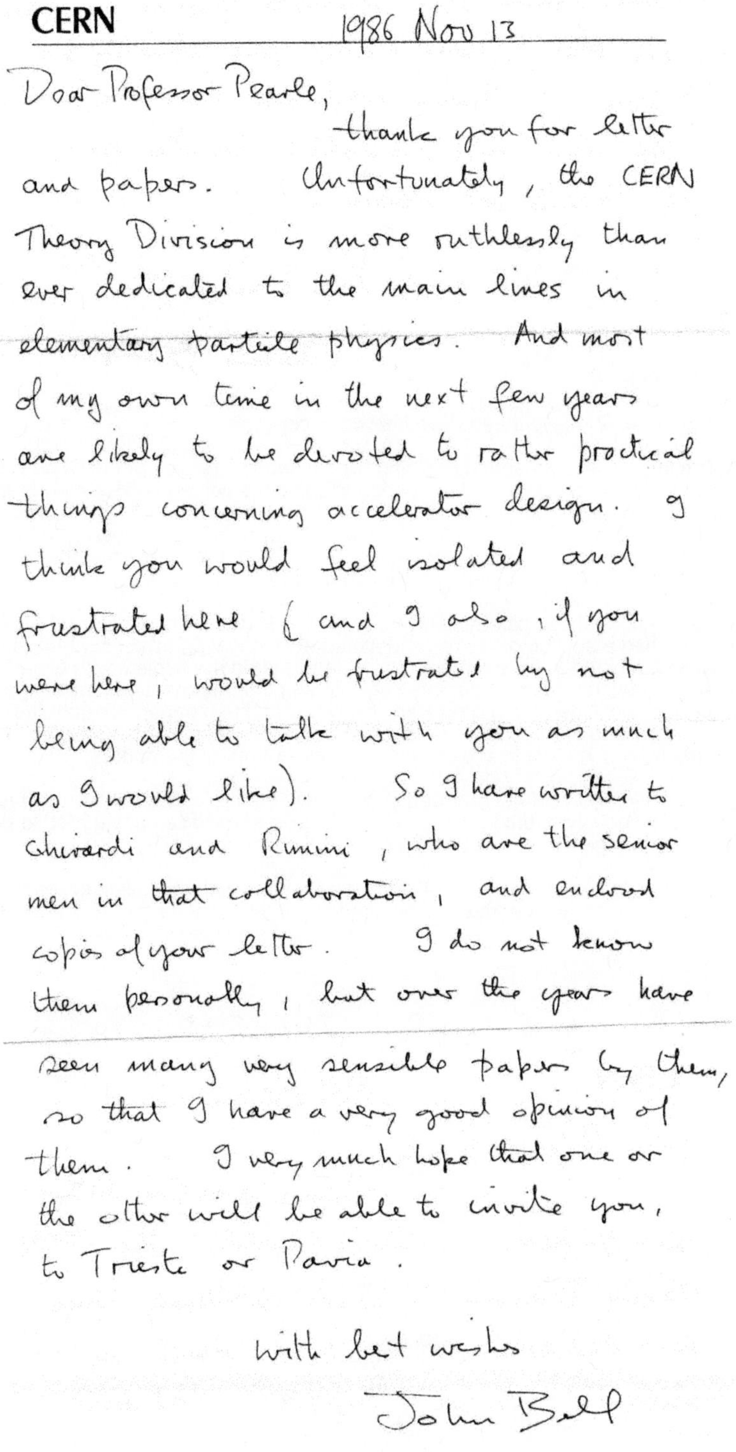

CERN 1986 Nov 13

Dear Professor Pearle,

thank you for letter and papers. Unfortunately, the CERN Theory Division is more ruthlessly than ever dedicated to the main lines in ~~elementary particle~~ physics. And most of my own time in the next few years are likely to be devoted to rather practical things concerning accelerator design. I think you would feel isolated and frustrated here (and I also , if you were here, would be frustrated by not being able to talk with you as much as I would like). So I have written to Ghirardi and Rimini , who are the senior men in that collaboration, and enclosed copies of your letter. I do not know them personally, but over the years have seen many very sensible papers by them, so that I have a very good opinion of them. I very much hope that one or the other will be able to invite you, to Trieste or Pavia.

With best wishes

John Bell

Figure 23.4 Letter from John Bell, November 1986.

UNIVERSITÀ DEGLI STUDI DI PAVIA

DIPARTIMENTO
DI FISICA NUCLEARE E TEORICA

Via Bassi 6 - 27100 PAVIA (Italy)

Pavia,

Prot. n.

Dr. John Bell
CERN
Geneva, Switzerland

Dear Dr. Bell,

I am sending under separate cover two enclosures.
~~In addition to this letter you will find~~ two enclosures.

The large one is a paper containing my continuous version of the GRW process that I tried to explain to you in Padua. I hope you find the written version clearer than the oral one - it ought to be clearer, since I understand it much better than I did then. GianCarlo, Alberto and I think that it is the framework for future development, because it is based on a (stochastic) Schrodinger-like equation that is *linear* in the wavefunction, and nice results are coming out of it.

Chief among these nice results is a very natural treatment of identical particles. I originally had a section of my paper devoted to identical particles, but I excised it because I decided to write a more extensive paper on the subject with GianCarlo and Alberto. However, I recall your interest in seeing the treatment of identical particles that GianCarlo and Alberto had produced. They finished that paper, but both feel that my teatment is superior, and GianCarlo urged me to tell you about it. So, I am enclosing also the section on identical particles that I had cut out of the paper.

GianCarlo and I did not succeed in producing a satisfactory relativistic theory of statevector reduction, but I am still hopeful because we have not yet tried to do it in the context of the theory in this paper.

I wish to thank you again for bringing me together with my Italian colleagues. I have made some good friends, this has been a very fruitful year for me, and I have learned a lot.

With very best wishes,

Phil Pearle

Phil Pearle

Figure 23.5 Letter to John Bell, May 1988.

without my knowledge, with the result that I was surprised on Monday to learn of their excitement about it, and also to learn of some excellent ideas of their own, so we decided to write another paper together [9]. The letter I wrote to John about this development appears as Figure 23.5.

(Incidentally, regarding the topic of identical particles in GRW's SL theory mentioned in this letter, a satisfactory treatment was found by Euan Squires and Chris Dove [10] and independently by Rodi Tumulka [7].)

I saw John Bell two more times.

At a conference in Erice in August 1989 entitled "Sixty-Two Years of Uncertainty," he gave a simply wonderful talk, "Toward an Exact Quantum Mechanics," but published under the bolder title "Against 'measurement'" [11]. In it, he incisively critiqued the texts of L.D. Landau/ E.M. Lifshitz, K. Gottfried, and a paper by N. G. van Kampen with regard to their treatments of the mysterious transition of the state vector from AND to OR, as he put it (i.e., collapse), condemning them for their imprecision by the acronym he introduced especially for the purpose, FAPP (for all practical purposes).

David Mermin was sitting with Viki Weisskopf up front, and David rose to defend orthodoxy. After listening patiently to a few minutes of this remonstration, Bell interrupted with the single loud word "FAPPtrap" (for those unfamiliar with this idiom, "claptrap" is an eighteenth-century word referring to theatrical techniques employed to garner applause, but which has come to mean "absurd or nonsensical talk or ideas"). This broke up the audience, and David sat down.

Afterwards, I spoke with John and expressed my appreciation, remarking that during his talk I had this vision of him as a knight, behind whose shield people like me who dared to challenge orthodoxy could safely work. While thanking me he commented, nonetheless, that he would be critical of me as well should he perceive the necessity. Apparently, the necessity arose a few days later.

Prior to the mandatory banquet, a number of us were standing around and talking about dynamical collapse. This included Abner Shimony. In my early attempts at constructing collapse dynamics, Abner had argued for the necessity of what is now called the "tails" criterion. This is that the state vector has to end up with 0 amplitude in a finite time for any macroscopic state but one. In my pre-CSL dynamical models, I had acquiesced and used this as a criteria to distinguish between some alternatives [12]. However, with the advent of CSL, for which other amplitudes become exponentially small but not 0, I have found a number of good reasons to disacquiesce.

Abner still maintained that position. So I raised the tails issue as if it was a concern of mine, thinking it would be nice for Abner to hear John's opinion. In a disquisition that began with the booming phrase for all to hear in his inimitable charming brogue, "Phil, don't be a ninny" (for those unfamiliar with this idiom, "ninny" is a sixteenth-century word referring to "innocent," but which has come to mean "a foolish or simple-minded person"), which made me feel rather warm, he stressed that one can unambiguously tell the difference between a really small-amplitude term compared with a very-much-larger-amplitude term. I had already believed this and, though tarted up, it is the main argument today [13], but Abner sticks by his guns.

At Erice, I also gave a talk, on an attempt to make a relativistic version of CSL [14], which culminated in a paper by Ghirardi, Grassi and myself [15]. While talking, I would look up every once in a while to see John sitting at the end of the row with his feet in the aisle and a grin on his face. Soon after the conference, I got a note from him (Figure 23.6), so I sent him as much of [14] as I had finished, and received a postcard (Figure 23.7) in an envelope in return, my last written communication from him.

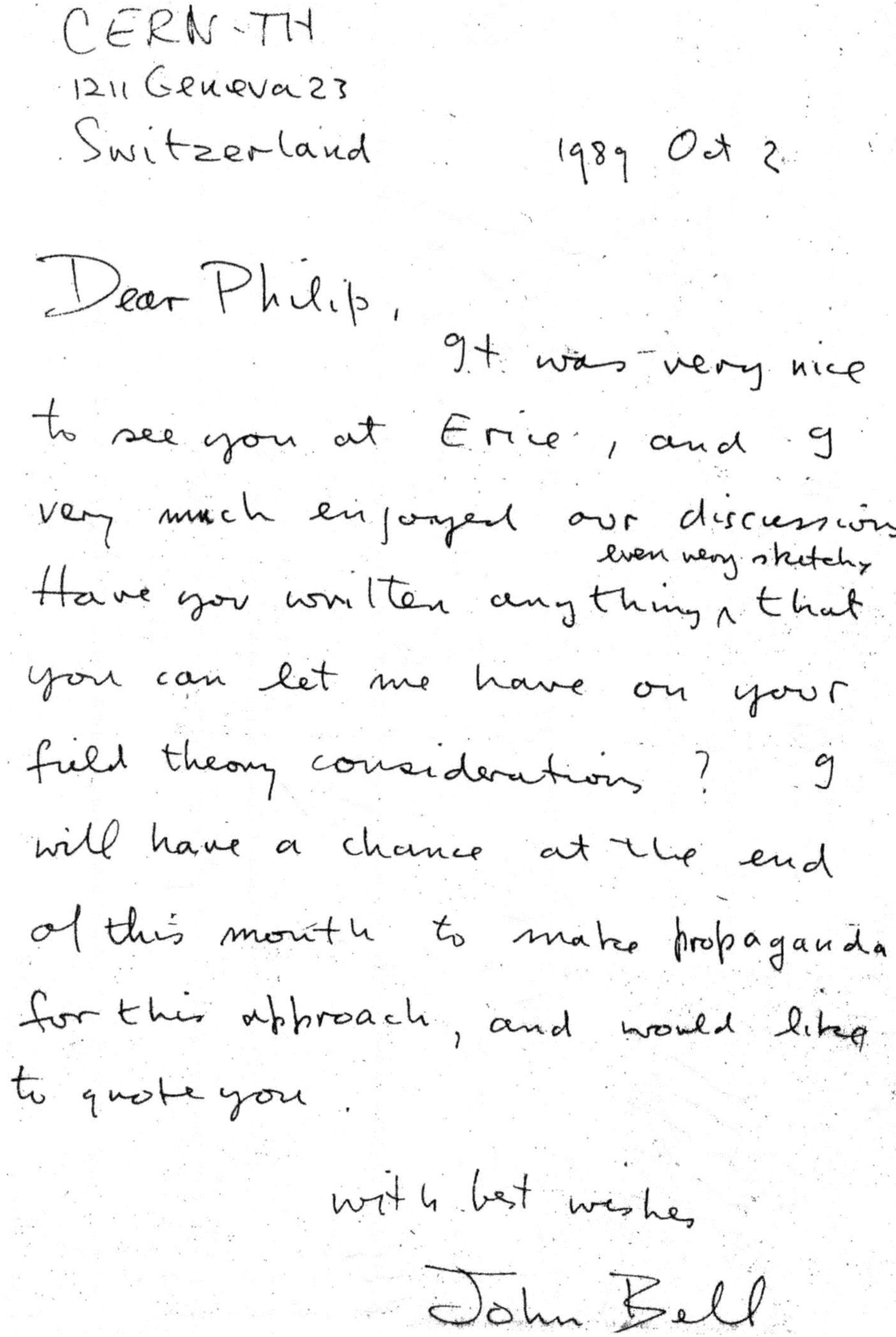

CERN-TH
1211 Geneva 23
Switzerland   1989 Oct 2

Dear Philip,

It was very nice to see you at Erice, and I very much enjoyed our discussions. Have you written anything ∧ that you can let me have on your field theory considerations? I will have a chance at the end of this month to make propaganda for this approach, and would like to quote you.

                    with best wishes

                            John Bell

Figure 23.6 Letter from John Bell, October 1989.

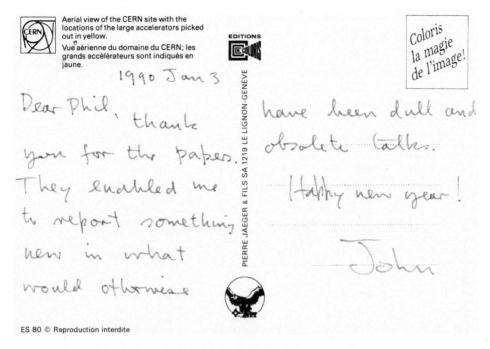

Figure 23.7 Letter from John Bell, January 1990.

However, on June 10–15 in 1990, at a Workshop on the Foundations of Quantum Mechanics convened at Amherst College, to all our delight, John showed up with his wife Mary, largely, I believe, because Mary had an old friend who lived in the vicinity. The format was very informal and congenial to private conversations, and Kurt Gottfried and John conversed privately a good deal. I thought to myself, here is a modern Einstein–Bohr debate, I can't wait to hear the outcome. But both men were too courteous for my expectations, and I guess they agreed to disagree, for nothing memorable that I recall emerged.

At the end of the conference we all sat around in a relaxed mood and were asked to say something we had learned. It came my turn, and I said that I had previously characterized the tails problem of CSL, rather poetically I had thought, that a little bit of everything that might have been coexists with what is (at which point I snuck a peek at a frowning John), but that I had learned from John that one should not express the ideas of a new theory in an old language (at which point John positively beamed!).

Three months later, we learned of our irreparable loss.

David Mermin, who had been at Amherst, too, Kurt Gottfried and I, perhaps feeling the need to do *something*, arranged to meet for private discussions concerning foundational issues at Cornell a couple of times. However, it is a matter of core beliefs, of religion if you will: which is holier to you, quantum theory or reality? If the former, one may muck up reality. If the latter, one may muck up quantum theory. None of us was converted [16].

23.2 CSL

Here is a brief introduction to the muck-up of quantum theory that is CSL, without proofs [17].

The state vector evolution in standard quantum theory is

$$|\psi, t\rangle = \mathcal{T} e^{-i \int_0^t dt' H(t')} |\psi, 0\rangle,$$

where the time-ordering operator $\mathcal{T}$ is necessitated if $[H(t), H(t')] \neq 0$ for some $t \neq t'$.

The CSL evolution is

$$|\psi, t\rangle = \mathcal{T} e^{-i \int_0^t dt' H(t') - \frac{1}{4\lambda} \int_0^t dt' [\mathbf{w}(t') - 2\lambda \mathbf{A}]^2} |\psi, 0\rangle. \qquad (23.1)$$

Here, λ is a collapse rate parameter. $\mathbf{w}(t') = (w_1(t'), w_2(t'), \ldots)$ is a vector whose components are independent functions of time of a white noise character; i.e., they can take on any value between $-\infty$ and ∞ at any time. $\mathbf{A} = (A_1, A_2, \ldots)$ are a set of mutually commuting "collapse generating" operators, so they have a joint complete eigenbasis. If $H = 0$, one of these basis vectors is the end result of collapse, i.e., is the state vector at $t \to \infty$.

Nature is supposed to provide a particular $\mathbf{w}(t')$ under which the state vector evolves. The probability of any $\mathbf{w}(t')$ is postulated to depend only upon the state vector $|\psi, t\rangle$ that has evolved under it:

$$\text{Probability} \sim \langle \psi, t | \psi, t \rangle. \qquad (23.2)$$

State vectors have different norms, since their evolution is not unitary: this *probability rule* says that the state vectors of largest norm are the most probable. The proportionality constant in (23.2) can be chosen so that the integrated probability is 1 (the integral is the product of $\int_{-\infty}^{\infty} dw_j(t')$ for each fixed j and each fixed t', $0 \leq t' \leq t$).

That is all there is to the formalism of CSL, Eqs. (23.1) and (23.2). It may be applied to many different situations. The most familiar is nonrelativistic CSL, where the indices j refer to the points of space $\mathbf{x}$ and $A_{\mathbf{x}}$ is chosen to be essentially proportional to the mass density operator at $\mathbf{x}$ [8, 9]. This makes a superposition of states such as a pointer in the state $\alpha |\text{here}\rangle + \beta |\text{there}\rangle$ collapse toward either $|\text{here}\rangle$ (which happens for a fraction $|\alpha|^2$ of the outcomes) or $|\text{there}\rangle$ (which happens for a fraction $|\beta|^2$ of the outcomes). The bigger the pointer mass, the faster the collapse proceeds. However, the CSL formalism may be applied otherwise; e.g., it has recently been applied to inflaton field fluctuation operators in the early universe [18] so that a particular universe (presumably ours) is chosen by collapse dynamics, instead of the superposition of universes given by the standard theory.

If one sets $H = 0$ to see the collapse mechanism operate without interference, $\mathcal{T}$ is no longer needed in Eq. (23.1), since the A_j commute. Upon expanding the exponent, one sees it may be written in terms of $B_j(t) \equiv \int_0^t dt' w_j(t')$:

$$-\frac{1}{4\lambda} \int_0^t dt' [\mathbf{w}(t') - 2\lambda \mathbf{A}]^2 = -\frac{1}{4\lambda} \int_0^t dt' \mathbf{w}^2(t') - \mathbf{B}(t) \cdot \mathbf{A} + \lambda \mathbf{A}^2$$

$$= \left[-\frac{1}{4\lambda} \int_0^t dt' \mathbf{w}^2(t') + \frac{1}{4\lambda t} \mathbf{B}^2(t) \right]$$

$$- \frac{1}{4\lambda t} [\mathbf{B}(t) - 2\lambda t \mathbf{A}]^2. \tag{23.3}$$

It may be shown that the term in the second line of Eq. (23.3) makes no dynamical or probabilistic contribution, so in this case the evolution (23.1) simplifies to

$$|\psi, t\rangle = e^{-\frac{1}{4\lambda t}[\mathbf{B}(t) - 2\lambda t \mathbf{A}]^2} |\psi, 0\rangle, \tag{23.4}$$

where (23.2) says that the probability that $\mathbf{B}(t)$ lies in the range $d\mathbf{B}(t)$ is

$$\text{Probability} = \prod_j \frac{dB_j}{\sqrt{2\pi \lambda t}} \langle \psi, t | \psi, t \rangle. \tag{23.5}$$

We shall use Eqs. (23.4) and (23.5) in the rest of this paper.

23.3 Nonlocality in CSL

The deBroglie–Bohm pilot wave theory and quantum theory yield the same predictions. In the former, spatially separated objects influence each other nonlocally through the quantum potential. Bell suspected the nexus of these two features, identical predictions and nonlocality, for *any* alternative to quantum theory, and was thus led to his inequality. Here we consider the nexus of these two features in a simplified version of nonrelativistic CSL [19].

Consider an EPR–Bohm state of two spin-1/2 particles, $\alpha |\Uparrow\rangle_L |\Downarrow\rangle_R + \beta |\Downarrow\rangle_L |\Uparrow\rangle_R$ ($|\alpha|^2 + |\beta|^2 = 1$), where L and R refer to two widely separated locations. At the left and right there are identical apparatus, each attached to a pointer containing N nucleons. The usual Hamiltonian evolution is such that each apparatus locally, faithfully and very rapidly measures the spin in the vertical direction, resulting in the associated pointers being either up (state $|\uparrow\rangle$) or down (state $|\downarrow\rangle$), correlated with the spin that is encountered at each location. The resulting quantum state, in usual quantum theory, evolves no further. However, this state,

$$|\psi, 0\rangle = \alpha |\uparrow\rangle_L |\downarrow\rangle_R + \beta |\downarrow\rangle_L |\uparrow\rangle_R, \tag{23.6}$$

is the initial quantum state for applying CSL dynamics. (Here we have neglected the spin particles as well as the rest of the apparatus, which are assumed to have negligibly different mass distributions in the different states, and therefore to have a negligible effect on the collapse dynamics.)

Simplifying nonrelativistic CSL, we define four collapse-generating operators that represent the number of nucleons (i.e., $\approx$ the mass of the pointer divided by the mass of a nucleon) at left or right in the relevant directions: $N_{L\uparrow} \otimes 1_R$, $N_{L\downarrow} \otimes 1_R$, $1_L \otimes N_{R\uparrow}$, $1_L \otimes N_{R\downarrow}$. They are defined by

$$N_{L\uparrow}|\uparrow\rangle_L = N|\uparrow\rangle_L, N_{L\uparrow}|\downarrow\rangle_L = 0; N_{L\downarrow}|\downarrow\rangle_L = N|\downarrow\rangle_L, N_{L\downarrow}|\uparrow\rangle_L = 0;$$

$$N_{R\uparrow}|\uparrow\rangle_R = N|\uparrow\rangle_R, N_{R\uparrow}|\downarrow\rangle_R = 0; N_{R\downarrow}|\downarrow\rangle_R = N|\downarrow\rangle_R, N_{R\downarrow}|\uparrow\rangle_R = 0.$$

Similarly, we introduce four random variables $B_{L\uparrow}, B_{L\downarrow}, B_{R\uparrow}, B_{R\downarrow}$ that fluctuate in each of the four regions Left–Right/Up–Down.

The apparatus on either side may be rotated so that spin in any direction may be measured, and so pointers may point in any direction. Therefore, we should have, in our expressions, collapse-generating operators and random variables associated with pointers pointing in any direction. But all the associated collapse-generating operators operating on the basis vectors in this example give 0, make no contribution to any physical result and may be ignored.

It only remains to apply Eqs. (23.4) and (23.5) and draw the consequences.

The state vector at time t is

$$|\psi, t\rangle = \alpha|\uparrow\rangle_L|\downarrow\rangle_R e^{-\frac{1}{4\lambda t}[(B_{L\uparrow}-2\lambda Nt)^2+B_{L\downarrow}^2+B_{R\uparrow}^2+(B_{R\downarrow}-2\lambda Nt)^2]}$$
$$+ \beta|\downarrow\rangle_L|\uparrow\rangle_R e^{-\frac{1}{4\lambda t}[B_{L\uparrow}^2+(B_{L\downarrow}-2\lambda Nt)^2+(B_{R\uparrow}-2\lambda Nt)^2+B_{R\downarrow}^2]} \qquad (23.7)$$

and the probability density of the random variables is

$$P = |\alpha|^2 e^{-\frac{1}{2\lambda t}[(B_{L\uparrow}-2\lambda Nt)^2+B_{L\downarrow}^2+B_{R\uparrow}^2+(B_{R\downarrow}-2\lambda Nt)^2]}$$
$$+ |\beta|^2 e^{-\frac{1}{2\lambda t}[B_{L\uparrow}^2+(B_{L\downarrow}-2\lambda Nt)^2+(B_{R\uparrow}-2\lambda Nt)^2+B_{R\downarrow}^2]}. \qquad (23.8)$$

Each term in (23.8) describes a Gaussian in the four-dimensional **B**-space: call them $G_\alpha(\mathbf{B}, t)$ and $G_\beta(\mathbf{B}, t)$. The peak of G_α, call it $\mathbf{B}_\alpha$, has components $B_{L\uparrow} = 2\lambda Nt$, $B_{L\downarrow} = 0$, $B_{R\uparrow} = 0$, $B_{R\downarrow} = 2\lambda Nt$. The peak of G_β, call it $\mathbf{B}_\beta$, has components $B_{L\uparrow} = 0$, $B_{L\downarrow} = 2\lambda Nt$, $B_{R\uparrow} = 2\lambda Nt$, $B_{R\downarrow} = 0$. Each Gaussian has standard deviation $\sigma = \sqrt{\lambda t}$, so its width spreads much more slowly than $2\lambda Nt$, the translation of the peaks. Thus, for $t \gg 1/\lambda N^2$, $G_\alpha(\mathbf{B}, t)$ and $G_\beta(\mathbf{B}, t)$ only have a very small overlap of their tails.

So the most probable **B**-values lie within a few standard deviations either of $\mathbf{B}_\alpha$ or of $\mathbf{B}_\beta$. The other **B** values have a negligible probability of occurring, so they need not be considered. If **B** lies anywhere within a few standard deviations of $\mathbf{B}_\alpha$, the associated state vector given by (23.7) is $|\psi, t\rangle \sim |\uparrow\rangle_L|\downarrow\rangle_R$ to an excellent approximation, as the other term is exponentially smaller (a similar statement holds for $\mathbf{B}_\beta$). This is the collapse: CSL has described something that standard quantum theory never does, the occurrence of an event.

(The norm of a state in CSL is as unimportant as the phase factor multiplying a state in standard quantum theory: One can normalize a state at any time. A state is defined by its direction in Hilbert space.)

If, using (23.5), one integrates the probability (23.8) over a few standard deviations of $\mathbf{B}_\alpha$, the result is $|\alpha|^2$ to an excellent approximation (similarly for $\mathbf{B}_\beta$). This is the identical prediction given by standard quantum theory for the outcome $|\psi, t\rangle \sim |\uparrow\rangle_L|\downarrow\rangle_R$.

Thus, in CSL, is achieved one-half of the nexus described above, identical predictions. What about the second half, nonlocality?

The state vector evolution is linear and local.

It is in the probability rule, nonlinear in the state vector, where the nonlocality resides. The two Gaussians, G_α and G_β each depend on four variables: $B_{L\uparrow}$, $B_{L\downarrow}$, which exist at the left, and $B_{R\uparrow}$, $B_{R\downarrow}$, which exist at the right. The probability rule therefore says that the left variables must be *correlated* with the far-removed right variables, in order for a high-probability **B** to occur, resulting in a high-probability outcome of the collapse.

In CSL, the probability rule, responsible for the nonlocal correlation, has the status of a law of nature, assumed but unexplained. As for any law of nature, it might some day be explained by something deeper. If the explanation involves nonlocal communication, this need not necessarily violate special relativity, since relativity allows tachyonic influences.

23.4 Bell's Inequality Violation

Bell's inequality [20] concerns a theory's expression, $P(\mathbf{a}, \mathbf{b})$, for the ensemble average of the spin correlation $\sigma_1 \cdot \mathbf{a}\sigma_2 \cdot \mathbf{b}$, where **a** and **b** are unit vectors making an angle θ with each other and the state is the angular momentum 0 (singlet) state, Eq. (23.6), with $\alpha = -\beta = 1/\sqrt{2}$. Quantum theory gives $P(\mathbf{a}, \mathbf{b}) = -\cos\theta$, and this does not satisfy the inequality.

As we have seen, CSL's probability rule has a nonlocal influence, correlating the probabilities of detection at left and right. Certainly, at $t = 0$, when the pointer state mimics the spin state, and at time $t = \infty$, when the outcomes of the collapses are precisely the same, as predicted by quantum theory's "collapse postulate," $P(\mathbf{a}, \mathbf{b}; 0) = P(\mathbf{a}, \mathbf{b}; \infty) = -\cos\theta$. But what is $P(\mathbf{a}, \mathbf{b}; t)$ for $0 < t < \infty$? While Bell showed that locality implies disagreement with quantum theory, nonlocality does not necessarily imply agreement.

For example, one might envision an experiment where the apparatus discussed above involves pointers with a small value of N, so that the associated collapse proceeds "slowly." By this is meant that the ensemble of evolving state vectors is still in a superposition of pointer states with comparable amplitudes at some experimentally accessible time. At this time, suppose one utilizes two other apparatus to make fast measurements of the orientation of the pointers. If the ensemble of state vectors at this time does not have the statistical distribution of pointer orientations initially predicted by quantum theory, the result will be different from that predicted by quantum theory. Such an experiment would be a test of CSL vs. quantum theory.

So we analyze the predicted value of $P(\mathbf{a}, \mathbf{b}; t)$ according to CSL, where $\mathbf{a} = \hat{\mathbf{k}}$, $\mathbf{b} = \hat{\mathbf{k}}\cos\theta + \hat{\mathbf{i}}\sin\theta$. With the left apparatus measuring spin in the vertical direction, but with the right apparatus rotated by an angle θ in the $x - z$ plane, one readily finds that the initial state vector (23.6) is to be replaced with

$$|\psi, 0\rangle = \frac{1}{\sqrt{2}}\left[\sin\frac{\theta}{2}|\uparrow\rangle_L| \nearrow\rangle_R + \cos\frac{\theta}{2}|\uparrow\rangle_L| \swarrow\rangle_R - \cos\frac{\theta}{2}|\downarrow\rangle_L| \nearrow\rangle_R + \sin\frac{\theta}{2}|\downarrow\rangle_L| \swarrow\rangle_R\right],$$
(23.9)

where $|\nearrow\rangle_R$, $\swarrow\rangle_R$ are the states of the pointer at the right pointing parallel or antiparallel to **b**. $N_{R\uparrow}$, $N_{R\downarrow}$ are to be replaced with $N_{R\nearrow}$, $N_{R\swarrow}$ satisfying

$$N_{R\nearrow}|\nearrow\rangle_R = N|\nearrow\rangle_R, \; N_{R\nearrow}|\swarrow\rangle_R = 0; \; N_{R\swarrow}|\nearrow\rangle_R = 0, \; N_{R\swarrow}|\swarrow\rangle_R = N|\swarrow\rangle_R,$$

and $B_{R\uparrow}$, $B_{R\downarrow}$ are to be replaced with $B_{R\nearrow}$, $B_{R\swarrow}$. Thus, the evolving state vector replacing (23.7) is

$$
\begin{aligned}
|\psi, t\rangle = &\frac{\sin\frac{\theta}{2}}{\sqrt{2}} |\uparrow\rangle_L |\nearrow\rangle_R e^{-\frac{1}{4\lambda t}[(B_{L\uparrow}-2\lambda Nt)^2+B_{L\downarrow}^2+(B_{R\nearrow}-2\lambda Nt)^2+B_{R\downarrow}^2]} \\
&+ \frac{\cos\frac{\theta}{2}}{\sqrt{2}} |\uparrow\rangle_L |\swarrow\rangle_R e^{-\frac{1}{4\lambda t}[(B_{L\uparrow}-2\lambda Nt)^2+B_{L\downarrow}^2+B_{R\nearrow}^2+(B_{R\swarrow}-2\lambda Nt)^2]} \\
&- \frac{\cos\frac{\theta}{2}}{\sqrt{2}} |\downarrow\rangle_L |\nearrow\rangle_R e^{-\frac{1}{4\lambda t}[B_{L\uparrow}^2+(B_{L\downarrow}-2\lambda Nt)^2+(B_{R\nearrow}-2\lambda Nt)^2+B_{R\swarrow}^2]} \\
&+ \frac{\sin\frac{\theta}{2}}{\sqrt{2}} |\downarrow\rangle_L |\swarrow\rangle_R e^{-\frac{1}{4\lambda t}[B_{L\uparrow}^2+(B_{L\downarrow}-2\lambda Nt)^2+B_{R\nearrow}^2+(B_{R\swarrow}-2\lambda Nt)^2]}. \quad (23.10)
\end{aligned}
$$

To express $P(\mathbf{a}, \mathbf{b}; t)$ in compact form, it is convenient to define Pauli operators $\mathbf{\Sigma}_L$, $\mathbf{\Sigma}_R$, for the pointers, where $\mathbf{\Sigma}_L \cdot \mathbf{a} = \Sigma_3$, $\mathbf{\Sigma}_R \cdot \mathbf{b} = \Sigma_3 \cos\theta + \Sigma_1 \sin\theta$ and

$$\mathbf{\Sigma}_L \cdot \mathbf{a}|\uparrow\rangle_L = |\uparrow\rangle_L, \; \mathbf{\Sigma}_L \cdot \mathbf{a}|\downarrow\rangle_L = -|\downarrow\rangle_L, \; \mathbf{\Sigma}_R \cdot \mathbf{b}|\nearrow\rangle_R = |\nearrow\rangle_R, \; \mathbf{\Sigma}_R \cdot \mathbf{b}|\swarrow\rangle_R = -|\nearrow\rangle_R.$$

Thus,

$$
\begin{aligned}
\langle\psi, &t|\mathbf{\Sigma}_L \cdot \mathbf{a}\mathbf{\Sigma}_R \cdot \mathbf{b}|\psi, t\rangle \\
=&\frac{\sin^2\frac{\theta}{2}}{2} |\uparrow\rangle_L |\nearrow\rangle_R e^{-\frac{1}{2\lambda t}[(B_{L\uparrow}-2\lambda Nt)^2+B_{L\downarrow}^2+(B_{R\nearrow}-2\lambda Nt)^2+B_{R\downarrow}^2]} \\
&- \frac{\cos^2\frac{\theta}{2}}{2} |\uparrow\rangle_L |\swarrow\rangle_R e^{-\frac{1}{2\lambda t}[(B_{L\uparrow}-2\lambda Nt)^2+B_{L\downarrow}^2+B_{R\nearrow}^2+(B_{R\swarrow}-2\lambda Nt)^2]} \\
&- \frac{\cos^2\frac{\theta}{2}}{2} |\downarrow\rangle_L |\nearrow\rangle_R e^{-\frac{1}{2\lambda t}[B_{L\uparrow}^2+(B_{L\downarrow}-2\lambda Nt)^2+(B_{R\nearrow}-2\lambda Nt)^2+B_{R\swarrow}^2]} \\
&+ \frac{\sin^2\frac{\theta}{2}}{2} |\downarrow\rangle_L |\swarrow\rangle_R e^{-\frac{1}{2\lambda t}[B_{L\uparrow}^2+(B_{L\downarrow}-2\lambda Nt)^2+B_{R\nearrow}^2+(B_{R\swarrow}-2\lambda Nt)^2]}. \quad (23.11)
\end{aligned}
$$

We must divide (23.11) by $\langle\psi, t|\psi, t\rangle$ to have the matrix element in (23.11) expressed in terms of the normalized state vector, and so obtain the correct expressions for the fractions of the various pointer directions associated with the state $|\psi, t\rangle$.

Then, for the complete ensemble of state vectors, we obtain

$$
\begin{aligned}
P(\mathbf{a}, \mathbf{b}; t) &= \int_{-\infty}^{\infty} \frac{dB_{L\uparrow}}{\sqrt{2\pi\lambda t}} \frac{dB_{L\downarrow}}{\sqrt{2\pi\lambda t}} \frac{dB_{R\nearrow}}{\sqrt{2\pi\lambda t}} \frac{dB_{R\swarrow}}{\sqrt{2\pi\lambda t}} \langle\psi, t|\psi, t\rangle \frac{\langle\psi, t|\mathbf{\Sigma}_L \cdot \mathbf{a}\mathbf{\Sigma}_R \cdot \mathbf{b}|\psi, t\rangle}{\langle\psi, t|\psi, t\rangle} \\
&= \sin^2\frac{\theta}{2} - \cos^2\frac{\theta}{2} = -\cos\theta. \quad (23.12)
\end{aligned}
$$

Thus, it does not matter at what time the ensemble of pointers is observed; the result will be the same as that predicted by quantum theory.

23.5 Conflict with Quantum Theory

Although, for the situation described above, the measurement of the ensemble of state vectors at a time between the onset of CSL dynamics and the completion of collapse gives no conflict with the predictions of quantum theory, this need not be true for other situations. Here we shall consider the general case of $\langle \psi, t | Op | \psi, t \rangle$, where Op is an arbitrary operator. We shall see that there will be a conflict if Op does not commute with the collapse-generating operators. Indeed, in the example above, they do commute as $\Sigma_L \cdot \mathbf{a} = N^{-1}[N_{L\uparrow} - N_{R\downarrow}]$, $\Sigma_R \cdot \mathbf{b} = N^{-1}[N_{R\nearrow} - N_{R\swarrow}]$, which is why there is no conflict.

In the general case, the initial state vector which CSL dynamics is to act upon (we are assuming the situation is such that the Hamiltonian is inconsequential) is

$$|\psi, 0\rangle = \Sigma_n \alpha_n |\mathbf{a}_n\rangle, \tag{23.13}$$

where $|\mathbf{a}_n\rangle$ are eigenstates of the collapse-generating operators, $A_j|\mathbf{a}_n\rangle = a_{jn}|\mathbf{a}_n\rangle$. It is assumed that there is no degeneracy; i.e., if $\mathbf{a}_n = \mathbf{a}_m$, then $m = n$. We are assuming that the $|\mathbf{a}_n\rangle$ are pointer states, so the prediction of quantum theory is that this will be the outcome state with probability $|\alpha_n|^2$.

Therefore, from (23.4), the matrix element to be considered is

$$\langle \psi, t | Op | \psi, t \rangle = \Sigma_{n,m} \alpha *_m \alpha_n \langle \mathbf{a}_m | Op | \mathbf{a}_n \rangle e^{-\frac{1}{4\lambda t}\{[\mathbf{B}(t) - 2\lambda t \mathbf{a}_n]^2 + [\mathbf{B}(t) - 2\lambda t \mathbf{a}_m]^2\}}, \tag{23.14}$$

and, from (23.5), the ensemble average is

$$\overline{\langle \psi, t | Op | \psi, t \rangle} = \Sigma_{n,m} \alpha *_m \alpha_n \langle \mathbf{a}_m | Op | \mathbf{a}_n \rangle e^{-\frac{\lambda t}{2}[\mathbf{a}_n - \mathbf{a}_m]^2}. \tag{23.15}$$

If Op is a function of $\mathbf{A}$, then $\langle \mathbf{a}_m | Op | \mathbf{a}_n \rangle = \langle \mathbf{a}_n | Op | \mathbf{a}_n \rangle \delta_{nm}$, so (23.15) becomes time-independent:

$$\overline{\langle \psi, t | Op | \psi, t \rangle} = \Sigma_n |\alpha_n|^2 \langle \mathbf{a}_n | Op | \mathbf{a}_n \rangle. \tag{23.16}$$

In any case, the right side of Eq. (23.16) also gives the ensemble average at $t \to \infty$, since $e^{-\frac{\lambda t}{2}[\mathbf{a}_n - \mathbf{a}_m]^2} \xrightarrow{t \to \infty} \delta_{nm}$. The right side of Eq. (23.16) is also the prediction of quantum theory for the value of this matrix element.

If Op does not commute with $\mathbf{A}$ then, according to (23.15), the ensemble average of the matrix element does not agree with the prediction (23.16) of quantum theory, and so its measurement at an intermediate time can serve as a test of CSL vs. quantum theory. For example, in nonrelativistic CSL, the collapse-generating operators are mass density operators, which do not commute with momentum density operators. So, if one can contrive to measure the matrix element when Op is a function of momentum, for example in an interference experiment, one can test nonrelativistic CSL [21].

References

[1] P. Pearle, How stands collapse, I, *J. Phys. A: Math. Theor.* **40**, 3189 (2007) contains a bit of an autobiographical account of how I came to care about this, a more detailed account of this paper, as well as other professional events in my life touched upon here. One other place I have written autobiographically is P. Pearle, Tales and tails

and stuff and nonsense, in R.S. Cohen, M. Horne, and J. Stachel (eds.), *Experimental Metaphysics: Quantum Mechanical Studies for Abner Shimony* (Kluwer, Dordrecht 1997), p. 143.

[2] P. Pearle, Alternative to the orthodox interpretation of quantum theory, *Am. J. Phys.* **35**, 742 (1967).

[3] P. Pearle, Reduction of the state vector by a nonlinear Schrödinger equation, *Phys. Rev. D* **13**, 857 (1976).

[4] R. Penrose, Time asymmetry and quantum gravity, in C. Isham, R. Penrose, and D.W. Sciama (eds.), *Quantum Gravity 2, A Second Oxford Symposium* (Clarendon, Oxford 1981), p. 244.

[5] J.S. Bell, Are there quantum jumps? in C.W. Kilmister (ed.), *Schrödinger: Centenary Celebration of a Polymath* (Cambridge University Press, Cambridge 1987), p. 41.

[6] C. Dove and E.J. Squires, A local model of explicit wavefunction collapse, quant-ph/9605047.

[7] R. Tumulka, A relativistic version of the Ghirardi–Rimini–Weber model, *J. Statist. Phys.* **125**, 825 (2006).

[8] P. Pearle, Combining stochastic dynamical state-vector reduction with spontaneous localization, *Phys. Rev. A* **39**, 2277 (1989).

[9] G.C. Ghirardi, P. Pearle, and A. Rimini, Markov processes in Hilbert space and continuous spontaneous localization of systems of identical particles, *Phys. Rev. A* **42**, 78 (1990).

[10] C. Dove and E.J. Squires, Symmetric versions of explicit wavefunction collapse models, *Found. Phys.* **25**, 1267 (1995).

[11] J.S. Bell, Against "measurement," in Arthur I. Miller (ed.), *Sixty-Two Years of Uncertainty* (Plenum, New York 1990), p. 17.

[12] P. Pearle, On the time it takes a state vector to reduce, *J. Statist. Phys.* **41**, 719 (1985).

[13] On tails, see Sect. 4.3 of [14]; A. Shimony, Desiderata for a modified quantum dynamics, in A. Fine, M. Forbes and L. Wessels (eds.), *PSA 1990*, Vol. 2 (Philosophy of Science Association, East Lansing 1991), p. 49; G.C. Ghirardi and T. Weber, An interpretation which is appropriate for dynamical reduction theories, in R.S. Cohen, M. Horne and J. Stachel (eds.), *Potentiality, Entanglement and Passion-at-a-Distance: Quantum Mechanical Studies for Abner Shimony Vol. 2* (Kluwer, Dordrecht 1997), p. 89; G.C. Ghirardi, R. Grassi and F. Benatti, Describing the macroscopic world: Closing the circle within the dynamical reduction program, *Found. Phys.* **25**, 5 (1995); P. Pearle, Tales and tails and stuff and nonsense, in R.S. Cohen, M. Horne and J. Stachel (eds.), *Experimental Metaphysics, Quantum Mechanical Studies for Abner Shimony Vol. 1* (Kluwer, Dordrecht 1997), p. 143; P. Pearle, How stands collapse, II, in W.C. Myrvold and J. Christian (eds.), *Quantum Reality, Relativistic Causality and Closing the Epistemic Circle: Essays in Honour of Abner Shimony* (Springer, New York, 2009), p. 257.

[14] P. Pearle, Toward a relativistic theory of statevector reduction, in Arthur I. Miller (ed.), *Sixty-Two Years of Uncertainty* (Plenum, New York) (1990), p. 193.

[15] G.C. Ghirardi, R. Grassi and P. Pearle, Relativistic dynamical reduction models: General framework and examples, *Found. Phys.* **20**, 1271 (1990).

[16] David Mermin's most recent belief is expressed in C.A. Fuchs, N.D. Mermin, and R. Schack, An introduction to QBism with an application to the locality of quantum mechanics, *Am. J. Phys.* **82**, 749 (2014).

[17] For a recent introduction, see P. Pearle, Collapse miscellany, in D. Struppa and J. Tollakson (eds.), *Quantum Theory: a Two Time Success Story. Yakir Aharonov Festschrift*, eds. D. Struppa, J. Tollakson (Springer, Milan 2013), p. 131.

[18] P. Cañate, P. Pearle and D. Sudarsky, Continuous spontaneous localization wave function collapse model as a mechanism for the emergence of cosmological asymmetries in inflation, *Phys. Rev. D* **87**, 104024 (2013).

[19] Bedingham has considered this in the context of a relativistic CSL-type model: D. Bedingham, Stochastic particle annihilation: A model of state reduction in relativistic quantum field theory, *J. Phys. A: Math. Theor.* **40**, 647 (2007): Relativistic state reduction model, *J. Phys.: Conference Series* **306**, 012034 (2011).

[20] J.S. Bell, On the Einstein–Podolsky–Rosen paradox, *Physics* **1**, 195 (1964).

[21] K. Hornberger, S. Gerlich, P. Haslinger , S. Nimmrichter and M. Arndt, Colloquium: Quantum interference of clusters and molecules, *Rev. Mod. Phys.* **84**, 157 (2012); J. Bateman, S. Nimmrichter, K. Hornberger and H. Ulbricht, Near-field interferometry of a free-falling nanoparticle from a point-like source, *Nature Commun.* **5**, 4788 (2014).

24

Gravitation and the Noise Needed in Objective Reduction Models

STEPHEN L. ADLER

Abstract

I briefly recall intersections of my research interests with those of John Bell. I then argue that the noise needed in theories of objective state vector reduction most likely comes from a fluctuating complex part in the classical spacetime metric; that is, state vector reduction is driven by *complex-number-valued* "space–time foam."

24.1 Introduction

My research interests have intersected those of John Bell three times. The first was when I found the forward lepton theorem for high-energy neutrino reactions, showing that for forward lepton kinematics, a conserved vector current (CVC) and a partially conserved axial vector current (PCAC) imply that the neutrino cross section can be related to a pion scattering cross section [1]. This led to an exchange of letters and discussions with Bell during 1964–5, which are described in the Commentaries for my selected papers [2]. The second was in the course of my work on the axial-vector anomaly [3], when I had further correspondence with John Bell, as described both in the Commentaries [4] and in the volume of essays on Yang–Mills theories assembled by 't Hooft [5]. The third time was a few years after Bell's death in 1990, when I became interested in the foundations of quantum theory and the quantum measurement problem, in the course of writing my book on quaternionic quantum mechanics [6]. Foundational issues in standard, complex quantum theory had preoccupied Bell for much of his career and led to his best known work. However, my correspondence with Bell in the 1960s never touched on quantum foundations, and I only read his seminal writings on the subject much later on. In this article I focus on this third area of shared interests.

24.2 Objective Reduction Models

There is now a well-defined phenomenology of state vector reduction, pioneered by the work of Ghirardi, Rimini, and Weber (GRW) and of Philip Pearle, and worked on by many others. John Bell was interested in this program from the outset. His 1987 essay, "Are there quantum jumps?" [7], is devoted to a discussion of the GRW model, and Bell states, "For myself, I see the GRW model as a very nice illustration of how quantum mechanics, to become rational, requires only a change which is very small (on some measures!)."

Current formulations of objective reduction models use not the discrete localizations of the original GRW paper but rather a nonlinear coupling of the Schrödinger equation to a stochastic noise variable, as introduced in the continuous spontaneous localization (CSL) model of Ghirardi, Pearle, and Rimini [8]). The structure of this model, as I noted in my book on emergent quantum theory [9], is uniquely fixed by two natural physical requirements, which might be expected to arise in an integral way from a more fundamental physical theory. The first is the requirement of state vector normalization – the unit norm of the state vector should be maintained in time. The second is the requirement that there should be no faster-than-light signaling – the density matrix averaged over the noise should satisfy a linear evolution equation of Lindblad form. The form of the stochastic equation that is fixed by imposing these two requirements has the satisfying feature that reduction to definite outcomes, with probabilities obeying the Born rule, can be proved. This proof was first given in [8], and was extended to include the Lüders rule for degenerate systems in Adler, Brody, Brun and Hughston [10].

For the noise in the CSL model to give state vector reduction, as opposed to a unitary evolution, it must be introduced as an anti-Hermitian Hamiltonian term that acts linearly on the wave function; norm preservation then requires the presence of a compensating quadratic term as well. To achieve localization, the noise is coupled to a local density operator. There is then an important constraint, since as shown by Pearle and Squires [11] and subsequent papers of Collett et al. [12] and Pearle et al. [13], the noise coupling must be mass-proportional to avoid conflicts with experiment. This suggests a noise coupling to the local mass density as the favored form of the CSL model. Thus we have, in the extension of the CSL model to nonwhite noises [14],

$$\frac{d|\psi(t)\rangle}{dt} = \left[-iH + \sqrt{\gamma} \int d^3x M(\vec{x})\Phi(\vec{x}, t) + O\right]|\psi(t)\rangle, \tag{24.1}$$

with $M(\vec{x})$ the mass density for particles of masses and coordinates m_i and $\vec{q}_i$,

$$M(\vec{x}) = \sum_i m_i \delta^3(\vec{x} - \vec{q}_i), \tag{24.2}$$

and with O denoting nonlinear terms that preserve state vector normalization. Here $\Phi(\vec{x}, t)$ is a classical noise field, with expectation values $\mathcal{E}$ giving the mean and autocorrelation,

$$\mathcal{E}[\Phi(\vec{x}, t)] = 0, \quad \mathcal{E}[\Phi(\vec{x}, t_1)\Phi(\vec{y}, t_2)] = D(\vec{x} - \vec{y}, t_1 - t_2), \tag{24.3}$$

where $D(\vec{x} - \vec{y}, t_1 - t_2)$ is the noise correlation function and $\sqrt{\gamma}$ is a coupling constant that can be absorbed into the definitions of Φ and of the correlation function $D(\vec{x} - \vec{y}, t_1 - t_2)$.

24.3 What Is the Physical Origin of the Noise?

We now turn to the crucial question of the physical origin of the noise. Some possible cosmological particle physics origins of the noise were discussed in the second paper of [14], but here I want to broach another possibility, that the noise arises from a rapidly fluctuating complex part of the classical gravitational metric $g_{\mu\nu}$. (In scalar–tensor theories of gravitation, the scalar field that accompanies the metric could also play a role.) Let us suppose that the classical metric has the form

$$g_{\mu\nu} = \bar{g}_{\mu\nu} + \phi_{\mu\nu}, \tag{24.4}$$

with $\bar{g}_{\mu\nu}$ the conventional real space–time metric, and with the line element given as usual by

$$(ds)^2 = \bar{g}_{\mu\nu} dx^\mu dx^\nu. \tag{24.5}$$

We assume that the extra part $\phi_{\mu\nu}$ is an irreducibly *complex* fluctuation term, with nonzero imaginary part, and with expectations $\mathcal{E}$ given by

$$\mathcal{E}[\phi_{\mu\nu}] = 0,$$
$$\mathcal{E}[\phi_{\lambda\sigma}(\vec{x}, t_1)\phi^*_{\mu\nu}(\vec{y}, t_2)] = D_{(\mu\nu,\lambda\sigma)},$$
$$\mathcal{E}[\phi_{\lambda\sigma}(\vec{x}, t_1)\phi_{\mu\nu}(\vec{y}, t_2)] = U_{(\mu\nu,\lambda\sigma)}. \tag{24.6}$$

From the definition of the matter stress–energy tensor $T^{\mu\nu}$, the variation of the matter interaction action δS_{int} produced by the fluctuating term in the metric is

$$\delta S_{\text{int}} = -\frac{1}{2} \int d^4x \left(^{(4)}\bar{g}\right)^{1/2} T^{\mu\nu} \phi_{\mu\nu}, \tag{24.7}$$

with $\left(^{(4)}\bar{g}\right)^{1/2}$ the square root of the determinant of $-\bar{g}_{\mu\nu}$. This action variation corresponds to minus one times the time integral of a variation in the matter interaction Hamiltonian of

$$\delta H_{\text{int}} = \frac{1}{2} \int d^3x \left(^{(4)}\bar{g}\right)^{1/2} T^{\mu\nu} \phi_{\mu\nu}. \tag{24.8}$$

Since T^{00} is proportional to the local mass density, in a flat space–time with $\bar{g}_{\mu\nu}$ the Minkowski metric, the coupling of the imaginary part of the ϕ_{00} term in the metric gives a real noise coupling corresponding to $\sqrt{\gamma}\Phi$ of Eq. (24.1) with D of Eq. (24.3) computable from $D_{(00,00)}$ and $U_{(00,00)}$ of Eq. (24.6).

In writing Eq. (24.4) I am assuming that both terms in the metric, $\phi_{\mu\nu}$ as well as $\bar{g}_{\mu\nu}$, are symmetric in the indices μ, ν. An alternative way of introducing a complex metric, the Kähler metric with Hermitian metric $g^*_{\mu\nu} = g_{\nu\mu}$, has an imaginary part that is antisymmetric in its indices, and so does not couple to the symmetric stress–energy tensor. The Ansatz I that am making for the metric is similar to the definition of so-called "space–time foam," except that the usual "foam" fluctuations are assumed to be real-valued if classical, or self-adjoint if of quantum origin. Instead, I am taking the fluctuation terms to be purely classical with a nonzero imaginary part.

Before proceeding to further discussion of the Ansatz of Eq. (24.4), I note that it differs substantively from previous proposals to relate state vector reduction to gravitation, which have been reviewed by Shan Gao [15] in an article that critiques the well-known proposal of Penrose and Diósi. (For a recent survey of the Diósi–Penrose proposal and references see L. Diósi [16].) Equation (24.4), augmented by requirements of state vector normalization and no faster-than-light signaling, gives the usual decorrelation function (in Diósi's term the "catness" function) of the CSL model. The Diósi–Penrose proposal, when incorporated into Diósi's universal position localization model [17], gives a different decorrelation function that is related [16] to the Newtonian gravitational potential. Shan Gao's review [15] also notes other articles suggesting a relation between state vector reduction and gravitation. Pearle and Squires [18] propose to relate the CSL noise to fluctuations in the Newtonian potential or the curvature scalar, but do not explicitly address the issue of Hermiticity properties of the noise when viewed as an addition to the Hamiltonian. Károlyházy and subsequent collaborators [19] try to relate wave function phase fluctuations induced by real-valued fluctuations in the metric to the localizations needed in the GRW model. However, within the framework of the CSL model, real-valued metric fluctuations lead to unitary state vector evolution and do not give state vector reduction.

24.4 Arguments for a Classical, but Complex-Valued Metric

Although much effort has been devoted to trying to quantize gravitation, there has been considerable discussion in the literature of whether gravity has to be quantized. Feynman [20], in his 1962–3 lectures on gravitation, notes the possibilities both that gravity may not have to be quantized, and that quantum theory may break down at large distances for macroscopic objects. Dyson [21] argues that the Bohr–Rosenfeld argument for quantization of the electromagnetic field does not apply to gravity, and moreover, by a number of examples, shows that it is hard (perhaps not possible) to formulate an experiment that can detect a graviton. A similar conclusion about detectability of gravitons has been given by Rothman and Boughn [22]. Dyson also notes that arguments that have been cited to show that gravity must be quantized really only show the inconsistency of a particular model for classical gravity coupled to quantized matter, the Møller and Rosenfeld semi-classical Einstein equation $G_{\mu\nu} = -8\pi G \langle \psi | T_{\mu\nu} | \psi \rangle$. In a recent paper [23], I argued that in trace dynamics prequantum theory, the metric should be introduced as a c-number in order for there to be an invariant volume element defined through the determinant of the metric. (For a generalization of [23] that allows a quantized metric in trace dynamics, see Appendix A below.) In the trace dynamics framework, a consistent coupling of classical gravity to prequantum matter that obeys the Bianchi and covariant conservation identities is obtained by writing the classical Einstein equation as

$$G_{\mu\nu} = -\frac{8\pi G}{\text{Tr}(1)} \mathbf{T}_{\mu\nu}, \tag{24.9}$$

with $\mathbf{T}_{\mu\nu}$ the covariantly conserved trace stress–energy tensor.

The possibility that the metric is complex-valued has been considered previously in the literature; see [24] for references. The formalism of general relativity, involving the construction of both the affine connection and the curvature tensor, is polynomial in the metric and its derivatives, and does not impose a restriction that the metric be real-valued. In fact, one could argue that just as in the analysis of polynomial algebraic and differential equations, an extension of the metric from the real number field to the complex number field is natural. Since macroscopic bodies have a real stress–energy tensor, they serve as a source only for the real part of the Newtonian gravitational potential $(g_{00} - 1)/2$, not for the imaginary part. Hence an imaginary part of the metric would not change gravitational astrophysics. An analysis that I carried out with Ramazanoğlu [25] of spherically symmetric Schwarzschild-like solutions in trace-dynamics-modified gravity shows that in polar coordinates the metric component g_{00} develops a square root branch cut and becomes complex below a finite radius. Although the branch cut turns out to be a coordinate singularity, and g_{00} is real-valued in isotropic coordinates, this calculation suggests that the metric $g_{\mu\nu}$ should be considered as a complex-valued classical field.

24.5 Classical Noise, not Quantum Noise, Is Needed for State Vector Reduction

In this section I argue that quantum noise, unlike classical noise, does not lead to state vector reduction. I begin by contrasting the kinematic structures of classical and quantum noise.

In the case of classical noise acting on a system S in a Hilbert space $\mathcal{H}_S$, the pure state density matrix $\rho_\alpha = |\psi_\alpha\rangle\langle\psi_\alpha|$ and its general matrix element $\rho_{\alpha;ij} = \langle i|\rho_\alpha|j\rangle$, are functions of a classical noise variable α. One can then form an order-n density tensor, with $\mathcal{E}$ the expectation over the noise variable,

$$\rho^{(n)}_{i_1 j_1, i_2 j_2, \ldots, i_n j_n} = \mathcal{E}[\rho_{\alpha;i_1 j_1} \rho_{\alpha;i_2 j_2} \cdots \rho_{\alpha;i_n j_n}], \qquad (24.10)$$

which satisfies various identities derived by contracting pairs of indices [26]. The hierarchy of all such tensors captures the properties of the noise acting on the system; in particular, the expectation $\mathcal{E}[V]$ of the variance $V = \mathrm{Tr}\rho A^2 - (\mathrm{Tr}\rho A)^2$ of an operator A can be rewritten as

$$\mathcal{E}[V] = \mathrm{Tr}(\rho^{(1)}A^2) - \rho^{(2)}_{i_1 j_1, i_2 j_2} A_{j_1 i_1} A_{j_2 i_2}. \qquad (24.11)$$

The proof of state vector reduction of [8] and [10] can be recast in terms of the Lindblad evolution satisfied by the density matrix $\rho^{(1)} = \rho$ and special properties of the density tensor $\rho^{(2)}$ for the nonlinear evolutions that obey the conditions of norm preservation and no faster-than-light signaling.

To define quantum noise, we consider a closed system, consisting of a system S and an environment E in an overall pure state, characterized by a density matrix ρ. Quantum noise consists of the environmental fluctuations acting on S that are averaged over when the environment is traced over. Labeling system states by $|i\rangle$, $|j\rangle$ and environmental states by $|e_a\rangle$, $a = 1, 2, \ldots$, we can define a density tensor $\rho^{(n)}$ by

$$\rho^{(n)}_{i_1 j_1, i_2 j_2, \ldots, i_n j_n} = \mathrm{Tr}_E \rho_{i_1 j_1} \rho_{i_2 j_2} \cdots \rho_{i_n j_n},$$

$$\mathrm{Tr}_E \mathcal{O} = \sum_a \langle e_a | \mathcal{O} | e_a \rangle, \tag{24.12}$$

with $\rho_{i_\ell j_\ell}$ the matrix acting on the environmental Hilbert space according to

$$(\rho_{i_\ell j_\ell})_{e_q e_r} = \langle e_q i_\ell | \rho | e_r j_\ell \rangle. \tag{24.13}$$

This quantum hierarchy does not obey all of the properties of the classical hierarchy of Eq. (24.10). In particular, the variance V defined by $V = \mathrm{Tr}(\rho A^2) - (\mathrm{Tr}\rho A)^2$, with Tr the trace over the full system plus the environment Hilbert space, cannot be put into a form analogous to Eq. (24.11), and one cannot construct a proof of reduction following the method used in the classical case. This should not be a surprise, since Bassi and Ghirardi have proved [27], using only linearity of the Schrödinger evolution, that quantum evolution by itself cannot give rise to state vector reduction.

Can this conclusion be evaded by decreeing that the system plus environment forms an open system, which is not in a pure state? I do not believe so, when the finite propagation of signals is taken into account [28]. Consider a system consisting of a Stern–Gerlach molecular beam apparatus, plus its environment, enclosed in a large container, which is at a distance from the apparatus much greater than 3×10^7 cm, and which is used to conduct the following thought experiment. Let the container consist of perfectly reflecting boundaries, which can be simultaneously opened by command from synchronized timers just inside. When the boundaries are closed, the interior is in a pure state and the nonreduction proof of [27] applies. When the boundaries are open, the interior forms an open system, since photons and other particles (generically, information) can get out. The typical time for a molecular beam experiment is 10^{-3}s, which is less than the time for the Stern–Gerlach apparatus inside to be informed whether the boundaries are open or closed. So I do not see how the state of the boundaries, and thus whether the system is closed or open, can influence the outcome of the Stern–Gerlach experiment being conducted inside.

24.6 Conclusion

I have argued that there is a natural confluence between the noise requirements of objective theories of state vector reduction and the noise that can be furnished by a fluctuating, classical, complex-valued component of the gravitational metric.

24.7 Acknowledgments

Completion of this work was supported in part by the National Science Foundation under Grant PHYS-1066293 and the hospitality of the Aspen Center for Physics. I wish to thank Shan Gao for conceiving the volume in which this essay appears, and for his editorial assistance.

Appendix A Generalization to a Quantized Metric

To extend the argument of [23] to a quantized metric, we note that under a general coordinate transformation from coordinates x_μ to x'_μ the metric changes according to

$$g'_{\mu\nu} = g_{\rho\sigma} \frac{\partial x^\rho}{\partial x'^\mu} \frac{\partial x^\sigma}{\partial x'^\nu}. \tag{A.1}$$

Since in trace dynamics the coordinates x_μ and x'_μ are c-numbers, when the metrics $g'_{\mu\nu}$ and $g_{\rho\sigma}$ are matrix-valued Eq. (A.1) applies separately to their traces and their trace-free parts. That is, writing

$$g'_{\mu\nu} = g'_{\text{CLASSICAL}\,\mu\nu} + g'_{\text{PREQUANTUM}\,\mu\nu},$$
$$g'_{\text{CLASSICAL}\,\mu\nu} = \frac{\text{Tr}(g'_{\mu\nu})}{\text{Tr}(1)},$$
$$\text{Tr}(g'_{\text{PREQUANTUM}\,\mu\nu}) = 0, \tag{A.2}$$

and

$$g_{\rho\sigma} = g_{\text{CLASSICAL}\,\rho\sigma} + g_{\text{PREQUANTUM}\,\rho\sigma},$$
$$g_{\text{CLASSICAL}\,\rho\sigma} = \frac{\text{Tr}(g_{\rho\sigma})}{\text{Tr}(1)},$$
$$\text{Tr}(g_{\text{PREQUANTUM}\,\rho\sigma}) = 0, \tag{A.3}$$

we have the transformation laws

$$g'_{\text{CLASSICAL}\,\mu\nu} = g_{\text{CLASSICAL}\,\rho\sigma} \frac{\partial x^\rho}{\partial x'^\mu} \frac{\partial x^\sigma}{\partial x'^\nu},$$
$$g'_{\text{PREQUANTUM}\,\mu\nu} = g_{\text{PREQUANTUM}\,\rho\sigma} \frac{\partial x^\rho}{\partial x'^\mu} \frac{\partial x^\sigma}{\partial x'^\nu}. \tag{A.4}$$

We see that the trace part of the matrix-valued metric $g_{\mu\nu}$, which we have defined here as $g_{\text{CLASSICAL}\,\mu\nu}$, has the same transformation law as the full matrix-valued metric, and so the determinant of the trace part can be used to define an invariant volume element. The classical metric $g_{\mu\nu}$ of the text of this article, and of [23], can thus be identified with the trace part $g_{\text{CLASSICAL}\,\mu\nu}$ of the matrix-valued metric.

The arguments given in [23] for the form of the effective action functional of the classical metric arising from prequantum fluctuations require that the action for the prequantum fields be Weyl-scaling-invariant. When a trace-free prequantum metric $g_{\text{PREQUANTUM}\,\mu\nu}$ is included, Weyl scaling of the metric takes the form

$$g_{\text{CLASSICAL}\,\mu\nu}(x) \rightarrow \lambda^2 g_{\text{CLASSICAL}\,\mu\nu}(x),$$
$$g^{\mu\nu}_{\text{CLASSICAL}}(x) \rightarrow \lambda^{-2} g^{\mu\nu}_{\text{CLASSICAL}}(x),$$
$$g_{\text{PREQUANTUM}\,\mu\nu}(x) \rightarrow \lambda^2 g_{\text{PREQUANTUM}\,\mu\nu}(x),$$
$$g^{\mu\nu}_{\text{PREQUANTUM}}(x) \rightarrow \lambda^{-2} g^{\mu\nu}_{\text{PREQUANTUM}}(x), \tag{A.5}$$

where indices on both the classical and the prequantum metric are raised using the classical metric,

$$g^{\rho\sigma}_{\text{CLASSICAL}} = g^{\rho\gamma}_{\text{CLASSICAL}} g^{\sigma\delta}_{\text{CLASSICAL}} g_{\text{CLASSICAL}\,\gamma\delta},$$

$$g^{\rho\sigma}_{\text{PREQUANTUM}} = g^{\rho\gamma}_{\text{CLASSICAL}} g^{\sigma\delta}_{\text{CLASSICAL}} g_{\text{PREQUANTUM}\,\gamma\delta}. \tag{A.6}$$

The requirement of Weyl scaling invariance of the prequantum gravitational action can now be satisfied by fourth-order trace actions such as

$$\mathbf{S}_{\text{GRAV PREQUANTUM}} = \int d^4x \left(-\text{Det}(g_{\text{CLASSICAL}\,\xi\eta}) \right)^{1/2} g^{\alpha\beta}_{\text{CLASSICAL}} g^{\gamma\delta}_{\text{CLASSICAL}}$$

$$\times \text{Tr} \left(\frac{\partial g_{\text{PREQUANTUM}\,\rho\sigma}}{\partial x^\alpha \partial x^\gamma} \frac{\partial g^{\rho\sigma}_{\text{PREQUANTUM}}}{\partial x^\beta \partial x^\delta} \right). \tag{A.7}$$

The classical part of the metric is taken, as in [23], to have a standard Einstein–Hilbert action,

$$S_{\text{GRAV CLASSICAL}} = \frac{1}{16\pi G} \int d^4x \left(-\text{Det}(g_{\text{CLASSICAL}\,\xi\eta}) \right)^{1/2} R[g_{\text{CLASSICAL}\,\xi\eta}]. \tag{A.8}$$

Because the gravitational constant G is dimensional, this action is not Weyl-scaling-invariant.

References

[1] S.L. Adler, Tests of the conserved vector current and partially conserved axial-vector current hypotheses in high-energy neutrino reactions, *Phys. Rev.* **135** (1964), B963–B966.

[2] S.L. Adler, *Adventures in Theoretical Physics: Selected Papers with Commentaries* (Singapore: World Scientific, 2006), section on "Forward lepton theorem," pp. 6–8.

[3] S.L. Adler, Axial-vector vertex in spinor electrodynamics, *Phys. Rev.* **177** (1969), 2426–38.

[4] S.L. Adler, *Adventures in Theoretical Physics: Selected Papers with Commentaries* (Singapore: World Scientific, 2006), section on "Chiral anomalies and $\pi^0 \to \gamma\gamma$ decay," pp. 30–36, and for Bell correspondence, p. 34.

[5] S.L. Adler, Anomalies to all orders, in G. 't Hooft (ed.), *50 Years of Yang–Mills Theory* (Singapore: World Scientific, 2005), pp. 187–228, and for Bell correspondence, pp. 192–3.

[6] S.L. Adler, *Quaternionic Quantum Mechanics and Quantum Fields* (Oxford: Oxford University Press, 1995). See especially pp. 520–25, which deal with the quantum measurement problem in the context of quaternionic quantum theory.

[7] J.S. Bell, Are there quantum jumps? In *Schrödinger, Centenary of a Polymath.* (Cambridge: Cambridge University Press, 1987). This article is reprinted in S. Bell, *Speakable and Unspeakable in Quantum Mechanics* (Cambridge: Cambridge University Press, 1987), pp. 201–12; M. Bell, K. Gottfried and M. Veltman (eds.), *Quantum Mechanics, High Energy Physics and Accelerators: Selected Papers of John S Bell (with Commentary)* (Singapore: World Scientific, 1995), pp. 866–86; M. Bell, K. Gottfried and M. Veltman (eds.), *John S. Bell on the Foundations of Quantum Mechanics.* (Singapore: World Scientific, 2001), pp. 172–92. For an

appreciation by G.C. Ghirardi of Bell's paper, see G.C. Ghirardi, John Stewart Bell and the dynamical reduction program, in R.A. Bertlmann and A. Zeilinger (eds.), *Quantum [Un]speakables: From Bell to Quantum Information* (Berlin: Springer, 2002), pp. 287–305.

[8] G.C. Ghirardi, P. Pearle and A. Rimini, Markov processes in Hilbert space and continuous spontaneous localization of systems of identical particles, *Phys. Rev. A* **42** (1990), 78–89.

[9] S.L. Adler, *Quantum Theory as an Emergent Phenomenon: The Statistical Mechanics of Matrix Models as the Precursor of Quantum Field Theory* (Cambridge: Cambridge University Press, 2004), pp. 165–8. This analysis draws on work of S.L. Adler and T.A. Brun, Generalized stochastic Schrödinger equations for state vector collapse, *J. Phys. A: Math. Gen.* **34** (2001), 1–13, and was extended to include couplings nonlinear in the noise by A. Bassi, D. Dürr and Günter Hinrichs, On the uniqueness of the equation for state vector collapse, *Phys. Rev. Lett.* **111** (2013), 210401.

[10] S.L. Adler, D.C. Brody, T.A. Brun and L.P. Hughston, Martingale models for quantum state reduction, *J. Phys. A: Math. Gen.* **34** (2001), 8795–820.

[11] P. Pearle and E. Squires, Bound state excitation, nucleon decay experiments, and models of wave function collapse, *Phys. Rev. Lett.* **73** (1994), 1–5.

[12] B. Collett, P. Pearle, F. Avignone and S. Nussinov, Constraint on collapse models by limit on spontaneous X-ray emission in Ge, *Found. Phys.* **25** (1995), 1399–412.

[13] P. Pearle, J. Ring, J.I. Collar and F.T. Avignone III, The CSL collapse model and spontaneous radiation: An update, *Found. Phys.* **29** (1999), 465–80.

[14] S.L. Adler and A. Bassi, Collapse models with non-white noises, *J. Phys. A: Math. Theor.* **40** (2007), 15,803–908; S.L. Adler and A. Bassi, Collapse models with non-white noises: II. Particle-density coupled noises, *J. Phys. A: Math. Theor.* **41** (2008), 395308.

[15] Shan Gao, Does gravity induce wavefunction collapse? An examination of Penrose's conjecture, *Stud. Hist. Philos. Mod. Phys.* **44** (2013), 148–51. In addition to a review of work suggesting a relation between state vector reduction and gravitation, this article gives a critique of the Diósi–Penrose proposal, as does Shan Gao, On Diósi–Penrose criterion of gravity-induced quantum collapse, *Int. J. Theor. Phys.* **49** (2010), 849–53.

[16] L. Diósi, Gravity-related wave function collapse: Mass density resolution, *J. Phys. Conf. Ser.* **442** (2013), 012001-(7).

[17] L. Diósi, Models for universal reduction of macroscopic fluctuations, *Phys. Rev. A* **40** (1989), 1165–73.

[18] P. Pearle and E. Squires, Gravity, energy conservation, and parameter values in collapse models, *Found. Phys.* **26** (1996), 291–305.

[19] F. Károlyházy, Gravitation and quantum mechanics of macroscopic objects, *Nuovo Cimento A* **42** (1966), 390–402; F. Károlyházy, A. Frenkel and B. Lukács, On the possibility of observing the eventual breakdown of the superposition principle, in A. Shimony and H. Feshbach (eds.), *Physics as Natural Philosophy* (Cambridge, MA: MIT Press, 1982), pp. 204–39; F. Károlyházy, A. Frenkel and B. Lukács, On the possible role of gravity in the reduction of the wave function in R. Penrose and C.J. Isham (eds.), *Quantum Concepts in Space and Time* (Oxford: Clarendon, 1986), pp. 109–46; A. Frenkel, Spontaneous localizations of the wave function and classical behavior, *Found. Phys.* **20** (1990), 159–88.

[20] R.P. Feynman, On the philosophical problems in quantizing macroscopic objects, In B. Hatfield (ed.), *Feynman Lectures on Gravitation* (Reading: Addison-Wesley, 1995), Sect. 1.4, pp. 11–15.

[21] F.J. Dyson, Is a graviton detectable? Poincarè Prize Lecture (2012), *Int. J. Modern Phys. A* **28** (2013), 1330041.

[22] T. Rothman and S. Boughn, Can gravitons be detected? *Found. Phys.* **36** (2006), 1801–25.

[23] S.L. Adler, Incorporating gravity into trace dynamics: The induced gravitational action, *Class. Quantum Grav.* **30** (2013), 195015, 239501 (Erratum).

[24] J. Munkhammar, Linearization of Moffat's symmetric complex metric gravity. arXiv:0909.3586.

[25] S.L. Adler and F.M. Ramazanoğlu, Spherically symmetric vacuum solutions arising from trace dynamics modifications to gravitation, arXiv:1308.1448.

[26] S.L. Adler, A density tensor hierarchy for open system dynamics: Retrieving the noise, *J. Phys. A. Math. Theor.* **40** (2007), 8959–90. See citations [5] and [6] of this paper for earlier related work of Mielnik, and Brody and Hughston.

[27] A. Bassi and G.C. Ghirardi, A general argument against the universal validity of the superposition principle, *Phys. Lett. A* **275** (2000), 373–81.

[28] S.L. Adler, Why decoherence has not solved the measurement problem: A response to P.W. Anderson, *Stud. Hist. Philos. Mod. Phys.* **34** (2003), 135–42.

25

Towards an Objective Physics of Bell Nonlocality: Palatial Twistor Theory

ROGER PENROSE

Abstract

In 1964, John Stewart Bell famously demonstrated that the laws of standard quantum mechanics demand a physical world that cannot be described entirely according to local laws. The present article argues that this nonlocality must be gravitationally related, as it comes about only with quantum state reduction, this being claimed to be a gravitational effect. A new formalism for curved space–times, *palatial* twistor theory, is outlined, which appears now to be able to accommodate gravitation fully, providing a nonlocal description of the physical world.

25.1 Nonlocality and Quantum State Reduction

Although *quantum entanglement* is a normal consequence of the unitary (Schrödinger) evolution U of a multiparticle quantum system, this evolution is nevertheless *local* in the sense that it is described as the continuous (local) evolution in the relevant configuration space. What the EPR situations considered by John Stewart Bell [1] demonstrated was that when widely separated quantum *measurements*, of appropriate kinds, are performed on such entangled states (a current record for distance separation being 143 km, for entangled photon pairs [2]), the results of such measurements (which may be probabilistic or of a yes/no character) cannot be mimicked by any local realistic model. Thus, in trying to propose a mathematical model of what is going on *realistically* in the physical world, one would need to face up to the actual physical process involved in quantum *state reduction* R, which is an essential feature of quantum measurement.

For many years, my own position on the state-reduction issue has been that R would have to involve a fundamental mathematical extension of current quantum theory, giving a description of something *objectively taking place* 'out there' in the physical world (OR: *objective* reduction), rather than R being the effect of some kind of 'interpretation' of the standard unitary quantum formalism. Moreover, the extended theory should be able to describe the *one* physical world that we all experience, rather than some sort of coexisting superposition of vast numbers of alternatives. To be more specific about my own viewpoint, it has long been my position that such necessary *deviation* from unitarity U in this OR would result from a correct melding of the quantum formalism with that of *general relativity*. Unlike standard attempts at a quantum gravity theory, this would entail some accommodation on the part of *both* quantum mechanics *and* gravitational theory, involving some kind of 'gravitisation' of quantum mechanics [3], in addition to the conventional viewpoint of there being some needed quantum modification of the classical picture of space–time.

In my opinion, there are several reasons for believing that standard U-evolution cannot remain completely true when the effects of Einstein's general relativity become significantly involved. For example, there is the so-called 'black-hole information paradox'. According to Hawking's 1976 analysis [4], information would be lost as an aspect of black-hole evaporation, entailing a deviation from standard unitary evolution U. Although Hawking subsequently reversed his opinion (see [5]), I believe that his original argument is the more cogent one (and, in my view, information loss is also a clear implication of an examination of the appropriate conformal diagrams; see [6], especially Fig. 30.14), so that U must in any case be violated in such extreme gravitational situations. Otherwise, one is led to unpleasant and improbable-sounding conclusions such as 'firewalls' instead of horizons [7]. Moreover, the very curious way in which the Big Bang must have been an extraordinarily special initial state, as implied by the second law of thermodynamics, also sheds doubt on the conventional presumption that that event can have been the result of a standard quantum-gravity type evolution, since *gravitational* degrees of freedom were enormously suppressed at the Big Bang, this applying *only* to gravitation, the matter and radiation degrees of freedom having apparently been almost completely thermalized (see [6, Chap. 27]).

In addition to these very large-scale phenomena, fundamental issues are raised when quantum superpositions of even very tiny gravitational fields are involved [8, 9]. It turns out that Einstein's foundational *principle of equivalence* (between a local gravitational field and an acceleration) is in conflict with the standard linear U-evolution even in the case of the Newtonian limit of general relativity. For it can be shown that such superpositions are, strictly speaking, *illegal* according to the quantum formalism, as the accelerating frames relevant to the two components of such a superposition refer to *different vacua* [3]. The specific type of scheme that I have been promoting, to handle this conflict, suggests a particular scale for an OR-type process. All this agrees fairly closely with an earlier proposal introduced by Lajos Diósi [10] (see also [11]), but now with clear motivations from foundational principles of general relativity, most particularly the Galilean limit of the equivalence principle (see [3]). According to this proposal, a macroscopic quantum superposition between two quantum states **A** and **B**, in the (Newtonian) $c = \infty$ limit, would undergo a spontaneous reduction (OR) of the superposed state to one of **A** *or* **B**, on a time scale $\tau \sim \hbar/E_G$, where E_G is the gravitational self-energy of the mass distribution (positive in some regions and negative in others) of the *difference* between the distributions in **A** and **B** separately [3, 8, 9]. Here we assume that each of **A** and **B** would, on its own, have been completely *stationary*. In space–time terms, we have a quantum superposition of two different space–time geometries that persists only for a time $\sim \tau$, where the total space–time separation (superposition) period, before reduction takes place, would be of order *unity*, in

Planckian units [3], the separation of the space–time geometries being given in terms of a symplectic measure that can be explicitly given in the linearized limit [8].

25.2 Twistor Nonlocality and Its Basic Algebra

Among the principal motivational ideas behind the original introduction of *twistor theory* [12] was the feeling that one should seek a description of the physical world that would be fundamentally *nonlocal*. I had hit upon twistor theory's initial notions in late 1963, before I had had the advantage of knowing John Bell's remarkable demonstration that the effects of conventional quantum mechanics cannot actually be explained in terms of a local realistic model. Yet I had already felt that there were some reasonably persuasive indications from the earlier work of Einstein, Podolski and Rosen (EPR) [13] and David Bohm [14] that some kind of spatial nonlocality must be true of the real world, although we were certainly not *forced* into this viewpoint until Bell's famous result appeared in 1964.

The kind of nonlocality exhibited in the original twistor viewpoint was, however, rather limited. The idea was that space–time points should not be thought of as primitive entities, but *secondary* to the nonlocal notion of a space–time *light ray* – henceforth referred to simply a *ray* – namely a complete null geodesic in the *classically* viewed space–time manifold $\mathcal{M}$. Clearly a ray *is* a *non*local entity, representing the entire history of an idealized freely moving classical massless (and spinless) particle. The 5-real-dimensional space $\mathbb{PN}$ of all such rays in $\mathcal{M}$ is Hausdorff, provided that we can assume that $\mathcal{M}$ is globally hyperbolic [15]. Any point $\mathbf{r}$ of $\mathcal{M}$ may be identified within $\mathbb{PN}$ (with perhaps some ambiguity, for certain very special $\mathcal{M}$s) as the sphere $\mathbf{R}$ of rays that pass through $\mathbf{r}$. This S^2 (basically the celestial sphere of an observer at $\mathbf{r}$) has the structure of a *conformal* sphere, and one of the defining aspects of twistor theory was to try to regard $\mathbf{R}$ as a *Riemann sphere* (complex projective line) within the structure of $\mathbb{PN}$.

However, for a general $\mathcal{M}$, the real 5-manifold $\mathbb{PN}$ does not possess anything directly of the nature of the *complex* structure that would enable such an interpretation of the 2-sphere $\mathbf{R}$ to be inherited from an ambient complex structure, of some sort, within $\mathbb{PN}$ itself. Nevertheless, the 6-manifold $\mathbb{PN}$ of *momentum-scaled* rays is actually a *symplectic* manifold, where a momentum scaling assigns a (space–time) *null* convector $\mathbf{p}$ at each point of the ray, written in index form as p_a, which points along the ray in the future direction and is parallel propagated along it. The symplectic structure of $\mathbb{PN}$ is defined by the closed symplectic 2-form Σ and a natural symplectic potential 1-form Φ, given by

$$\Sigma = \mathrm{d}p_a \wedge \mathrm{d}x^a = \mathrm{d}\Phi, \ \text{ where } \ \Phi = p_a \mathrm{d}x^a$$

(abstract indices being used throughout [16]). In conventional coordinate notation, $\mathrm{d}x^a$ would simply stand for the coordinate 1-form basis; in abstract indices $\mathrm{d}x^a$ is just a Kronecker delta translating the abstract index on p_a to a conventional 1-form notation. We shall be seeing in Sect. 25.3 that, in some sense, it is the process of *canonical quantisation* (in the guise of geometric quantisation [17]), when applied to the symplectic structure

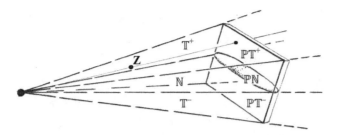

Figure 25.1 A positive-norm twistor **Z** lies in the portion $\mathbb{T}^+$ of the nonprojective twistor vector space $\mathbb{T}$, this being the disjoint union of $\mathbb{T}^+$, $\mathbb{N}$, and $\mathbb{T}^-$. The projective versions of these three spaces are $\mathbb{PT}^+$, $\mathbb{PN}$, and $\mathbb{PT}^-$, respectively.

Σ (via its associated Poisson-bracket structure Σ^{-1}), that gives quantum significance to this Riemann sphere interpretation. But we shall find that there are considerable subtleties about such a quantisation procedure, which, as it turns out, cannot be applied globally in the appropriate way to the whole 6-space $\mathbb{PN}$ when $\mathcal{M}$ is conformally curved. The global inconsistency of this quantisation process is the key to a new notion in twistor theory known as *palatial twistor theory* [18], which will be described in outline in Sect. 25.6 and which offers some hope for a fundamentally nonlocal description of general curved 4-dimensional space–times.

When $\mathcal{M}$ is *conformally flat*, and most particularly in the case $\mathcal{M} = \mathbb{M}$, where $\mathbb{M}$ is Minkowski 4-space, we do not need to appeal to quantisation procedures, and we find that the required complex structure is *already* provided by classical theory, when looked at in the appropriate way. This interpretation is explicit, as we shall see below, and (at least locally) $\mathbb{PN}$ can then be identified as a 5-dimensional real hypersurface $\mathbb{PN}$ in a complex space $\mathbb{PT}$ (which is a complex projective 3-space, $\mathbb{CP}^3$) referred to as *projective twistor space*. Thus, $\mathbb{PN}$ becomes what is referred to as a *CR-manifold* (Cauchy–Riemann or complex–real manifold [19]).

The real 5-space $\mathbb{PN}$ divides $\mathbb{PT}$ into two halves, $\mathbb{PT}^+$ and $\mathbb{PT}^-$. In each case, the prefix '$\mathbb{P}$' refers to 'projective' (i.e., all nonzero complex multiples being projected to a single point), and there is a *non*projective version, respectively $\mathbb{N}$, $\mathbb{T}^+$, and $\mathbb{T}^-$ of each (see Fig. 25.1). The space $\mathbb{T}$, referred to simply as 'twistor space' (or sometimes as '*non*projective twistor space'), is a 4-dimensional complex vector space (zero included) with *pseudo-Hermitian metric form*, of split signature $(+,+,-,-)$. The subregions $\mathbb{T}^+$, $\mathbb{T}^-$ and $\mathbb{N}$ of $\mathbb{T}$ are defined, respectively, by the metric form taking positive, negative and zero values. A *twistor* **Z**, sometimes written in abstract-index form as Z^α, is an element of $\mathbb{T}$, and in standard coordinates (Z^0, Z^1, Z^2, Z^3) for $\mathbb{T}$, the metric form is

$$||\mathbf{Z}|| = Z^0\overline{Z^2} + Z^1\overline{Z^3} + Z^2\overline{Z^0} + Z^3\overline{Z^1} = Z^\alpha\overline{Z}_\alpha = \mathbf{Z}\cdot\overline{\mathbf{Z}},$$

where the complex conjugate $\overline{\mathbf{Z}}$ of the twistor **Z** is a *dual* twistor, i.e., an element of the dual twistor space $\mathbb{T}^*$, and written in abstract-index form as $\overline{Z}_\alpha$, and in standard coordinates, its components are $(\bar{Z}_0, \bar{Z}_1, \bar{Z}_2, \bar{Z}_3) = (\overline{Z^2}, \overline{Z^3}, \overline{Z^0}, \overline{Z^1})$.

If $||\mathbf{Z}|| = 0$, we call $\mathbf{Z}$ a *null* twistor, and it represents a *ray* in $\mathbb{M}$. To see this explicitly, we need the fundamental *incidence relation* between $\mathbb{M}$ and $\mathbb{T}$ given by

$$\begin{pmatrix} Z^0 \\ Z^1 \end{pmatrix} = \frac{i}{\sqrt{2}} \begin{pmatrix} t+z & x+iy \\ x-iy & t-z \end{pmatrix} \begin{pmatrix} Z^2 \\ Z^3 \end{pmatrix},$$

where (t, x, y, z) are standard Minkowski space–time coordinates for $\mathbb{M}$ (with $c = 1$) for a point $\mathbf{r}$, either in $\mathbb{M}$ or in its *complexification* $\mathbb{CM}$. When $\mathbf{Z}$ is a *null* twistor, the real points $\mathbf{r}$ which are incident with $\mathbf{Z}$ (i.e., whose coordinates satisfy the incidence relation with Z^α) are simply the points that constitute a *ray* $\mathbf{z}$ in $\mathbb{M}$ (or at infinity, in the conformal compactification $\mathbb{M}^\#$ of $\mathbb{M}$, if $Z^2 = Z^3 = 0$). When $\mathbf{Z}$ is not null, there are no real points incident with $\mathbf{Z}$, but there are *complex* solutions for (t, x, y, z), giving points of $\mathbb{CM}$. We may see, from the incidence relation, that the rays determined by the null twistors $\mathbf{Y}$ and $\mathbf{Z}$ *intersect* (possibly at infinity, in $\mathbb{M}^\#$), if and only if $\bar{\mathbf{Z}} \cdot \mathbf{Y} (= \bar{Z}_\alpha Y^\alpha)$ vanishes. If we fix $\mathbf{r}$ as a point of $\mathbb{CM}$, then the solutions for $\mathbf{Z}$ of the incidence relation give us a complex projective line (a Riemann sphere) in the projective 3-space $\mathbb{PT}$. If $\mathbf{r}$ is a real point (point of $\mathbb{M}$, or even of $\mathbb{M}^\#$), then this Riemann sphere lies in $\mathbb{PN}$, in accordance with the above comments concerning $\mathbb{PN}$. For more details, see [20–22].

25.3 Spinor Parts and Physical Interpretation of a Twistor

The four components of a twistor Z^α, in a standard frame, may be identified as the pairs of components of two *2-spinors* ω^A and $\pi_{A'}$, according to

$$Z^0 = \omega^0, \, Z^1 = \omega^1, \, Z^2 = \pi_{0'}, \, Z^3 = \pi_{A'},$$

so we can write

$$Z^\alpha = (\omega^A, \pi_{A'}) \text{ and } \bar{Z}_\alpha = (\bar{\pi}_A, \bar{\omega}^{A'}),$$

or more simply $\mathbf{Z} = (\boldsymbol{\omega}, \boldsymbol{\pi})$ and $\bar{\mathbf{Z}} = (\bar{\boldsymbol{\pi}}, \bar{\boldsymbol{\omega}})$. The incidence relation with a space–time point $\mathbf{r}$ becomes

$$\omega^A = ir^{AA'}\pi_{A'}, \text{ which we can write } \boldsymbol{\omega} = i\mathbf{r}\cdot\boldsymbol{\pi},$$

where we are now regarding $\mathbf{r}$ as standing for the position vector of that point with respect to a given origin point $\mathbf{O}$. Under a change of origin $\mathbf{O} \mapsto \mathbf{Q}$, where q^a is the position vector $\overrightarrow{\mathbf{OQ}}$, so that the position vector $\mathbf{r}$ must undergo $r^a \mapsto r^a - q^a$, the spinor parts of $\mathbf{Z}$ must accordingly undergo

$$\omega^A \mapsto \omega^A - i\,q^{AA'}\pi_{A'}, \, \pi_{A'} \mapsto \pi_{A'},$$

this preserving the incidence relation $\boldsymbol{\omega} = i\mathbf{r}\cdot\boldsymbol{\pi}$. For a dual twistor $\mathbf{W}$, with $W_\alpha = (\lambda_A, \mu^{A'})$, we correspondingly have

$$\lambda_A \mapsto \lambda_A, \, \mu^{A'} \mapsto \mu^{A'} + iq^{AA'}\lambda_A.$$

The inner product of a twistor $\mathbf{Z} = (\omega, \pi)$ with a dual twistor $\mathbf{W} = (\lambda, \mu)$ (i.e., $W_\alpha = (\lambda_A, \mu^{A'})$) is

$$\mathbf{W}{\cdot}\mathbf{Z} = \lambda_A \omega^A + \mu^{A'} \pi_{A'} = \lambda{\cdot}\omega + \mu{\cdot}\pi.$$

In terms of ω and π, a *physical* interpretation of a twistor $\mathbf{Z} = (\omega, \pi)$, can be provided, up to the phase freedom $\mathbf{Z} \mapsto e^{i\theta} \mathbf{Z}$ (θ real), even in the *non*-null case. The twistor, up to this phase freedom, gives us the 4-momentum $\mathbf{p}$ (p_a in index form) and 6-angular momentum M^{ab} of a massless particle with spin,

$$p_a = \pi_{A'} \overline{\pi}_A, \, M^{ab} = i\omega^{(A}\overline{\pi}^{B)}\varepsilon^{A'B'} - i\overline{\omega}^{(A'}\pi^{B')}\varepsilon^{AB}$$

(where I use abstract indices, so that $a = AA'$, $b = BB'$, etc., the skew ε 'metric' spinors defining the space–time metric via $g_{ab} = \varepsilon_{AB}\varepsilon_{A'B'}$; see [16, Sect. 3.1; 22, Sect. 6.3]). Provided that $\pi \neq 0$, this p_a and M^{ab} automatically satisfy all the conditions

$$p_a p^a = 0, \, p_0 > 0, \, M^{(ab)} = 0, \, \frac{1}{2}\varepsilon_{abcd} p^b M^{cd} = sp_a$$

required for a free massless particle. Conversely, the twistor $\mathbf{Z}$ (with $\pi_{A'} \neq 0$) is determined, uniquely up to a phase multiplier $e^{i\theta}$, by p_a and M^{ab}, subject to these conditions. (When $\pi = 0$, we get a limiting situation, where the particle is at infinity, but with zero momentum and nonzero angular momentum.) We shall be seeing shortly that the *phase* also has a key geometrical role to play in the geometric (pre)-quantisation procedure mentioned in Sect. 25.2, which is relevant to the *palatial* twistor theory sketched in Sect. 25.6.

The quantity s is the *helicity* of the particle, i.e., its *spin*, but with a *sign*, positive for right-handed and negative for left-handed. The helicity s finds the very simple (and fundamental) twistor expression

$$2s = \omega^A \overline{\pi}_A + \pi_{A'}\overline{\omega}^{A'} = Z^\alpha \overline{Z}_\alpha = ||\mathbf{Z}||.$$

When $s = 0$, the complete interpretation of $\mathbf{Z}$, up to phase, is the momentum-scaled ray it determines (possibly at infinity), as given earlier. But when $s \neq 0$, there are no such real points, and there is no real world line that can be associated with $\mathbf{Z}$ in a Poincaré-invariant way (although there is a spatially nonlocal real interpretation that can be given in terms of a twisting configuration referred to as a 'Robinson congruence' [22, Sect. 6.2], from which the term 'twistor' was originally derived). There is also an interpretation of $\mathbf{Z}$ in terms of the family of *complex* points $\mathbf{x}$ satisfying the incidence relation $\omega = i\mathbf{x}{\cdot}\pi$, constituting what is referred to as an α–plane in $\mathbb{CM}$ (or $\mathbb{CM}^\#$, if we include points $\mathbf{x}$ at complex infinity).

In curved space–time $\mathcal{M}$, these notions are not so well defined. Most particularly, the concept of an α–plane disappears for a complexified (real analytic) space–time $\mathbb{C}\mathcal{M}$ for which $\mathcal{M}$ is not conformally flat. Nonetheless, we can go some small way towards defining a hypothetical twistor space $\mathcal{T}$ for $\mathcal{M}$ (analogous to $\mathbb{T}$ defined for $\mathbb{M}$), in that the nonprojective 7-space $\mathcal{N}$ can be defined from $\mathcal{M}$ by extending the $\mathbf{p}$–scaling of rays that gave us the 6-space $\mathbb{P}\mathcal{N}$ of Sect. 25.2 to a π-scaling for rays γ. Thus, the 7-space $\mathcal{N}$ is a circle-bundle over $\mathcal{N}$, where the circle is simply the *phase freedom* $\pi_{A'} \mapsto e^{i\theta}\pi_{A'}$ (θ real) referred to earlier, to give us a $\mathbb{C}^*$-family of π-*scaled* rays Γ for each ray γ in $\mathcal{M}$, where $p_a = \pi_{A'}\bar{\pi}_A$, as above, the 2-spinor $\pi_{A'}$ being taken as parallel propagated along each γ. Yet we do *not* get a canonically defined full complex twistor space $\mathcal{T}$ for a conformally curved $\mathcal{M}$.

As noted above, this circle bundle is, in fact, just what is needed for the procedure of *geometric quantisation*, when applied to the symplectic 6-manifold $\mathbb{P}\mathcal{N}$ (see [17]). What is first required for this (in the preliminary procedure of prequantisation) is a circle-bundle *connection* for which the curvature is ($\hbar$ times) the symplectic 2-form Σ (of Sect. 25.2). This connection is directly given by the 1-form $i\hbar\Phi$, where Φ is the symplectic potential referred to in Sect. 25.2. We find that to proceed to a full *quantisation* procedure, we run into issues of nonuniqueness, these being actually *central* to the nonlocality that comes about in palatial twistor theory, for which a tentative description will be given in Sect. 25.6.

25.4 Local Twistors and the Einstein Λ-Equations

Despite these ambiguities, there is, however, a *local* notion of a twistor (*not* requiring $s = 0$), defined at *each point* $\mathbf{q}$ of any $\mathcal{M}$ and which also can be carried over to each entire *ray* γ in $\mathcal{M}$, i.e., to each point of the associated ray space of $\mathbb{P}\mathcal{N}$, and thence to each point Γ of the π-scaled-ray space $\mathcal{N}$. This provides us with a flat twistor space $\mathbb{T}_{\mathbf{q}}$, canonically and conformally invariantly defined for each $\mathbf{q} \in \mathcal{M}$, and also such a space $\mathbb{T}_\gamma$, for each $\gamma \in \mathbb{P}\mathcal{N}$, where $\mathbb{T}_\gamma$ may be interpreted as a kind of complex pseudo-tangent space to $\mathcal{N}$ at each corresponding point $\Gamma \in \mathcal{N}$. These are obtained via the notions of *local twistor* and *local twistor transport* [22, Sect. 6.9].

A local twistor is a quantity $Z^\alpha = (\omega^A, \pi_{A'})$, defined at a point $\mathbf{q}$ of the space–time $\mathcal{M}$, which transforms as

$$\omega^A \mapsto \omega^A, \pi_{A'} \mapsto \pi_{A'} + i\omega^A \Omega^{-1}\nabla_{AA'}\Omega,$$

under a conformal rescaling of $\mathcal{M}$'s metric, according to $g_{ab} \mapsto \Omega^2 g_{ab}$ (Ω being a smooth positive-valued function on $\mathcal{M}$). To get an exact correspondence with the twistor concept introduced in Sect. 25.2, we must think of ω^A (and $\pi_{A'}$) as not being defined with respect to a *fixed* origin point $\mathbf{O}$, as in Sect. 25.3, but now taken with respect to a *variable* point $\mathbf{q} \in \mathcal{M}$. Recall that in $\mathbb{M}$, when the origin $\mathbf{O}$ is displaced to a *general* point $\mathbf{q}$ (with position vector q^a with respect to $\mathbf{O}$) in $\mathbb{M}$, the twistor $(\omega^A, \pi_{A'})$ defined with respect to $\mathbf{O}$ becomes $(\omega^A - iq^{AA'}\pi_{A'}, \pi_{A'})$ with respect to $\mathbf{q}$. The local twistor perspective on this is that $(\omega^A, \pi_{A'})$, defined at $\mathbf{O}$, when carried to $\mathbf{q}$ by local twistor transport, becomes $(\omega^A - iq^{AA'}\pi_{A'}, \pi_{A'})$ at $\mathbf{q}$. This enables us to extend this concept, in a conformally invariant way, to a general $\mathcal{M}$. When $\mathcal{M} = \mathbb{M}$ (or, indeed, when $\mathcal{M}$ is any simply connected conformally flat space–time) the notion of local twistor transport is *path-independent*, so that the local twistor concept extends to a global one, but this is not true in general.

The definition of *local twistor transport*, along a smooth curve γ in $\mathcal{M}$ with tangent vector t^a, is

$$t^a\nabla_a\omega^B = -it^{BB'}\pi_{B'}, t^a\nabla_a\pi_{B'} = -it^{AA'}P_{AA'BB'}\omega^B,$$

where

$$P_{ab} = \frac{1}{12}Rg_{ab} - \frac{1}{2}R_{ab}, \text{ with } R_{ac} = R^b_{abc}.$$

(sign conventions as in [16, 22]). Taking γ to be a *ray* – which is simply connected, with topology $\mathbb{R}$ (by $\mathcal{M}$'s global hyperbolicity) – we use local twistor transport to propagate (ω, π) uniquely all along γ, thereby providing us with our canonical twistor space $\mathbb{T}_\gamma$, assigned to γ. Correspondingly, we shall have spaces $\mathbb{PT}_\gamma$, $\mathbb{N}_\gamma$, and $\mathbb{PN}_\gamma$, just as in Sect. 25.2. When $\mathcal{M}$ is *conformally flat* (and simply connected), these spaces are all independent of the choice of any curve γ connecting a pair of points in $\mathcal{M}$, owing to the integrability of local twistor transport, so the local twistor spaces are all canonically identical and may be referred to simply as spaces $\mathbb{T}$, $\mathbb{PT}$, $\mathbb{N}$, and $\mathbb{PN}$, respectively, but this does not hold if $\mathcal{M}$ is conformally curved.

We must raise the question of the relation between each $\mathbb{PN}_\gamma$ and the global space $\mathbb{PN}$ of rays in a general $\mathcal{M}$. Within each $\mathbb{T}_\gamma$, for a ray γ, this ray, when π-scaled to Γ, can itself be unambiguously represented by $(0, \pi_{A'})$ all along γ, this being unchanged by local twistor transport along γ (since $t^{AA'} \propto \bar{\pi}^A \pi^{A'}$ and $\pi^{A'}\pi_{A'} = 0$). When $\mathcal{M}$ is *conformally flat* (and simply connected), the integrability of local twistor transport allows us to achieve this globally for the whole of $\mathbb{PN}$, where a π-scaled ray η in $\mathcal{M}$ that meets γ at a point $\mathbf{q}$ would be represented at $\mathbf{q}$ by the local twistor $(0, \eta_{A'})$, in both $\mathbb{T}_\gamma$ and $\mathbb{T}_\eta$, where $\eta_{A'}$ provides the direction and π-scaling for η. In fact, the spaces $\mathbb{N}_\gamma - \{0\}$ are all canonically isomorphic with each other, and (locally) with $\mathcal{N}$ itself, so it makes sense to identify $\mathcal{N} \cup \{0\}$ with each $\mathbb{N}_\gamma$. (at least locally). However, this close association does not apply when $\mathcal{M}$ is *not* conformally flat.

When $\mathcal{M}$ is $\mathbb{M}$ or the *de Sitter* 4-space $\mathbb{D}$ with (positive) cosmological constant Λ (or the *anti*-de Sitter space if $\Lambda < 0$), it *is* conformally flat, and the local twistor spaces can all be identified, as can their vector-space tensor algebras. More particularly, they have a specific structure defined by antisymmetric 2-valent twistors referred to as *infinity twistors* [22], which fix the *metric* structure of the space–time. These are $I^{\alpha\beta}$ and $I_{\alpha\beta}$, taken to be both complex conjugates and *duals* of one another:

$$I_{\alpha\beta} = \overline{I^{\alpha\beta}}, I^{\alpha\beta} = \overline{I_{\alpha\beta}},$$

$$I_{\alpha\beta} = \frac{1}{2}\varepsilon_{\alpha\beta\rho\sigma}I^{\rho\sigma}, I^{\alpha\beta} = \frac{1}{2}\varepsilon^{\alpha\beta\rho\sigma}I_{\rho\sigma},$$

where $\varepsilon_{\alpha\beta\rho\sigma}$ and $\varepsilon^{\alpha\beta\rho\sigma}$ are Levi–Civita twistors fixed by their antisymmetry and $\varepsilon_{0123} = 1 = \varepsilon^{0123}$ in standard twistor coordinates. In standard 2-spinor descriptions (Sect. 25.3), we have

$$I_{\alpha\beta} = \begin{pmatrix} \frac{\Lambda}{6}\varepsilon_{AB} & 0 \\ 0 & \varepsilon^{A'B'} \end{pmatrix}, I^{\alpha\beta} = \begin{pmatrix} \varepsilon^{AB} & 0 \\ 0 & \frac{\Lambda}{6}\varepsilon_{A'B'} \end{pmatrix}.$$

For the de Sitter space $\mathbb{D}$, the infinity twistors provide a *complex symplectic structure* (not to be confused with the *real* symplectic structure of Sect. 25.2) defined by the 2-form

$$\mathcal{J} = I_{\alpha\beta} dZ^\alpha \wedge dZ^\beta; \ d\mathcal{J} = 0.$$

Also, there is a *symplectic potential* 1-form

$$\mathcal{J} = I_{\alpha\beta} Z^\alpha dZ^\beta, \ \text{where} \ \mathcal{J} = d\mathcal{J}.$$

When $\Lambda = 0$, this symplectic structure becomes degenerate, the matrices for $I_{\alpha\beta}$ and $I^{\alpha\beta}$ becoming singular. When $\Lambda \neq 0$, they are essentially inverses of one another,

$$I_{\alpha\beta} I^{\beta\gamma} = -\frac{\Lambda}{6} \delta_\alpha^\gamma,$$

but they annihilate each other if $\Lambda = 0$. For a given Λ, the structure afforded to the twistor space $\mathbb{T}$ by $I_{\alpha\beta}$ (or equivalently $I^{\alpha\beta}$, where $\varepsilon_{\alpha\beta\rho\sigma}$ and $\varepsilon^{\alpha\beta\rho\sigma}$ are assumed given) will be called its *I-structure* (or $\mathbf{I}_\Lambda$*-structure*).

A significant feature of local twistor transport is that the satisfaction of Einstein's Λ-vacuum equations $R_{ac} = \Lambda g_{ab}$ is *equivalent* to the fact that $I_{\alpha\beta}$ (or $I^{\alpha\beta}$) is *constant* under local twistor transport. (See [22, p. 376] for the case $\Lambda = 0$. When $\Lambda \neq 0$ this fact can be directly established; see also [23].) Moreover, the local twistors $\varepsilon_{\alpha\beta\rho\sigma}$ and $\varepsilon^{\alpha\beta\rho\sigma}$ can be seen to be *automatically* local-twistor constant, independent of the Einstein equations. Accordingly, the Einstein Λ-vacuum equations can be phrased in terms of the existence of an **I**-structure that holds globally for all the local twistor spaces $\mathbb{T}_\gamma$ for the ray space $\mathcal{N}$.

25.5 Twistor Quantisation and Cohomological Wave Functions

Up to this point, I have been concerned only with classical twistor theory. For the *quantized* theory, we need to introduce the *commutation laws* [22, Sect. 6.10]:

$$Z^\alpha \overline{Z}_\beta - \overline{Z}_\beta Z^\alpha = \hbar \delta_\beta^\alpha$$

and

$$Z^\alpha Z^\beta - Z^\beta Z^\alpha = 0, \overline{Z}_\alpha \overline{Z}_\beta - \overline{Z}_\beta \overline{Z}_\alpha = 0,$$

where now Z^α and $\overline{Z}_\alpha$ are taken to be *linear operators* generating a noncommutative algebra $\mathbb{A}$, acting on some appropriate 'ket space' [24]. We should think of that space as a quantum state space of some kind, which I shall sometimes refer to as '$| \ldots \rangle$', but it is best not to be too specific about this for the time being. (In the language of standard quantum mechanics, a ket space may be thought of as a complex linear space with a basis that is a complete set of commuting variables.) In fact, it will be a key part of the arguments in Sect. 25.6 that for *conformally curved* space–times, the ket space will *not* be provided in a globally consistent way, though having a *local* (but nonunique) existence in appropriately defined 'sufficiently small' regions of the ray space $\mathbb{P}\mathcal{N}$. The proposal, according to palatial twistor theory, is that in the case of a curved $\mathcal{M}$ we obtain an algebra $\mathcal{A}$ that generalizes the role that $\mathbb{A}$ plays for $\mathbb{M}$ (or for $\mathbb{D}$), where $\mathcal{A}$ is defined *completely globally* for $\mathcal{M}$, even though there would be *no* globally consistent ket space.

There are, however, issues concerning the nature of $\mathbb{A}$ (and certainly of $\mathcal{A}$) that are not completely resolved at the time of writing. We would certainly require that $\mathbb{A}$ contain polynomial expressions in Z^α and $\bar{Z}_\beta$, but, as we shall be seeing, expressions that are *analytic* in these quantities must also play a role. In a (conformally) *curved* space–time $\mathcal{M}$, we would have a *deformed* such algebra $\mathcal{A}$ that is, in some appropriate *local* sense, the same as $\mathbb{A}$, but whose global structure would encode the entire (conformal) geometry of a given curved space–time $\mathcal{M}$. The algebra $\mathbb{A}$ itself is to be thought of as, in some sense, the algebra of linear operators acting on (germs of?) holomorphic entities of some kind defined on $\mathbb{T}$, but the precise notion of what is required has not yet become completely clear. In basic terms, $\mathbb{A}$ is to be taken as the algebra generated by Z^α and $\partial/\partial Z^\beta$, but where infinite series in these (noncommuting) operators would also need to be considered as belonging to $\mathbb{A}$. This raises issues of *convergence* and *locality* that will need to be sorted out in due course, but for our present purposes I shall ignore these subtleties and merely explain the general idea of what is required.

In the case of $\mathbb{M}$, the above commutation laws are *almost* implied by the standard quantum commutators for position and momentum,

$$p_a x^b - x^b p_a = i\hbar \delta_a^b,$$

but there appears to be an additional input related to the issue of helicity. By direct calculation, we may verify that the twistor commutation laws reproduce exactly the (considerably more complicated-looking) commutation laws for p_a and M^{ab} that arise from their roles as translation and Lorentz-rotation generators of the Poincaré group. In this calculation, we take note of the fact that there is no factor-ordering ambiguity in the expressions for p_a and M^{ab} in terms of the spinor parts of Z^α and $\bar{Z}_\alpha$ (because of the symmetry brackets in the spinor expression for M^{ab}). However, when we examine the calculation for obtaining the helicity s, we do not retrieve the classical expression $2s = Z^\alpha \bar{Z}_\alpha$ (or $2s = \bar{Z}_\alpha Z^\alpha$) but, specifically (writing the helicity *operator* as **s**, in bold type),

$$\mathbf{s} = \frac{1}{4}(Z^\alpha \bar{Z}_\alpha + \bar{Z}_\alpha Z^\alpha).$$

In analogy with the standard quantum-mechanical procedures, if we wish to consider what the *wave function* for a massless particle should be in twistor terms, we need to think of functions of Z^α that are "independent of $\bar{Z}_\beta$." This means "annihilated by $\partial/\partial\bar{Z}_\beta$," in other words *holomorphic* in Z^α (by the Cauchy–Riemann equations). Thus, a twistor wave function (in the **Z**-description) must be holomorphic in **Z** and we have the operators representing Z^α and $\bar{Z}_\alpha$:

$$Z^\alpha \rightsquigarrow Z^\alpha \times, \bar{Z}_\alpha \rightsquigarrow -\hbar \frac{\partial}{\partial Z\alpha}.$$

We can alternatively consider wave functions expressed in terms of the conjugate variables $\bar{Z}_\alpha$, which are *dual* twistors, and relabelling $\bar{Z}_\alpha$ as W_α, we have commutation laws

$W_\alpha \overline{W}^\beta - \overline{W}^\beta W_\alpha = -\hbar \delta_\alpha^\beta, W_\alpha W_\beta - W_\beta W_\alpha = 0, \overline{W}^\alpha \overline{W}^\beta - \overline{W}^\beta \overline{W}^\alpha = 0$, leading to a *dual*-twistor **W**-description of wave functions.

For purposes of being definite, I just adopt the **Z**-description here. It should be remarked that, in the **Z**-description, the above quantisation procedure allows the removal of all operations that involve $\overline{\mathbf{Z}}$, replacing them completely by operations in **Z**, and thereby providing us with entirely *holomorphic* descriptions. This will prove to be of *central* importance to the theory.

If we are asking that a wave function describe a (massless) particle of *definite helicity*, then we need to put it into an eigenstate of **s**, which, by the above, is

$$\mathbf{s} = -\frac{1}{2}\hbar \left(Z^\alpha \frac{\partial}{\partial Z^\alpha} + 2 \right).$$

This is simply a displaced *Euler homogeneity operator* $Z^\alpha \frac{\partial}{\partial Z^\alpha}$, so that for a helicity eigenstate, with eigenvalue s, we need a twistor wave function $f(\mathbf{Z})$ that is not only holomorphic but also *homogeneous* of degree

$$n = -2s - 2,$$

where, for convenience, I henceforth choose $\hbar, =, 1$. Then $2s$ is an integer (odd for a fermion and even for a boson).

We need to see the relation between such a twistor wave function and the space–time description, in terms of the zero-mass field equations in flat space–time $\mathbb{M}$ – or in a conformally flat $\mathcal{M}$ – for each helicity s. These field equations are expressed in the 2-spinor form

$$\nabla^{AA'} \psi_{AB...E} = 0, \text{ or } \psi = 0, \text{ or } \nabla^{AA'} \tilde{\psi}_{A'B'...E'} = 0,$$

for the integer $2s$ satisfying $s < 0, s = 0$, or $s > 0$, respectively, where we have *total symmetry* for each of the $|2s|$-index quantities

$$\psi_{AB...E} = \psi_{(AB...E)}, \tilde{\psi}_{A'B'...E'} = \tilde{\psi}_{(A'B'...E')}.$$

These equations give the spinor form of (the anti-self-dual and self-dual parts of) the free Maxwell equations (if $|s| = 1$) and of the source-free linearized free gravitational field (if $|s| = 2$).

The required relation between $f(\mathbf{Z})$ and the appropriate ψ (or $\tilde{\psi}$) can be achieved by a simple contour integral expression (see [25] and [22, Sect. 6.10]), where the contour lies within the Riemann sphere **R**, representing a point **r** of complex Minkowski space $\mathbb{CM}$, **R** being the locus of projective twistors $\mathbb{P}\,\mathbf{Z}$ that are incident ($\omega = i\mathbf{r}\cdot\pi$) with **r**, where $\mathbf{Z} = (\omega, \pi)$. A free wave function $\psi(\mathbf{r})$ (or $\tilde{\psi}(\mathbf{r})$) should be of *positive frequency*, and this is achieved if $\psi(\mathbf{r})$ (or $\tilde{\psi}(\mathbf{r})$) remains holomorphic when we allow **r** to be any complex point lying in the *forward tube* $\mathbb{M}^+$. This is the part of $\mathbb{CM}$ consisting of points whose imaginary parts are timelike past-pointing, and is represented, in projective twistor space $\mathbb{PT}$, by the lines **R** lying entirely in $\mathbb{PT}^+$. See Fig. 25.2 for a picture of this arrangement (the shaded region indicating where f is free of singularities).

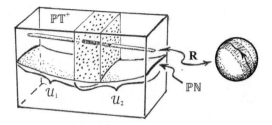

Figure 25.2 The contour integral arrangement for a twistor wave function. The two open sets $\mathcal{U}_1$ and $\mathcal{U}_2$ (regions of $\mathbb{PT}^+$ depicted above those letters) together cover the whole of $\mathbb{PT}^+$, the function f being defined on their intersection. The complex point $\mathbf{r}$ is represented by a Riemann sphere (complex projective line) $\mathbf{R}$ in $\mathbb{PT}^+$.

Figure 25.3 The 'impossible tribar' illustrates the 1st cohomology. This arises from a local ambiguity in the distance of the depicted object from the viewer.

The details are best not entered into here, but what we find is that the twistor wave function f is not really to be thought of as 'just a function' in the ordinary sense, but as a representative of an element of the *1st cohomology* (actually the 1st *sheaf* cohomology) of the space $\mathbb{PT}^+$. It is at this stage that we begin to realize the deeper and subtler aspects of the *nonlocality* of the twistor picture of physical reality, this nonlocality finding expression in the essential nonlocality of cohomology [26]. A good intuitive way of appreciating the idea of 1st cohomology is to contemplate the 'impossible tribar' depicted in Fig. 25.3. Here we have the picture that for each local region, there is an interpretation provided, of a 3-dimensional structure that is unambiguous, except for an uncertainty as to its distance from the viewer's eye. As we follow around the triangular shape, our interpretation remains consistent until we return to our starting point, only to find that it has actually become *inconsistent*! The element of the 1st cohomology that is expressed by the picture is the measure of this *global* inconsistency [27], where *locally* there is no inconsistency, but merely a mild-seeming ambiguity of the distance from the viewer's eye to the pictured object.

In the case of a twistor wave function there is an additional subtlety, in that the global inconsistency arises from the 'rigidity' of *holomorphic functions* rather than from that of the solid structures conjured up by the local parts of Fig. 25.3. To be more specific, we may think of $\mathbb{PT}^+$ as being broken into pieces – and simply the *two* overlapping open regions $\mathcal{U}_1$ and $\mathcal{U}_2$ depicted in Fig. 25.2 will do – which together cover the whole of $\mathbb{PT}^+$, where there is a region $\mathcal{R}$ of intersection of $\mathcal{U}_1$ and $\mathcal{U}_2$ which is where our holomorphic function f is actually defined:

$$\mathbb{PT}^+ = \mathcal{U}_1 \cup \mathcal{U}_2, \mathcal{R} = \mathcal{U}_1 \cap \mathcal{U}_2$$

(see Fig. 25.2). We can think of this as being analogous to the tribar of Fig. 25.3 by imagining a splitting of the tribar picture into two overlapping parts, a left-hand one ($\mathcal{U}_1$) and a right-hand one ($\mathcal{U}_2$), where the (disconnected) overlap region provides us with instructions as to how to glue the two parts together. These instructions are the analogue of the holomorphic function f, and it is the *rigidity* of holomorphic functions (as expressed in the local uniqueness of analytic continuation) that provides the analogy with that of the rigid-body structures depicted (locally) in Fig. 25.3.

The nonlocality of twistor 1st cohomology that is illustrated here reflects a *physical* nonlocality that is exhibited in one-particle wave functions that had worried Einstein, way back in 1927 (*Einstein's boxes*; see [28]), though much milder and elementary than the two-particle nonlocality that Bell established in 1964. We may imagine that a photon source, aimed at a photosensitive screen some distance away, emits a single photon towards the screen, the state of this photon being described by a wave function ψ. As soon as the screen registers reception of the photon at any one of its points – say $\mathbf{x}$ – this detection event instantly forbids every other point of the screen from detecting the photon, despite the fact that at another point $\mathbf{y}$ on the screen, the wave function itself may have had, a moment previously, a ψ-value comparable with that at $\mathbf{x}$. The probability of detection at each point of the screen is determined in the same way (by some form of Born rule), but as the detection probability refers to a *single* particle, it cannot be detected at more than one point. It is a global thing, quite unlike the situation with a water wave, say, whose effect on each point on a cliff face is *locally* determined, being independent of its effect at other points on the cliff.

Of course, taken on its own, such a situation can be explained in accordance with the point of view that the wave function is simply a kind of 'probability wave' with no actual *reality* attached to it. This simple picture cannot be maintained, however, when interference effects are involved, and for such reasons I would myself insist on attributing some measure of physical reality to the wave function. Yet this 'reality' has to acquire some kind of nonlocality, as this simple example demonstrates.

For an extreme illustration of this, we may imagine that an astronomer detects a photon from a galaxy some millions of light years away. Prior to detection, one would consider that the photon's wave function had spread, enormously diluted, over a region several millions of light years across. Yet the astronomer's detection of the photon at once forbids its detection at any other place in that vast region (where I here ignore complicating matters of Bose statistics and quantum field theory). The twistor picture of this nonlocality is like that of a vast impossible tribar, whose impossibility is removed once it is broken at any *one* place, this breaking being the analogy of a quantum state reduction (R) occurring at that place.

Of course, as we know, a local-realistic model (i.e., a 'Bertlmann's socks' type of explanation; see [29]) can easily be provided for this kind of single-particle nonlocality, and we need to pass to multiparticle situations in order to provide instances of genuine Bell nonlocality. The twistor description of an *n*-particle wave function would require *n*th cohomology, so the twistor picture becomes more complicated (and not yet adequately discussed, as far as I am aware). I am more concerned, in this article, with how one might attempt to address, in twistor terms, the issue of the *state reduction* R that occurs when a *measurement* is applied to a (twistor) wave function. Since, according to Sect. 25.1, I adopt the view that this is an *objective* physical process (OR) which is gravitational in nature, we shall need to see how genuine curved-space geometry might be incorporated into the twistor formalism. The basic ideas for this are outlined in the next section, and we see a new kind of nonlocality arising in the twistor picture.

25.6 Palatial Twistor Theory

Before turning to the proposal of 'palatial twistor theory' for a general curved space–time $\mathcal{M}$, it will be helpful to give a brief outline of the earlier procedure referred to as the 'nonlinear graviton' construction [30] (and see also [31] for the case of nonzero Λ), whereby the general (complex) solution of Einstein's Λ-vacuum equations, in the anti-self-dual case, may be expressed in twistor terms. The gist of this construction can be gleaned from Fig. 25.2, where we now think of the two overlapping open regions $\mathcal{U}_1$ and $\mathcal{U}_2$ as being separate portions of twistor space that are to be glued together over the shaded intersection region. Now, rather than thinking of f as just being painted onto the overlap region $\mathcal{R}$, we can think of f as playing a more *active* role, whereby it effects a sliding of one patch over the other, to obtain some form of curved twistor space, achieved by a process of coordinate patching. In the infinitesimal case, this can be expressed as a shift in $\mathcal{U}_1$ as matched to $\mathcal{U}_2$ along the vector field

$$I^{\alpha\beta}\frac{\partial f}{\partial Z^\alpha}\frac{\partial}{\partial Z^\beta}$$

(defined on $\mathcal{R}$), where f is homogenous of degree 2 (corresponding to helicity $s = -2$). Exponentiating this, we get a genuinely curved twistor space of the required type (with a globally defined **I**-structure). The *points* of the complex 4-manifold $\mathcal{M}$ arise as completed Riemann spheres that are deformed versions of the line **R**, in Fig. 25.2. A remarkable theorem due to Kodaira [32] tells us that such deformed **R**s do indeed form a complex four-parameter family (at least for deformations that are not too large). It is a striking fact that the nontrivial local structure of the resulting $\mathcal{M}$, which involves a nonzero anti-self-dual Weyl curvature (Ψ_{ABCD}; see [16, 30]), arises from the *global* structure of the deformed twistor space $\mathcal{T}$, itself having merely a *local* **I**-structure, identical with that of $\mathbb{M}$, or of $\mathbb{D}$ (with the corresponding Λ).

The idea behind palatial twistor theory [18] is that we try to mimic this nonlinear-graviton procedure, but instead of matching the complex-manifold structure of $\mathcal{U}_1$ to that of $\mathcal{U}_2$ in order to get a curved twistor space $\mathcal{T}$, we try to match their respective twistor *quantum algebras* $\mathbb{A}_1$ and $\mathbb{A}_2$ to give a curved quantum algebra $\mathcal{A}$, but without involving actual spaces, such as the regions of $\mathcal{U}_1$ and $\mathcal{U}_2$, or $\mathcal{T}$. Had we maintained a notion of a twistor *manifold* for the regions $\mathcal{U}_1$, $\mathcal{U}_2$, this would have forced us into the *left*-handed space–time framework, rather than allowing both helicities. Matching the twistor *algebras* rather than the twistor *spaces* of the nonlinear graviton construction enables us to resolve a long-standing ($\sim$40-year-old) conundrum known as the 'googly problem' (see [22, footnote on p. 164]). That problem demands finding a suitable procedure which would yield (complex) space–times possessing *self-dual* conformal curvature ($\tilde{\Psi}_{A'B'C'D'}$), i.e., *right*-handed ($s = 2$), for which the linearized version would require a twistor function f of the awkward-looking homogeneity degree -6, which has proved very problematic. Of course, that could be addressed if the *dual* (**W**-description) were adopted, but then the same problem would arise for the *anti*-self-dual part, which does not help, as a single formalism has to be found, able to cope with *both* parts at once. This is what palatial twistor theory is proposed to achieve.

Since these twistor quantum algebras are noncommutative, we are led into the kind of picture provided by the ideas of *noncommutative geometry* [33], and ordinary spaces such as $\mathcal{U}_1$ and $\mathcal{U}_2$ are not determined uniquely by such algebras. Nevertheless, we need some notion of locality (or 'topology') in order to express the concept of building up the entire structure out of 'flat' pieces. This is achieved by appealing to the light-ray spaces $\mathbb{PN}$, $\mathbb{P\bar{N}}$, and $\mathcal{N}$, described in Sects. 25.2 and 25.3. We can imagine that $\mathbb{PN}$ is divided into, say, n partially overlapping regions $\mathbb{PN}_1$, $\mathbb{PN}_2$, ..., $\mathbb{PN}_n$, providing an open covering $(\mathcal{N}_1, \mathcal{N}_2, ..., \mathcal{N}_n)$ of $\mathcal{N}$. The idea is that if the individual spaces $\mathcal{N}_k$, together with their intersections, are, in an appropriate sense, 'simple', then they can be (noncanonically) assigned respective flat twistor quantum algebras $\mathbb{A}_1, \mathbb{A}_2, ..., \mathbb{A}_n$. On the various overlaps $\mathcal{N}_j \cup \mathcal{N}_k$ the algebras needs to match appropriately, but the resulting patched algebra will *not* be flat in its total structure if $\mathcal{M}$ is conformally curved.

I need to explain some of these terms a little more fully. Basically, the algebra will be considered to be *flat* if it has a consistent ket space $|...\rangle$, and the idea is to find such a flat algebra for any simple region $\mathcal{U}$ in $\mathcal{N}$. Here, an open subregion of $\mathcal{N}$ is called *simple* if it is topologically and holomorphically trivial – by which I mean that it has Euclidean topology and is in some appropriate sense *convex*. To construct a (flat) twistor quantum algebra suitably assigned to any such simple region, we first consider the twistor space $\mathbb{T}_\gamma$, for each ray γ in $\mathbb{P}\mathcal{U}$ (from local twistor transport Sect. 25.4) and then construct the twistor quantum algebra $\mathbb{A}_\gamma$ from each in the standard way (as in Sect. 25.4). We have no reason to expect a canonical isomorphism between these algebras for different rays γ in $\mathbb{P}\mathcal{U}$ when $\mathcal{M}$ is conformally curved. However, it is to be expected that we *can*, for a simple region $\mathcal{U}$ in $\mathcal{N}$, *deform*, continuously and holomorphically, the various $\mathbb{A}_\gamma$s at the different points of $\mathcal{U}$, to obtain a holomorphic trivialisation of the bundle of $\mathbb{A}_\gamma$s over $\mathcal{U}$, thereby obtaining (by no means uniquely) a single algebra $\mathbb{A}_\mathcal{U}$ continuously and holomorphically isomorphic to each $\mathbb{A}_\gamma$ for $\gamma \in \mathbb{P}\mathcal{U}$ (and therefore isomorphic to $\mathbb{A}$) consistently over the whole of $\mathcal{U}$. This is to be expected because the relation between the various local twistor spaces $\mathbb{T}_\gamma$ and their immediate neighbours (i.e., $\mathbb{T}_{\gamma'}$ where γ' is neighbouring to γ) becomes a *holomorphic* one between the corresponding $\mathbb{A}_\gamma$s, by virtue of the holomorphic nature of the twistor quantisation process (and by virtue of the prequantisation connection of Sect. 25.3. This notion of consistency (though still not fully understood in mathematical details) demands that there be a consistent ket-space $| \ldots \rangle$ for $\mathbb{A}_\mathcal{U}$, isomorphic to $\mathbb{A}$, over each region $\mathcal{U}$.

Each such flat $\mathbb{A}_\mathcal{U}$ is to be thought of in the spirit of a coordinate patch. Over the *intersection* $\mathcal{U}_i \cap \mathcal{U}_j$ of two simple regions $\mathcal{U}_i$ and $\mathcal{U}_j$ we require consistency of the algebras $\mathbb{A}_{\mathcal{U}_i}$ and $\mathbb{A}_{\mathcal{U}_j}$, in the sense of having a continuous and holomorphic deformation of one to the other, retaining a consistent ket space on the intersection, but we do not require a common ket-space to be present for the whole of their *union* $\mathcal{U}_i \cup \mathcal{U}_j$. Such consistency would not generally be possible globally. Instead, our fully patched together algebra $\mathcal{A}$ would *not* have a consistent ket space (unless $\mathcal{M}$ was conformally flat). The idea would be that a measure of the *departure* from global consistency of a ket space, over the whole of $\mathcal{N}$, would be something of the nature of a (nonlinear) 1st cohomology element (as with the inconsistency expressed in Fig. 25.3) and which, in space–time terms, would express the presence of a nonzero *Weyl conformal tensor*, i.e., conformal curvature for $\mathcal{M}$.

We need to be able to identify the *points* of $\mathcal{M}$ in terms of the algebra $\mathcal{A}$. These have to arise from *nonlocal* considerations (as was the case with the nonlinear graviton construction). Corresponding to any particular point $\mathbf{r}$ of $\mathcal{M}$ there would be a locus $\mathbf{R}$ in $\mathbb{P}\mathcal{N}$ representing $\mathbf{r}$, namely the family of all rays through $\mathbf{r}$, which is topologically S^2. The idea is that the consistency (i.e., trivialisation, in the above strong sense of having a consistent ket-space) of the $\mathbb{A}$-bundle over $\mathbf{R}$ is what determines such an S^2 locus as representing a point of $\mathcal{M}$. For this to work, as a sufficiently restrictive proposal for locating $\mathcal{M}$'s points in terms of such a twistorial construction, we need to establish the validity of various technical issues that have been skimmed over in the previous two paragraphs. Moreover, to ensure that the construction outlined above actually provides us with a 4-dimensional $\mathcal{M}$, we would certainly need some suitable generalisation of the Kodaira theorem [32] that was central to the nonlinear graviton construction.

None of this yet encodes the formulation of Einstein's equations. It is perhaps remarkable, therefore, to find that Einstein's Λ-vacuum equations are themselves very simply encoded into this structure. For these equations provide *precisely* the necessary and sufficient condition that the local twistor spaces $\mathbb{T}_\gamma$ possess an **I**-structure (for given Λ), so all we now require that is that the needed continuous and holomorphic deformations of the $\mathbb{T}_{\mathcal{U}}$ algebras preserve their nature as algebras on $\mathbb{T}_{\gamma}$ with this **I**-structure. If all these procedures (or something like them) indeed work as intended (with generalisations to the Yang–Mills equations and other aspects of physics), then there would appear to be significant openings for twistor theory in a nonlocal basic physics, not envisaged before.

Nevertheless, we have still not addressed the issue raised in Sect. 25.1 of the need for a (nonlocal) physics capable of describing the **R**-process as a realistic gravitational phenomenon. Indeed, the formalism as so far described cannot yet be taken as a 'quantum-gravity' theory, if only for the simple reason that no place has been found for the Planck length l_P or equivalently the Planck time t_P. A tempting way to incorporate such dimensional quantities might be to modify the commutators of Sect. 25.5 as follows:

$$Z^{\alpha}\overline{Z}_{\beta} - \overline{Z}_{\beta}Z^{\alpha} = \hbar\delta^{\alpha}_{\beta}$$

and

$$Z^{\alpha}Z^{\beta} - Z^{\beta}Z^{\alpha} = \varepsilon I^{\alpha\beta}, \overline{Z}_{\alpha}\overline{Z}_{\beta} - \overline{Z}_{\beta}\overline{Z}_{\alpha} = \overline{\varepsilon}I_{\alpha\beta},$$

where ε is a very small (complex?) constant related to the Planck length. These commutator equations have not yet been significantly explored, and it cannot yet be said whether or not they supply anything of the kind of quantum gravity framework that might be needed.

Acknowledgments

I am grateful for valuable early input from Michael Atiyah, and also for useful discussions with Michael Enciso, Andrew Hodges, and Carlo Rovelli. The majestic ambiance of *Buckingham Palace* no doubt also had its role to play, as it was there that a brief discussion with Atiyah initially took place, which inspired the thought that noncommutative twistor algebra might be the key to the subsequent developments that are described in this article.

References

[1] Bell, J.S. (1964), On the Einstein Podolsky Rosen paradox, *Physics* **1**, 195–200.
[2] Xiao, S.M., Herbst, T., Scheldt, T., Wang, D., Kropatschek, S., Naylor, W., Wittmann, B., Mech, A., Kofler, J., Anisimova, E., Makarov, V. Jennewein, Y., Ursin, R., and Zeilinger, A. (2012), Quantum teleportation over 143 kilometres using active feedforward, *Nature Lett.* **489**, 269–73.
[3] Penrose, R. (2014), On the gravitization of quantum mechanics: 1. Quantum state reduction, *Found. Phys.* **44**, 557–75.
[4] Hawking, S.W. (1976), Breakdown of predictability in gravitational collapse, *Phys. Rev. D* **14**, 2460–73.
[5] Hawking, S.W. (2005), Information loss in black holes, *Phys. Rev. D* **72**, 084013–6.
[6] Penrose, R. (2004), *The Road to Reality: A Complete Guide to the Laws of the Universe* (Jonathan Cape, London).

[7] Almheiri, A., Marolf, D., Polchinski, J. and Sully, J. (2013), Black holes: complementarity or firewalls? *J. High Energy Phys.* **2013** (2), 1–20.

[8] Penrose, R. (1993), Gravity and quantum mechanics, in R.J. Gleiser, C.N. Kozameh, and O.M. Moreschi (eds.), *General Relativity and Gravitation 13. Part 1: Plenary Lectures 1992* (Inst. of Phys. Publ., Bristol/Philadelphia), pp. 179–89.

[9] Penrose, R. (1996), On gravity's role in quantum state reduction, *Gen. Rel. Grav.* **28**, 581–600.

[10] Diósi, L. (1984), Gravitation and quantum mechanical localization of macro-objects, *Phys. Lett. A* **105**, 199–202.

[11] Diósi, L. (1987), A universal master equation for the gravitational violation of quantum mechanics, *Phys. Lett. A* **120**, 377–81.

[12] Penrose, R. (1987), On the origins of twistor theory, in W. Rindler and A. Trautman (eds.), *Gravitation and Geometry: A Volume in Honour of I. Robinson* (Bibliopolis, Naples).

[13] Einstein, A., Podolsky, P. and Rosen, N. (1935), Can quantum-mechanical description of physical reality be considered complete? *Phys. Rev.* **47**, 777–80.

[14] Bohm, D. (1951) *Quantum Theory* (Prentice–Hall, Englewood Cliffs, NJ), Chap. 22, Sects. 15–19.

[15] Geroch, R. (1970), Domain of dependence, *J. Math. Phys.* **11**, 437.

[16] Penrose, R. and Rindler, W. (1984), *Spinors and Space–Time, Vol. 1: Two-Spinor Calculus and Relativistic Fields* (Cambridge University Press, Cambridge).

[17] Woodhouse, N.M.J. (1991), *Geometric Quantization,* 2nd ed. (Clarendon Press, Oxford).

[18] Penrose, R. (2015), Palatial twistor theory and the twistor googly problem, *Phil. Trans. R. Soc. A* **373**, 20140237.

[19] Tomassini, S. and Dragomir, G. (2006), Differential geometry and analysis on CR manifolds, *Progress in Mathematics*, Vol. 246 (Birkhäuser, Basel).

[20] Huggett, S.A. and Tod, K.P. (1985), *An Introduction to Twistor Theory,* London Mathematical Society Student Texts, Vol. 4 (Cambridge University Press, Cambridge).

[21] Ward, R.S. and Wells, R.O., Jr. (1989), *Twistor Geometry and Field Theory* (Cambridge University Press, Cambridge).

[22] Penrose, R. and Rindler, W. (1986), *Spinors and Space–Time, Vol. 2: Spinor and Twistor Methods in Space–Time Geometry* (Cambridge University Press, Cambridge).

[23] Frauendiener, J. and Sparling, G. (1992), On the symplectic formalism for general relativity, *Proc. R. Soc. (London) A* **436**, 141–53.

[24] Dirac, P.A.M. (1947), *The Principles of Quantum Mechanics,* 3rd ed. (Clarendon Press, Oxford).

[25] Penrose, R. (1969), Solutions of the zero rest-mass equations, *J. Math. Phys.* **10**, 38–9.

[26] Wells, R.O., Jr. (1991), *Differential Analysis on Complex Manifolds* (Prentice-Hall, Englewood Cliffs, NJ).

[27] Penrose, R. (1991), On the cohomology of impossible figures, *Structural Topology* **17**, 11–16.

[28] Norsen, T. (2004), Einstein's boxes, *Am. J. Phys.* **73**, 164–76.

[29] Bell, J.S. (1966), *Speakable and Unspeakable in Quantum Mechanics* (Cambridge University Press, Cambridge). [Reprint 1987]

[30] Penrose, R. (1976), Non-linear gravitons and curved twistor theory, *Gen. Rel. Grav.* **7**, 31–52.

[31] Ward, R.S. (1980), Self-dual space–times with cosmological constant, *Comm. Math. Phys.* **78**, 1–17.

[32] Kodaira, K. (1963), On stability of compact submanifolds of complex manifolds, *Am. J. Math.* **85**, 79–94.

[33] Connes, A. and Berberian, S.K. (1995), *Noncommutative Geometry* (Academic Press, San Diego).

26

Measurement and Macroscopicity: Overcoming Conceptual Imprecision in Quantum Measurement Theory

GREGG JAEGER

26.1 Introduction

John S. Bell, of course, is best known for the result now referred to simply as "Bell's theorem," which removed from consideration the class of so-called local hidden-variable theories that at the time of its publishing appeared to be the most natural class of theories among those that would render quantum mechanics a form of statistical mechanics. If, as this and other results suggest, quantum theory is to serve as a truly *fundamental* theory, conceptual precision in its interpretation is not only desirable but paramount. John Bell was accordingly concerned about what he viewed as conceptual imprecision, from the physical point of view, within the standard approaches to the theory. He saw this as most acute in the case of their treatment of *measurement at the level of principle*. His concerns were strongly expressed in the article "Against measurement" [1]. This was published in 1990 in a volume of proceedings of the 1989 Erice meeting "Sixty-Two Years of Uncertainty," during which it was my pleasure to meet and eat with Bell, and to listen to him present this paper. He pointed out that this conceptual imprecision is reflected in the *terminology* of the foundations of quantum theory, a great deal of which he explicitly deemed worthy of banishment from discussions of principle, because it corresponds to a set of what he saw as vague and, in some instances, outright destructive concepts. His concern was thus not one regarding mathematics so much as regarding basic concepts used in contemporary quantum physics, which he viewed as failing the needs of natural philosophy and so of fundamental physics, despite their practical adequacy. Here, I consider John Bell's critique of standard quantum measurement theory and some alternative treatments wherein he saw greater conceptual precision, and make further suggestions as to how to improve conceptual precision, as he advocated.

That the source of difficulties is to be understood specifically as a problem of *imprecision of physical concepts* that stands in the way of the achievement of an exact fundamental mechanical theory is pointed out at the outset of "Against measurement": Bell wished to make it clear "at once that it is not mathematical precision, but physical" that caused him such great concern. As he saw it, one should have had by the time of its writing "an exact formulation of a serious part of mechanics," where by "exact" he means "only that the theory should be fully formulated in mathematical terms, with nothing left to the discretion

419

of the theoretical physicist," with nonrelativistic 'particle' quantum mechanics and that of the electromagnetic field constituting a sufficiently "serious part" [1]. Bell also made it immediately clear that he saw physics as part of the long tradition of *natural philosophy*, and that his concerns about physical precision were, in effect, concerns regarding the *precision of concepts* of natural philosophy.

In the analysis offered in "Against measurement," a key distinction is made with regard to theoretical treatments compatible with experimental data obtained between those sufficiently precise to be accepted as fundamental physics and those good enough "for all practical purposes," for which he supplies the memorable acronym 'FAPP' (which, following his usage, continues to be employed in this sense now, two and one-half decades later). Bell's exploration begins by pointing out that there is a lack of precision in the traditional "proper treatments" one finds published in respectable and frequently consulted sources. He recalls and answers the often-asked rhetorical question of why one should bother making quantum mechanics more precise than it already is: "Why not look it up in a good book? But *which* good book? Usually the good unproblematic formulation is still in the head of the person in question . . . For the good books known to me are not much concerned with physical precision." His verdict on the various available treatments set firmly within quantum theory involving the standard, unmodified dynamical laws is that they are useful for practical predictions of the statistics to be found in experiments but fall far short of what physics ought to be *at the level of principle*: "The orthodox approaches, whether the authors think they have made derivations or assumptions, are just fine FAPP" but ultimately fail to fully describe the physical world [1].

Bell provides a lengthy laundry list of standard quantum physical terms, reflection upon which shows that the lack of physical precision in the then current thinking – from which it should pointed out we have yet to significantly advance – is due to *conceptual* imprecision, and suggests that physics reject a considerable amount of its standard terminology.

Here are some words which, however legitimate and necessary in application, have no place in a *formulation* with any pretension of physical precision: *system, apparatus, environment, microscopic, macroscopic, reversible, irreversible, observable, information, measurement.* The concepts of 'system', 'apparatus', 'environment', immediately imply an artificial division of the world, and an intention to neglect, or take only schematic account of, the interaction across the split. The notions of 'microscopic' and 'macroscopic' defy precise definition. So also do the notions of 'reversible' and 'irreversible'. Einstein said that it is theory which decides what is 'observable'. I think he was right – 'observation' is a complicated and theory-laden business. Then that notion should not appear in the *formulation* of fundamental theory. [1]

The ages-old philosophical question of the relationship of observation to reality is relevant to the issues engaged in the article, but is itself not engaged in the text in any detail, beyond a general rejection of subjectivism. It suffices here to note that in his writings generally, Bell sides with realism and cites one of its great physicist champions. His primary concern instead is more specifically the relationship of *physical theory* to reality, and his position is that, at a minimum, physical theory should explain to the physicist what can and cannot be measured, something given by the answer to the question of how measurements are and can be made without themselves being considered *fundamental* to physical theory.

26.2 The Negative Influence of Inappropriate Terminology

As Bell sees it, "On this list of bad words from good books, the worst of all is 'measurement'." Again, he does not reject the term in general, particularly not its use in practice. He mentions, as an example, its use in the command "measure the mass and width of the Z boson," viewing such use as entirely acceptable; he objects, rather and most specifically, to "its use in the fundamental interpretive rules of quantum mechanics." When Bell considers as an example of the problems that arise through the use of 'measurement' in the foundations of quantum theory, he comments and answers rhetorically that, when reading Dirac's "good book" *Quantum mechanics*, one gets the sense that "the theory is exclusively concerned about 'the results of measurement', and has nothing to say about anything else. What qualifies some physical systems to play the role of 'measurer'?" [1].

He suggests that the role of the notion of measurement be taken over by the more neutral term 'experiment': "Even in a low-brow practical account, I think it would be good to replace the word 'measurement', in the formulation, by the word 'experiment'. For the latter word is altogether less misleading." But, for him, this term also has limitations and, so, he continues,

However, the idea that quantum mechanics, our most fundamental physical theory, is exclusively about the results of experiments would remain disappointing . . . To restrict quantum mechanics to be exclusively about piddling laboratory operations is to betray the great enterprise [of natural philosophy]. A serious formulation will not exclude the big world outside the laboratory. [1]

Thus, he finds use of the term 'experiment' tolerable in the formulation of quantum mechanics, but best avoided if possible.

'Measurement,' however, is "entirely inappropriate." Bell makes two specific charges against the term:

The first charge against 'measurement', in the fundamental axioms of quantum mechanics, is that it anchors there the shifty split of the world into 'system' and 'apparatus'. A second charge is that the word comes loaded with meaning from everyday life, meaning which is entirely inappropriate in the quantum context . . . In other contexts, physicists have been able to take words from everyday language and use them as technical terms with no great harm done . . . Would that it were so with 'measurement'. But in fact the word has had such a damaging effect on the discussion, that I think it should now be banned altogether in quantum mechanics. [1]

Beyond the general difficulties of its use of this most problematic term, Bell sees the traditional, "orthodox" treatment of measurement-like processes as reinforcing the imprecision of the enterprise of quantum natural philosophy in several ways. He also objects to any distinction between systems based on imprecise reference to *physical scale*, where the term 'macroscopic' is brought into play.

The kinematics of the world, in this orthodox picture [with probabilities of obtaining outcomes], is given by a wavefunction (maybe more than one?) for the quantum part, and classical variables – variables that *have* values – for the classical part: $(\Psi(t, q, \ldots), X(t \ldots \ldots))$. The Xs are somehow macroscopic. This is not spelled out very explicitly. They dynamics is not very precisely formulated either. [1]

In this way, another to-be-banned term, 'macroscopic,' is indicated. Notably, Bell had expressed concern with the term 'macroscopic' in previous years as well, although he had previously seen a sharpening of the concept as still viable, particularly in that he commented that in regard to the 'EPR correlations' that violated his equality that he had "very little understanding of the position of . . . Bohr," which depended on restrictions on what was to be considered possible in measurements and made use of the term.

For example, in his 1981 article entitled "Bertlmann's socks and the nature of reality," he indicated as one possibility for progress in foundations of quantum theory that "it may be that Bohr's intuition was right – in that here is no reality below some 'classical' 'macroscopic' level. Then fundamental physical theory would remain fundamentally vague, until concepts like 'macroscopic' could be made sharper than they are today" [2]. In his comment to this paper, made directly after its presentation, Bell's contemporary and friend Abner Shimony, who, along with John Clauser, Michael Horne, and Richard Holt, provided a directly experimentally testable form of Bell's inequality [3], may have influenced Bell's thinking regarding measurement, in that he remarks there that

perhaps I can help to focus on the source of the difficulty [in understanding Bohr's answer to E.P.R.]. In any measuring process, Bohr insists upon a sharp distinction between object and subject. The apparatus is considered to be situated on the subject's side of this division. Hence it is characterized in terms of the concepts of everyday life (of which the concepts of classical physics are refinements). One may ask, however, whether it is possible to investigate the physical behavior of the apparatus . . . Bohr's answer is that . . . [it] is possible but then other apparatus will be employed in the investigation. The boundary between the object and the subject has shifted. [2]

By the time of "Against measurement," Bell was referring to this division exactly as "the shifty split."

The focus of the critique in "Against measurement" is in fact most specifically on the conceptual imprecision involved in the treatment of *state evolution* during measurement that depends on the above (as Bell sees it) problematic system–apparatus division, which is typically made by having the apparatus, in one way or the other, qualify as 'macroscopic.' He notes that a range of different, often incompatible assumptions as to how a system can be considered to be macroscopic have been used in the standard approaches to quantum measurement [4]. In "Against Measurement," Bell analyzes the traditional treatments, which invoke sudden changes of quantum state during experiments, making use of the following distinction: "It will be convenient later to refer to . . . the *spontaneous* jump of a macroscopic system [S] into a definite configuration, as the [Landau–Lifschitz] LL jump. And the *forced* jump of a quantum system as a result of 'measurement' – *an external intervention* – as the Dirac jump." A "jump" placed in the same location as the latter appears in the formulation of von Neumann; the postulate of state collapse according to von Neumann is also noted in the text: "what vN actually *postulates* is that 'measurement' – an external intervention by [the rest of the world] R on S – causes the state $\sum_n c_n \phi_n$ to jump, with various probabilities into ϕ_1 *or* ϕ_2 *or* . . . From the 'or' here, replacing the 'and', as a result of external intervention, vN infers that the density matrix, averaged over the several possibilities, has no interference terms between states of the system which correspond to different measurement results." It is this invocation of measurement at the level of postulates that Bell finds extremely objectionable.

Von Neumann was clearly forced to postulate such a process, which he called "Process 1," pointing out its exceptional nature by noting that, on physical grounds, one would rather expect the more usual Process 2 to be the only one needed.

[O]ne should expect that [Process 2] would suffice to describe the intervention caused by a measurement: Indeed, a physical intervention can be nothing else than the temporary insertion of a certain energy coupling into the observed system, i.e., the introduction of an appropriate time dependency of $\hat{H}$. [5]

Von Neumann had argued that the boundary between the measuring and measured systems should also be "arbitrary to a very large extent" [5, p. 420], because whether the collapse happens to the measured system alone or to the joint system of measuring apparatus together with the measured system, the statistics of outcomes will be the same from the point of view of any physical system, such as a human being, separate from them that becomes correlated with them in the same way a measurement is assumed to become during an experiment. Although this is true, it does not aid our understanding of what takes place during measurement, but instead leaves its details obscure. This is very clearly a case of a theory working, but working only "FAPP."

Bell then surveys other traditional treatments in other "good books," indicating the various instances of physical imprecision within them. In the case of the treatment of Landau and Lifschitz (LL), which "derive[s] the Dirac jump from the LL jump," he says, "In the LL formulation . . . the theory is ambiguous in principle, about exactly when and exactly how the collapse occurs, about what is microscopic and what is macroscopic, what quantum and what classical." These, of course, are the most critical questions involved in a fuller understanding of the quantum mechanics of experimentation. For this reason, imprecision in relation to these obscures the problem itself, and makes it all the more difficult to solve.

Bell also considers the treatment of Kurt Gottfried, and offers natural suggestions for its missing details as part of an exploration of its more realistic character. Formally, he takes this as a treatment in which the density matrix ρ for the joint system of system $S' = S + A$, where A is the measurement apparatus system, is replaced with another density matrix $\hat{\rho}$, in which all nondiagonal elements are zero in the Hilbert space basis in which measured values and the apparatus "pointer" variable values are to be perfectly correlated, something which is a prerequisite of accurate measurement often postulated by realist (as well as operationalist) interpretations of quantum theory. To him, this appears to be the case because there is conceptual drift "away from the 'measurement' (. . . external intervention) orientation of orthodox quantum mechanics towards the idea that systems, such as S' above, have intrinsic properties – independently of and before observation. In particular, the readings of external apparatus are supposed to be really there before they are read" in that explication of measurement in which "KG derives, FAPP, the LL jump from assumptions at the shifted split R'/S', which include a Dirac jump there," where $R' = R - A$. This is seen as having the advantage that some

'macroscopic' 'physical attributes' *have* values at all times, with a dynamics that is related somehow to the butchering of ρ into $\hat{\rho}$ – which is seen as somehow not incompatible with the international Schrödinger equation of the system. Such a theory, assuming intrinsic properties, would not need external intervention, would not need the shifty split, but the retention of the vague word 'macroscopic' would reveal limited ambitions as regards precision. [1]

One might, he finally notes, avoid this term by introducing variables that have values even at small scales, as in the de Broglie–Bohm approach.

Bell had previously viewed this approach as important, in that it showed that "the subjectivity of the orthodox version, the necessary reference to the 'observer,' could be eliminated" [7]. This indicates a potentially promising direction for increased conceptual precision, one that he sought to "publicize." In this earlier discussion, he drew three morals from the existence of the de Broglie–Bohm model: (1) "always test your general reasoning against simple models," (2) "in physics the only observations we must consider are position observations, if only the positions of instrument pointers," and (3) one concerning terminology that was to be the main theme of "Against measurement." In the paper where these morals were drawn, "On the impossible pilot wave," Bell notes regarding (3) that "serious people" were likely "misled by the pernicious misuse of the word 'measurement'," which "strongly suggests the ascertaining of some preexisting property of some thing, any instrument involved playing a purely passive role. Quantum experiments are just not like that, as we learned especially from Bohr" [7].

However, as seen below, by the end of the 1980s, Bell found a different approach more promising, one that deviates from standard quantum mechanics at the level of law: Indeed, he had concluded already concluded by the time of his 1986 article "Are there quantum jumps?" that "If, with Schrödinger, we reject extra variables, then we must allow that his equation is not always right ... it seems to me inescapable ... a recent idea [of Ghirardi, Rimini, and Weber (Ghirardi, Rimini and Weber, 1985)], a specific form of spontaneous wave function collapse, is particularly simple and effective" [8]. (The cited work by Ghirardi, Rimini and Weber was actually published 1986; see [9].)

26.3 Modified Quantum Dynamics

26.3.1 The Desiderata and Superpositions at Large Scales

In addition to their common use of notions of measurement and macroscopicity, standard analyses of quantum mechanical situations, as viewed from the perspective of data production, suffer from what has been called the "opportunistic employment of the superposition principle." That is, one is tempted to allow the superposition principle to operate whenever convenient and not operate whenever inconvenient, as opposed to understanding via specific basic quantities precisely when it may or may not be in force. This issue was taken up by Shimony in his article "Desiderata for a modified quantum dynamics," presented in a memorial session for Bell, where he *also* noted that "At a workshop at Amherst College in June *Bell* remarked that the stochastic modification of quantum dynamics is the most important new idea in the field of foundations of quantum mechanics during his professional lifetime" [10].

It is clear in "Against measurement" (AM) that Bell viewed the modification of standard quantum dynamics and the presence or absence of state superposition as important in that it is a move that provides an opportunity to correct at least some of the forms of imprecision noted above, by providing objectivity to the circumstances under which measurement-like events would or would not take place. It is, therefore, worth looking more closely at the context in which such theories can be developed. This is just want Shimony does just in "Desiderata . . . ," by spelling out four assumptions "concerning the interpretation of the quantum mechanical formalism have the consequence of making the [measurement problem and the problem of Schrödinger's cat] so serious that it is difficult to envisage their solution without some modification of the formalism itself." These assumptions, variously sanctioned in AM, are generally "strongly supported by physical and philosophical considerations, and therefore a high price would be paid by sacrificing one of them in order to hedge standard quantum mechanics against modifications." They are the following [10]:

(i) "The quantum state of a physical system is an objective characterization of it." (As Bell puts it in AM, a "serious formulation will not exclude the big world outside the laboratory" and will not be concerned exclusively with "piddling laboratory operations.")

(ii) Connected with Bell's theorem, "The objective characterization of a physical system by its quantum state is complete, so that an ensemble of systems described by the same quantum state is homogeneous, without any differentiations stemming from differences in 'hidden variables.'"

(iii) "Quantum mechanics is the correct framework theory for all physical systems, macroscopic as well as microscopic, and hence it specifically applies to measuring apparatuses." (About which, however, it should be noted that "The main consideration in favor of [it being] the incompatibility proved by Bell (1987, pp. 14–21 and 29–39 [7]) between quantum mechanics and local hidden variables theories, but Bell himself emphasizes that there is still an option of non-local hidden variables theories, which he does not regard as completely repugnant." Furthermore, this assumption has the implication that all variants of the Copenhagen interpretation are "ruled out.")

(iv) "At the conclusion of the physical stages of a measurement (and hence, specifically, before the mind of an observer is affected), a definite result occurs from among all those possible outcomes (potentialities) compatible with the initial state of the object." (Bell is skeptical even of having *biology* pertinent to measurement induction – hence his comment regarding collapse: "Was the wavefunction of the world waiting to jump for thousands of years until a single-celled living creature appeared . . . or some better qualified system . . . with a PhD?")

He then sets out a list of eight well supported desiderata for such a dynamics, the last pertaining critically to the proposal Bell looked to, namely, that of GRW. It is

The modified dynamics should be capable of accounting for the occurrence of definite outcomes of measurements performed with actual apparatus, not just with idealized models of apparatus. The Spontaneous Localization theory of [9] has been criticized for not satisfying this desideratum... Albert and Vaidman (Albert 1990, 156–8) [11] note that the typical reaction of a measuring apparatus in practice is a burst of fluorescent radiation, or a pulse of voltage or current, and these are hard to subsume under the scheme of measurement of the Spontaneous Localization theory. [9]

Shimony also notes difficulties in this approach – and others pertaining to Bell's concern about *irreversibility* – which Bell says also defies a precise conceptual basis: "a stochastic modification of quantum dynamics can hardly avoid introducing time-asymmetry. Consequently, it offers an explanation at the level of fundamental processes for the general phenomenon of irreversibility, instead of attempting to derive irreversibility from some aspect of complexity (which has the danger of confusing epistemological and ontological issues)."

Now, Bell was not the first to notice the imprecision of the traditional approach to quantum measurement. Notably, in 1970, Wigner – who had shown the limitations of von Neumann's arbitrariness of the location of the division involved in his own measurement schema by showing that if a cognitive system is used as a measurement apparatus contradictions can appear [12] – pointed this out very clearly in his critique of one highly developed standard treatment: In what was perhaps the most sophisticated treatment within that approach, that of Danieri, Loingier and Prosperi, the three authors were said by him to be "using phrases such as 'macroscopic variables' and 'macroscopic objects' without giving a precise definition of these terms" [13], so that, for example, their premises could not be rigorously formulated.

[M]ost quantities which we believe to be able to measure, and surely all the very important quantities such as position, momentum, fail to commute with all the conserved quantities so that their measurement cannot be possible with a microscopic apparatus. This raises the suspicion that the macroscopic nature of the apparatus is necessary in principle and reminds us that our doubts concerning the validity of the superposition principle for the measurement process were connected with the macroscopic nature of the apparatus. [14]

And this nature, now sometimes referred to by the term 'macroscopicity', is not rigorously characterized within that approach. Wigner noted that "the theory of the interaction of a quantum system with a classical (macroscopic) system has not been formulated so that the mathematical meaning of the arrows [indicating the change of joint-system state-vector upon measurement] is not clear" [13, p. 7]. In remarks on Prosperi's paper in a key meeting of measurement theorists at the outset of the 1970s, which continued along the lines of the DLP approach, Wigner noted specifically the inappropriateness of making use of "something as inadequately defined as is the macroscopic nature of something" in serious physical discussions (his remarks immediately follow the article of Prosperi [15]) [13].

In the years between Wigner's critique and Bell's later criticisms of traditional quantum measurement theory, Anthony Leggett had considered performing tests for quantum effects in "macroscopic systems," preferably large material systems, to better illuminate the question of whether there is a clear role for macroscopicity in measurement. Leggett still wishes to find "evidence of a breakdown of the quantum mechanical scheme of the physical world [in] that which connects the world of atoms and electrons, for which it was originally developed, with the everyday world of our immediate experience," where quantum mechanically complementary properties appear compatible [16]. In particular, he wishes to find superposition and so interference effects, which DLP had argued should not occur during measurements (cf. [17]). For this purpose, Leggett has suggested studying superconducting devices (SQUIDs) and the Josephson effect, in which states of a current of electrons could, in principle, enter a superposition of states of clockwise and/or counter clockwise circulation [18]. Since the early work of Leggett, 'macroscopic' has become defined increasingly in terms of large values of specific observable quantities, generalizing Bohr's original belief that heft and rigidity or those of others simply that a sufficiently large number of degrees of freedom are essential, and rather than being identified via criteria related to the resolution of the naked eye, as the most direct understanding of the meaning of the term would suggest.

Leggett has proposed a measure he calls the disconnectivity D, a "semi-quantitative" and "qualitatively defined" notion, claiming that "the quantum states important in the discussion of the [cat] paradox are characterized by a very high value of . . . 'disconnectivity'; by contrast, the states necessary to explain so-called 'macroscopic quantum phenomena' in superfluids and superconductivity have only low disconnectivity, so that they are irrelevant to our question . . . " [19]. Rather, these center on the "most promising area to look [for high disconnectivity states being that of] phenomena where quantum tunneling plays an essential role" [19]. The GRW approach can be viewed as having participated in this trend as well.

26.3.2 *Continuous Spontaneous Localization and Beables*

It was the work of GRW, shared by other workers such as Philip Pearle and conveniently called continuous spontaneous localization (CSL), that Bell saw as most clearly offering an alternative and "explicit model allowing a unified description of microscopic and macroscopic systems" with the starting point of "a modified quantum dynamics for the description of macroscopic objects" in which systems of many components have wave functions that frequently spontaneously localize to small regions, claiming that with it "most features of the behavior of macroscopic objects are accounted for by quantum mechanics in a natural way, due to the irrelevant spreads of wave packets for macroscopic masses" [9]. Regarding the behavior of their model, GRW remark that "If one assumes for simplicity that the localization frequencies λ_i of all microscopic (e.g., atomic) constituents of a macroscopic body are of the same magnitude . . . , the center of mass is affected by the same process with a frequency $\lambda_{\text{macro}} = N\lambda_{\text{micro}} \ldots$ where [the "macroscopic number"] N is of the order of Avogadro's number" [9].

In response to this work, Bell noted that

In the GRW scheme this vagueness [regarding wavefunction collapse] is replaced by mathematical precision... departures of the Schrödinger equation show up very rarely and very weakly in few-particle systems. But in macroscopic systems, *as a consequence of the prescribed equations,* pointers very rapidly point, and cats are very quickly killed *or* spared. [3]

Bell saw this aspect of the GRW approach in marked, positive contrast to examples of 'solutions of the measurement problem involving infinite limits that had appeared – in particular, the 1972 model of Coleman and Hepp, where a solution for the dynamics of a model apparatus consisting of a semi-infinite array of spin-1/2 particles was given, which was viewed by some as a sort of solution to the measurement problem. He was critical of the Coleman–Hepp model, noting that, for it, "the rigorous reduction does not occur in physical time but only in an unattainable mathematical limit... the distinction is an important one" [20].

One difficulty subsequently encountered by the CSL approach is finding a set of parameters that allow it to describe what is observed. This difficulty is connected with what Bell called "beables," those quantities that could be understood realistically and that could correspond with what is actually observed. Those beables associated with local space–time regions are "local beables." Bell viewed contemporary quantum mechanics textbooks as failing to focus on these quantities.

What you may find there are the so-called 'local observables'. It is then implicit that the apparatus of the 'observation', or, better, of experimentation, and the experimental results are real and localized. We will have to do the best we can with these rather ill-defined local beables, while hoping always for a more serious reformulation of quantum mechanics where the local beables are explicit and mathematical and rather than implicit and vague. [21]

For their part, CSL theories have mainly followed what has been called the 'mass density ontology' [22], as evidenced, for example, by the use of the parameter N above, to which the mass density would be proportional for systems built from a given sort of fundamental subsystem. For a state in a superposition of states with differing mass densities for which there is an operator, the larger the difference of the mass density distribution of the states is, the more quickly a collapse will take place. Thus, the collapse rate for superpositions of states of microscopic systems is low because the mass density differences are likewise small, and for superpositions of macroscopic states it is large because the mass density differences are likewise large.

However, there is a problem of persistent "tails" for any collapse process that completes in finite time: State functions' correspondence to perfectly sharp values in position is not obtained in finite time in cases where sharp values are obtained, and it is admitted that "[f]or a macrosystem, the precisely applied eigenstate–eigenvalue link does not work" [23]. At least one of the longest and most active advocates of CSL, Philip Pearle, does not see this as precluding the success of the theory, despite Shimony's desideratum, which regards tails specifically, namely,

d. If a stochastic dynamical theory is used to account for the outcome of a measurement, it should not permit excessive indefiniteness of the outcome, where "excessive" is defined by considerations of sensory discrimination. This desideratum tolerates outcomes in which the apparatus variable does not have a sharp point value, but it does not tolerate 'tails' which are so broad that different parts of the range of the variable can be discriminated by the senses, even if very low probability amplitude is assigned to the tail. The reason for this intolerance is implicit in Assumption (iv) . . . If registration on the consciousness of the observer of the measurement outcome is more precise than the 'tail' indicates, then the physical part of the measurement process would not yield a satisfactory reduction of the initial superposition, and a part of the task of reducing the superposition would thereby be assigned to the mind. For this reason, I do not share the acquiescence to broad 'tails' that Pearle advocates (1990, pp. 203–4) . . . [10]

Pearle has argued more recently, as he had once in Bell's presence – at the same Amherst conference mentioned above – that "one should not express a new theory in an old theory's language," a comment "at which he beamed" [23]. In particular, Pearle argues that "a collapse theory is different from standard quantum theory and . . . therefore requires a new language, conceptual as well as terminological" (Pearle). Emphasis is put, for example, on "near possessed" rather than "possessed" values of physical quantities. In his explication of CSL, Pearle argues that

CSL retains the classical notion that the physical state of a system corresponds to the state vector. Corresponding to a random field $w(\mathbf{x}, t)$ whose probability of occurrence is non-negligible, the dynamics always evolves a realizable state. Therefore, one is freed from requiring the (near) eigenstate–eigenvalue link criterion for the purpose of selecting the realizable states. I suggest that the eigenstate–eigenvalue link criterion be subsumed by a broader concept. It must be emphasized that this new conceptual structure is only applicable for a theory which hands you macroscopically sensible realizable states, not superpositions of such states. In the new language, corresponding to a quantum state, every variable possesses a distribution of values . . . [23]

The notions of this new 'language' is to be given meaning by considering ways in which it is to be used in context. The distribution here is not to be understood as a probability distribution, despite possessing all the defining properties of one, because in classical physics, unlike in the case of the fundamental quantum context, argues Pearle, probabilities are understood as due to *ignorance*. One

may give the name 'stuff' to a distribution's numerical magnitude at each value of the variable, as a generalization of Bell's quasi-biblical characterization, 'In the beginning, Schrödinger tried to interpret his wavefunction as giving somehow the density of the stuff of which the world was made.' One is encouraged to think of each variable's stuff distribution as something that is physically real. The notion allows retention of the classical idea that, for a physical state, every variable possesses an entity. What is different from classical ideas is that the entity is not a number. [23]

Bell had used the term 'stuff' in the context of stochastic localization theory in AM:

The GRW-type theories have nothing in their kinematics but wavefunctions. It gives the density (in a multi-dimensional configuration space!) to *stuff*. To account for the narrowness of that stuff in macroscopic dimensions, the linear Schrödinger equation has to be modified, in the GRW picture by a mathematically prescribed spontaneous collapse mechanism. [1]

On this view, *every variable* possesses such a distribution, so that "complementarity here means that variables whose operators do not commute do not possess joint distributions, but they do jointly possess distributions" (Pearle, 2009). As an example, one can consider a state that is the quantum superposition of two states, one with state amplitude $\sqrt{1-\epsilon}$ and the other with amplitude $\sqrt{\epsilon}$. Under the above interpretation, the smaller "tail" state is to considered to represent "an unobservably small amount of stuff which allows describing the state vector by (qualified) possessed values assigned to macroscopic variables, consistent with the dominant state" [23].

As noted above, a central problem for CSL is finding parameter ranges for which it would have experimental predictions deviating from those of standard quantum mechanics. Interference experiments are archetypal and would serve to differentiate the two, because CSL tends to destroy interference in that it naturally destroys one of the necessary pair of distinct states in the case of massive systems, with an interference visibility that decreases with the increase in system mass. Thus, one might test the theory by considering, for example, two-slit experiments on each of a range of sorts of systems differing in their masses: photons, electrons, neutrons, atoms, and molecules. At one limit of this range, the photons, of course, are massless, and CSL would have no additional effect on them. In the upper range of the experiments that have been performed, one finds the C-60 molecule, which has $N = 720$ nucleons. The value 720 is too small for a great impact, and is very much smaller than the Avogadro number, often taken as a value one can say is clearly "macroscopic." Because collapse narrows wavepackets, it also leads to a momentum increase and hence to an energy increase, requiring collapse rates that differ not only with particle number, but also with particle mass [23]. Unfortunately, experiments capable of testing this hypothesis are not of the sort commonly performed and currently await testing.

26.4 Toward the Removal of Conceptual Imprecision

In the quantum theory of measurement, experiments are typically understood schematically as follows. A system S is initially prepared in a quantum state T through a series of physical interactions, after which it is measured through interaction with an apparatus A that is required, in the process, to enter a state, the value of the "pointer" property, which itself becomes perfectly correlated with the value of the measured property E of S. A minimal requirement placed on a measurement is that a certain "calibration condition" be satisfied, namely, that if a property to be measured is a real one, then it should exhibit its value unambiguously and with certainty; cf. [24]. For so-called "sharp observables," that is, properties represented by Hermitian operators, this calibration condition is equivalent to a probability reproducibility condition, namely, that a probability measure E_T for a property be "transcribed" onto that of the corresponding apparatus pointer property. In addition, measurement is taken to include the reading of registered value in addition to the above process of registration of the measured property by apparatus A. The question of how this pointer "objectification" is achieved, in view of the nonobjectivity of the measured operator, is the first part of the so-called "objectification problem."

The second part of the objectification problem is that of "value objectification." A pointer reading refers to the property value of the object system prior to measurement only if the measured observable was objective *before* the measurement. When the observable is nonobjective, the question arises of what happens to the system in the course of the measurement. In general, some state change is unavoidable. The attempts to minimize this irreducible 'disturbance' then naturally lead to the concept of ideality of a measurement. Ideality requires another characteristic, namely, repeatability: A *repeatable* measurement will put the system in a state in which the pointer reading X refers to an objective value of the measured observable. This is taken to show that the existence of repeatable measurements is necessary for realistic interpretations of quantum mechanics [25]. For such measurements, pointer objectification entails value objectification via a strong value correlation. Such an operational approach to measurement, however, threatens to mask the objective, physical nature of measurements in themselves, with which Bell was so concerned.

In his 1981 article, "Quantum theory for cosmologists," Bell asked the following rhetorical questions about quantum measurements as understood within such a scheme.

If [quantum] theory is to apply to anything but idealized laboratory operations, are we not obliged to admit that more or less 'measurement-like' processes are going on more or less all the time more or less everywhere? . . . The concept of measurement becomes so fuzzy that it is quite surprising to have it appearing in physical theory at the most fundamental level . . . [D]oes not any analysis of measurement require concepts more fundamental than measurement? And should not the fundamental theory be about these . . . ? [26]

There is a stark contrast between the everyday "classical" measurements and quantum measurements. In classical physics, it *is* the case that situations that are more or less measurement-like are going on all the time everywhere. The difficulty in the quantum case is that similar processes should be happening, yet the experimental outcomes found by us are still consistent with the predictions made using the Schrödinger state evolution, although the superposition principle should not apply when such measurement-like processes are taking place, if human beings are treated just like other physical entities. If it is indeed the case that measurements, as distinct from subjective acts of observation, are nonetheless an integral part of physics and *not* artificially introduced, the special physical circumstances appearing in measurements must be circumscribed. In the search for physical clarity, we can remove anthropocentric elements from our conception of quantum measurement by finding the set of radically influential objects corresponding to this natural kind rather than a generic *apparatus* for measurements, and thus remove impediments to progress in isolating the physical conditions underlying measurement as objective. If human beings or other, larger sets of, for example, biological entities precipitate such physical conditions, then the special role of these entities will have been objectively grounded and natural philosophy will have been advanced.

It may be helpful to consider the possibility that the set of related entities is of a *natural kind*, because when successful, classes employed by science do correspond to natural kinds, as in the cases of the sorts of chemical element, subatomic particle, star, and galaxy. This does not require that these objects be treated differently from others, but only assists in our comprehending the implications of their sets of common characteristics. In addition to having particular sets of natural properties in common with one another, these tokens should be subject to laws of nature relevant to these properties. One can seek a set of conditions for being a member of the kind 'radically influential object' as a way of making progress toward an improved realist physics. From the formal point of view, such influencers are typically assumed to have these characteristics in common: they (i) induce nonunitary state change and (ii) satisfy the conditions on the systems for providing a robust record of measurement outcomes. The former relates to the ability to disentangle the joint state of the influencer–target system and the latter corresponds to the production of Einsteinian elements of reality. However, Bell rightly pointed out that more fundamental properties should be present, of which these properties are *consequences*. It is these configurations of fundamental properties that should play the key role in describing measurement via truly fundamental physical principles.

Some models of measurement, discussed further below, assume that measurements involve complex measuring systems with a large number of degrees of freedom prepared in metastable states. Some natural systems are known to us to measure – for example, our eyes when connected with our nervous systems are such systems. Consider the human optical system taken to comprise all the material from the eyes to the brain inclusively, while asking whether previously assumed characteristics are essential. We must look to our understanding of the behavior of macromolecules of the optical nervous system when light is incident upon it. Shimony has pointed out that the photoreceptor protein of the rod cells, known as rhodopsin, absorbs a photon, followed by a biochemical cascade that is then followed by an electrical pulse in the optic nerve. Rhodopsin has two components,

retinal, which can absorb a photon, and opsin, which acts as an enzyme that effects the binding of about five hundred mediating molecules when it is triggered by the excited retinal . . . [W]hat if the unitary dynamics of evolution of the photon and the retinal produces a superposition of the cis and the trans conformations? . . . Would not such a superposition produce an indefiniteness of seeing or not seeing a visual flash, unless, of course, a reduction occurred further along the pathway from the optic nerve to the brain to the psyche? [9] [10]

The distinct, alternative physical states corresponding to different conformations of a molecule that can superpose and then enter a specific state when in contact with the remainder of the nervous system are central to the functioning of this light detection process in a biochemical and electric realm. The presence of this larger subsystem beyond rhodopsin has an effect of "amplification." Some artificial systems can also mimic the behavior of such natural radical influencers, and so are exploitable by designers of experiments, e.g., avalanche photodiodes (APDs) plus electronics. These systems, for example, are rather complex and involve many degrees of freedom and metastable initial states. Their effect appears to us to be completed in a way that, for example, a Stern–Gerlach magnet alone without a downstream beam-occupation detector is not. Nonetheless, it has been argued by some, such as Asher Peres [27], that amplification is not required for measurement to take place, an argument considered below in greater detail.

It is the resolution of questions regarding requirements such as these that could provide the classification of the measurers as a natural kind and would be a helpful element of any realist treatment of quantum state change, such as that called for by Bell. Although, like Bell, I am concerned about the notion of measurement being given an unusually prominent place in physics, I am much more concerned about notions associated with other terms that he criticizes in less detail in "Against measurement," such as 'observable' and 'observation,' which are clearly laden with the influence of theory and more importantly with a directly *subjective*, that is, nonphysical aspect. This should be kept in mind when asking the question of which characteristics of the above example are *necessary* to an objective understanding of measurement, in contrast to those which may be systematically present in the considerations of physicists only because they themselves are subjects.

Although at least one human knower is always *eventually* present who is witness in any successful experiment – or, in less contrived cases, simply observes ongoing natural events – that happens also to be large, this should not be allowed to beg the question of whether largeness is a necessary characteristic of a radical influencer in all data-yielding situations. The matter of the necessity of amplification, which aids in the delivery of signals perceptible to the human senses, is similarly subtle. Amplification has often been considered to lead to the objectivity of data produced, because from the statistical point of view it makes it is very difficult to undo, providing many systems and observers to be affected by the resulting signal. Despite the presence of these characteristics in many or most familiar situations wherein one learns about the state of a part of the universe, one must ask whether there is an objective reason for requiring them in data-yielding situations *in general*.

Another characteristic mentioned above that is often selected out for special status is *complexity*. Notably, it was indicated by DLP, who required in their treatment that interference "be absent by virtue of the complexity of the considered system," with complexity taken to refer to the number of degrees of freedom of the joint system being large [28]. It is also strongly indicated as a requirement in the above-mentioned work of Peres that "benefited from comments by J.S. Bell," entitled "Can we undo quantum measurements?" [27]. Peres introduces a simple measurement model and with it demonstrates that systems of many degrees of freedom may effectively obey superselection rules, because it is impossible to measure any phase relationship between two quantum states in the limit of an *infinite number* of degrees of freedom. He does so without requiring another common candidate requirement, namely, a measurer prepared in a metastable state. The demonstration is a sensible on FAPP, but only FAPP, like many others. Its value is thus that it shows that amplification and the presence of metastable states of the measuring system are *not* necessary for measurement.

All the commonly assumed characteristics of measurers might be thought by us to be natural simply because they reflect a subtle and unnoticed anthropocentrism, seeming natural to us only because we scientists are human. Just because humans are comparative large physical systems and perform measurements in the experimental context does not mean that we must have access to all measurement results: measurement and the experience of a measurement are distinct, despite their going together in our human experience. As Bell suggests, we must ask ourselves what might be the purely objective properties of measurement processes that could be solely responsible for, or most significant to, the occurrence of every measurement.

We see above, for example, that having entered the field of a Stern–Gerlach magnet alone is insufficient for the measurement of an appropriate particle's spin and that, following its action, there is a sufficient set of elements present for a successful measurement to take place when detector plus electronics capture the output beams, which can later be viewed by an experimenter. What is happening in this case? According to the Schrödinger evolution, the effect of the magnetic field is only to entangle the particle's spin with its direction; it is usually understood that it is the detector suite that allows the spin to be identified after the above spin–path correlation has occurred. The detector, unlike the magnetic field created by the magnet, is complex, as is the human nervous system. The one characteristic that appears to survive our removal of unnecessary conditions is what appears to be something similar to physical *complexity*. It is helpful now to recall, as Shimony noted, that a benefit of stochastic modification of quantum dynamics was to offer an explanation ("for the general phenomenon of irreversibility") at the level of fundamental processes, something clearly in harmony with Bell's call for a more fundamental notion than measurement to account for the emergence of experimental data. We should, of course, note Bell's warning that looking to "some aspect of complexity" has "the danger of confusing epistemological and ontological issues." Proceeding cautiously, then, we can ask in precisely what sense the ontology of above situations might be complex.

The commonly recognized measurement-like situations involve a significant number of degrees of freedom and may also involve a number of distinct parts. (Incidentally, the notion of natural kinds can also assist in distinguishing between internal and external degrees of freedom, via the notion of fundamental entities.) An important, well-known example relevant to our considerations here is the Schrödinger's cat experiment. In this thought experiment, Schrödinger considered an unstable atom A, the decay of which would release a hammer H that would break a vial containing poison and allowing the cat C, which is taken as otherwise isolated from the rest of the universe, to be exposed to the poison [29]. Aside from the "absurdity" (as he calls it) of the appearance there of two equiprobable distinct states of the cat's 'health' at a particular moment in the corresponding overall state

$$|\Psi\rangle = 1\sqrt{2}(|\text{undecayed}\rangle_A|\text{unreleased}\rangle_H|\text{alive}\rangle_C + |\text{decayed}\rangle_A|\text{released}\rangle_H|\text{dead}\rangle_C) \quad (26.1)$$

of the joint system, this situation has been viewed (in the terminology to be replaced in a more precise conception of measurement), as linking the atom to the 'macroscopic domain (cf. [4]). One can equally well, as Bell's position would relatively favor, view this as a situation involving complex physical circumstances. Clearly, the joint system of

A+H+C is more complex in several respects than that of C taken alone. For example, it has more subsystems and more degrees of freedom and involves an interaction of C with H. The same can be said for A+H relative to A+H+C, which, although A+H already involves an interaction involves additional interactions. But C itself involves internal interactions. One can ask, for example, whether these overlooked internal interactions are significant in relation to the complexity of the situation or otherwise influence measurement, because no one has ever observed that sort of alive–dead circumstance in a cat or similar being.

Finally, although the above example continues to be of conceptual importance, one is now in need of further, and more practical examples to make progress. On the side of practical, rather than thought experiments, let us consider as an example of what is now available to be explored, some experiments performed relatively recently by Gerlich et al. [30]. These experiments have been taken by their creators to "prove the quantum wave nature and delocalization of compounds composed of up to 430 atoms, with a maximal size of up to 60 Ångstroms, masses up to $m = 6,910$ AMU." To us, they can be seen to involve something somewhat similar to the surprising predictions of quantum mechanics in the Schrödinger cat thought experiment. These quantum systems involve thousands of internal degrees of freedom.

An important point here is the factor in these experiments that differentiates the detector used from conventional detectors operating in the visible and near infrared, such as avalanche photodiodes and photomultiplier tubes of past quantum optics experiments, namely, that although those may be single-photon sensitive, they could not reliably determine the *number* of photons in a pulse of light like the photon-number-resolving detectors used in this experiment. Such determination is made possible as follows. The experiments used a calorimetry-based photon detector in which energy is deposited in an absorber whose energy content was determined via an observed change in temperature. Tungsten transition-edge sensors were used and understood to operate in such a way that tungsten electrons acted as both energy absorbers and thermometers, and were prepared in such a way as to keep the tungsten electrons on the edge of a superconducting-to-normal-conduction transition; a dependence of resistance on temperature was set up to allow precise thermometry. The change of current in the voltage-biased detector was measured with a superconducting quantum interference (SQUID) array and analyzed. The results demonstrated quantum state superpositions of states of certain properties of large entities, suggesting that the number of degrees of freedom involved does not correspond to the sort of physical complexity required for induced behavior that would differ from that given by the standard quantum mechanical description.

The detailed study of such experiments, in particular, the array of instrumentation and its necessity for successfully providing data describing quantum phenomena, alongside the study of human sensory systems themselves, can be expected to yield a more refined understanding of measurement-like processes within quantum theory and should provide novel insights allowing more precise vocabulary and concepts to be introduced to improve upon its basic principles, as John Bell recommended. Novel approaches to measurement in quantum optics, much as in the testing of the Bell CHSH inequality itself in the past, can also aid us in transcending the current limitations, illustrated in the weakness that he identified in quantum physical terminology deployed in measurement situations. After such work, one will be in a better position to consider specific modifications of the fundamental laws of quantum physics or the quantum state description itself, both of which should help us progress in the direction that Bell suggested we should go in order to advance natural philosophy.

Acknowledgments

This work was partly supported by funding from the DARPA QUINESS program through U.S. ARO Award W31P4Q-12-1-0015.

References

[1] Bell, J.S. (1990), Against measurement, in A.I. Miller (ed.), *Sixty-Two Years of Uncertainty: Historical, Philosophical, and Physical Inquiries into the Foundations of Quantum Mechanics*, New York: Plenum Press, pp. 17–31.

[2] Bell, J.S. (1981), Bertlmann's socks and the nature of reality, *Journal of Physics 42*, C2, supplément au No. 3, p. 41.

[3] Clauser, J., Horne, M., Shimony, A. and Holt, R. (1969), Proposed experiment to test local hidden-variable theories, *Physical Review Letters* **23**, 880–84.

[4] Jaeger, G.S. (2014), What in the (quantum) world is macroscopic? *American Journal of Physics* **82**, 896–905.

[5] Von Neumann, J. (1932), *Mathematische Grundlagen der Quantenmechanik*, Berlin: Julius Springer.

[6] Von Neumann, J. (1955), *Mathematical Foundations of Quantum Mechanics*, Princeton, NJ: Princeton University Press, Ch. V, Sect. 4.

[7] Bell, J.S. (1982), On the impossible pilot wave. *Foundations of Physics* **12**, pp. 989–99.

[8] Bell, J.S. (1987), Are there quantum jumps? In Schrödinger: *Centenary Celebration of a Polymath*, Cambridge: Cambridge University Press.

[9] Ghirardi, G.C., Rimini, A. and Weber, T. (1986), Unified dynamics for microscopic and macroscopic systems, *Physical Review D* **34**, 470–91.

[10] Shimony, A. (1991), Desiderata for a modified quantum dynamics, in *PSA 1990*, Chicago: Philosophy of Science Association. Reprinted in Shimony, A. (1993), *Search for a Naturalistic World View*, Cambridge: Cambridge University Press, Chapter 4.

[11] Albert, D. (1990), On the collapse of the wavefunction, in A.I. Miller (ed.), *Sixty-Two Years of Uncertainty: Historical, Philosophical, and Physical Inquiries into the Foundations of Quantum Mechanics*, New York: Plenum Press, pp. 153–65.

[12] Wigner, E.P. (1963), The problem of measurement, *American Journal of Physics* **131**, 6.

[13] Freire, O. (2005), Orthodoxy and heterodoxy in the research on the foundations of quantum mechanics: E.P. Wigner's case, in B.S. Santos (ed.), *Cognitive Justice in a Global World*, Madison, WI: University of Wisconsin Press.

[14] Wigner, E.P. (1971), The subject of our discussions, in *Foundations of quantum mechanics. Proceedings of the International School of Physics "Enrico Fermi,"* London: Academic Press, p. 5.

[15] Prosperi, G.M. (1971), Macroscopic physics and the problem of measurement, in *Foundations of Quantum Mechanics. Proceedings of the International School of Physics "Enrico Fermi,"* London: Academic Press, pp. 97–126.

[16] Leggett, A. (2002), Testing the limits of quantum mechanics: Motivation, state of play, prospects, *Journal of Physics: Condensed Matter* **14**, R415–R451.

[17] Home, D. and Whitaker, A. (2007), *Einstein's Struggles with Quantum Theory: A Reappraisal*, New York: Springer, p. 310.

[18] Leggett, A. (2000), The order parameter as a macroscopic wavefunction, in J. Berger and J. Rubinstein (eds.), *Lecture Notes in Physics* m62, Heidelberg: Springer, pp. 230–238.

[19] Leggett, A. (1980), Macroscopic quantum systems and the quantum theory of measurement, *Progress in Theoretical Physics Supplement* **69**, 80–100.

[20] Bell, J.S. (1975). On wave packet reduction in the Coleman–Hepp model, *Helvetica Physica Acta* **48**, pp. 93–8.

[21] Bell, J.S. (1990), La nouvelle cuisine, in A. Sarlemijn and P. Kroes (eds.), *Between Science and Technology*, Dordrecht: Elsevier Science Publishers.

[22] Allori, V., Goldstein, S., Tumulka, R. and Zanghì, N. (2008), On the common structure of Bohmian mechanics and the Ghirardi–Rimini–Weber theory, *British Journal for the Philosophy of Science* **59**(3), 353–89.

[23] Pearle, P. (2009), How stands collapse, II, in *Quantum Reality, Relativistic Causality, and Closing the Epistemic Circle*, The Western Ontario Series in Philosophy of Science, Vol. 73, New York: Springer, pp. 257–92.

[24] Busch, P., Lahti, P. and Mittelstaedt, P. (1991), *The Quantum Theory of Measurement*, Heidelberg: Springer.

[25] Busch, P. and Jaeger, G. (2010), Unsharp quantum reality, *Foundations of Physics* **40**, 1341.

[26] Bell, J.S. (1981), Quantum mechanics for cosmologists, in C. Isham, R. Penrose and D. Sciama (eds.), *Quantum Gravity 2*, Oxford: Clarendon Press, pp. 611–37.

[27] Peres, A. (1980), Can we undo quantum measurements? *Physical Review D* **69**, 879–83.

[28] Danieri, A., Loinger, A. and Prosperi, G.M. (1962), Quantum theory of measurement and ergodicity conditions, *Nuclear Physics* **33**, 297–319.

[29] Schrödinger, E. (1935), Die gegenwärtige Situation in der Quantenmechanik, *Die Naturwissenschaften* **23**, 807–12, 823–28, 844–49.

[30] Gerlich, S., Eibenberger, S., Tomandl, M. et al. (2011), Quantum interference of large organic molecules, *Nature Communications* **2**, 263.

Index